THE
DREAM MACHINE

ALSO BY M. MITCHELL WALDROP

Complexity: The Emerging Science at the Edge of Order and Chaos

Man-Made Minds: The Promise of Artificial Intelligence

THE
DREAM MACHINE

⁓

J. C. R. Licklider and the Revolution
That Made Computing Personal

⁓

M. MITCHELL WALDROP

VIKING

VIKING
Published by the Penguin Group
Penguin Putnam Inc., 375 Hudson Street, New York, New York 10014, U.S.A.
Penguin Books Ltd, 27 Wrights Lane, London W8 5TZ, England
Penguin Books Australia Ltd, Ringwood, Victoria, Australia
Penguin Books Canada Ltd, 10 Alcorn Avenue, Toronto, Ontario, Canada M4V 3B2
Penguin Books (N.Z.) Ltd, 182–190 Wairau Road, Auckland 10, New Zealand

Penguin Books Ltd, Registered Offices:
Harmondsworth, Middlesex, England

First published in 2001 by Viking Penguin,
a member of Penguin Putnam Inc.

1   3   5   7   9   10   8   6   4   2

Grateful acknowledgment is made for permission to reprint excerpts from
History of Programming Languages by Thomas J. Bergin and Richard G. Gibson.
© 1996 ACM Press. Reprinted by permission of Pearson Education, Inc.

PHOTO CREDITS: U.S. Department of Defense: p. 1; courtesy of Tracy R. Licklider: p. 7;
MIT Museum: pp. 24, 66, 196, 411; The Computer Museum History Center: pp. 99, 142, 259, 333.
Jacket photo courtesy of Frances Antupit, Koby-Antupit Photographers,
Cambridge, Mass., and MIT Museum.

LIBRARY OF CONGRESS CATALOGING-IN-PUBLICATION DATA
Waldrop, M. Mitchell.
The dream machine : J. C. R. Licklider and the revolution that made computing personal /
M. Mitchell Waldrop.
p.   cm.
ISBN 0-670-89976-3 (alk. paper)
1. Licklider, J. C. R. 2. Microcomputers—History. I. Title.
QA76.17.W35 2001
004.16'092–dc21
[B]   2001017985

This book is printed on acid-free paper. ∞

Printed in the United States of America
Set in Garamond BE
Designed by Nancy Resnick

To A.F.F

THE SLOAN TECHNOLOGY SERIES

PREFACE TO THE SLOAN TECHNOLOGY SERIES

Technology is the application of science, engineering, and industrial organization to create a human-built world. It has led, in developed nations, to a standard of living inconceivable a hundred years ago. The process, however, is not free of stress; by its very nature, technology brings change in society and undermines convention. It affects virtually every aspect of human endeavor: private and public institutions, economic systems, communications networks, political structures, international affiliations, the organization of societies, and the condition of human lives. The effects are not one-way; just as technology changes society, so too do societal structures, attitudes, and mores affect technology. But perhaps because technology is so rapidly and completely assimilated, the profound interplay of technology and other social endeavors in modern history has not been sufficiently recognized.

The Sloan Foundation has had a long-standing interest in deepening public understanding about modern technology, its origins, and its impact on our lives. The Sloan Technology Series, of which the present volume is a part, seeks to present to the general reader the stories of the development of critical twentieth-century technologies. The aim of the series is to convey both the technical and the human dimensions of the subject: the invention and effort entailed in devising the technologies, and the comforts and stresses they have introduced into contemporary life. As we begin the new century, it is hoped that the series will disclose a past that might provide perspective on the present and inform the future.

The Foundation has been guided in its development of the Sloan Technology Series by a distinguished advisory committee. We express deep gratitude to John Armstrong, Simon Michael Bessie, Samuel Y. Gibbon, Thomas P. Hughes, Victor McElheny, Robert K. Merton, Elting E. Morison (deceased), and Richard Rhodes. The Foundation has been represented on the committee by Ralph E. Gomery, Arthur L. Singer, Jr., Hirsh G. Cohen, and Doron Weber.

—Alfred P. Sloan Foundation

CONTENTS

TRACY'S DAD

Tricycles.

That's what Tracy would always remember most about it: tricycles in the Pentagon.

It was a Saturday afternoon in late 1962 or maybe early 1963—sometime in the Kennedy administration, anyway, not too long after Tracy's family had moved down from the Boston area so his father could go to work for the Defense Department. The air in Washington was electric in those days, with all the energy and drama of a new, young administration. The Cuban Missile Crisis, the Berlin Wall, the Civil Rights marches—it was heady stuff for a fifteen-year-old. So when his dad had offered to take him along to his office that afternoon while he picked up some papers he'd forgotten to bring home, Tracy had jumped at the chance. He was still slightly awestruck at the very thought of the Pentagon.

Well, it *was* an awesome place, especially when you saw it up close. The Pentagon was nearly a thousand feet on a side and sat up on a little rise like a walled city. Tracy and his father had left their car in an immense parking lot and hiked to the front door. They had gone through a very impressive security procedure

The mightiest bureaucracy in the world: the Pentagon, with a panorama of Washington, D.C.

at the guard station, where Tracy had been signed in, vouched for, and given a badge. They had walked on in and started down the hallway into the nerve center for the defense of the Free World. And the first thing Tracy had seen was some earnest-looking young soldiers moving up and down the halls to deliver the interoffice mail—pedaling grown-up-sized tricycles.

It was absurd. But the soldiers on the tricycles seemed very serious about their work. And Tracy had to admit that the tricycles made a certain amount of sense for getting around: the Pentagon had some *very* long hallways. He thought he and his father were going to walk forever.

Actually, Tracy thought it was strange that his father worked in the Pentagon at all. You couldn't imagine anyone less like a general, or a bureaucrat, or a politician. His dad was more like an overgrown kid. He looked ordinary enough: a tall, slightly chubby guy in a tweed sports coat and dark-rimmed glasses. But he had this mischievous expression on his face, as if he were always looking for a way to stir things up. Take dinnertime, for example, which was never what you would call casual where Dad was concerned. Unless his Pentagon job took him out of town, he would always come home to eat with the family before going back to work in the evening. It was fun: Dad would tell stories and make awful puns, and sometimes he'd crack himself up before he could even get to the punch line. He'd laugh so hard you just had to laugh with him. But the very first thing he'd always do was turn to Tracy and his thirteen-year-old sister Lindy and ask, "What have you done today that was altruistic, creative, or educational?" And he meant it. Tracy and Lindy would have to think through all the things they had done that day to find something they could fit into one of those categories.

Eating out could be pretty intense, too. Mom and Dad both loved to try new restaurants. But while they were waiting for the food to arrive, Dad would put Lindy and Tracy to work on some brainteasers like "If a train is going west at forty miles per hour and an airplane is flying overhead at . . ." Tracy got to be pretty good at these, to the point where he could generally work out the answers in his head. But Lindy would just develop this resigned thirteen-year-old's expression.

"OK, Lindy," Dad would ask, "if a bike wheel is rolling along the ground, do all the spokes move at the same speed?"

"Of course, Dad."

"No, they don't," he'd say, and then he'd explain how the spokes on the bottom would actually stand still for an instant while the spokes on top went twice as fast as the bicycle—all the while drawing on his place mat, making diagrams that would have done Leonardo proud. (Once, at a convention, a guy had leaned over Dad's shoulder and offered fifty dollars for a page of his doodles.)

And then there were the art galleries. On the weekends Mom liked to have a little time for herself, so Dad would take Tracy and Lindy out to look at paintings, usually at the National Gallery of Art down on the Mall. It was mostly the Impressionists that Dad liked: Gauguin, Monet, Pissarro, Cézanne. He loved the light, the glow that seemed to come through their paintings. But Dad had also come up with a technique for how you should look at a painting, based on "color displacements" (he had been a psychologist up at Harvard and MIT). He claimed that if you covered one eye with your hand and stood back about five feet, and then suddenly dropped your hand so you could look at the picture with both eyes, the flat surface would seem to leap into three dimensions. And it worked! He and Tracy and Lindy would go wandering through the National Gallery for hours, each of them looking at the paintings with a hand over one eye.

They got some strange looks. But then, they'd always been a pretty strange family, in a good sort of way. Compared to their school friends, Tracy and Lindy felt different somehow. More special. *Worldlier*. Dad loved to travel, for example. So Tracy and Lindy had grown up thinking it was perfectly natural to go traipsing through Europe or California for weeks or months at a time. In fact, their parents would rather spend money on trips than on furniture, which was why their big old Victorian house up in Massachusetts had a sort of orange-crate-and-boards decor. Between them, too, Dad and Mom managed to fill up that house with actors, directors, writers, artists, and all kinds of wonderfully oddball people—not to mention Dad's students, who would drop by at all hours. Mom would just send them straight on up to Dad's office on the third floor, where he had a desk surrounded by endless piles of paper. Dad never filed anything. On his desk, however, he did keep a bowl of Ayds diet candy, which was supposed to curb your appetite and which Dad ate like, well, candy.

So no, Dad was definitely not the kind of guy you'd expect to find working at the Pentagon. Except that here he was, walking with Tracy down these *looong* hallways.

By the time they got to his father's office, Tracy figured they must have walked the length of several football fields. And when he finally saw it, he felt . . . disappointed. It was just one more door in a hallway lined with doors. And the inside was nothing more than a drab little room painted drab military green, with a desk, a couple of chairs, and a bunch of filing cabinets. There was a window, but all you could see was a wall full of other windows. Tracy hadn't been quite sure what to expect, but this wasn't it.

In fact, Tracy wasn't even sure what his father *did* in that office all day. The work wasn't secret, but it was still Defense Department business—which Dad took pretty seriously, so he didn't talk about it much at home. And to be honest, at age fifteen, Tracy didn't pay much attention when he did. About the only things he knew for sure were that his father was on the road a good deal, that he spent a lot of time trying to get people to do things, and that it all had something to do with computers.

No surprise there. His father was crazy about computers. Back home in Cambridge, at a company called Bolt Beranek and Newman, the members of Dad's research group had had a computer that they'd modified themselves. It was a fairly big thing, about the size of a couple of refrigerators. But it also had a keyboard, a screen that would show you what you typed, a light pen—everything you could want. It even had some special software so that several people could work simultaneously from remote terminals. Dad played with the thing day and night, writing programs. And on the weekends, he took Tracy and Lindy in so *they* could play with it. (Afterward they'd go for hamburgers and fries at the Howard Johnson's across the street; it got to the point where the waitresses wouldn't even bother to take their orders anymore, they'd just bring out the burgers as soon as they saw them coming.) Dad even wrote a computerized language tutor for them. If you got a word right, it would respond "Commendable." And if you got it wrong, it would respond "Dumbkopf!" (It would be years before anyone pointed out to Tracy's dad that there's no *b* in the German word *Dummkopf*.)

Tracy had taken to all of this like a natural; he'd even learned how to program the machine himself. But now, in hindsight, looking back forty years from the perspective of the new millennium, he thinks maybe that's also why he didn't pay too much attention to what his father was doing in the Pentagon. He'd been spoiled. He was like one of those kids today who fool around with 3-D computer graphics, play games off of DVD-ROMs, go blithely surfing through the Internet—and take all of that for granted. Since his father's joyfully interactive computer was the only one he'd ever seen, Tracy just assumed that computing was like that for everyone. As a teenager, he had no way of knowing that for most people the word *computer* still meant a big, mysterious mainframe sitting off in a back room somewhere, an ominous, implacable, pitiless machine of the sort that served only large institutions—*them*—and seemed to be reducing the rest of us to numbers on a punch card. Nor did Tracy have any way of understanding that his father was one of the very few people in the world who had looked at that technology and seen the possibility of something profoundly different.

Always the big dreamer, the kind of guy who was forever asking "What if?," Tracy's dad had come to believe that one day all computers would be like his machine up in Cambridge. They would be humane and intimate. They would respond to people and help them as individuals. They would serve as a new medium of expression. They would democratize access to information, foster wider communities, and build a new global commons for communication, commerce, and collaboration. Ultimately, in fact, they would enter into a kind of *symbiosis* with humans, forming a cohesive whole that would think more powerfully than any human being had ever thought and process data in ways that no machine could ever do by itself.

Now that he was at the Pentagon, moreover, Tracy's dad was doing everything he could to turn that vision into a reality. Up at MIT, for example, he was already in the process of establishing Project MAC, the world's first large-scale experiment in personal computing. The project managers there couldn't hope to give anyone a stand-alone personal computer, of course, not with the cheapest machines still costing hundreds of thousands of dollars. But they could scatter dozens of remote terminals around the campus and in people's homes. And then, through the technology of "time-sharing," they could tell their big, central machine to dole out little slices of processing time very, very rapidly, so that each user would feel as if the thing were responding to him or her as an individual. This scheme would work surprisingly well. Indeed, within just a few years, Project MAC would not only introduce hundreds of people to the joys of interactive computing but also evolve into the world's first on-line community, complete with on-line bulletin boards, E-mail, a "freeware" exchange—and hackers. It would pioneer essentially all the social phenomena that would later be found in the on-line communities of the Internet era. Along the way, moreover, its remote terminals would plant the notion of a "home information center," an idea that would circulate in the technological community's collective consciousness until the 1970s, when it would inspire a slew of young hobbyists with names such as Jobs and Wozniak to market something called a microcomputer.

Meanwhile, Tracy's dad was also gambling on a soft-spoken, rather lonely guy who had approached him on practically his first day at the Pentagon, and whose ideas on "augmenting the human intellect" had proved to be identical to his own notion of human-computer symbiosis. Douglas Engelbart had been a voice in the wilderness until then; his own bosses at SRI International, in what would soon become Silicon Valley, thought he was an absolute flake. But once Tracy's father had given him his first real funding—and vigorously defended him to his higher-ups—Engelbart, with his group, would go on to invent the mouse, on-screen windows, hypertext, full-screen word processing, and a host of other innovations. Engelbart's December 1968 presentation at a computer meeting in San Francisco would blow about a thousand minds at once—and later be remembered as one of the turning points in computer history, the moment when the rising generation of computer professionals at last began to *understand* what interactive computing could do. By no coincidence, this was also the rising generation whose members had had their graduate educations supported by Tracy's dad and his successors at the Pentagon—and a talented portion of which would soon be gathering at PARC, the Xerox Corporation's legendary Palo Alto Research Center. There they would put Dad's "symbiosis" vision into the form we are still using more than three decades later: a stand-alone personal computer equipped with a graphics display screen and a mouse. A graphical user interface with windows, icons, menus, scroll bars, and all the rest. A laser printer to print things out. And the Ethernet local-area network to tie it all together.

And finally, there was networking. Now that he had come to work for the Pentagon, Tracy's father was actually spending most of his time on airplanes, constantly seeking out isolated research groups that were already doing work consistent with his vision of human-computer symbiosis; his goal was to forge them into a nationwide community, a self-sustaining movement that would carry on with the dream long after he himself left Washington. On April 25, 1963, in a memo to "the members and affiliates of the Intergalactic Computer Network," he would outline a key part of his strategy: to connect all their individual computers and time-sharing systems into a single computer network spanning the continent. True, networking technology was still too primitive to enable the creation of such a system—for now, anyway. But Dad's mind was already racing ahead. He would soon be talking about the Intergalactic Network as an electronic commons open to all, "the main and essential medium of informational interaction for governments, institutions, corporations, and individuals." This electronic commons would support electronic banking, E-commerce, digital libraries, "investment guidance, tax counseling, selective dissemination of information in your field of specialization, announcement of cultural, sport, and entertainment events"—and on and on. By the late 1960s such visions would inspire Dad's handpicked successors to implement his Intergalactic Network, now known as the Arpanet. By the 1970s, moreover, they would begin to expand the Arpanet even further, into the network of networks known today as the Internet.

In short, Tracy's dad was setting in motion the forces that would give rise to essentially all of modern computing: time-sharing, personal computing, the mouse, graphical user interfaces, the explosion of creativity at Xerox PARC, the Internet—all of it. Of course, not even he could have imagined such an outcome, not in 1962. But it would have delighted him no end. After all, it was why he had uprooted his family from the home they loved, and why he had come to Washington to work in the sort of bureaucracy he hated: because he believed in his dream.

Because he was determined to see it become real.

Because the Pentagon—though some of the higher-ups didn't quite seem to understand this yet—was putting up the money to *make* it real.

Once Tracy's dad had gotten all his papers together and was ready to go, he grabbed a handful of green plastic tags. This was how you kept the bureaucrats happy, he explained. Every time you left your office for the day, you were supposed to label all your file drawers and all the stacks of paper on your desk with the appropriate-colored tags: green for nonclassified material, then yellow, red, and so forth through all the other levels of security. It was kind of silly, considering that he almost never needed anything besides green. But those were the rules. *Soooo* . . .

Tracy's dad dealt his green tags around the office like so many playing cards, just enough to make it look as if someone had thought carefully about where each one should go. "OK," he told his son, "we can leave now."

Tracy and his father shut the door behind them—

ADVANCED RESEARCH PROJECTS AGENCY
Command and Control Research
J. C. R. Licklider, Director

—and began to walk back down the long, long halls of the Pentagon, while all around them earnest young men on tricycles carried information to and fro through the mightiest bureaucracy in the world.

CHAPTER I

MISSOURI BOYS

Joseph Carl Robnett Licklider did tend to make an impression on people. Even in the early days, long before he got involved with computers, he had a way of making everything look so *easy*.

"Lick was probably the most gifted intuitive genius I have ever known," declared the late William McGill in an interview recorded shortly before his death, in 1997. McGill explained in that interview that he had first met Lick when he entered Harvard University as a psychology graduate student in 1948: "Whenever I would finally come to Lick with the proof of some mathematical relation, I'd discover that he already knew it. He hadn't worked it out in detail, he just . . . knew it. He could somehow envision the way information flowed, and see relations that people who just manipulated the mathematical symbols could not see. It was so astounding that he became a figure of mystery to all the rest of us: How the hell does Lick do it? How does he see these things?

"Talking with Lick about a problem," added McGill, who would later serve as president of Columbia University, "amplified my own intelligence about thirty IQ points."

Lick made a similarly deep impression on George A. Miller, who first worked

Louise's favorite snapshot: Lick as a young psychologist

with him at Harvard's Psycho-Acoustics Laboratory during World War II. "Lick was the All-American Boy—tall, blond, and good-looking, good at everything he tried," Miller would write many years later. "Extremely intelligent, intensely creative, and hopelessly generous—when you made a mistake, Lick persuaded everyone that you had just brought off the cleverest possible joke. He loved jokes. Many of my memories have him telling of some amusing absurdity, usually at his own expense, while he gestured with a Coca-Cola bottle in one hand."

It wasn't that he cracked people up, exactly. It was just that when Lick held forth with that laconic Missouri twang and lopsided grin, his listeners somehow found themselves smiling. He came at the world with a sunny, open-faced friendliness that made it seem as though everyone he met was going to be just great. And somehow, for him, everyone usually was.

He was a Missouri boy from way back. The name itself (pronounced LICK-lye-der) had originated generations earlier in Alsace-Lorraine, along the French-German border, but his family on both sides had lived in Missouri since before the Civil War. His father, Joseph Licklider, had been a farm boy from the middle of the state, near the town of Sedalia. Joseph also seems to have been a determined and resourceful young man. In 1885, after his own father died in an accident involving a horse, the twelve-year-old Joseph at once took on responsibility for his family. Realizing that he, his mother, and his sister couldn't hope to manage the farm on their own, he moved them all to St. Louis and went to work for one of the local railroads until he had put his sister through high school and college. Once that was done, Joseph apprenticed himself to an advertising firm to learn the crafts of writing and design. And once he had mastered those skills, he switched to insurance, eventually becoming a prizewinning salesman and head of the St. Louis Chamber of Commerce.

Meanwhile, at a Baptist youth revival meeting, Joseph Licklider had caught sight of a Miss Margaret Robnett. "I took one look at her," he later said, "and heard her sweet voice singing in the choir, and I knew I'd found the woman I loved." He immediately started taking the train out to her family's farm every weekend, determined to win her as his bride. He succeeded. Their only child was born in St. Louis on March 11, 1915. He was named Joseph for his father, and Carl Robnett for his mother's late brother.

The baby's sunny outlook was understandable. Joseph and Margaret were rather old for first-time parents in those days—he was forty-two, and she was thirty-four—and they could be quite strict in matters of religion and good behavior. But they were also a warm, loving couple who adored their little boy and doted on him constantly. Everyone did: young Robnett, as they called him at home, was not just an only child but the only grandchild on either side of the family. As he grew, moreover, his parents encouraged him in piano lessons, tennis, and whatever else he wanted to try, especially if the activity was at all intellectual. And Robnett did not disappoint them, maturing into a bright, energetic boy with a lively sense of fun, an insatiable curiosity, and an abiding love of all things technological.

When he was twelve, for example, he and just about every other boy in St. Louis conceived a passion for building model airplanes. Maybe it was the burgeoning aircraft industry in their hometown. Maybe it was Lindbergh, who had just made his solo flight across the Atlantic in an airplane named the *Spirit of St. Louis*. Or maybe it was simply that airplanes were the technological wonders of the age. No matter: the boys of St. Louis were model-airplane-mad. And nobody could build them better than Robnett Licklider. With his parents' permission, he turned his room into something resembling a lumberyard for balsa wood. He bought directions and plans, and drew up detailed designs of his own. He cut out balsa wood members and struts with painstaking care. And he stayed up all night putting the pieces together, covering the wings and body with cellophane, painting it all just so, and doubtless getting a little high on airplane glue. He was so good at it that one of the kit-manufacturing companies paid his way to a national air show in Indianapolis so he could show the visiting fathers and sons how it was done.

And then, with the approach of that all-important sixteenth birthday, his interest shifted to cars. It wasn't just that he wanted to drive them around; he wanted to *understand* them, inside and out. So his parents allowed him to buy an old junker, on the condition that he drive it no farther than the end of their long, curving driveway. Young Robnett happily took that car apart and rebuilt it again and again, starting up the engine each time he added a piece just to see what happened: "OK, so that's how *that* works." Margaret Licklider, fascinated by this technological prodigy she had raised, would stand by as he worked underneath the car and hand down the tools as he needed them. Her son got his license on March 11, 1931, the day he turned sixteen. And for years thereafter, he refused to pay more than fifty dollars for a car; whatever shape it was in, he could fix it up and make it go. (Faced with the ravages of inflation, he was eventually forced to raise his limit to $150.)

The sixteen-year-old Rob, as he was now known to his classmates, had grown up tall, handsome, athletic, and outgoing, with sun-bleached hair and blue eyes that gave him a notable resemblance to Lindbergh himself. He played a fiercely competitive game of tennis (and would continue to do so until his early twenties, when a back injury took just enough of the edge off his game that he quit). And of course, he had impeccable southern manners. He had to: he was surrounded by impeccable southern ladies. The Lickliders' home in University City, the suburb around Washington University, was a huge old house that they shared with Joseph's mother; Margaret's married sister and her husband; and Margaret's other, unmarried sister. Every evening from the time Robnett was five, it had been his duty and honor to take the arm of his maiden aunt, escort her to the dinner table, and hold out her chair for her like a gentleman. Even as an adult, Lick would be known as a remarkably courteous and considerate man who rarely raised his voice in anger, who nearly always wore a coat and tie even at home, and who found it almost physically impossible to remain seated when a woman entered the room.

But Rob Licklider was also maturing into a young man whose mind was very much his own. When he was a very small boy, according to a story he often told in later years, his father had served as a lay minister in their local Baptist church. Whenever Joseph preached, it was his son's job to crouch under the keyboard of the organ and operate the instrument's foot bellows for the church's elderly lady organist, who couldn't quite manage them by herself. One sleepy Sunday evening when Robnett was just about to nod off down there, he heard his father exhort the congregation, "Those of you who seek salvation, stand up!" So he instinctively leapt to his feet—and smashed his head against the bottom of the keyboard. Instead of finding salvation, he saw stars.

This experience, Lick would say, gave him an instant insight into the scientific method: Always be extremely careful in your work—and in your proclamations of faith.

Three quarters of a century after the fact, of course, it is impossible to know whether young Robnett really learned those lessons by slamming into a keyboard. But to judge from his attitudes in later life, he definitely learned them somewhere. Underlying his meticulous craftsmanship and insatiable curiosity was a complete lack of patience for sloppy work, easy solutions, or glib answers. He refused to be satisfied with the ordinary. The young man who would later talk of the "Intergalactic Computer Network" and publish professional papers with titles like "The System System" and "The Gridless, Wireless Rat-Shocker" possessed a mind that was constantly probing, and constantly at play.

He also possessed a streak of mischievous anarchy. When confronted by officious stupidity, for example, he would never challenge it directly; the belief that a gentleman never caused a scene was bred into his bones. But he loved to subvert it. When he pledged the Sigma Chi fraternity during his freshman year at Washington University, he was informed that pledges had to carry two kinds of cigarettes with them at all times so that upperclassmen could demand a smoke at any hour of the day or night. Not being a smoker himself, he promptly went out and bought the foulest Egyptian cigarettes that St. Louis had to offer. Nobody bothered him for a smoke more than once.

Meanwhile, that restless refusal to be satisfied with the ordinary was leading him on a roundabout quest for his purpose in life. He was already changing his identity to an extent. He had been "Robnett" at home and "Rob" to his schoolmates, but now, perhaps as a symbol of his new status as a college man, he started introducing himself by his fraternity nickname: "Call me 'Lick.' " From then on, only his very oldest friends would have the slightest idea who "Rob Licklider" was.

As for what the college man was going to *do*, however—well, he would have happily majored in everything if he could have; whenever Lick heard someone wax enthusiastic about a new field of study, he wanted to try it, too. So in his freshman year he majored in art for a while, then switched to engineering. Then it was physics and math. And for one disconcerting stretch, he even majored in the real world: at the end of Lick's sophomore year, embezzlers gutted his father's insurance company, and the firm collapsed, leaving Joseph out of a job and his son without the money for tuition. Lick had to drop out for a year and go to work as a carhop at a drive-in restaurant, one of the very few jobs he could find during the Depression. (Joseph Licklider, going stir-crazy just sitting around the house with all those southern ladies, went out and found a rural Baptist congregation in need of a minister; he and Margaret would spend the rest of their days serving in one country church after another, happier than they had ever been in their lives.) But when Lick finally returned to school, bringing with him a renewed enthusiasm for higher education—plus a convertible—one of his part-time jobs was to take care of the experimental animals in the psychology department. And once he began to understand the kind of research the professors there were up to, he realized that his quest was over.

What he had stumbled across was "physiological" psychology, a line of research that was then in the midst of a remarkable period of ferment. Today it would be known as neuroscience: the precise, detailed study of the brain and how it functions.

It was a discipline with roots going well back into the nineteenth century, when scientists such as Thomas Huxley, Darwin's most forceful advocate, had begun to assert that behavior, experience, thought, and even awareness had material bases in the brain. This was a fairly radical position to be taking in an age that was considerably more literal about religion than our own. Indeed, many scientists and philosophers in the early part of the nineteenth century tried to assert that the brain was not even made of ordinary matter, but was instead the realm of the mind and the seat of the soul, transcending all physical law. The evidence, however, soon began to show otherwise. As early as 1861, a systematic study of brain-damaged patients led the French physiologist Paul Broca to make the first association of a particular mental function—language—with a specific region of the brain: a patch of the left cerebral hemisphere now known as Broca's area. By the early twentieth century, it was known that the brain is an electrical organ, with impulses propagated by billions of thin, cablelike cells known as neurons. By the 1920s, it had been established that the regions of the brain that govern motor control and the sense of touch are located in two parallel bands of neural tissue running up the sides of the brain. It was likewise known that the centers governing vision are sited at the very back of the brain—ironically, at about the farthest point from the eyes themselves—whereas the hearing centers are right where they logically ought to be: in the temporal lobe, just under the ear.

But even this work had been comparatively crude. By the time Lick encountered the field, in the 1930s, researchers had begun to make use of the increasingly sophisticated electronics technology being spun off from the radio and telephone industries. Through electroencephalography, or EEG, they could eavesdrop on electrical activity in the brain by taking precise readings from detectors placed around the outside of the head. They could also go inside the skull and apply very precisely defined stimuli to the brain itself, and then measure how the neural response propagated anywhere else in the nervous system. (By the 1950s, in fact, they would be able to stimulate and record the activity of individual neurons.) And in the process, they could begin to map out the neural circuitry of the brain in unprecedented detail. In short, the physiological psychologists had gone from the early-nineteenth-century vision of the brain as something mystical, to a twentieth-century notion of the brain as something knowable. It was a system of immense complexity, to be sure. But it was nonetheless a *system*—not so very different in its fundamentals from the increasingly sophisticated electronics systems that physicists and engineers were building in their laboratories.

Lick was in heaven. Physiological psychology had everything he loved: mathematics, electronics, and the challenge of deciphering that ultimate gadget, the brain. He threw himself into it—and in the process, though he certainly couldn't have known it at the time, he took his first giant step toward that office in the Pentagon. Considering all that happened later, Lick's youthful passion for psychology might seem like an aberration, a sideline, a twenty-five-year-long diversion from his ultimate career in computers. But in fact, his grounding in psychology would prove central to his very conception of computers. Virtually all the other computer pioneers of his generation would come to the field in the 1940s and 1950s with backgrounds in mathematics, physics, or electrical engineering, technological orientations that led them to focus on gadgetry—on making the machines bigger, faster, and more reliable. Lick was unique in bringing to the field a deep appreciation for human beings: our capacity to perceive, to adapt, to make choices, and to devise completely new ways of tackling apparently intractable problems. As an experimental psychologist, he found these abilities every bit as subtle and as worthy of respect as a computer's ability to execute an algorithm. And that was why to him, the real challenge would always lie in adapting computers to the humans who used them, thereby exploiting the strengths of each.

For now, however, Lick's course was clear. In 1937, he finished his undergraduate career at Washington University with a triple degree in physics, mathematics, *and* psychology. He stayed on for another year to earn a master's degree in psychology. (The master's diploma awarded to "Robnett Licklider" was one of the last times that name ever appeared in print.) And in 1938 he entered the doctoral program at the University of Rochester in New York, one of the nation's premier centers for research on the auditory regions of the brain, the parts that dictate how we hear.

Lick's departure from Missouri wasn't just a change in address, as it happened. For the first two decades of his life he had been a model son to his parents, conscientiously attending Baptist services and prayer meetings three and four times a week. After he left home, however, he almost never set foot in a church again. He couldn't bring himself to tell his parents about it, knowing they would be terribly hurt by his abandonment of the faith they loved. But he found the strictures of Southern Baptist life unbearably oppressive. More important, he couldn't profess a belief he did not feel. As he later remarked when asked about all that time spent in prayer meetings, "It didn't take."

If many things changed, though, at least one thing stayed the same: Lick had been a star in the Washington University psychology department, and he was a star at Rochester. For his Ph.D. dissertation he made what may well have been the first maps of neural activity on the auditory cortex. In particular, he pinpointed the regions that seemed critical for distinguishing different sound frequencies, a key to our ability to hear musical pitch. And along the way he became such an expert in vacuum-tube electronics—not to mention such a creative wizard at designing experiments—that even his professors started coming to him for advice.[1]

Lick also excelled at Swarthmore College, outside Philadelphia, where he took a temporary appointment as a postdoctoral student after receiving his Ph.D. degree in 1942. In his short time there he proved conclusively that contrary to Gestalt theories of perception, magnetic coils placed in an asymmetrical pattern around the back of a subject's skull would *not* cause distorted vision—though they *would* make the subject's hair stand on end.

In general, however, 1942 was not a good year for lightheartedness. Lick's career, like the careers of a great many other researchers, was about to take a much more serious turn.

IN THE DUNGEON

Wars are noisy, to put it mildly. And by the eve of the United States' entry into World War II, it was clear to the Army Air Corps that war had become intolerably noisy in military aircraft. The roar of the engines and the concussion of anti-aircraft fire could be deafening—literally. The din made voice communication difficult, if not impossible, and the incessant vibration was thought to be a major factor in pilot fatigue: American B-17 crews flying for Britain under the Lend-Lease program were returning from their missions over Germany in such a state of exhaustion that they could barely land safely.

So in the late summer of 1940, with the Battle of Britain already raging over London, the Air Corps decided that its increasingly urgent preparations for war should include a major research program in practical acoustics. After a short delay wherein certain big-name researchers in the field proved to be much more interested in fighting one another for control than in doing the work, leadership of the project fell to Leo Beranek, a very young Harvard physicist whose only credential was a brand-new doctor of science degree in acoustics. Beranek, feeling more than a little overwhelmed by his sudden responsibility, quickly talked the Air Corps into funding a separate, allied laboratory to study the psychological and physiological impact of noise. The two laboratories would function like "the two ends of a dumbbell," as he put it. To head this Psycho-Acoustics Laboratory, as it came to be known, Beranek selected a thirty-three-year-old Harvard psychologist named Stanley Smith "Smitty" Stevens.[2]

Good choice. The formidable Stevens was a muscular, rough-hewn Mormon who had grown up in the polygamous frontier household of his grandfather. He was blunt. He was demanding. And he was the kind of man who did exactly what he said he would do—which in this case meant creating the finest psychology laboratory in the world. Stevens's strategy was simple: hire the most brilliant experimenters in the country, give them the best equipment money could buy, inspire them to the highest possible standards of intellectual clarity and experimental precision—and work them fourteen hours per day. With the Air Corps's comparatively lavish funding behind him, moreover, Stevens proceeded to do just that; by the end of the war, his lab would be a beehive of almost fifty young researchers.

Among the first, of course, was an affable twenty-seven-year-old named J. C. R. Licklider. Indeed, Stevens seems to have taken all of about five minutes to make up his mind: "I have a disciple here," he declared after Lick came up from Swarthmore for an interview in the spring of 1942. "I need this guy!"

The feeling was mutual. From Lick's point of view, the work going on in Stevens's lab was not only vitally important to the war but fantastically interesting. He would be working with state-of-the-art equipment of the sort that most psychologists could only dream about, *and* he would be interacting on a daily basis with the physicists and electronics engineers in Beranek's laboratory. Lick accepted Stevens's offer immediately and started full-time at the Psycho-Acoustics Lab in the summer of 1942.

It has to be said that the laboratory environment itself wasn't much of an attraction, especially in those early days. The staff called it the Dungeon, and with good reason. Harvard, having very little space that was suitable for classified war research, and recognizing that no one wanted a high-decibel sound-research laboratory as a next-door neighbor, had decided to house the Psycho-Acoustics Lab just outside Harvard Yard, in the basement of a bizarre Victorian-Gothic structure named Memorial Hall. (Try to imagine a redbrick cathedral with a mansard roof and castle battlements.) The outside basement door was almost impossible to find unless you knew just where to look, while the basement windows were set in wells below grade and were blacked out anyway. Inside, moreover, the basement was in the throes of renovation, much of it done by Stevens himself, with his own hands. On his first day there, Lick was shown through a dingy warren of soot-covered pipes, cramped offices, and hastily opened up laboratory spaces, all littered with the dusty clutter of construction.

It also has to be said that life under Stevens was not exactly low-key. "When [Stevens] was seriously interested in a problem, he could move forward only at full speed," recalled George Miller in a memoir written after Stevens's death, in 1973. "Sometimes he ran over you."[3] In truth, Stevens's title as "director" of the Psycho-Acoustics Lab didn't quite capture the reality. He was its patriarch.

"Stevens was a primitive—he had in him the force of Nature," wrote Miller, who first encountered the director in August 1943, when he arrived to start his graduate studies at Harvard. "When the clouds gathered and thunder rolled forth, he was as little concerned as Nature for who might be caught in the storm."[4]

And yet Miller, Lick, and many others somehow managed to thrive in that atmosphere, Miller noted. It was largely a matter of understanding Stevens's ways. From time to time he might appear at the door of your office and bark, "Know what you're doing?" But once you realized that this was just a basically shy person's way of saying "Good morning," the panic subsided, and you could begin to see a man who cared deeply about his laboratory and everyone in it. "The laboratory was his family, and members were given the duties and privileges of siblings, nephews, or cousins," wrote Miller. "The head of this extended family was concerned for the welfare of his kindred, and he rewarded them or disciplined them for their own good and the good of the group." The patriarch even looked after the social life of his family, taking his charges to dinner at the Faculty Club, on a group foray to Boston's Chinatown, and for weekends at his "farm" in New Hampshire—trips that included doing maintenance work in the summers and skiing in the winters.[5]

Lick, in particular, was in his element. He was just as committed to clarity and excellence as Stevens was. And besides, playing around in the lab until the wee hours was his idea of fun. "The first thing I remember Lick doing at the laboratory was to borrow from the petty cash fund—with Smitty's approval," says Karl Kryter, who had been Lick's roommate and best friend when they were both in graduate school at the University of Rochester, and who arrived at the Psycho-Acoustics Lab not long after him. "Then he went out to the clothing stores and bought up all these snaps, the kind that you used to make shirts and dresses. He would stay in the lab until one or two in the morning putting together boards and wires with snaps on them. He called this setup the Snappiac. He knew that if you asked the university shop to build you a gizmo for your experiment, it would require two or three months. But with the Snappiac, he could rearrange components very easily, so he could put together any kind of amplifier or electronic components he wanted."

Lick's research at the Psycho-Acoustics Lab focused mainly on the process of speech understanding, and how distortion in a radio or telephone signal can affect a listener's ability to make sense of what is being said. To everyone's surprise, including his own, he discovered that at least one kind of distortion could actually make words *easier* to understand. "Peak-clipping," as he called it, had the effect of selectively emphasizing the consonants in each word relative to the vowels: CoN-So-NaNT-S. This helped the words punch through any background noise. That discovery in turn led to a major research effort within the Psycho-Acoustics Lab, to which many members contributed—though with Lick always at the forefront.

Indeed, in his soft-spoken way, Lick was proving to be a natural-born leader. A classic illustration of this quality came at the culmination of the peak-clipping project, when Lick, Kryter, and several other researchers from the Psycho-Acoustics Lab went down to Eglin Army Air Force base in Florida to carry out field tests using military radio equipment. It later became one of Lick's favorite stories to tell on himself. When they checked in at the base, each member of the group was asked his equivalent rank. Not understanding the significance of the question, they all answered "Lieutenant" and were shown to cramped little rooms. All, that is, but Lick, who declared that his equivalent rank was "General" and was shown to a fine suite of rooms in the Officers Club. But then the others simply made his suite their headquarters and partied every night. Lick never did get any sleep (or so he claimed).[6]

Back in Cambridge, meanwhile, Lick had attracted the somewhat perplexed attention of Louise Carpenter Thomas, a twenty-three-year-old Kansan who worked as one of the only two secretaries (and only two females) in the lab. She was married to a graduate student over in the main psychology department, several blocks away. She was also blond, vivacious, and stunningly pretty, reminding many people of Ginger Rogers. It was Louise Thomas who had shown Lick around the Dungeon on his first day there, and she had recognized his accent immediately. "So you're from Missour*ah*," she'd teased him, deliberately mispronouncing it. Lick had been able to respond only with a laugh. In the movies, the actresses he liked best had always been the healthy, wholesome, all-American girls; now he suddenly found himself standing right next to one.

Actually, Louise wasn't quite as apple-pie perfect as all that. True, she had grown up in the tiny, tiny town of Pleasanton, Kansas, in the remote rural county of Linn. But when she left there, she *left:* "I rejected everything—including the Republican party," she recalls. At the University of Kansas she became passionately involved in the theater and found herself drawn to the generally left-wing arts crowd. "I was very liberal in college and marched in the May Day parade, and all that silliness that was so important to us back then," she says. Every morning she would buy the socialist magazine *PM* as she walked to work through Harvard Square. Then she would leave it on the corner of her desk so the guys could pick it up and read it when they went to the men's room.

She had never been able to figure out exactly what Smitty Stevens expected her to do at the lab, especially since he had been more than a little bit drunk when he asked her to join his staff there. (He'd cornered her at the psychology department's Christmas party, shortly after Pearl Harbor, and ordered her to report for an interview the next day.) She assumed that Stevens just liked having pretty girls around to impress the visiting colonels. So, lacking any better job description, she simply made herself into an indispensable gal-of-all-trades, editing papers, putting out research contracts, helping build the experimental apparatus, and doing whatever else needed to be done.

Louise Thomas had some trouble at first making up her mind about the new guy. He was warm and pleasant, of course. Very smart. Lovely smile. Obviously one of Smitty's favorites. And yet she was struck by the fact that whenever Lick went to the soda machine—which was constantly; he drank Coca-Cola morning and night—he'd come back with an armload of Cokes for everybody. Whenever he passed somebody's desk on his way to the pencil sharpener—and he insisted on very sharp pencil points—he'd offer to sharpen that person's pencils, too. As she told the other woman in the office one day, about a week after Lick arrived, "I can't figure it out. He's too . . . *nice.*"

It can't have been too long after that that Louise and the other woman went out to lunch and Louise came back to find a fresh gardenia in her desk drawer. In fact, the gardenias got to be a regular thing, she says, though "it took me a long time to find out where they were coming from!"

In any case, the Psycho Acoustics Lab was much too small and claustrophobic for staffers to remain strangers for long. So by and by, Louise, Lick, and Karl Kryter became a threesome, frequently going out to lunch together. "He and Karl were like a comedy team," she remembers. "They were very different people, but such very, very good friends." But after a while she couldn't help but notice how often Karl would suddenly remember some errand: "Got to go get my shoes," he would say as they made their way across Harvard Yard, and then he'd veer off to leave the other two to go to lunch on their own.

Lick, Louise was discovering, was the most amazingly unspoiled person she had ever met. "He could strike up a conversation with *anybody*—waitresses, bellhops, janitors, gardeners. It was a facility I marveled at. No one was a stranger to him, not if that person seemed interesting enough to talk to. And he was also the most creative person I'd ever known. Everything he did—whether it was the puns, which were constant and horrible, or the stories he would tell—everything he did was multilayered. He would start laughing at something and then see another reason to laugh, and then another. He was always having fun, always challenging conventional ideas."

So as time went by, she says, there were many lunches with Lick. A stroll through the Fogg art museum on campus. A walk along the Charles River. When she learned that his childhood name had been Robnett, she teasingly started to call him Robin. And then her husband, Garth Thomas, learned that his student draft deferment was about to expire and hurriedly took a draft-deferred research job with a firm down in Florida, while she stayed in Cambridge. . . .

In the end, when what was inevitable became unavoidable, it was all handled about as amicably as such things can be. Louise's divorce was finalized in late 1944. She and Lick were married in Harvard's Appleton Chapel on January 20, 1945. And every Saturday, the weekly anniversary of their first real date, she was still getting a fresh gardenia.

It wasn't easy. But somehow, even in the midst of late-night lab sessions, a surreptitious love affair, and marriage, Lick found time to indulge his endlessly restless curiosity. "Lick put his oar in almost everywhere," recalls Leo Beranek, who was working with him regularly during this period. Indeed, as Louise Licklider had come to know very well by that point, her new husband almost couldn't help himself: "He had a very hard time *not* grabbing on to the latest wave of ideas," she remembers, "or to anything that sounded interesting or that he didn't know much about. He wanted to learn about physics. He wanted to learn about abstract math. And when he wanted to learn something, he went after the people at the top—the ones who were most imaginative."

Once again, Lick had no way of knowing that he was taking yet another giant step on his long route to computers. What he did know, however, was that when he went after the people at the top, the ones who could really teach him something, he was rarely finding them at Harvard. More and more, he was finding them two miles away, on the other side of Cambridge—at the Massachusetts Institute of Technology.

THE *WIENERWEG*

Aesthetically, at least, the drive from Harvard to MIT was a real comedown in those final days of the war. Harvard, founded more than three hundred years earlier on the western side of Cambridge, still inhabited an Ivy League world of trees, town houses, radical bookstores, and cerebral good taste. But MIT, established in 1861 amid the industrial flats on the eastern side of town, looked more than ever like a redbrick engineering school. The main campus was handsome enough, with its imposing central dome, broad sweep of lawn running down to the Charles River, and fine view of Boston on the far side. But since the beginning of the war, MIT had been the home of a top-secret crash program known as the Radiation Laboratory, in which some four thousand physicists and engineers were feverishly working to perfect the technology of radar. And to house them, the university had expanded into the nearby neighborhoods with row after row of stark, ugly, prefabricated laboratory buildings. Even before the war, the area around the campus had been a dreary mix of warehouses, power plants, and light industry. Now it was even drearier.

Intellectually, however, the trip from Harvard to MIT was another story entirely. Once upon a time the institute *had* been just an engineering school, and not much more. But in the 1930s, under the inspired presidency of the physicist Karl Compton, MIT had begun to transform itself into a world-class center for physics, chemistry, mathematics, architecture, and much else besides. Then the war had transformed the place yet again, as the federal government poured in some $117 million for the Radiation Lab, the Servomechanism Lab, the High Voltage Lab, and innumerable smaller projects—a huge sum of money in those days, and far more than went to any other single contractor. By 1945, the level of intellectual energy in East Cambridge was unmatched anywhere else in the country, with the possible exception of Los Alamos. The war years had seen physicists working side by side with psychologists; psychologists learning from mathematicians; and mathematicians immersing themselves in electrical engineering. For four solid years, people had talked, argued, fought, created. Now they were alive with new ideas, new techniques, new ways of looking at the world.

And right there at the intellectual epicenter was a portly little man who could often be spotted wandering through the MIT campus in a natty suit and vest, his stubby fingers gripping an enormous cigar. He wasn't an imposing figure, to put it mildly. His awkward, splay-footed way of walking made him waddle like a duck, while his graying goatee and thick, black-rimmed glasses made him look, well, distinctive: the reporters of the day invariably compared him to Santa Claus, but a later generation would have instantly thought of Kentucky Fried Chicken.

However, that faintly ridiculous impression never lasted past the first encounter. "How's it going?" Norbert Wiener would invariably ask as he dropped into someone's office unannounced. Then, without waiting for an answer, he would launch into a discourse on whatever he was thinking about at the moment. A mathematical analysis of neurological disease, perhaps, or the abstract essence of communication, or the question of how an inanimate machine could embody something as ineffable as purpose. The conversation was usually a monologue and almost always brilliant, even when it bordered on the delphic. Robert Fano can vividly remember Wiener's doing this to him again and again in the years immediately after the war, when Fano was working toward his Ph.D. in electrical engineering. "You know," Wiener would announce each time, "information is entropy." And then he would walk out of Fano's office without saying another word. By the time Fano managed to figure out just what Wiener was getting at, he had also managed to turn his own doctoral dissertation into a crucial contribution to the newly emerging field of information theory.

Wiener's impromptu wanderings were known around campus as the *Wiener-weg*, German for "Wiener's path," and might include conversations with anyone from janitors on up. "He would just walk right in—whoever was there, he would interrupt and start talking," marveled a later MIT president, Julius Stratton, a former student of Wiener's who actually rather enjoyed the interruptions.[7] "Whatever was on his mind, Norbert Wiener's visit was one of the high points of the day at MIT for me and many others," agreed Jerome Wiesner, then a rising star within the Radiation Laboratory and later an MIT president himself. "[He] drew the communications sciences together at MIT . . . by personal contact, by stimulating meetings, and particularly by writing and speaking."[8]

Indeed, Wiener had long since become the single most influential figure at MIT, especially when it came to generating new ideas in unexpected places. He didn't just cross disciplinary boundaries, noted another commentator; he never even noticed their existence.

But then, he'd had plenty of practice at that sort of thing. His academic career had been under way for forty years by that point, ever since he entered college, in 1905—at the age of eleven. Norbert Wiener was that rarest of creatures, a former child prodigy who had actually made good.

Maybe it was his goatee and cigar, or the influence of his student days overseas, but somehow Wiener looked European, as if he ought to be speaking with a mysterious Continental accent and pronouncing his name in the Viennese manner: VEE-ner. In fact, he pronounced it with a perfectly ordinary American accent: WEE-ner, like the hot dog. Technically speaking, he was another Missouri boy, having been born there on November 26, 1894, when his Russian-immigrant father, Leo, was teaching German at the state university in Columbia. For all practical purposes, however, his true home was the rarefied air of Cambridge, Massachusetts, where his family had moved in 1895. Leo Wiener was a prodigiously talented man who was familiar with a broad range of science and mathematics, and who reputedly was able to read some forty different languages; he eventually became the first professor of Slavic languages at Harvard. But Leo was also a firm believer in nurture over nature: genius, he felt, could be systematically engineered. He therefore took his son's education in hand when Norbert was age six, and shortly became a harshly perfectionist taskmaster—surprisingly so, since he was a loving parent otherwise. Norbert was often reduced to tears over his lessons. Yet when he went back to public school two years later, he was seven years ahead of his age group—thus his entry into nearby Tufts University at eleven. In 1913 he received his Ph.D. in mathematics from Harvard, at age eighteen. From then until 1917, when the United States entered World War I, he did postdoctoral work at Cornell, Columbia, Cambridge, Göttingen, and Copenhagen universities, studying with such figures as the philosopher-mathematician Bertrand Russell and the great German mathematician David Hilbert. In 1918, as a clumsy, hopelessly nearsighted, but intensely patriotic twenty-four-year-old, he put his mathematical skills to use as an army private at the Aberdeen Proving Ground in Maryland, where he calculated artillery trajectories by hand. And in 1920, after briefly working as a journalist for the *Boston Herald* to tide himself over between jobs, he joined the mathematics faculty at MIT.

It was *not* a prestigious appointment. MIT's transformation still lay in the future, and the mathematics department existed mainly to teach math to the engineering students. The school wasn't oriented toward research at all. However, no one seems to have informed Wiener of that fact, and his mathematical output soon became legendary. The Wiener measure, the Wiener process, the Wiener-Hopf equations, the Paly-Wiener theorems, the Wiener extrapolation of linear times series, generalized harmonic analysis—he saw mathematics everywhere he looked. He also made significant contributions to quantum theory as it developed in the 1920s and 1930s. Moreover, he did all this in a style that left his more conventional colleagues shaking their heads. Instead of treating mathematics as a formal exercise in the manipulation of symbols, Wiener worked by intuition, often groping his way toward a solution by trying to envision some physical model of the problem. He considered mathematical notation and language to be necessary evils at best—things that tended to get in the way of the real ideas.

Wiener's eccentricities were meanwhile becoming equally legendary. There was his remarkable clumsiness, for example, which he attributed in part to near-sightedness: his eyeglass lenses were so thick that he couldn't judge where things were, which meant that he tended to trip over them. There was his disconcerting habit of loudly snoring through other people's lectures, only to wake up at the end with a series of penetrating questions and insights. And then there was his phenomenal absentmindedness. Legend has it that he was once wandering along a hallway, distractedly running his finger along one wall, when he encountered the open doorway of a classroom. Lost in thought, he followed his finger around the doorjamb, around all four sides of the room, and out the door again without ever once realizing that a lecture was in progress. On another famous occasion—an incident recounted so often that it probably happened more than once—Wiener ran into a friend as he was walking down a campus sidewalk. After they chatted for a moment, he asked the friend to remind him which way he had been going when they met. When told that he'd been heading *away* from the Faculty Club, Wiener said, "Oh—that means I've already had lunch."

Finally, there was the fact that he could never remember such mundane details as people. "I must have met Wiener a half dozen times, and I was a stranger each time," recalls George Miller, who would spend several years at MIT during the 1950s. Once, in fact, Miller escorted a visiting journalist to Wiener's office, in-tending to introduce the two. No need: Wiener and the journalist immediately fell to conversing in an obscure dialect of Chinese (having inherited his father's linguistic prowess, Wiener had a working knowledge of thirteen languages). After two or three minutes of this, Wiener stopped and looked at Miller: "And who are you?" he asked. Miller quietly left.

People, in short, were not Wiener's forte. He was as famous for his feuds as he was for his mathematics. He would quarrel with lifelong friends over other-wise innocuous scientific disagreements, which he tended to take as personal at-tacks. And then he would quarrel with other friends if they didn't back him up enthusiastically enough in the first fight. He could also be suspicious to the point of paranoia. While studying at Gottingen before World War I, Wiener had claimed (apparently without much basis in fact) that his mathematical ideas were stolen by Richard Courant, the administrative head of the university. Wiener later wrote an unpublished novel in which a professor who bore a strik-ing resemblance to Courant built his reputation on ideas appropriated from young geniuses.

"He was fundamentally a gentle and humane soul," wrote Wiener's former colleague Pesi Masani in a generally adoring biography. "He was, however, given to recurrent manifestations of petulance, egotism, emotional instability, irra-tional insecurity, and anxiety."[9] A childhood spent as a prodigy, not to mention Leo Wiener's tutoring methods, had clearly taken their toll.

For all his shortcomings, however, Wiener had transformed MIT mathemat-ics by the 1940s, even as the institute was transforming itself. And now, amid all that explosive intellectual energy in wartime Cambridge, Wiener was doing as much as anyone to provide a focus for it, a forum where people could get together and talk until all hours about crazy, visionary ideas—especially *his* kind of ideas.

The venue was Wiener's monthly series of "supper seminars," which he began shortly before the end of the war. "Dinnertime talk was always very animated and, though a bit rambling, tended to explore the special interests of the evening's guests," recalled Jerry Wiesner. After dinner, he added, someone would try to give a reasonably formal talk about his work, though it wasn't always easy. "These talks were expected to be short, but interruptions and diversionary arguments were so frequent that the speaker would often still be trying to finish his outline when the group gave up at midnight. There were often heated arguments, many of which would seem naive today; for example, the analogies between the functioning of digital computers and the brain were debated almost endlessly."[10]

Lick was usually there, of course, along with Stevens, Miller, and many others from the Psycho-Acoustics Lab. Jerry Wiesner remembered Lick as the tall fellow who was always in the thick of those arguments, Coke bottle in hand, having a grand time. Lick had obviously thought a lot about the relationship between humans and machines, thanks to his work in Memorial Hall. And now, like all the others, he listened avidly as Norbert Wiener articulated a breathtaking vision of where that relationship was taking mankind.

"The thought of every age is reflected in its technique," Wiener asserted shortly after the war. During the Scientific Revolution of the seventeenth and eighteenth centuries, for example, when figures such as Galileo, Kepler, and Newton were laying the foundations of modern science, the most advanced and evocative technology had been that of the clock, whose gears seemed to move with the timeless, pristine perfection of a planet orbiting the sun. It was no coincidence that seventeenth-century philosophers such as René Descartes had described even plants and animals as organic clockwork mechanisms. Then, during the Industrial Revolution of the nineteenth century, the defining technology had been that of the steam engine, which was capable of converting vast amounts of energy and heat into work. And again it was no coincidence that scientists of *that* era had conceived of living organisms as biological heat engines, mechanisms that burned food to do useful physiological work.

But now, said Wiener, in the twentieth century, we could perceive the beginnings of a new revolution. Unlike clocks and steam engines, he argued, the emerging technologies of the modern age didn't just operate blindly, forging ahead without any reference to the world around them. Rather, they operated responsively, taking in information from their surroundings to guide their future actions. "The machines of which we are now speaking are not the dream of the sensationalist nor the hope of some future time," he wrote. "They already exist as thermostats, automatic gyrocompass ship-steering systems, self-propelled missiles—especially such as seek their target—anti-aircraft fire-control systems, automatically controlled oil-cracking stills, ultra-rapid computing machines, and the like. . . . Scarcely a month passes but a new book appears on these so-called control mechanisms, or servomechanisms, and the present age is as truly the age of servomechanisms as the nineteenth century was the age of the steam engine or the eighteenth century the age of the clock."[11]

By the same token, said Wiener, the dominant sciences of the past had largely been physical sciences, dealing with matter, energy, motion, and intensity. But the central sciences of the modern era would increasingly deal with systems-level issues such as communication, control, organization, and information—whether in "the animal or the machine," as he put it. The older sciences were already homing in on the fundamental nature both of matter and of the universe itself, two of the deepest mysteries of human existence. But the new sciences would point the way toward the fundamental nature of life and the mind, mysteries that the physical sciences had never been able to touch. Indeed, Wiener believed that these new sciences had already brought mankind to the brink of a kind of Grand Unified Theory of behavior, encompassing biological systems and artificial systems alike—a theory based on the underlying themes of *communication* and *control*.

Out of all the developments that had led him to this conviction, Wiener would explain as Lick and the others happily listened, perhaps the first and most important was the automatic computer. In fact, he had been following this technology ever since the mid-1920s, when his wanderings through the MIT campus had led him to the electrical engineering department and a professor there named Vannevar Bush.

CHAPTER 2

THE LAST TRANSITION

Actually, Vannevar Bush would have been a hard man to miss. He was only in his midthirties when Norbert Wiener first encountered him in the 1920s, but he was already a New England classic: a descendant of sea captains, a Yankee inventor, a lanky, gruff, straight-talking individual who looked like Uncle Sam without the beard. He radiated competence and authority. And he had an instinctive rapport with hardware that left the hopelessly fumble-fingered Wiener in awe. "One of the greatest apparatus men that America has ever seen," Wiener later called him, "[a man who] thinks with his hands as well as with his brain."[1]

Best of all, for Wiener, was that Bush shared his fascination with computing devices. True, this was a distinctly unmathematician-like interest on Wiener's part: most other members of his tribe were contemptuous of computation, arguing that a real mathematician gained insight by abstract reasoning, not by reckoning. Slide rules were acceptable for scientists and engineers, but brute number crunching was just arithmetic, a task for desktop adding machines—*women's* work (the word *computer* was still a job description in the 1920s, carrying much the same pink-collar connotation as *typist*). The building of computing machinery was, by extension, a job for mere tinkerers.

The greatest apparatus man in America: Vannevar Bush contemplates the Differential Analyzer

Wiener, however, was among the few great mathematicians who had actually worked as computers, and his patriotic stint calculating artillery trajectories during World War I had taught him all too well that numerical calculation was *not* just a matter of punching numbers into an adding machine. A professional computer had to understand the whole complex process of computation—knowing which calculations to do in which order, for example, or whether to do *this* sequence of calculations or *that* sequence based on the result of, say, step C. That insight in turn resonated with Wiener's lifelong interest with the workings of the brain and with the fundamental underpinnings of intelligence (in 1909 he had actually started his graduate studies at Harvard in the zoology department, until his hopeless clumsiness at the lab bench sent him back to mathematics, he had hoped to make brain science his career). The notion of a machine that could automate computation was deeply compelling to Wiener: here was a device that was purely mechanical yet could carry out a task that seemed to embody intelligence itself.

And now here was Vannevar Bush, taking that automation to a whole new level.

For Bush, the inspiration had come early in the 1920s, when he tackled one of the most vexing technical problems of the day, the instability of electric power networks. The equation that described such networks was straightforward in principle but horrendous in practice, and all but impossible to solve by hand. The result had been a plague of brownouts and blackouts as power companies struggled to meet soaring demand with high-voltage lines designed on the basis of the rule of thumb and guesswork. Bush was convinced that there had to be a better way.

He was right. By 1927, after experimenting with a number of preliminary devices, Bush and his students started construction on the Differential Analyzer, an elaborate system of gears and pulleys and parallel rods that took up most of a large room. Completed in 1930, the analyzer was an *analog* computer, meaning that it represented mathematical variables not by numbers, the way a digital computer would, but by measuring the rotation of various rods. So to solve a power network, for example, Bush and his students would assign the variable *time* to one rod, the variable *current* to another, and so on. Then they would "program" the analyzer to solve that horrendous network equation by connecting all the rods with gears and pulleys in a way that mimicked the structure of the equation. In effect, they would turn the analyzer into a physical model of the network.

Granted, that last step was the tricky part; setting up the analyzer for something as complicated as the network could take as long as two days. But no matter: once all the connections were complete, solving the equation was as simple as starting the analyzer's motor. Gears would mesh, pulleys would pull, rods would rotate—and a plotter pen connected to the appropriate rod would neatly trace out the solution on a sheet of graph paper, accurate to 2 percent. Calculations that had once required days or weeks could now be completed in minutes.

For scientists and engineers used to endless sweaty hours with slide rules and calculators, this kind of performance was a godsend. By the mid-1930s, researchers from all over the world were flocking to the analyzer, using it not just for electrical engineering but for atomic physics, seismology, and astrophysics—including one epic occasion that had five staffers and eighteen graduate students working for thirty days to calculate the motion of cosmic rays in Earth's magnetic field. Replicas of the machine were either under construction or already completed at nearly a dozen sites in the United States and abroad. And the original analyzer had undergone numerous upgrades—including several suggested by Norbert Wiener, whose instinct for understanding mathematical concepts in physical terms had produced a steady stream of ideas for "analogy machines," as he called them. Indeed, over the years Wiener had become a regular visitor at Bush's lab and a close friend of its director, who later wrote that he'd never realized "a mathematician and an engineer could have such good times together."

Bush himself, meanwhile, had even bigger plans. In the spring of 1935, once the Depression had eased a bit, he got the Rockefeller Foundation to contribute eighty-five thousand dollars for the development of a second-generation Differential Analyzer based entirely on high-speed electronic circuits. Of course, this new machine would still be an analog computer, in the sense that it would still have to be "programmed" to create a physical model of each equation; but because the modeling would be done with circuits and voltages instead of with rods, gears, and pulleys, it would be far faster, more precise, and more versatile than the original. And indeed, after the "Rockefeller Analyzer" was finally dedicated, in 1942, following numerous delays due to its complexity, it would go on to do yeoman computational work during World War II.

Even more ambitious was Bush's scheme to consolidate the Rockefeller Analyzer project and several smaller development efforts into a world-class "Center of Analysis," which would systematically pursue the creation of innovative new calculating machines based on cutting-edge technologies such as microfilm, photocells, and digital electronics. For better or for worse, however, this proposal was a harder sell; Bush's formidable powers of persuasion didn't bear fruit until January 1939, when the Carnegie Corporation at last agreed to put up forty-five thousand dollars in seed money for the new center. And by that point, it was clear that the project would have to proceed without Vannevar Bush at the helm. Just as the money was arriving at MIT, Bush was leaving.

In truth, Bush had become increasingly distracted from his computer work ever since 1932, when MIT president Karl Compton asked him to serve as the first dean of MIT's newly formed engineering school, as well as vice president of the institution itself. As the decade wore on, moreover, his preoccupation with administrative duties had given way to an even greater anxiety over the looming threat of a new world war, and a conviction that the United States would almost inevitably be drawn into that conflict. Bush saw an urgent need for the country to prepare itself technologically as well as militarily. No one else seemed to be stepping up to meet that challenge, so in late 1938 he accepted an offer to become director of the Carnegie Institution of Washington, a major private research foundation that just happened to be located at the seat of power, where he hoped he could have some influence.

Bush departed for Washington in January 1939 and immediately started educating himself about the eternal game of politics, an activity that he found distasteful at best but that he knew to be essential to his larger goal of preparedness. He also struggled with the Carnegie Institution itself, which proved to be badly in need of revitalization. And in the meantime, he tried to tie up loose ends from Cambridge—including one idea that had been niggling in his brain for more than half a decade.

Bush's idea had first appeared in print in "The Inscrutable 'Thirties," a semiserious speculative article he published in January 1933. Toward its end, almost in passing, he fantasized about being able to fit an "unabridged dictionary on a square foot of film," and about "the contents of a thousand volumes [being] located in a couple of cubic feet in a desk, so that by depressing a few keys one could have a given page instantly projected before him."[2]

Bush was talking about microfilm, of course, which would allow all the world's literature to be stored page by page as tiny photographs. He may well have gotten the idea from librarians, some of whom had been touting the virtues of that medium since 1926. But whatever had inspired him, Bush had taken the "library problem" to heart, declaring it to be the most critical bottleneck in modern research: "The investigator is staggered by the findings and conclusions of thousands of other workers," he wrote a few years later, "[yet] the means we use for threading through the consequent maze to the momentarily important item is the same as was used in the days of square-rigged ships."[3]

The Yankee Inventor had continued to refine his desk-library notion during the rest of his time at MIT; at least two of the projects in his lab, the Navy Comparator and the Rapid Selector, were actually conceived as early prototypes of the system. And now, in Washington, Bush felt it was time for the idea to reach a wider audience. So on December 7, 1939, having prepared an article that he hoped would be accessible to lay readers, he submitted the forty-five-page typescript of "Mechanization and the Record" to Eric Hodgins, the publisher of *Fortune* magazine.

Hodgins loved it. "One of the most exciting events of all my life in the publishing business," he called the manuscript. "If anyone had told me that the future of filing and reference systems could be made the subject of such a bold . . . attack, I would have laughed at him prior to reading [Bush's] essay."[4]

Exciting indeed—and that was in 1939, before anyone could know just how prescient the article really was. Granted, from today's perspective, Bush's technology looks hopelessly quaint: his desk library was still very much an analog device, grounded in the microfilm and photocell technologies of the 1930s. In a larger sense, however, his vision was thoroughly modern, so that reading his 1939 draft now is like finding a fragment of turn-of-the-millennium sensibility that has somehow fallen sixty years backward in time.

The essence of that sensibility lay in two ideas that were introduced for the first time in the 1939 draft. First, Bush no longer described his desk library as a tool for trained librarians, as he had in previous discussions, but rather characterized it as a device that *anyone* could use. More than that, he stressed that the desk library wouldn't just help those nonprofessional users do the same old filing and retrieving a little faster; it would also support and improve their very "processes of thought."

Of course, Bush was writing here for a general audience, so perhaps his emphasis on nonprofessionals shouldn't be a complete surprise. Nonetheless, his vision of high technology's enhancing and empowering the individual, as opposed to serving some large institution, was quite radical for 1939—so radical, in fact, that it wouldn't really take hold of the public's imagination for another forty years, at which point it would reemerge as the central message of the personal-computer revolution.

Even more radical, however, was Bush's second new idea, which was the mechanism for implementing his vision. Instead of storing those countless microfilmed pages alphabetically, or according to subject, or by any of the other indexing methods in common use—all of which he found hopelessly rigid and arbitrary—Bush proposed a system based on the structure of thought itself. "The human mind . . . operates by association," he noted. "With one item in its grasp, it snaps instantly to the next that is suggested by the association of thoughts, in accordance with some intricate web of trails carried by the cells of the brain. . . . The speed of action, the intricacy of trails, the detail of mental pictures [are] awe-inspiring beyond all else in nature."[5] By analogy, he continued, the desk library would allow its user to forge a link between any two items that seemed to have an association (the example he used was an article on the English long bow, which would be linked to a separate article on the Turkish short bow; the actual mechanism of the link would be a symbolic code imprinted on the microfilm next to the two items). "Thereafter," wrote Bush, "when one of these items is in view, the other can be instantly recalled merely by tapping a button. . . . It is exactly as though the physical items had been gathered together from widely separated sources and bound together to form a new book. It is more than this, for any item can be joined into numerous trails."

Such a device needed a name, added Bush, and the analogy to human memory suggested one: "Memex." This name also appeared for the first time in the 1939 draft.

In any case, Bush continued, once a Memex user had created an associative trail, he or she could copy it and exchange with others. This meant that the construction of trails would quickly become a community endeavor, which would over time produce a vast, ever-expanding, and ever more richly cross-linked web of all human knowledge.

Bush never explained where this notion of associative trails had come from (if he even knew; sometimes things just pop into our heads). But there is no doubt that it ranks as the Yankee Inventor's most profoundly original idea. Today we know it as hypertext. And that vast, hyperlinked web of knowledge is called the World Wide Web.

And yet for all its prescience, and for all the initial enthusiasm at *Fortune,*

Bush's vision of the Memex wouldn't actually reach the public for another six years: Bush, and the rest of the world, were simply too distracted in the early 1940s. The Memex article we know today is a revised version that appeared in the July 1945 issue of the *Atlantic Monthly* under the title "As We May Think." Even then, if the measure of such an article is its direct, technological impact, the piece must be judged a failure. The initial public response was quite favorable, much to Bush's satisfaction. Yet in hindsight, he had managed to be both too far behind the times—his analog microfilm approach was hopelessly unworkable—and too far ahead of them: the newly emerging digital technology was hopelessly immature. *Indirectly*, however, the article's influence would be profound. In effect, Bush's larger vision of individual empowerment and creative enhancement was taken into the computer community's collective unconscious, lurking in certain brains like a time bomb that wouldn't explode for another twenty or thirty years, when the technology to realize that vision was finally ready.

But again, that was later. For now, Bush had ample reason to be preoccupied. On September 1, 1939, three months before Bush mailed off his typescript to *Fortune*, Hitler's invasion of Poland had ignited the long-dreaded war in Europe. Six months afterward, following an intensive backstage lobbying effort, Bush walked into the White House bearing a one-page proposal for a National Defense Research Council (NDRC), a high-level committee that would coordinate all the country's war-related technology development—with Bush, naturally, as its chairman. The lobbying paid off. Within fifteen minutes, that sheet of paper bore an inscription: "O.K.–FDR."

The date was June 12, 1940. Two days later, the German army marched into Paris.

A MEMO TO VANNEVAR

Norbert Wiener wasted no time. Indeed, he'd felt the coming crisis even more keenly than Vannevar Bush had. For years now, he'd found the Nazi persecution of his fellow Jews in Europe to be the stuff of nightmares—literally: at one point he'd been getting so little sleep that he was forced to consult a psychiatrist. Likewise the Japanese depredations in China: in 1935–1936 Wiener had spent one of the happiest years of his life as a guest professor at Tsing Hua University, in what was then known as Peking. For him to stand idle at such a time was unthinkable; he had to do *something*. So in the late summer of 1940, when Bush asked researchers all over the country to propose R&D projects that would help prepare the nation for its likely entry into the war, Wiener answered quickly.

Build computers, he advised his friend, thinking of their many conversations on the subject and the technological possibilities that were thick in the air by 1940. Build computers that used digital calculations for accuracy (as opposed to analog calculations); that used binary mathematics for simplicity; that used magnetic tape for mass storage; that used vacuum-tube electronics for speed;

and that used start-to-finish program control for still more speed (thus getting humans out of the loop). "If machines of this sort can be devised," Wiener concluded, "they will be of particular use in many domains in which the present theory is computationally so complex as to be nearly useless."[6]

Wiener mailed off his twelve-page memorandum on September 21, 1940, fully expecting a yes from Washington. In due course, however, the answer came back: no. Bush gently suggested that his friend turn his talents toward something more, um, *practical.*

That hurt. And indeed, since Wiener had outlined a machine that was remarkably close to the electronic, digital computer as we know it today,* Bush's rejection of the idea now seems like a monumental lost opportunity, not to mention a monumental blind spot in a man who understood the value of computing machinery as well as anyone in the country. What happened?

In fairness, Bush does seem to have been tempted by Wiener's scheme; his final no was actually several months in coming. But in the end, with research dollars scarce, Bush clearly felt he had to make choices. With the Luftwaffe's making nightly bombing runs over London, and with German U-boats' wreaking havoc in the North Atlantic, his most urgent priorities had to be the development of radar, antiaircraft fire control, and antisubmarine warfare. (In 1941 the list would expand to include a supersecret crash program known as the Manhattan Project, which was intended to exploit a newly discovered phenomenon called nuclear fission.) Computers just didn't seem to be that critical in 1940. Besides, the Yankee Inventor couldn't help but look at Wiener's proposal and see a theoretical design by a clumsy nonengineer. As Wiener himself later lamented, "[Bush's] estimate of any work which did not reach the level of actual construction was extremely low."[7] And in the absence of any such working device, Bush had no way of knowing whether Wiener's approach to computing was even the right one.

Once again, that last objection loomed larger in 1940 than we can easily understand now. Today we speak of "the computer" as if it were a single thing that had to be invented only once. But as Wiener's list of features suggests, the modern digital computer is actually a combination of at least half a dozen separate inventions, most of which involved not just another gadget but a shift in the way people *thought* about computing. At the time of Wiener's memo, moreover, it was far from clear whether he or anyone else had put the individual pieces together in the right way; those conceptual transitions were still very much works in progress.

Take the shift from analog computing to digital computing, for example. This was essentially the shift from measurement to arithmetic—from machines that

---

* Actually, Wiener had outlined a machine that would have been somewhat less flexible than a modern computer, in that much of its hardware would have been specialized for solving one kind of problem: so-called partial differential equations. That might not, however, have been such a big limitation in practice: partial differential equations encompass almost everything in physics, chemistry, and engineering. And his machine still would have been an enormous step forward.

made a physical model of the problem at hand, as in the Differential Analyzer, to machines that could manipulate numbers as discrete chunks of information. Even in 1940, of course, digital calculation had the advantage of accuracy: the lowliest desktop calculator could add 2 plus 2 and get 4, whereas the fanciest version of the Differential Analyzer would get 4 plus or minus 2 percent. But analog machines still had the huge advantage of speed, especially when they were grappling with the kinds of large, complex problems the analyzer had been built for. Furthermore, it seemed as though all-electronic machines such as the Rockefeller Analyzer would only increase that speed.

Better yet, for Vannevar Bush and for many others, was that analog machines had a wonderfully *evocative* quality. They didn't just calculate an answer; they invited you to go in and make a tangible model of the world with your own hands, and then they acted out the unfolding reality right before your eyes. For anyone watching that process, Bush wrote, "one part at least of formal mathematics will become a live thing."[8] Compared to that, digital computers seemed static and dead, nothing but electrons zipping invisibly through wires. That may have been why Bush himself later seemed to feel such a sense of loss as digital computing swept the world, starting in the 1950s. Certainly he never wavered in his own commitment to the analog approach. Doggedly, and without success, the Best Apparatus Man in America kept on trying to come up with a workable analog design for his Memex until his death, in 1974. And until the end, his colleagues could hear him grumbling about the "damn digital computer."

All of which just goes to show that Fate does have a sense of irony. The qualities that would give digital computers their ultimate advantage over analog machines—their vastly greater flexibility and programmability—were still only dimly perceived in 1940 (indeed, it would be several years yet before anyone even demonstrated a programmable digital computer). But one of the biggest single steps in that direction had already been taken—thanks to Vannevar Bush's own Differential Analyzer.

Actually, all Bush had been looking for was a bright young electrical engineering graduate who would be willing to work his way toward an MIT master's degree. With the lab increasingly preoccupied by the new Rockefeller Analyzer and several other projects, Bush's idea was to bring in a kid who could keep the original Differential Analyzer in repair and help visiting scientists get their problems set up on the machine. So in the spring of 1936, he'd sent out postcards advertising the position to universities across the country. And in due course, he'd settled on a young man who had just happened to see one of those postcards posted on a bulletin board at the University of Michigan. At age twenty, he was a shy, slender, jug-eared boy whose dark hair, pale-blue eyes, and quirky half-smile somehow combined to give him an elfin quality. His name was Claude Shannon.

From Shannon's point of view, Bush's postcard couldn't have appeared at a better time. With his graduation day approaching, he'd been debating whether to get a job or go on to graduate school; now he could do both. Besides, as he would recall in a 1987 interview in *Omni* magazine (perhaps thinking back to all the model planes and radio circuits he had put together as a boy in his tiny hometown of Gaylord, Michigan), the job seemed perfect for him: "I was always interested in building things with funny motions[, though] my interest later shifted to electronics." And now here was the Differential Analyzer. "The main machine was mechanical with spinning discs and integrators," he told *Omni*, "and there was a complicated control circuit with relays. I had to understand both of these. The relay part got me interested."

Shannon could see that the relays were actually little switches, each of which could be automatically opened and closed by an electromagnet. But as he traced his way through the control circuits, he was also struck by how closely the relays' mechanical operation resembled the workings of symbolic logic, a quasi-mathematical system developed in the nineteenth century to model human reasoning—and a subject he had just taken a course in during his senior year at Michigan. In the analyzer, each relay had a binary choice: it was either *closed* or *open*. And likewise, in logic, any given assertion had a binary choice: it was either *true* or *false* ("Socrates was a man" is *true*, to take the classic example, whereas "The moon is made of green cheese" is *false*).

Furthermore, as Shannon quickly remembered from his elementary physics courses, relays combined in circuits could physically embody all the standard ways that assertions were combined in logic. If two relays were lined up in series, for example, as in

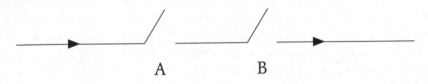

they would embody the logical relationship *and*—that is, current would flow through the circuit only if relay A *and* relay B were closed. Opening either one would break the circuit. Similarly, if the two relays were wired up in parallel—

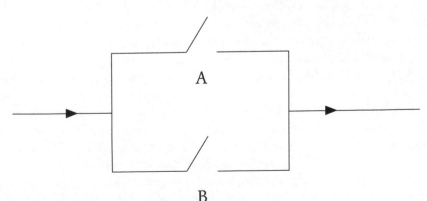

—they would embody the logical relationship *or,* meaning that the current would flow if either relay A *or* relay B was closed. And if a relay was wired up backward, so that it snapped closed when it normally would have opened, and vice versa, it would embody the logical operation of *not.*

Indeed, as Shannon told *Omni,* the analogy worked perfectly—and no one had ever noticed it before. So he made it the topic of his master's thesis and spent most of 1937 working out the implications. "I think I had more fun doing that than anything else in my life, creatively speaking," he mused.

It showed. "A Symbolic Analysis of Relay and Switching Circuits" has just the kind of cerebral exuberance you'd expect from a very bright twenty-one-year-old. Shannon's thesis is downright fun to read—and strangely compelling, given what's happened in the six decades since it was written. In an aside toward the end, for example, Shannon points out that *true* and *false* could equally well be denoted by the digits *1* and *0,* in which case the operation of the relays could be associated with what was then an arcane and unfamiliar form of arithmetic—namely, *binary* arithmetic. The digit *1* would mean that the relay was closed, while the digit *0* would mean that it was open. Thus, Shannon wrote, "it is possible to perform complex mathematical operations by means of relay circuits. Numbers may be represented by the positions of relays and stepping switches, and interconnections between sets of relays can be made to represent various mathematical operations."[9] As an illustration, he showed the design of a relay circuit that could add binary numbers.

But more than that, Shannon noted, a relay circuit could make comparisons and thus take alternative courses of action according to specified circumstances, as in, "*If* the number X equals the number Y, *then* do operations P, Q, and R." To illustrate the point, he showed that relays could be arranged to produce a combination lock that would open if and only if the user pressed a series of buttons in the proper order. Of course, the implications were much more general. Any desktop adding machine could do simple addition and subtraction; a relay circuit could *decide.* *

Even now, more than sixty years later, that remains a profound and startling insight. In effect—though he didn't emphasize it in his thesis—Shannon had shown that a relay circuit could physically embody that ineffable whatever-it-was in a human computer's head that allowed her to know *what* mathematical operations to perform *when,* and whether to perform operation A or B depending on the result of operation C. That ability, in turn, is ultimately what makes a modern digital computer so much more than just an adding machine: it can work its way through a sequence of such decisions automatically. In a word, it can be *programmed.* And that's why "A Symbolic Analysis of Relay and Switching Circuits" is arguably the most influential master's thesis of the twentieth century: in it Claude Shannon laid the theoretical foundation for all of modern

---

* It isn't even very hard. The assertion "If A then B" turns out to be logically equivalent to "(Not A) or B" So any circuit that can embody *not* and *or* can also embody *if-then.*

computer design, nearly a decade before such computers even existed. In the years since, switching technology has progressed from electromechanical relays to microscopic transistors etched on silicon. But thanks to Shannon, microchip designers still talk and think in terms of a device's internal logic.

Of course, neither Vannevar Bush nor anyone else had any way of recognizing all this in 1937. And in truth, the immediate impact of Shannon's work was remarkably small. It was obvious to everyone that this shy young man from Michigan possessed a "profound originality," as Wiener put it; indeed, Bush soon made him a full-fledged team member on one of the group's advanced development projects. But MIT's commitment to analog computation was already too great to be derailed by one piece of theory, however striking. Nor was Shannon himself in any position to build real logic circuits; with Bush's encouragement, he had decided to go on for his Ph.D. in mathematics, which effectively took him out of circulation for several years. And perhaps most important, no one else was waiting around—certainly not for any theoretical guidance. All across the United States (and the world), in fact, hands-on tinkerers were already plunging ahead with their own ideas for computing, and in the process achieving most of the other transitions on Wiener's list. For example:

From decimal math to binary math. The *1*s and *0*s of the binary number system weren't exactly an obvious choice in the 1930s. Human beings had been counting on their fingers and using base-10 arithmetic for millennia, after all. Manufacturers had perpetuated that base-10 system in the beautifully intricate gears and linkages of their desktop calculators. And the first instinct of many of the computer pioneers had been to follow suit. Nevertheless, binary math not only had deep ties to logic, as Shannon had pointed out, but also greatly simplified the design of switching circuits—which is probably why it was independently rediscovered on several different occasions.

In late November 1937, for instance, a young mathematical engineer named George Stibitz had just been given an assignment by his boss at the Bell Telephone Laboratories in New York City: he was to take a look at the telephone company's electromechanical relays, which were widely used within the Bell system for routing telephone calls, and think about ways to make them more reliable and efficient.

Well, as Stibitz would recall in a 1979 symposium on the history of computing, he had never had any experience with relays before. And he certainly didn't know anything at that point about Shannon's work with switching circuits, since Shannon was still writing up his results for publication. But being a mathematician, he was immediately struck by the same *on-off, true-false, 1-0* equivalence that Shannon had perceived. So one night he borrowed a few spare relays from the Bell Labs junk pile and took them home for what he thought of as a play project. "With a scrap of board, some strips of metal from a tobacco can, two relays, two flashlight bulbs, and a couple of dry cells," he said, "I assembled an [adding circuit] on the kitchen table of our home."

Stibitz's adding circuit was about as simple as it could be. Each strip of tin was a switch that inputted a single-digit binary number: press it down for *1*, leave it up for *0*. The two light bulbs served as an output display showing the two-digit sum of the inputs: *on* for the digit *1*, *off* for the digit *0*. Simple or not, however, it worked: press both strips down for *1 + 1*, say, and the lightbulbs would flash *1* and *0*, the binary equivalent of 2. What Shannon had achieved in theory, Stibitz had independently accomplished in hardware.

The next morning, he took his circuit in to Bell Labs to show his colleagues, who were more amused than impressed. Even to Stibitz, the thing still seemed like a toy. But it was proving to be a strangely addictive toy. Unable to leave it alone, he said, "I sketched a schematic for a multidigit adder and pointed out that a relay machine could do anything a desk calculator could do." To prove that point, he went on to design binary circuits that could subtract, multiply, and divide. Moreover, since it seemed unlikely (then) that users could ever be persuaded to learn binary notation, Stibitz designed a kind of user interface: a circuit that would take a number entered in ordinary decimal notation and automatically convert it to binary.

Before long, Stibitz's original task was forgotten, and it had become clear to everyone at Bell Labs that his relay circuits were much more than toys. In the early summer of 1938, his boss asked him if he could build a calculator that would do arithmetic with complex numbers—the kind involving "imaginary" quantities based on the square root of −1. Those quantities had turned out to have some very real applications in the design of AT&T's new coast-to-coast system of long-distance lines, and Bell Labs' computer division—a small team of women armed with desk calculators—was being swamped by complex arithmetic.

Stibitz, in collaboration with a switching engineer named Samuel B. Williams, went to work. The result, completed in October 1939 at a then-breath-taking cost of twenty thousand dollars, was the Complex Computer. Equipped with some 450 relays and three modified Teletype machines for entering problems and printing out the answers, the Complex Computer wasn't particularly fast even by 1939 standards; indeed, it took about a minute just to multiply two eight-digit decimal numbers. But it was very reliable and easy to work with. By the summer of 1940, a refined version was in routine use at Bell Labs. And that September, the Complex Computer made its public debut at a meeting of the American Mathematical Society, held at Dartmouth College in Hanover, New Hampshire.

It's hard to say which impressed the attending mathematicians more: the Complex Computer itself or the fact that they were communicating with it through the world's first computer network. "Williams designed an interface so that the number signals could be serialized and transmitted over a standard telegraph line," explained Stibitz. "[Attendees at Dartmouth] were invited to transmit problems from a Teletype in McNutt Hall to the computer in New York. Answers returned over the same telegraph connection and were printed out on the Teletype."[10]

Either way, the Complex Computer was the hit of the meeting. Even Norbert Wiener seemed to find it a revelation. After he'd spent quite a bit of time playing with the Teletype and getting totally exasperated (for some reason he kept trying to make the remote computer divide a number by 0 and produce infinity—an impossible task for any real machine, and one that Stibitz and Williams had explicitly guarded against), Wiener told a colleague that he was now convinced: binary math was the future of computing. Indeed, the Dartmouth presentation may have been what inspired the memo he sent to Vannevar Bush only a few weeks later.

From mechanical to electromechanical to fully electronic. Wiener's emphasis on high-speed electronics in that memo certainly didn't come as a surprise to Bush. Electronics technology had evolved rapidly in the 1930s, thanks to the fast-growing radio and telephone industries, and the advantages for computing were obvious. Electromechanical relays could open and shut many times per second, thus providing a significant speedup over purely mechanical calculators, and vacuum tubes could switch currents on and off as much as a thousand times faster still—*if*, that is, anyone felt heroic enough to use them.

The problem was that vacuum tubes were fragile devices that tended to burn out like light bulbs. This was tolerable enough in radio circuits, which used only a comparative handful of tubes, but it was felt to be a showstopper for digital computers, which would need hundreds or even thousands of the things in order to do anything useful. Thousands of tubes meant a constant string of failures, went the argument. And when even a single burned-out tube could bring the entire system to a halt, how would you ever finish a calculation?

This is why most of the early machines were electromechanical. The first person to successfully take on the challenge of creating a fully electronic computer was a physics professor named John V. Atanasoff, of Iowa State University.

For years, as Atanasoff later told the story, he had been determined to build a computing machine that could help his students; he found it disturbing to watch them waste most of their effort on mindless calculation, as opposed to real thinking. But with all his teaching responsibilities, he'd had very little time to focus on the problem. Finally, however, on a bitterly cold winter night in late 1937, he just couldn't stand it anymore; he had to get away to *concentrate*. So he jumped into his car in Ames, Iowa, and drove east through the subzero temperatures at more than eighty miles per hour. Almost three hours later, after he crossed the Mississippi River into Illinois, he stopped at a roadhouse to warm up. And there, somewhere between his first and second bourbons, he conceived four crucial ideas to make his computer work.

First, just as George Stibitz and his Bell Labs colleagues were independently doing at the same time, Atanasoff decided to do his calculating with base-2 binary numbers; the circuitry would be *so* much simpler that way. Second, Atanasoff decided that he would go ahead and use vacuum tubes. He was well aware of the burn-out problem, but he reasoned that relays failed, too, and the tubes weren't *that* unreliable. In any case, the headaches would be worth it for the sake of speed.

Third, Atanasoff came up with a serial, step-by-step calculation scheme that would allow his computer to deal with larger numbers than it could have done otherwise. And last of all, he devised a circuit that could store binary numbers in electronic form until they were needed, so that the computer could have an internal "memory" (in fact, he may well have been the first person to use that word in this context).

Upon returning to campus, Atanasoff went to work. By 1939 he had his designs worked out and was seeking funding. And by mid-1941, with the collaboration of his student Clifford Berry, he had the machine up and running. The Atanasoff-Berry Computer, or ABC, as he later insisted it be called, was a relatively small device about the size of an office desk, comprising some three hundred vacuum tubes and a data-storage device that consisted of a rotating drum studded with capacitors. Moreover, with the exception of a balky data-input unit—an IBM-style card reader that the designers never could get to function without error—the ABC worked reliably.

Atanasoff didn't develop his invention any further, as it happened. Soon after the United States entered the war, in December 1941, he went to the Naval Ordnance Laboratory in Washington, D.C., where he supervised the acoustic testing of mines. He never returned to computing. Nonetheless, he always considered the ABC to be the first demonstration of a fully electronic computer.

From operator-supervised to programmable computers. Fully electronic though it was, the ABC was still a very long way from Wiener's ideal of getting humans out of the loop: using it meant standing at the console, pressing buttons, watching meters, loading and unloading punch cards, and in general intervening in every step of the computation. But then, Bell Labs' Complex Computer was even further away from that ideal, since it wasn't programmable, couldn't make decisions, and had no internal memory; it actually functioned more like a desktop calculator, responding whenever the operator entered numbers for addition or whatever, but otherwise just sitting there. And on the Differential Analyzer, of course, every new problem required days of preparation to get all the gears and pulleys set up.

To eliminate such bottlenecks, as Wiener had pointed out, computers would have to be given the power to carry out long sequences of operations on their own, under the control of some sort of *program*. In particular, they would have to be able to make their own decisions about what operation to perform next— the kind of decisions that programming languages now express with statements such as "*If* variable X = 0, *then* do operation Y, *else* do operation Z." Shannon had shown that a machine could carry out such operations in theory. But at the time of Wiener's memo, in 1940, the only computer capable of doing so in practice—the Automatic Sequence Controlled Calculator at Harvard University, later called the Mark I—was still under construction.

The calculator was the obsession of a Harvard applied-mathematics instructor named Howard H. Aiken. Like Bush and most of the other computer pioneers, Aiken had been intensely frustrated by the massive, mind-numbing calculations required to solve so many of the most worthwhile problems in mathematics, and had been driven to find a way to automate those calculations. Unlike most of the others, however, Aiken was familiar with the work of the nineteenth-century English inventor Charles Babbage, who had designed a series of computational devices that culminated in the 1830s with his "Analytical Engine," a general-purpose, programmable calculator that could automatically carry out arithmetic operations specified on a series of punched cards—and that could indeed make some decisions about what operation to do next. Babbage's marvelously intricate array of brass gears was never built, unfortunately, largely due to a lack of funding and the difficulty of fabricating precise enough parts. Nonetheless, his patroness Augusta Ada Byron, Countess of Lovelace (and daughter of Lord Byron), wrote a number of programs for the machine to demonstrate its potential power. Moreover, Babbage's surviving designs show that his Analytical Engine would indeed have worked; today, Ada Lovelace is widely honored as the first programmer, and Charles Babbage is likewise revered as the spiritual father of modern computing. It is certainly true that Howard Aiken, who had come across Babbage's biography in the mid-1930s, when almost no one in the United States had ever heard of him, considered himself Babbage's spiritual heir.

Aiken began design work on his automatic calculating machine in 1937, when he was thirty-seven (he had worked for more than ten years as an electrical engineer at Westinghouse before going back to school to get a doctorate in physics). He had no intention of using vacuum tubes in his machines, since he was one of the many engineers who were still suspicious of their reliability. He likewise had no inclination to use binary arithmetic; conventional decimal arithmetic seemed fine to him. And in practice, his designs actually had little in common with Babbage's engine. Instead of creating a new machine from the ground up, he basically planned to lash together a series of existing calculators, tabulators, and sorters, with modifications. Nonetheless, Aiken's system, like Babbage's, was going to be programmable: it would read off its commands from holes punched in spools of paper tape.

Once his design was on paper, Aiken needed backing. He therefore approached the biggest names in office calculators: Marchant, Monroe, National Cash Register. He definitely made an impression. At six feet two inches tall, he was a formidable figure, with arched eyebrows and long, pointed ears that gave him a distinctly Mephistophelian look, and an energetic, forceful, self-assured manner that allowed him to dominate any room he entered. All he got for his efforts, however, was a great many compliments on his fascinating idea. It wasn't until 1939 that he finally caught the fancy of Thomas B. Watson, Sr., the patriarch of the IBM Corporation. This was long before the name IBM became synonymous with computers; in those days it was identified with the ubiquitous IBM punch cards and the very sophisticated and expensive machines that could sort and tabulate them. Nonetheless, Watson was interested in the scientific applications of IBM equipment, and he agreed to support Aiken's proposal, provided that Aiken agreed to build his calculator out of IBM statistical tabulators and to let IBM's own engineers handle the actual construction. Aiken gritted his teeth and decided that he could live with that.

Construction of Aiken's machine started in 1939 at the IBM plant in Endicott, New York, and went on for more than four years, as the engineers there struggled to turn Aiken's rather abstract designs into real, working hardware. Indeed, the completed machine wasn't delivered to Harvard until early 1944. By that point it had already been commandeered for the war effort, of course: it would henceforth be devoted to urgent calculations for the navy under the supervision of Naval Reserve Commander Howard Aiken, who had been called to active duty in 1941 and who now became the first naval officer in history to command a computer. But Harvard was thrilled to have the machine anyway and promptly began laying plans for a gala debut on August 1, 1944.

It was quite an occasion, apparently. Aiken, who had never had a very high opinion of Thomas Watson's engineers, had managed to approve a press release that barely mentioned IBM's contributions to the project. Watson was furious; as his son, Thomas Watson, Jr., later recalled, "If Aiken and my father had had revolvers they would both have been dead." Nonetheless, the machine made a tremendous impression on the reporters who gathered for the event. Not only was it huge, being eight feet tall, fifty-one feet long, and two feet thick, but it had a sleek, shiny, sci-fi look; at the insistence of Watson, who was a past master at public relations, the machine had been encased in a futuristic stainless-steel-and-glass skin. The reporters instantly dubbed it "the electronic brain," a phrase that Aiken despised. But for better or worse, the name stuck, and the American public had its first introduction to the computer age.

The Automatic Sequence Controlled Calculator acquired its "Mark I" designation a year later, when Aiken and his team began work on an upgraded Mark II for the navy. (There would eventually be a Mark III and a Mark IV as well.) None of these machines would have much impact on the development of computer hardware per se; Aiken's insistence on features such as base-10 arithmetic was just too idiosyncratic. Yet the Mark I, especially, would be profoundly important as a proof of principle, the first demonstration that a complex calculating machine could operate continuously and reliably while carrying out a complicated sequence of actions under the control of a program.

Indeed, Howard Aiken may well have been the first person to realize that programming would eventually become a profession in its own right. Not only would he later convince Harvard to start the first master's degree courses in what would now be called computer science, but he insisted from the beginning that the Mark I project be staffed with trained mathematicians—among them, most notably, a thirty-seven-year-old naval reserve lieutenant named Grace Murray Hopper, formerly a mathematics professor at Vassar College. ("Where the hell have you been?" was Aiken's greeting to her on the day she arrived, in 1944.)

Hopper would later gain fame both as a teacher and as a pioneer in the development of high-level programming languages. Yet perhaps her best-known contribution came in the summer of 1945, when she and her colleagues were tracking down a glitch in the Mark II and discovered a large moth that had gotten crushed by one of the relay switches and shorted it out. She taped the dead moth into the logbook with the notation "First case of an actual bug being found."

From Hopper's wording, it's clear that the term "bug" wasn't new in this context (it's been traced back to the time of Edison). But even so, the logbook and the moth are now on display in the Smithsonian Institution.

So there it was: the history of the computer in the 1930s was a history of conceptual groping on the part of many individuals, each one wrestling with recalcitrant hardware and each one solving a piece of the conceptual puzzle. But only at the end of the decade were a few people such as Norbert Wiener beginning to put all the pieces together—and even they were only *just* beginning.* Ultimately, in fact, it would take the war itself to forge those pieces into a unified whole.

The war, that is, plus a small group of young men with insufficient respect for the wisdom of their elders—and another world-famous mathematician on an insatiable quest for calculating power.

THE WIZARD

On a summer's evening in early August 1944, as U.S. Army Lieutenant Herman H. Goldstine was standing on the railway platform just outside the Aberdeen Proving Ground in Maryland, waiting for the train back to Philadelphia, he hap-

---

* Actually, this isn't quite true, though few people would realize it for decades. In what has to be one of the greatest ironies of modern computer history, the earliest, most fully thought out, and most farsighted of the pioneering projects of the 1930s was also the most isolated and poorly funded. Working virtually alone and at his own expense, a German civil engineer named Konrad Zuse (T'SOO-sa) not only started experiments with digital computing well before the Americans or British, but went much further. Beginning with his Z1 computer in 1934, Zuse built four increasingly sophisticated prototypes that utilized binary arithmetic, stored programs, and virtually all the rest. (Only his last machine, the Z4, survived the Allies' bombing of Berlin.) He was prevented from building a fully electronic vacuum-tube computer only because it was essentially impossible for a civilian to acquire the tubes in wartime Germany. Along the way, moreover, he invented the first programming language—the "Plan Calculus," or *Plankalcül*—and began to analyze methods by which a computer could play chess.

Zuse told his story in 1979, at the same history-of-computing conference that George Stibitz attended. His account is published as "Some Remarks on the History of Computing in Germany" in *A History of Computing in the Twentieth Century,* edited by N. Metropolis, J. Howlett, and Gian-Carlo Rota (New York: Academic Press, 1980), 611–27. Zuse died at his home in Hünfeld, Germany, on December 18, 1995.

pened to catch sight of another man stepping onto the platform—a short, plump, rather ordinary-looking fellow with an expansive forehead, a round, cheerful face, and the air of a meticulously dressed banker.

Von Neumann! Goldstine was awestruck. Before his current incarnation—he was liaison officer for the army's computing substation at the University of Pennsylvania's Moore School of Engineering—Goldstine had been a Ph.D. mathematics instructor at the University of Michigan. So he already knew the legends. At age forty, John von Neumann (pronounced fon NOY-man) held a place in mathematics that could be compared only to that of Albert Einstein in physics. In the single year of 1927, for example, while still a mere instructor at the University of Berlin, von Neumann had put the newly emerging theory of quantum mechanics on a rigorous mathematical footing; established new links between formal logical systems and the foundations of mathematics; and created a whole new branch of mathematics known as game theory, a way of analyzing how people make decisions when they are competing with each other (among other things, this field gave us the term "zero-sum game").

Indeed, Janos Neumann, as he was known in his native Budapest, had been just as remarkable a prodigy as Norbert Wiener. The oldest of three sons born to a wealthy Jewish banker and his wife, he was enthralled by the beauty of mathematics from the beginning. He would discuss set theory and number theory on long walks with his childhood friend Eugene Wigner, himself destined to become a Nobel Prize–winning physicist. Once, when he noticed his mother pause in her crocheting and gaze contemplatively into space, Janos asked her, "What are you calculating?" His own powers of mental calculation were legendary. A certain renowned mathematician once struggled all night to solve a problem with the help of a desk calculator; the next morning von Neumann solved the same problem in his head—in six minutes.

His memory was equally prodigious. Asked to quote the opening chapter of *A Tale of Two Cities*, which he had read many years before, von Neumann continued to recite without a pause or a mistake for more than ten minutes—until his audience begged him to stop. He was likewise gifted in languages: by the age of six he was making jokes with his father in classical Greek. And he had a knack for fitting in with the culture around him. In Budapest his family had been prosperous enough to earn the Hungarian honorific "Margattai" from the Austro-Hungarian emperor, so in Göttingen and Berlin, he transmuted both name and honorific into an aristocratic sounding "Johannes von Neumann." And then in 1930, when the lack of tenured professorships in Europe led him to immigrate to Princeton University in the United States, he quickly became Johnny von Neumann, a cheerful, party-going lover of funny hats, jokes, puns, and limericks, the racier the better.

The Americans responded in kind. Von Neumann's brilliance was obvious to everyone; it took him precisely one year to achieve tenure at Princeton. Then, when the Institute for Advanced Study was established there in 1933, he became one of its founding members. Indeed, with the inadvertent assistance of Adolf

Hitler, whose obsession with "racial purity" was forcing many of Europe's best mathematicians into exile, von Neumann helped to make Princeton a world-class mecca for mathematics. And by the eve of the war, he reigned alongside Norbert Wiener as one of the two most visible and innovative mathematicians in the country, if not in the world.

The two men were good friends, as it happened. They frequently expressed admiration for each other's work—indeed, they frequently found themselves working in the same fields of mathematics—and they maintained a warm correspondence for many years. And yet they were strikingly different, both in style and in personality. Wiener was impulsive and intuitive in his mathematics, a Beethoven who could bring his concepts to life only through struggle, trial and error. Von Neumann, by contrast, was a consummate craftsman, a Mozart who could pour out his work so effortlessly that almost everything he produced seemed perfect the moment he set it down. Colleagues described how von Neumann might write out five pages of letter-perfect and breathtakingly original mathematics before breakfast, while sitting at his kitchen table in a bathrobe.

The differences also extended to the two men's demeanor in the lecture hall. Whereas Wiener's human foibles were abundantly apparent, von Neumann seemed to have no human failings at all. His lectures were brilliantly conceived, beautifully organized, and masterfully delivered (with just the barest trace of an accent), albeit so rapid-fire that only the most alert listeners could keep up. Indeed, the effect was enough to make von Neumann seem rather chilly and intimidating to those who had only heard him lecture and did not know him personally. And since that was a group that happened to include Lieutenant Herman H. Goldstine, Ph.D., the younger man was justified in feeling timid on that evening in August 1944.

"It was therefore with considerable temerity that I approached this world-famous figure, introduced myself, and started talking," Goldstine later recounted. "Fortunately for me von Neumann was a warm, friendly person who did his best to make people feel relaxed in his presence. The conversation soon turned to my work. When it became clear to von Neumann that I was concerned with the development of an electronic computer capable of 333 multiplications per second, the whole atmosphere of our conversation changed from one of relaxed good humor to one more like the oral examination for a doctor's degree in mathematics."[11]

Suddenly von Neumann was demanding details, details, details—and with good reason. He had just spent eight months covertly scouring the country for serious computational power and had come up short. His purpose was one that he certainly couldn't talk about on a railway platform, but it was of the highest national priority imaginable. Starting in April 1943, von Neumann had become a frequent visitor to the high country of New Mexico, where he was serving as a consultant to a new mesa-top laboratory so secret that it officially didn't exist. Its name was Los Alamos. There, along with his fellow Hungarian Edward Teller, the German refugee Hans Bethe, and many others in the Manhattan Project's theory division, von Neumann had been struggling to carry out elaborate calculations for the design of a fundamentally new type of explosive, based on the energy of nuclear fission. The team's particular challenge was to figure out how a bomb could be forged from a recently discovered, man-made element known as plutonium-239. Unless they could get the specifications for the plutonium device just right, it would not explode. And the determination of those precise specifications had turned out to require an overwhelming number of numerical calculations. Los Alamos was well supplied with IBM's finest punch-card tabulators, of course (as was the whole U.S. military effort), but these machines still had to be operated by mortal human beings, who simply could not keep up with the number-crunching demands being placed on them.

In January 1944, von Neumann had accordingly written to Warren Weaver, the head of applied-mathematics research at Vannevar Bush's National Defense Research Council: Who is working on high-speed automatic computing machines? he asked. Weaver had given him three contacts: Howard Aiken at Harvard, George Stibitz at Bell Labs, and the Astronomical Computing Center at Columbia University, another group supported by IBM.

Frustratingly, however, none of the three had panned out. The Columbia group was still working with conventional IBM tabulators, and von Neumann was unimpressed with the results. The Bell Labs work was more to his liking; Stibitz and Williams were now breaking in Model 3. But building a duplicate of the Bell Labs machine at Los Alamos would have taken the better part of two years, and time was of the essence (it was still considered very possible that Hitler's scientists would create a workable atomic bomb before the Manhattan Project). Von Neumann also thought well of Harvard's Mark I, but Aiken's machine didn't have quite the power and flexibility that the physicists needed, and in any case, Harvard was already fully committed to navy projects.

The upshot was that von Neumann and his colleagues at Los Alamos had resigned themselves to making do with improved versions of the IBM tabulators—which explains his sudden, intense interest on that evening in August 1944 when Goldstine happened to mention that he and the Moore School group were building a computer they called the Electronic Numerical Integrator and Calculator, or ENIAC. Von Neumann had never heard of it. Warren Weaver, strangely enough, hadn't said a word about ENIAC the previous January, though he certainly knew about the machine.

Or maybe it wasn't so strange. The most benign interpretation is that Weaver had simply been guiding von Neumann to projects that might help him right away, not two years down the road. In January 1944 the ENIAC group had barely finished designing its machine and had not yet started construction. However, it's also possible that Weaver didn't mention ENIAC because the project was widely considered to be a boondoggle—a hopelessly overambitious endeavor that Bush's NDRC wouldn't begin to waste money on. The fact that it had funding at all was considered a measure of the ignorance and desperation of the U.S. Army.

The desperation arose from those same artillery trajectories that Private Norbert Wiener had worked on back in 1918–a computing challenge that may not have been as world-shaking as von Neumann's but was just as daunting. By its nature, in fact, it was a job that could never be finished. An artillery gun is useful only if it can hit its target. And it can hit its target only if it is pointed in just the right direction at just the right elevation, taking account of factors such as muzzle velocity, wind speed, and air temperature. Gunnery crews were therefore supplied with little books full of firing tables, with each table's listing exactly how they should aim for a given combination of factors. Each book had hundreds or thousands of tables. Each new type of gun needed its own book. And new, improved guns were pouring off the wartime assembly lines all the time.

This made for a lot of calculation: someone had to evaluate a nasty differential equation for each entry of each table in each book, over and over again. Shortly before Pearl Harbor, in fact, the army had constructed two replicas of Bush's Differential Analyzer for that very purpose, one at the Ballistics Research Laboratory at Aberdeen, and the other–now supervised by Goldstine–at the Moore School. In practice, however, the analyzers' lengthy setup time meant that they could be used only to calculate the basic trajectories. All the multitudinous variations were left to the Aberdeen computer division, which had quickly grown to comprise some two hundred WACs armed with desk calculators. But human fingers can punch keys only so fast. By the spring of 1943 the Aberdeen computers were falling disastrously behind.

Enter John Mauchly, then thirty-five years old. Until the eve of the war Mauchly had been chairman of the physics department at tiny Ursinus College outside Philadelphia, where his research had originally focused on improving weather prediction. Like virtually all of the other computer pioneers, though, he had found himself stymied by massive calculations that simply couldn't be accomplished by short-lived human beings. Mauchly had accordingly spent more and more of his time in the late 1930s experimenting with electronics, looking for a way to make high-speed electronic computing feasible. (That he didn't actually try to build a computer was due largely to a lack of money.) By the time of Pearl Harbor, in December 1941, this had led him into a full-time career in electrical engineering at the Moore School, thanks to an army-funded summer course there designed to transform scientists from other fields into electrical engineers for the coming war effort. By August 1942, moreover, Mauchly was circulating a five-page memo roughing out his concept for a vacuum-tube calculating machine, and urging that the Moore School build it. With a fully electronic device, he argued, computations that had once taken years would take just days.

His crusade got a decidedly chilly reception. Most of Mauchly's colleagues still thought that vacuum tubes were wildly impractical for anything except small-scale radio circuits and the like. In fact, the frustrated former physicist got precisely nowhere with his idea until March 1943, when Goldstine learned of it during a casual conversation with one of the mechanics who worked on the analyzer. Being a mathematician and not an engineer, Goldstine didn't know that a vacuum-tube computer was impossible. What he *did* know was that the artillery calculations were falling further behind every day. If Mauchly's gadget had even half a chance of working, he reasoned, then the army ought to take a chance on it. Goldstine went to Mauchly and asked him to draw up a more detailed proposal while he went to talk to his superiors at the army's ordnance department.

Goldstine was able to arrange for Mauchly to give a presentation at Aberdeen on April 9, 1943. Mauchly spent the night before the meeting feverishly getting his proposal down on paper. Working alongside him was a Philadelphia blue blood named J. Presper Eckert, who had been a laboratory instructor for the summer course that turned Mauchly into an electrical engineer. Eckert was bright, talented, and imaginative (as a boy he had built a crystal radio set on a pencil). And by no coincidence, he had been Mauchly's one real supporter at the Moore School, the only person there who had been excited by the idea of an electronic computer and had encouraged Mauchly to keep at it. Eckert was also still a graduate student; the following day, April 9, would be his twenty-fourth birthday.

Mauchly and Eckert staggered into the meeting at Aberdeen having had no sleep and no breakfast. But they did have their proposal. And somehow, they and Goldstine together managed to make their presentation persuasive. Listening closely were Colonel Leslie Simon, director of the army's Ballistic Research Lab, and Professor Oswald Veblen, director of the Institute for Advanced Study and chairman of the Ballistic Research Lab's advisory board. Nearly thirty years later, Goldstine would remember the moment of decision: Veblen, he wrote, "after listening for a short while to my presentation and teetering on the back legs of his chair[,] brought the chair down with a crash, arose, and said, 'Simon, give Goldstine the money.' "[12] That very day, contracts were signed that backed the ENIAC project with an astounding $150,000 in seed money—a sum that would eventually grow to an even more astounding $400,000. Project PX, as the classified program was called, was officially launched on June 1, 1943.

From the beginning, Mauchly, Eckert, and their colleagues felt tremendous pressure, not only because their reputations were at stake, along with the Moore School's, but because friends and relatives were risking their lives on the front lines while they were home safe in Philadelphia. Working hours at the Moore School were not short. Nonetheless, the final design for ENIAC wasn't completed until late 1943. And the result was ambitious, to say the least. ENIAC was to be built around three walls of a room, in the shape of a horseshoe, with arms eighty feet long. It would stand eight feet tall and weigh thirty tons. It would contain 17,468 vacuum tubes of six different types, as well as 10,000 capacitors, 70,000 resistors, 1,500 relays, and 6,000 manual switches—an array of electronics so vast that large blowers would be required to dissipate the heat it produced. (When ENIAC finally became operational in 1946, temperatures in the surrounding room would reach 120° F.) And every one of those vacuum tubes would be switching on and off a hundred thousand times per second. As Goldstine later noted, this worked out to roughly 1.7 billion chances for failure every second.

Mauchly and Eckert had devised a number of safeguards, of course, including such common-sense measures as testing the tubes beforehand and running them well below their rated voltage. But even so, Bush, Aiken, and Stibitz were all known to regard ENIAC as a total waste of government funds that would be better spent on the proven technologies of relay calculators and differential analyzers. (The Moore School group returned the contempt; after visiting the new Rockefeller Analyzer at MIT, Goldstine wrote, "It was, I think, a pretty sad spectacle of what the supermen at NDRC can do.") As von Neumann listened to Goldstine's account on the railroad platform at Aberdeen, however, he was not in a nay-saying frame of mind; he had just spent eight months looking at "proven" technologies that couldn't do the job. When the two men parted, von Neumann had Goldstine's promise that he would arrange a visit for him.

Getting the right security clearances for Professor von Neumann proved to be no trouble at all. He arrived at the Moore School on September 7, 1944.

There actually wasn't much for von Neumann to see. Construction on ENIAC was just getting under way, and it was clear that the machine wasn't going to be ready in time to help Los Alamos. Nevertheless, von Neumann was impressed by the young team, which had grown to include a dozen engineers, and so fascinated by their work that—at a point when he was already swamped by his commitments to Los Alamos and other war research—he agreed to become a regular consultant for the Moore School group. Somehow he would make room in his schedule to visit the project every few weeks or so.

Mauchly, Eckert, and Goldstine, for their part, were overjoyed to have him. From now on, who cared if the supermen of the NDRC scoffed at vacuum-tube computing? *They* had von Neumann on their side. And besides, von Neumann was already beginning to make a real contribution to the project, especially when it came to their efforts to go beyond ENIAC.

The need to do so was clear even then, Mauchly later noted: "Considering the pressure to produce the fastest calculator as quickly as we could, we had to do the best with whatever seemed the most available methods." As a result, the original ENIAC suffered from two major drawbacks. First, it had to store its input data and its various intermediate calculations in a clumsy hodgepodge of specialized memories, ranging from very fast electronic "flip-flops" to very slow IBM-style punch cards. "The fast memory was not cheap, and the cheap memory was not fast," Mauchly noted wryly.

Second, ENIAC was almost as hard to "program" as the Differential Analyzer. In principle, at least, ENIAC was supposed to be a general-purpose computer capable of handling any type of calculation. But in practice, the ENIAC technicians would have to rearrange a massive tangle of wires for each new problem, fitting them into holes on a big plugboard in much the same way that 1940s-era telephone operators made connections on a switchboard. "Cables ran from place to place over the entire eighty feet of panels that housed various arithmetic and functional devices," said Mauchly. These connections "provided a program sequence that was as rapid as the arithmetic and functional units required. But it allowed no flexibility whatsoever." This wasn't going to be a fatal drawback for the army's artillery calculations, since each trajectory would be set up once and then computed many times without change. Nonetheless, something better was clearly needed if electronic computing was ever going to reach its full potential. Thus the effort to plan a next-generation machine: the Electronic Discrete Variable Automatic Computer, or EDVAC.

Mauchly and Eckert's basic goals for EDVAC were more memory, easier setup, and fewer tubes. They had any number of ideas about how to achieve those goals, including Eckert's solution to the expensive memory problem: store the data as electronic pulses circulating through glass tubes filled with liquid mercury. Given the pressure of getting ENIAC up and running, however, they had only just begun to work out their ideas in detail. Von Neumann accordingly became an active participant in their discussions about the new machine. Indeed, it seems to have become one of his central intellectual interests. Building on his Los Alamos experience but looking past the end of the war, he had already come to the conclusion that very fast numerical calculation could revolutionize scientific research in fields ranging from fluid dynamics to astrophysics. And this, he realized, would require very fast, very flexible calculating machines like EDVAC. In March or April of 1945, in fact, after some particularly fruitful sessions on the new machine at the Moore School, von Neumann even volunteered to write up a report. It would summarize the group's thinking to date, he said, but would also expand and develop those ideas according to some notions of his own. Mauchly, Eckert, and rest of the EDVAC staff accepted the offer enthusiastically; after all, they still had ENIAC to worry about.

So von Neumann went off to Los Alamos, where the first atomic bombs were now being assembled for testing that summer, as well as for eventual use against Japan. And there in the clear, cool air that he found so invigorating, with vistas of the mountains on every side, he began to write the first draft of his report on EDVAC.

A CONFIRMED SOLITARY

He had plenty to mull over. The previous ten years had been a remarkably fertile period not only for experiments in digital-computer hardware—work that von Neumann was now as familiar with as anyone—but also, and even more so, for developments in the theory of computing. Like the hardware efforts, the theoretical insights had been scattered and apparently unrelated at first, which is why it took a von Neumann to see the connections. Yet these efforts were, in fact, laying the foundations for what would one day be called computer science. Witness Claude Shannon's recognition of the links between switching circuits and logic, for example. Or witness the ideas of another young man at the very start of his career—an English mathematician-in-training who had simultaneously followed the trail of logic to an even stranger destination.

Actually, Alan Turing had gotten a bit of a head start on Shannon; by the time his American contemporary started tracing the Differential Analyzer's relay circuits in 1936, Turing had already spent a year at Cambridge University obsessing about a single question: What are the ultimate powers and fundamental limits of a computer?

Turing was using the word in its 1930s sense, of course—the one in which the "computer" was a human sitting down with a pencil and paper to calculate his taxes, say, or the area of a circle that was four inches across. Yet his question was also one of the most profound in mathematics, going straight to the heart of reason itself: How far could logic actually take us? Was it all-powerful? Could it encompass all conceivable truth? Or were there limits? Were there numbers that couldn't be computed? Theorems that couldn't be proved? Deductions that couldn't be made?

Such issues had preoccupied mathematicians since the turn of the century. But by the time Turing learned about the situation in the spring of 1935, during a series of lectures given by one of his professors, M. H. K. Newman, that preoccupation had become something more akin to panic and despair. In 1931, Turing heard Newman explain, the Czech-born mathematician Kurt Gödel had shown that mathematical logic did indeed have limits—that there actually did exist numbers that could not be computed and theorems that could not be proved. Furthermore, it wasn't simply a matter of mathematicians' not being clever enough; Gödel had shown that the proofs literally would not exist.

Gödel's theorem had been an astonishing and unexpected result, to put it mildly, and it had unsettled mathematicians in much the same way that the discovery of the quantum uncertainty principle had unsettled physicists in the 1920s. But it had also pointed them toward a host of new challenges. For example, said Newman, Gödel had shown that unprovable assertions existed—somewhere. But he hadn't been able to point to one or even show how to find one. They were just out there, lurking in the intellectual dark. So what mathematicians clearly needed was a test, some diagnostic procedure that they could apply to any assertion they came across, and that would tell them whether it was provable or not. Such a diagnosis would proceed automatically—or, as Newman expressed it, "by a mechanical process"—and would output a simple yes or no. Thus the "decidability problem," which required mathematicians either to find this diagnostic procedure or else to show that it could not exist.

Hmm. Turing was intrigued. And so all by himself, without a word to Newman or anyone else, he took on the decidability problem as his own personal challenge.

This was typical. Alan Turing was a very solitary young man. Born in 1912, the second son of parents who had felt it best to leave their boys in foster care while they were away from England during their various tours of duty with the Indian civil service, he had been a shy, awkward, rather excitable child with an obsession with self-sufficiency: one of his favorite pastimes was what he called the desert-island game, in which he tried to synthesize exotic (and often dangerous) chemicals from everyday substances found around the house. When he ar rived at Cambridge in 1931, moreover, he proved to be an eccentric among eccentrics. He always wore a jacket and tie, of course, just as everyone did in those days. But he managed to make it look as though he'd bought his clothes at a rummage sale. He often neglected to shave, since he might nick himself and he would invariably faint at the sight of blood. He was rude, the kind of guy who thought it redundant to acknowledge your existence if he'd already said hello to you once that morning. And he had this . . . *voice*. It was a high-pitched stammer when he talked, a nervous crowing when he laughed, and a kind of random squeak when he was lost in thought.

True, Turing's classmates found him to be a witty and lively companion at the dinner table. But even so, their fellowship was rarely sufficient to breach that terrible solitude. Beyond everything else, Turing had realized that he was homosexual, a secret reality that he managed to hide from almost everyone, at the price of constant internal agonizing. "Gross indecency," as it was officially known, was a criminal offense in England. As Turing once lamented to a young (straight) friend from boarding school, he had the constant sense of living in a looking-glass world in which all the conventional ideas were the wrong way about.

As if by compensation, however, Turing's loneliness had also given rise to a strange and wonderful originality. He did things in a way that seemed obvious and straightforward enough to him but that no one else would ever have thought of. When his bicycle chain developed a habit of falling off after a certain number of revolutions, for example, he didn't get it fixed; he simply kept count of the revolutions and got off the bike in time to readjust the chain by hand.

Turing went at his mathematics in much the same strange spirit. Indeed, when he tackled a problem, he often wouldn't even glance at the mathematical literature to see what others had accomplished before him; he preferred to reinvent the wheel himself. And so it was with the decidability problem: he simply turned it over and over in his mind until finally the answer was clear to him.

"According to my definition," Turing proclaimed at the outset of his manuscript, "a number is computable if its decimal can be written down by a machine." This was taking Newman at his word about a "mechanical process." But still, as Turing's biographer Andrew Hodges points out, proper mathematical arguments were supposed to be couched in terms of abstract symbols and formal reasoning. In that context, Turing's statement sounded almost shockingly *industrial*.

But things quickly got even stranger: as he began to describe the structure of his imaginary machine, Turing seemed to veer off into a kind of abstract psychology. Since our intuitive picture of computation is of someone working with a pencil and paper, he suggested, why not refine that picture down to its absolute, irreducible essence, until we get to something simple enough to analyze rigorously? In particular, he said, let's replace the two-dimensional sheet of paper with an infinitely long tape divided into squares, like a roll of postage stamps. Each square will either be blank or have a symbol written on it. (The symbols could be anything—numbers, letters, colors, pictures, whatever—so long as there was just one per square.)

Next, said Turing, let's replace the mathematician's hand, eye, and pencil with a scanning device that can move backward and forward along the tape one square at a time, reading and writing symbols as it goes. (A modern analogy would be the read/write head of a tape recorder or a VCR.) Let's also replace the mathematician's "state of mind," his minute-to-minute sense of what's going on in the calculation, with a kind of internal scratch pad that records the current "state" of the machine. (Example: "I am currently in the process of carrying a *4* to the next column of addition.") Each different state can also be labeled by a symbol.

Finally, he said, let's replace the mathematician's knowledge, his sense of what to do next in the computation, with a table of elementary rules. There would be one table entry for each combination of tape symbol and machine state. In words, for example, one such entry might read "If the symbol *1* is in the current square of the tape, and the machine is in state CQT, then change the symbol to *0*, change the machine state to PPL, and move one square to the left." True, these rules would also have to be encoded in some fashion. But in practice, Turing noted, there was no reason that they couldn't be put on the same strip of tape being used for the data. That way you could just hard-wire the read/write head with a minimal set of rules that would respond the same way to any tape that was fed in, thus creating a *universal* machine. But however you encoded them, he asserted, an appropriate set of rules would allow the machine to generate precisely the same sequence of symbols as the human mathematician—which meant that the human and the machine would be computing the same thing.

Today, of course, Turing's machine is instantly recognizable as a (nonhuman) computer: his read/write head is just the modern computer's central processor, while his table of rules is a program residing in the computer's memory. And indeed, Turing may already have been thinking about how to build his imaginary machine "in the metal," as he later put it. For now, though, his mathematical blueprints would suffice; having defined his machine in rigorous detail, Turing was ready to begin his assault on the decidability problem.

His first result was not so much a proof as a definition: he declared that a number was computable if and only if it could be generated by his machine in a finite number of steps. By providing demonstrations of and sample programs for a wide variety of computations, Turing then made the case that his machine could compute anything that a human mathematician could. For most calculations, it was true, the very simplicity of the machine's design made it horribly inefficient. It might require a very long time to come up with the answer. But if a human mathematician could get there, the machine could get there. By extension, moreover, the same arguments applied to any *other* computing machine, then or in the future. No matter how clever its design, no machine could do more than Turing's. Or to state it in modern terms, all computers are fundamentally the same: given enough time and memory capacity, the lowliest handheld PC can do anything the mightiest supercomputer on the planet can.

But Turing also went further, showing that there were certain things that even the most powerful machine *could not* compute. In particular, he pointed out, no computer could predict its own behavior. Say you feed the machine a tape and come back to find that it's been chugging along for twenty-four hours. Say it's been chugging along for a week, or a year, or even ten years: read, write, left, right. There's still an infinite supply of tape remaining, but is this thing ever going to write down, say, a *0*? Is it ever going to *finish*? There's no way to tell, Turing showed, not in general. The "halting problem," as it's known, is hopeless.

Nowadays, of course, very few people wonder whether their computers will eventually output a *0*. But Turing's argument applies to *any* nontrivial behavior: except in very special cases, the fastest way to find out if your program will take a given action is to run it and see. And that, in turn, means that the old saying about the mindlessness of computers is true but also irrelevant. A computer does only what its programmers tell it to do—but the programmers can't really know the consequences of their commands until they see their program running. (That's one big reason software vendors have to spend so much time debugging their products: there is no universal testing program that can guarantee another program's correctness.) Or to say it still another way, the imaginary machine that Turing modeled on human mathematicians had some of the same unpredictability as the human mind.

Finally, Turing arrived at the decidability problem itself. And his conclusion was bleak. Give it up, he told his fellow mathematicians—it can't be solved. Just as there was no universal test for whether a computer would ever write down a *0*, there was no universal test for the computability of a number or the provability of a theorem. Such a test would be a bit like universal acid, the mythical stuff that supposedly dissolves everything it touches—what would you ever keep it in? More formally, Turing pointed out that if a universal decision procedure *did* exist, then it could be encoded on a tape and applied to itself. Is decidability decidable, yes or no? Then he showed that this quickly led to some ghastly logical contradictions along the lines of "This statement is false" (if the statement is true, then it's false; if it's false, then it's true—round and round in an infinite circle of futility). The only conclusion was that the universal decidability procedure could not exist.

He'd done it. So in April 1936, having completed a fifty-page typescript filled with German Gothic characters and other obscure symbols, Turing handed over the entirety to M. H. K. Newman, the professor who had started him on his journey. Newman was startled, since this was the first he'd heard about his student's interest in the decidability problem. But he quickly grasped the significance of Turing's achievement and encouraged him to submit his paper to the prestigious *Proceedings of the London Mathematical Society:* this, he told his much-relieved pupil, was first-class work.

All of which makes it easy to imagine Turing's horror when his penchant for reinventing the wheel finally caught up with him. In mid-May 1936, even as he was polishing his paper for publication, the mail brought a copy of an article just published by the American logician Alonzo Church of Princeton University—an article in which Church proved exactly the same result. True, Church had attacked the decidability problem using a different and much more conventional method: no machines here. But by the canons of mathematics, the differences didn't matter; research journals published original results, not also-rans. Worse, it turned out that Church had already published many of the steps leading up to his proof. If Turing had deigned to consult the research literature before he started, he would have known that.

Fortunately for Turing's peace of mind, however, Newman wasn't about to let such brilliant work go to waste. He prevailed upon the journal editors to accept Turing's paper anyway, on the grounds that his approach was so original and his result so important. And he likewise prevailed upon Alonzo Church to accept Turing as his student for the coming academic year at Princeton. Newman advised his colleague that given Turing's predilection for working without supervision or criticism from anyone, it would be best for him to "come into contact as soon as possible with the leading workers in this line, so that he should not develop into a confirmed solitary."

Those gestures of support presumably reassured the twenty-four-year-old Turing that he hadn't made a total fool of himself. But still, once he arrived at Princeton at the end of that September, he seemed to slip into something of a depression. By rights, of course, he should have been in heaven. In those days the Princeton mathematics department was still housed in the same building as the august Institute for Advanced Study, so the place was swarming with the finest mathematical minds in the world; Albert Einstein himself sometimes wandered the hallways. And yet Turing remained the same confirmed solitary he had been at Cambridge. To someone of his semi-upper-crust English background, America felt harsh and pushy. Besides, his long obsession with the decidability problem was over—and now what? In December he gave his first talk at Princeton; attendance was poor since he was a total unknown. In January

1937 his paper finally appeared in print, but only two people bothered to write and ask for offprints (a good measure of interest in the days before Xerox machines). The fact was that very few mathematicians seemed to understand or care what he had done.

Turing's sense of futility soon lifted, fortunately. Church published a glowing review of his paper. He found other problems to pique his interest. And by the spring of 1937 he had cheered up enough that he decided to remain with Church for another academic year and get his doctorate. Nonetheless, Turing was still achingly homesick; he had *never* learned to like America. And even more urgently, he could see by 1938 that his homeland would soon be at war. He desperately wanted to help. As he wrote to a friend that spring, "I hope Hitler will not have invaded England before I get back." So in July of 1938, with his brand-new doctorate in hand, Turing set sail for home, heading back to Cambridge, to England, and to the war.

THE PHYSICAL EMBODIMENT OF PURPOSE

Seven years later, as he was drafting the EDVAC report on the Los Alamos mesa top, John von Neumann almost certainly had Turing's machine in mind. Unfortunately, we have no way of knowing precisely how it influenced his thinking, since his draft nowhere mentions Turing explicitly. But we do know that von Neumann had had a very high regard for the Englishman ever since meeting him at Princeton. He had even tried to keep him from leaving there in 1938 by offering to make him his assistant—a compliment indeed, considering that the country was full of refugee mathematicians who would have been overjoyed to work with the great von Neumann. We also know that von Neumann still raved to colleagues about the young man's "brilliant ideas" even after Turing turned him down. After the war, in fact, von Neumann's advocacy would probably do more than anything else to win Turing's work the attention it deserved.

And besides, even if he had forgotten Turing's machine during the press of his wartime research—not that John von Neumann ever forgot *anything*—he'd had a much more recent reminder thanks to yet another chain of unexpected insights set in motion by his old friend Norbert Wiener.

However mitted Wiener may have been in the autumn of 1940, when Vannevar Bush rejected his "impractical" computer proposal, he certainly hadn't let it keep him out of war research. Within weeks, in fact, Wiener was already deep into a problem of desperate practicality: antiaircraft fire control. All across the southern tier of England, the speedy new warplanes of the Luftwaffe were proving entirely too agile for the artillery on the ground. The English gunners simply could not react fast enough to the German pilots' twists and turns; too many of their shells were hitting empty air. And the consequences of that failure were being felt every night in central London.

The obvious solution was some kind of high-speed electronic fire-control system, one that would take in radar data and then automatically give the gunners a pointing vector. The obvious problem, unfortunately, was that the system would have to allow for the shells' time of flight, which meant extrapolating the enemy pilot's trajectory at least a short time into the future—including all those random twists and turns.

Impossible? Not quite, Wiener realized. No pilot could ever make his trajectory completely random. He was constrained by the limits of his aircraft (how fast could it climb or turn?); by his own physiological limits (how fast could he react? how fast could he maneuver without blacking out?); and even by his flying style and skill level. In some statistical sense, Wiener reasoned, those constraints might make his trajectory predictable. Mathematically, in fact, it would be like extracting a meaningful "signal" that was embedded in the "noise" of the aircraft's random motion. And indeed, Wiener was quickly able to sketch out a set of mathematical techniques that would do just that—formulating them so broadly, in true Wiener style, that they would later prove invaluable across campus at the Radiation Lab, not to mention in electrical and communications engineering generally. In effect, he had shown how best to separate signal from noise in *any* kind of data stream. (After the war, Wiener's dense treatise on the subject, bound in canary-colored cloth and entitled *Extrapolation, Interpolation, and Smoothing of Stationary Time Series,* would become known to a generation of communications engineers as The Yellow Peril.)

In any case, by year's end Wiener had a contract from Bush's research council to develop hardware that could carry out his mathematical transformations automatically. And by January 1941, with the able assistance of an electrical engineer named Julian Bigelow—a "quiet, thorough New Englander"—he was hard at work.

In terms of immediate practical payoff, unfortunately, the Wiener-Bigelow effort proved to be another bust. By 1941 they had a prototype fire-control system that could follow a simulated target quite well. But as they themselves were the first to admit, it was too complex for use in the field. The system that was actually deployed relied on a simpler approach developed at Bell Laboratories in New York City.

In its influence on Wiener's own thinking, however, his work with Bigelow proved to be seminal—not least because it provoked in him a strong sense of déjà vu: "It will be seen that for the second time I had become engaged in the study of a mechanico-electrical system which was designed to usurp a specifically human function—in the first case, the execution of a complicated pattern of computation, and in the second, the forecasting of the future," he would later note.[13]

Extrapolating a series of data points with a mathematical equation was one thing, but seeing it done by an inanimate machine was quite another, and still a cause for astonishment in 1941. Indeed, it went straight to the heart of perhaps the greatest conundrum in natural philosophy: the mind-body problem. Ordinary physical matter ("body") is inherently passive, goes the classic argument. It simply responds to outside forces. An apple falls from its tree only if it is pulled by gravity. A baseball abruptly heads for the center-field bleachers only if its trajectory is intercepted by a bat. Things happen only if they are *made* to happen. In the physical world, every effect requires a cause—and the cause always comes first.

However, the argument continues, human beings (and other living things) are not passive. Even though our bodies are obviously made of physical matter, we don't just respond to outside forces. We have autonomy. We can take action. We have goals, expectations, desires, *purpose*. Our "causes" are not in the past but in the future. And therein lies the conundrum: how can that possibly be unless we possess some ineffable quality of "mind," or "spirit," or "soul" that transcends the physical matter of our bodies?

Good question—except that Wiener and Bigelow had just produced a purely physical device that took action based on a prediction. In however rudimentary a way, their fire-control apparatus had "causes" that lay in the future. Furthermore, they realized, it had that capacity for a very deep reason: *feedback*.

The concept of feedback is one of those brilliantly simple ideas that seem utterly obvious once someone has pointed them out. Consider a thermostat, for example: Whenever the surrounding room gets too cold, a strip of metal bends, closes a contact, and turns the furnace on; when the room warms up again, the strip straightens out, releases the contact, and cuts the furnace off. The system automatically counteracts the changes in temperature, and the room stays pretty close to comfortable.

In a world where no device is ever perfectly accurate or reliable, the two men recognized, some such self-correcting mechanism was essential for any kind of effective operation. And indeed, many specific examples of feedback mechanisms were already well known. In the late eighteenth century, for example, the Scottish inventor James Watt had equipped his new steam engine with a simple "governor" for safety: if the engine ever started going too fast, the governor device would automatically cut back the power and force it to slow down. By 1868 such governor devices had become common enough that the Scottish physicist James Clerk Maxwell published a pathbreaking mathematical analysis of them. And of course, in Wiener and Bigelow's own fire-control system, the attacking aircraft's *predicted* trajectory was constantly being updated through feedback from its *actual* trajectory.

But did feedback also apply to voluntary action? Definitely, Wiener and Bigelow argued. Consider the process of picking up a pencil. Your mind has a goal: Pick up the pencil. But as you reach out for it, the motion of your arm is never perfectly accurate. So your brain must constantly make corrections using information from your eyes, muscles, and fingertips. A smooth, coordinated action is possible if and only if your brain can complete that feedback loop.

And if it can't complete the loop? Then voluntary action becomes impossible, the two men reasoned. And indeed, when they consulted with Wiener's close friend Arturo Rosenblueth, a Mexican-born neurophysiologist who was then working at the Harvard Medical School, they learned that such breakdowns were common. In purpose tremor, for example, a condition often associated with injury to the cerebellum, a patient would try to pick up a pencil, overshoot the mark, and then go into uncontrollable oscillation—exactly as the mathematical theory of feedback predicted.

That confirmation was so striking that Wiener and Bigelow were immediately convinced, as was Rosenblueth: feedback was required for *any* voluntary action. Moreover, the implications were so profound that the three men kept exploring the issue even in the midst of their more urgent war research and the national mobilization after Pearl Harbor. In 1942, in fact, they even made time to lay out their conclusions publicly at a neurophysiology meeting in New York sponsored by the Josiah Macy Foundation.

First, Wiener, Bigelow, and Rosenblueth explained at that meeting, a deep understanding of *control* requires a deep understanding of *communication;* the two are inseparable. In the action of picking up that pencil, for example, information about the position of your hand must be communicated from your eye to your brain, and the corrections then have to be communicated through the motor nervous system to your hand. Given an appropriately general definition of "communication," moreover, much the same story can be told about any other feedback process.

Second, they pointed out, the processes of communication and control are based on the much more fundamental notion of *message,* "whether this should be transmitted by electrical, mechanical, or nervous means." The same principles of feedback apply in every medium, they said, just as the same principles of aerodynamics apply to the flight of a sparrow and the flight of a B-29. This meant, ultimately, that the human nervous system no longer had to be viewed as something utterly mysterious, a kind of organic black box possessing powers of responsiveness and will that science could never understand. Quite the opposite: viewed from an engineering perspective, the nervous system was a perfectly comprehensible array of feedback loops in active communication with its environment.

Conversely, they said, if it was valid to think of the nervous system in engineering terms, then it was just as valid to think of machines in biological terms. Look at the fire-control system, correcting its aim after every shot with feedback from radar. The gun and its fire-control system operated in a completely automatic fashion, with no humans in the loop anywhere. And yet the gun seemed guided by a grimly determined intelligence.

Through feedback, said Wiener, Bigelow, and Rosenblueth, a mechanism could embody *purpose.*

Even today, more than half a century later, that assertion still has the power to fascinate and disturb. It arguably marks the beginning of what are now known as artificial intelligence and cognitive science: the study of mind and brain as information processors. But more than that, it does indeed claim to bridge that ancient gulf between *body* and *mind*–between ordinary, passive matter and active, purposeful spirit. Consider that humble thermostat again. It definitely embodies a purpose: to keep the room at a constant temperature. And yet there is nothing you can point to and say, "Here it is–this is the psychological state called *purpose.*" Rather, purpose in the thermostat is a property of the system as a whole and how its components are organized. It is a mental state that is invisible and ineffable, yet a natural phenomenon that is perfectly comprehensible.

And so it is in the mind, Wiener and his colleagues contended. Obviously, the myriad feedback mechanisms that govern the brain are far more complex than any thermostat. But at base, their operation is the same. If we can understand how ordinary matter in the form of a machine can embody purpose, then we can also begin to understand how those three pounds of ordinary matter inside our skulls can embody purpose–and spirit, and will, and volition. Conversely, if we can see living organisms as (enormously complex) feedback systems actively interacting with their environments, then we can begin to comprehend how the ineffable qualities of mind are not separate from the body but rather inextricably bound up in it.

For those who had the time to pay attention to such things during the war years, Wiener, Bigelow, and Rosenblueth's argument at the Macy conference was electrifying. The anthropologist Gregory Bateson would look back on that moment more than thirty years later and still marvel: "The central problem of Greek philosophy–the problem of purpose, unsolved for 2,500 years–came within range of rigorous analysis."

Certainly the trio's presentation had quite an effect on one member of the audience. Then in his forty-third year, Warren McCulloch was officially a professor of psychiatry at the University of Illinois Medical School in Chicago. But in reality, there wasn't a job description on earth that could define him. Enigmatic, lyrical, a devout Quaker possessed of a mind that wandered freely over a vast intellectual landscape, McCulloch was a keen student of philosophy, neurophysiology, mathematics, logic–and poetry. He filled his scientific articles with allusions to everyone from Shakespeare to Saint Bonaventura and then gave them titles such as "Where Is Fancy Bred?," "Why the Mind Is in the Head," and "Through the Den of the Metaphysician." He was endlessly enthusiastic about the quest after the inner workings of thought. And he came away from the Macy meeting an inspired man. If one simple feedback loop was enough to endow a machine with purpose, McCulloch mused, then how much more could millions or billions of feedback loops accomplish? How much more might happen inside a web of connections as dense as that in, say, the cerebral cortex?

McCulloch was soon hard at work on the problem in collaboration with Walter Pitts, an eighteen-year-old mathematics prodigy at the University of Chicago. For simplicity's sake, they assumed that the brain as a whole could be modeled as a vast, interconnected electrical circuit, with neurons serving as both the wires and the switches. That is, each neuron would receive electrical input from dozens or hundreds of other neurons. And if the total stimulation passed a certain threshold, that neuron would then "fire" and send an output pulse to dozens or hundreds more.

The result—today it would be known as a "neural network" model—was admittedly a gross oversimplification of reality. But McCulloch and Pitts argued that it did capture the abstract essence of brain physiology. Moreover, it was mathematically tractable, which meant that they could begin to ask deeper questions about neural networks in general. What could such networks actually do, for example? What was their computational power? What kind of logical operations were they capable of carrying out? And what kind of operations could they *never* carry out, even in principle?

If these sound like the same questions that Alan Turing asked of his imaginary machine, it was no accident. By some miracle, the obscure McCulloch and Pitts had come across the decidability paper written by the equally obscure Turing, understood its significance, and taken Turing's work as one inspiration for their own. Moreover, McCulloch and Pitts now found themselves arriving at much the same destination as Turing. Their own paper, published in 1943 as "A Logical Calculus of the Ideas Immanent in Nervous Activity," was essentially a demonstration that their idealized neural networks were functionally equivalent to Turing machines. That is, any problem that a Turing machine could solve, an appropriately designed network could also solve. And conversely, anything that was beyond a Turing machine's power—such as the decidability problem—was likewise beyond a network's power. As the science historian William Aspray has written, "With the Turing machines providing an abstract characterization of thinking in the machine world and McCulloch and Pitts's neuron nets providing one in the biological world, the equivalence result suggested a unified theory of thought that broke down barriers between the physical and biological worlds."[14] Or, as McCulloch himself would put it in his 1965 autobiography *Embodiments of Mind,* he and Pitts had proved the equivalence of all general Turing machines, whether "man-made or begotten."

John von Neumann was deeply impressed by McCulloch and Pitts's neural-network ideas from the moment he saw their paper ("Johnny," was Norbert Wiener's message to his frequent correspondent: "You've got to read this thing!"). Brain science, logic, philosophy, computation, the fundamental nature of mind—they were all here, in one electrifying synthesis. Nor had von Neumann forgotten about neural networks under the pressure of Los Alamos and his long quest for computational power. Quite the opposite. It was essentially impossible to think about computers in those days without thinking about brains as well. Here, for the first time in history, was a machine that could *think*— or at least do arithmetic, compile data, integrate differential equations, and accomplish all manner of things that had once required human intelligence. Some researchers, including Howard Aiken, found the computer-brain analogy simplistic, misleading, and even dangerous; others, such as von Neumann and Wiener, found it subtle, illuminating, and provocative. But nobody could ignore it.

Indeed, von Neumann had found the analogy so compelling that in late 1944 he had joined forces with Wiener and Aiken to organize a conference on the subject. Held at Princeton on January 6–7, 1945, the meeting was a tiny affair that included only the three organizers plus McCulloch, Pitts, Goldstine, and a number of like-minded neurophysiologists—fewer than a dozen people, all told. But it definitely had an impact. With a lecture on computing machines given by von Neumann, one on communications engineering by Wiener, and still others on the structure of the brain by McCulloch and Rafael Lorente de No of Rockefeller University, plus some extremely lively discussions, it was later cited by Wiener as one of the defining moments in his emerging science of communication and control. Given the presence of von Neumann, Goldstine, and Aiken, moreover, it symbolized the emergence of the modern computer-research community, which today numbers in the millions. In those war years, the scattered computer pioneers were finally beginning to meet one another face to face, where they could trade ideas directly and start to learn from their respective experiences.

And not incidentally, that Princeton meeting in January 1945 meant that McCulloch and Pitts's neural-network ideas were fresh in von Neumann's mind a few months later as he began to draft his report on the EDVAC.

THE LAST TRANSITION

In his discussions with the Moore School team, von Neumann had agreed to base his report on two great abstractions, corresponding roughly to hardware and software. On the hardware side, the idea was to focus on the forest instead of the trees. Rather than get bogged down in a discussion of specific vacuum tubes and circuit elements (technology that was evolving rapidly in any case), the report should discuss what is now known as the abstract *architecture* of the computer—that is, the overall structure of the machine, the function of each part, and the way all those parts would interact to perform a computation (J. Presper Eckert, in particular, had always emphasized the "logical" design of his computers). Once this abstract design was in place, engineers would later be free to implement the various functions in any way they chose, based on the best technology available at the time.

As von Neumann began to sketch out what this architecture should *be*, however, his thinking was also heavily influenced by McCulloch and Pitts, whom he cited by name at a number of points. Von Neumann was particularly intrigued by their idealized neurons, which had precisely the kind of *on-off* behavior required for the logical elements of a computer. Indeed, McCulloch and Pitts themselves had emphasized that each of their model neurons actually behaved like a small logic circuit of the sort that could be built with a set of electro-mechanical relays or vacuum tubes. And that, if nothing else, meant that von Neumann could draw his own circuit designs using McCulloch and Pitts's neural-net notation, which would eventually evolve into the standard circuit notation used by computer engineers today.

Even more telling, however, was the fact that as von Neumann went on to describe the five functional units of his abstract computer, he referred to them as "organs" and went out of his way to make provocative comparisons with biological functions. Within the first page or two of his report, for instance, he was comparing the *input* units of the machine with the sensory neurons of the brain, the ones that allow us to perceive the world. He likewise compared the *output* units with the motor neurons of the brain, the ones that govern bone, muscle, and action. And then he compared the remaining three functions with the brain's associative neurons, which are devoted to abstract thought. The *central arithmetic* unit, for example, would be the computer's own internal calculating machine, the high-speed, electronic equivalent of an accountant's desktop calculator. This was the portion of the computer that would do the actual work of computation, carrying out additions, subtractions, multiplications, divisions, square roots, and the like, as needed. (All calculations, of course, were to be done in binary arithmetic.)

Next, said von Neumann, the *memory* unit would be the computer's electronic scratch pad, the place where it stored data, programs, intermediate results, and the final answer. Following a careful analysis of typical problems in fields such as statistics and fluid dynamics, von Neumann estimated that the memory unit would need a capacity of at least 256,000 binary digits—just thirty-two kilobytes in modern parlance, but a number that he clearly found daunting in 1945. After much additional discussion, he concluded that such a memory could be built, though it would be at the outer edge of technological feasibility. (He was right: memory would continue to be the biggest single constraint on computer performance until the microchip revolution of the 1970s.)

Finally, said von Neumann, the *central control* unit would be the heart of the computer, the part that decided what to do next. (Today this is usually known as the central processing unit, or CPU.) Its decisions would in turn be governed by a program stored in the memory unit. This left just one critical question: How was the central control unit supposed to go about executing those programs?

Von Neumann had quite a bit of leeway in his answer, thanks to Alan Turing. As he was undoubtedly well aware, his abstract architecture was logically equivalent to a Turing machine, with the memory, input, and output units collectively corresponding to the tape, and the central arithmetic and central control units collectively corresponding to the read/write head.* In particular, von Neumann knew that his architecture was universal in precisely the same sense as Turing's, which meant that he (like the generations of computer designers to come after him) was free to focus his attention on real-world factors such as cost, speed, efficiency, and reliability, without having to worry about his machine's fundamental ability to compute: there was nothing one design could do that another couldn't, given enough time and storage capacity.

In any case, von Neumann opted for the simplest scheme possible. He envisioned the central controller as going through an endless cycle: fetch the next chunk of instructions or data from the memory unit, execute the appropriate operation, and then send the results back for storage in memory. Fetch, execute, return. Fetch, execute, return. At one level, of course, this kind of serial, step-by-step processing was clearly absurd. It was a bit like trying to make a dinner salad the hard way: go out and buy the lettuce, bring it home, and chop it up. Go out and buy the carrots, bring them home, and chop them up. Go out and buy the celery, etc., etc. But von Neumann argued that any other approach, such as trying to coordinate many operations simultaneously, would make the central controller's circuitry much more complicated and thereby increase the chances of failure. At least in terms of 1940s technology, he said, serial operation was the price to be paid for reliability. Moreover, he had a point: it was only in the late 1970s, with the availability of reliable and inexpensive microchips, that computer scientists would begin serious experimentation with "parallel" computers that could carry out many operations simultaneously. To this day, the vast majority of computers in the world—including essentially all personal computers—are still based on the serial, step-by-step "von Neumann" architecture.

Von Neumann mailed off his handwritten manuscript to Goldstine at the Moore School in late June 1945. He may well have felt rushed at that point, since the Trinity test of the plutonium bomb was less than three weeks away (it would take place on July 16). But in any case, he left numerous blank spaces for names, references, and other information that he planned to insert after his colleagues had had a chance to comment. Presumably he intended that the final version of the report would give explicit credit to the originators of its key ideas—which in many cases meant Mauchly and Eckert.

Even as it stood, however, the draft was compelling. After a decade in which the computer pioneers had struggled to master the vagaries of hardware, barely able to see the forest for the trees, von Neumann had laid out the fundamental principles of computer design with breathtaking clarity. His paper was so good,

---

* Or more precisely, von Neumann's architecture *would* be equivalent if the memory unit had an infinite capacity.

in fact, that Goldstine, in what was apparently a burst of innocent enthusiasm, had it typed up verbatim as "First Draft of a Report on the EDVAC" and circulated it within the group for discussion—under von Neumann's name alone. Goldstine thereby gave outsiders their first inkling of what was going on at the Moore School, since it wasn't long before copies reached the hands of other groups working on high-speed computers. But alas, that very fact would soon give rise to a feud that would tear the Moore School group apart.

Initially, at least, Mauchly and Eckert had been quite pleased by von Neumann's paper. But as the months went by and they began to realize what was happening, their reaction gradually shifted from pleasure to annoyance and then to fury. They had been constrained by wartime secrecy rules from talking about their work. Yet now, because the "First Draft" was circulating under the name of the world-famous von Neumann, they were in danger of getting no credit whatsoever; people just assumed that all the ideas in it were his. And worse, the paper was turning out to have serious patent implications.

At issue was the *second* great abstraction in the "First Draft": the "stored program" concept. Up until that point, most of the pioneering computers had embodied at least part of the problem-solving process in their actual physical structure. The classic example was Bush's Differential Analyzer, in which the problem was represented by the arrangement of gears and shafts. The brilliant idea behind the stored-program concept was to make a clean split, to separate the problem-solving sequence from the hardware entirely. The act of computation thus became an abstract process that we now know as *software,* a series of commands encoded in a string of binary *1*s and *0*s and stored in the computer's memory.

In theoretical terms, of course, this idea was equivalent to Turing's inspired notion of encoding the instruction table on the tape of his imaginary machine instead of hard-wiring it into the read/write head. And it had precisely the same implications for universality: a stored-program computer, like a Turing machine, could treat its program instructions as just another kind of data.

In practical engineering terms, meanwhile, the stored-program approach had the obvious advantage of convenience. By storing the instructions electronically, you could change the function of the computer without having to change the wiring. With the advent of ultrafast electronic computing, moreover, it had become an increasingly urgent necessity. From the beginning, the Moore School group had been acutely aware that it was pointless to build a machine that could add up thousands of numbers per second if it had to wait seconds or minutes for a human operator to type in the next number or give it the next command. For an electronic computer to make sense at all, its access to data and commands had to be at least as fast as its arithmetic. And that meant storing all the data and all the commands in some form of electronic memory.

To von Neumann, in fact, the stored-program notion seemed so obvious and straightforward—assuming that you could provide the computer with enough memory to handle a full-scale program—that his paper treated it almost in passing. It seemed equally straightforward to Eckert and Mauchly, who had actually discussed the idea in the early months of 1944, half a year before von Neumann joined the EDVAC project.

Straightforward or not, however, the stored-program idea was the key technical advance in EDVAC. Eckert and Mauchly were anxious to patent it, arguing with some justice that they had conceived of it independently and that von Neumann's report had simply summarized the thinking of the group as a whole. As time went on, however, von Neumann proved just as eager to keep them from monopolizing the idea. He refused to acknowledge them as the originators of the stored-program concept, insisting that so much had been said during their early discussions that it was impossible to assign specific credit to anyone. As a result, what had started out as annoyance quickly gave way to bitterness, with Eckert and Mauchly on one side versus Goldstine and von Neumann on the other. Indeed, it soon became a three-sided argument, since the University of Pennsylvania also felt that it had some claim to the idea: Eckert and Mauchly, after all, were university employees and had done their work on university time. So for the time being, at least, the university forbade Eckert and Mauchly to apply for any patent in their own names.

The group managed to keep a lid on the situation—barely—until ENIAC was completed. The machine was demonstrated for a suitably awed press corps on Valentine's Day, February 14, 1946. And it worked just as Eckert and Mauchly had advertised, adding up thousands of multidigit numbers before the reporters could even blink. It was a striking demonstration that an ultrafast, fully electronic computer could indeed operate reliably. (It may also have been the first time that the word *computer* was applied to a machine; Mauchly and Eckert wanted to dramatize ENIAC's capabilities versus those of human computers.)

As gratifying as that performance was, however, the triumph was almost lost in the increasingly nasty patent dispute. Shortly after ENIAC's debut, the University of Pennsylvania imposed a new rule explicitly laying claim to all patents on all inventions produced by university employees. Eckert and Mauchly refused to go along with it. On March 31, 1946, they resigned in protest and went off to found the Eckert-Mauchly Corporation, thereby throwing the EDVAC project into limbo (the machine would not be completed until 1952). But within weeks, before Eckert and Mauchly could stake their claim, von Neumann made a preemptive strike by filing his own patent application on the stored-program idea.

The mess promptly landed in the lap of the army, whose role as the funding agency for both ENIAC and EDVAC gave it final authority in the matter. The whole unedifying saga would drag on for another year, ending only in April 1947, when exasperated army attorneys at last threw out everybody's patent claims on the ground that von Neumann's "First Draft" paper represented prior public disclosure. They decreed that the stored-program idea rightfully belonged in the public domain. And there it has remained.

That was probably just as well. However fierce the controversy surrounding its birth, the stored-program concept now ranks as one of the great ideas of the computer age—arguably *the* great idea.

By rendering software completely abstract and decoupling it from the physical hardware, the stored-program concept has had the paradoxical effect of making software into something that is almost physically tangible. Software has become a medium that can be molded, sculpted, and engineered on its own terms. Indeed, as the Yale University computer scientist David Gelernter has pointed out, the modern relationship between software and hardware is essentially the same as that between music and the instrument or voice that brings it to life. A single computer can transform itself into the cockpit of a fighter jet, a budget projection, a chapter of a novel, or whatever else you want, just as a single piano can be used to play Bach or funky blues. Conversely, a spreadsheet file can (with a little effort) be run on a Microsoft Windows machine, a Macintosh, or a Unix workstation, just as a Bach fugue can be performed on a pipe organ or by an ensemble of tubas. The bits and bytes that encode the spreadsheet obviously can't function without the computer, any more than a page full of notes can become music without a performer. And yet the spreadsheet also transcends the computer, in exactly the same way that the Bach fugue transcends any given performance of it. Everything important about that spreadsheet—its on-screen appearance, its structure, its logic, its functionality, its ability to respond to the user—exists, like the harmonies and cadences of Bach, in an abstract, platonic world of its own.

Metaphorically, at least, this abstraction is probably about as close as science and technology have ever come to the pagan notion of animation, spirit, enchantment. It is certainly a big part of what gives computers their emotional clout. Anyone who has ever switched on a personal computer has felt it: watching the programs fill the screen is disconcertingly like watching a dead thing come alive. Without software, the glowing glass box is just a glass box. With software, it becomes what the MIT sociologist Sherry Turkle has dubbed the first psychological machine—active, surprising, goal-driven, and capable of responding to us in ways that no ordinary machine could ever do. "A new mind that is not yet a mind," Turkle called the computer in her 1984 book, *The Second Self,* "a new object, betwixt and between, equally shrouded in superstition as well as science."

More recently, of course, software has also become the basis of a whole new abstraction, that vast network of interlinked computers known variously as the Internet, the Web, or cyberspace. After all, it is only when software becomes independent of hardware that we can even think about sending files and programs over a telephone line or a high-speed data network. Indeed, cyberspace now seems set to raise the software abstraction to yet another level, where programs won't even be tied to one place anymore. Instead, software "agents" searching for data will be able to leave their home computers and fan out through the network at will, merging, spawning, communicating, collaborating, and leaping from machine to machine like a society of tame computer viruses.

In any case, the stored-program concept articulated in von Neumann's 101-page "First Draft" marked the beginning of the last great transition in the development of computers, a transition that is still going on today. It was the shift in focus from structure to behavior, from how computers were made to what they could do. Before von Neumann's paper, people viewed electronic digital computers merely as fancier and fancier adding machines. The things were immensely useful, to be sure, but they were still just crunchers of numbers. After von Neumann's paper—though it would take the better part of a generation before this really became clear—people could begin to conceive of these machines as something fundamentally new. Computing machines had become *computers*, the devices that implemented software.

CHAPTER 3

NEW KINDS OF PEOPLE

Logic, computability, high-speed electronics, stored programs, neural networks, the physical embodiment of purpose–J. C. R. Licklider ate it all up, with gusto. Indeed, Norbert Wiener's Tuesday-night supper seminar was the high point of his week in those postwar years. Lick and the rest of the Psycho-Acoustics Lab contingent would hang around long after the plates were cleared away, jumping from table to table and happily fighting the battle of ideas with forty or fifty of their contemporaries from all over Cambridge. They would keep it up as they drove back to Harvard, arguing the fine points with one another in the car. And when he finally got back home to Louise, Lick would still be floating high, his blue eyes bright with intellectual intoxication.

It was an intoxicating time for everyone, really, especially with the almost explosive release of tension and anxiety that had accompanied the end of the war. "Cambridge was like an anthill," explains Louise Licklider. "Everybody was getting involved with everybody else–finding different challenges, taking up different ideas. I use the word *cross-fertilization* because there was an awful lot of that going on. And quite a lot of socializing, too."

The Lickliders, of course, were in the thick of it. Following their marriage, in

An imaginative forward glance: Norbert Wiener in his element

January 1945, their tiny upstairs apartment on Massachusetts Avenue near Harvard Square had become a frequent destination for both Louise's theater crowd and the bright young men of the Psycho-Acoustics Lab. "Louise was an absolutely stunning girl," remembered William McGill. "She was gifted aesthetically and poetically, Lick was gifted intellectually, and they made a magnificent couple." McGill also recalled one infamous evening when he and his wife knocked on the Lickliders' door for a cocktail party and their host greeted them with a pitcher in each hand. "Louise likes this, and I like *this*," Lick said, pouring them two martinis apiece. Everybody at the party was blitzed inside of fifteen minutes, said McGill with a laugh.

Of course, as George Miller points out, "Lick was the kind of guy you found off in a corner at a party arguing about Fourier transforms." It was just that Lick, being Lick, had trouble keeping a straight face. Miller recalls the weekly Wednesday-night "Pretzel Twist," in which the young experimental psychologists at Harvard would sit around in someone's apartment talking about research while consuming pretzels by the fistful and quite a lot of beer. "W. R. 'Tex' Garner and I were trying to learn something from those meetings," Miller says, "while James P. Egan was intensely serious about behavioristic dogma and Lick sat back laughing gently most of the time." Miller sometimes had to wonder how serious Lick really was about psychology. "But then," he adds, "I was never quite sure how serious Lick was about *anything*."

It *was* hard to tell sometimes. For example, there was the paper in which he presented an improved apparatus for animal conditioning experiments, entitled "The Gridless, Wireless Rat-Shocker." He claimed forever afterward that it was the most popular and often-requested paper he ever wrote. And then there was the paper by Licklider and Licklider, conceived in the hot and steamy summer of 1947, when Louise was suffering through the pregnancy that would soon produce their son, Tracy. She and her husband would frequently abandon their stifling apartment to spend their evenings in the air-conditioned depths of Memorial Hall. One night Lick suddenly said, "Let's do an experiment!" And so they did, wrapping cardboard tubes with foil and stringing together an elaborate maze. The result was "Observations on the Hoarding Behavior of Rats," published in 1950.

In truth, Lick was very serious about psychology, as was everyone in the Dungeon, not least because they all had to jump-start careers that had been stalled by the war. Granted, the place was in turmoil. The wartime Psycho-Acoustics Lab had abruptly shrunk after VJ Day, when its funding was cut. And then those who survived had been joined in their underground warren by the remnants of the main psychology department, which had just undergone a bitter split between the social psychologists and the experimenters. Nonetheless, the release of pent-up intellectual energy produced a torrent of research papers, which quickly gave the Memorial Hall basement a national reputation in psychology. Lick alone published at the rate of four or five papers per year during this period, including two chapters (one coauthored with Miller) in the landmark *Handbook of Experimental Psychology*, edited by Smitty Stevens.

However, for at least a handful of the Memorial Hall crew—Lick, Miller, and McGill, among others—that seriousness of purpose was more than just a return to academic business as usual. The supper-seminar discussions, the lectures and classes they were attending at MIT, the cross-fertilization of new ideas—that whole postwar ferment was beginning to push them in new directions. And none of them was changing faster than J. C. R. Licklider. Indeed, said McGill, by the time he got there, in 1948, Lick's efforts to apply engineering concepts to the brain were downright Wienerian—and so foreign-seeming in the context of Memorial Hall that the rest of them sometimes wondered if he was still doing psychology at all. "We found that talking with Lick was rather like talking with a very smart electrical engineer who knew more about your work than you did," said McGill, "but who also had the advantage on you because he knew all the engineering techniques, too."

In effect—though he himself probably wouldn't have expressed it this way—Lick was trying to do for the human auditory cortex what von Neumann had done for computers in his "First Draft": he was trying to understand the brain in architectural terms, rather than focus on the specific neural details. "Lick approached any problem we were dealing with as a systems problem," said McGill. For example, he might sit someone down in front of an oscilloscope where a dot was moving erratically on the screen, and ask him or her to keep a circle centered on the dot. Since nobody could do it perfectly, Lick would record the tracking errors and then use that record to try to reconstruct the machinery of hand-eye coordination, asking himself what kind of neural circuitry would produce exactly those errors and no others. Likewise, in his "Duplex Theory of Pitch Perception," Lick came up with a McCulloch-and-Pitts-style neural-network model to explain the pitchlike quality that musicians call chroma or tonality.

"You have to realize that back then these ideas were mind-boggling in psychology," said McGill. "They were right on the cutting edge. They were far more complex and sophisticated than standard psychological learning theory. What Lick was doing came out of systems engineering, the construction of servomechanisms, the construction of equipment for human operators in advanced aircraft—Lick must have developed these ideas by talking to electrical engineers and aircraft designers during the war."

But just as influential was the intellectual turmoil surrounding Wiener and company. As George Miller told the psychologist Bernard J. Baars for the latter's book *The Cognitive Revolution in Psychology,* "Norbert was around, and we read his books, and we all were impressed with things like . . . the Rosenblueth-Wiener-Bigelow paper on feedback systems that set their own goals. Those things had quite a freeing effect." Miller also remembers being deeply impressed by McCulloch and Pitts's wartime work on neural-network theory. "We all thought it was wonderful," he says. "Lick and I taught about it in our seminars. And right after the war we got Walter Pitts to come over from MIT to explain it to us."

Miller, of course, was very different from Lick in style and interests. Whereas

Lick was trying to understand the circuitry of the brain in systems-engineering terms, Miller was interested in higher-level cognitive issues: learning, the structure of memory, and most especially language. Yet the two of them were very much alike in their restless exploration of new directions in psychology, in their love of mathematics, and in their shared fascination with Norbert Wiener's vision of a new science. "It gave me these enthusiastic notions that any good idea could be mathematized," says Miller. "Just after the war, for example, I went down to MIT to take Wiener's course in time-series analysis [the techniques he had developed for the antiaircraft fire-control problem], except that it turned out to be taught by his protégé Y. W. Lee. So I just stayed and scribbled notes as fast as I could. And when I looked up, there was another guy next to me doing exactly the same thing—it was Jerry Wiesner!"

So yes, those years were incomparably exhilarating for the Memorial Hall natives—or rather, they *would* have been exhilarating if it hadn't been for two increasingly irksome problems. The first was Smitty Stevens. In his own mind Stevens was still the patriarch of the Psycho-Acoustics Lab. But Lick and Miller, in particular, were no longer children. Stevens could feel them growing up intellectually and slipping away—a feeling that did nothing to improve his legendary temper. "Miller and Lick were precious to him in a way that the rest of us weren't," explained McGill. "Everybody regarded them as the brightest people around. But you were highly regarded in Smitty's eyes only if you were working on something *he* thought was important. And he didn't believe in systems analysis or any of that. So he wound up arguing, and in some cases openly fighting, with both of them. I can remember him saying, almost verbatim, 'George Miller is working on words. Can you believe that? *Words!*' "

Miller himself today tends to downplay the tension with Stevens, preferring to remember the Psycho-Acoustics Lab as a close-knit if not always harmonious family: "Smitty didn't care if you liked him," he says. "But he didn't want you to ignore him. He would come up and tell you that you were full of shit just to get your attention. That was simply the nature of life with Smitty." And perhaps it was. But McGill, who watched the shouting matches as an admittedly impressionable graduate student, remembered wondering at the time how Miller and Lick could stand it.

Meanwhile, there was the second problem. Although Lick, Miller, and a few of the other young Turks were striking out in new directions, almost no one else in the Memorial Hall basement was following them. Without fully realizing it themselves, they were beginning to travel forbidden paths, to think forbidden thoughts, to question the received dogma in psychology. And a wise young psychologist simply didn't do that in the 1940s, particularly not at Harvard, where the Way of Truth was embodied in the short, dynamic figure of Burrhus Frederic Skinner. The renowned behaviorist was forty-five years old in 1948, the year that Harvard recruited him from Indiana University. He was already a familiar face in Memorial Hall, having earned his Ph.D. from Harvard some two decades earlier and returned for frequent visits ever since. But when he arrived to stay, he *arrived:* Skinner, his students, and his postdocs descended into the basement like an army of religious fanatics.

The Memorial Hall denizens knew all about behaviorism, of course. They could hardly have avoided it: in the United States, at least, it had long since become psychology's received dogma. Most fundamentally, behaviorism forbade any consideration of "mental states" such as happiness, fear, expectation, purpose, or even consciousness. Whatever those states were, went the argument, they were completely inaccessible to outside observers. They were locked inside an individual's head, subjective by definition. Thus they could have no place in objective science. The mind was effectively a black box; the only thing that was observable about a given organism was its behavior. And that meant, in turn, that the only way for scientists to understand that organism was to catalog observable inputs and outputs, to determine which stimulus yielded which response.

This attitude strikes present-day sensibilities—as it struck many nonpsychologists even then—as bizarre. Psychology, supposedly the science of the mind, had ruled the mind off-limits. And yet countless researchers embraced the approach as a kind of salvation: no sooner had the doctrine been defined—in 1913, by the brilliant iconoclast John B. Watson of Columbia University—than it swept through American psychology departments like the proverbial wildfire.

Watson, like many a fundamentalist reformer before him, had been trying to cleanse the mother church of its sins, which in this case meant the vague and hopelessly subjective "introspection" techniques that had preoccupied many of his colleagues since the late nineteenth century. In effect, those predecessors had looked at the classic mind-body duality and come down firmly on the "mind" side. If consciousness (or soul, or spirit) was literally something separate from the body—an unexplainable essence that was somehow *in* the brain but not *of* it—then they would study it in the only way possible: by analyzing their own thought processes from the inside.

Watson, by contrast, came down on the "body" side of the duality, with a fervor that may well have had something to do with his strict Baptist upbringing in South Carolina. Human beings were automatons, he declared, describable in purely natural terms. Their behavior was nothing more than a series of learned responses to outside stimuli; the "mind," assuming that it even existed, was irrelevant. As Watson boasted in his 1930 book *Behaviorism,* "Give me a dozen healthy infants, well-formed[,] and my own specified world to bring them up in and I'll guarantee to take any one at random and train him to become any type of specialist I might select—doctor, lawyer, artist, merchant-chief and yes, even beggar-man and thief, regardless of his talents, penchants, tendencies, abilities, vocations, and [the] race of his ancestors." Indeed, he continued, we might even imagine a utopian future built around "behavioristic freedom," in which children would be scientifically conditioned to produce a rationally ordered culture: "Will not these children . . . with their better ways of living and thinking, replace us as a society, and in turn bring up their children in a still more scientific way, until the world finally becomes a place fit for human habitation?"

Today, of course, that passage sounds profoundly cynical and manipulative, not to mention Orwellian. It seems somehow appropriate that after 1920, when he was forced out of academia by a messy divorce scandal, Watson made a lucrative second career for himself as vice president of the advertising firm J. Walter Thompson. Condition people's responses properly, he seemed to be saying, and you can make them do anything. And yet in the 1920s such ideas seemed just the opposite of cynical. Not only did behaviorism resonate with Americans' egalitarian belief that any child could grow up to become president, but in the bloody aftermath of the Great War it seemed like the last, best hope of saving the world. Certainly it allowed a great many psychologists to conduct their research in a spirit of idealism—and, not incidentally, with a rigor rivaling that employed in physics and chemistry. Behaviorist psychology would produce generations of well-conditioned pigeons and mice that would peck at lights or run through mazes with all the precision of planets orbiting the Sun. And that, in the status-conscious world of science, meant that psychologists could at last hold their heads high: they, too, were practicing a "hard" science.

Rigor, of course, is not a bad thing in research (nor, for that matter, is idealism). The problem was that by the 1940s, the hard-science doctrine of behaviorism had hardened into concrete. "The chairmen of all the important departments would tell you that they were behaviorists," George Miller explained to Baars. "Membership in the elite Society of Experimental Psychology was limited to people of behavioristic persuasion. . . . The power, the honors, the authority, the textbooks, the money, everything in psychology was owned by the behavioristic school. Those who didn't give a damn, in clinical or social psychology, went off and did their own thing. But those of us who wanted to be scientific psychologists couldn't really oppose it. You just wouldn't get a job." Indeed, young researchers quickly learned not to use the word *mind* at all, even in casual conversation, lest they be declared "unscientific" and thereby undermine their own careers.

If the mainstream behaviorists were rigorous, B. F. Skinner was a fanatic. Not only was the mind unobservable, Skinner proclaimed; the mind did not exist. It was all an illusion: there was nothing inside the black box but stimulus and response.

It is impossible to know precisely why Skinner was so rabid on this subject, though it's telling that his own religious upbringing had been just as strict as John B. Watson's, and his rejection of it just as visceral. Fanatical or not, though, Skinner had recruited an impressive number of followers. His scientific credibility was enormous: in the early 1930s he had single-handedly established the field of "operant" conditioning, in which animals were trained to exhibit almost any desired behavior through a system of reward and punishment. And he was a wonderfully charismatic figure, with an undeniable flair for publicity and the dramatic gesture. He had raised his own infant daughter partially inside a baby-sized, Plexiglas "Skinner box" rigged for the appropriate rewards and punishments; the reporters had eaten it up. He arrived at Harvard having just published the utopian novel *Walden Two,* in which he described an idyllic community governed by behaviorist principles; he made it sound so attractive that he immediately became the guru for hundreds of disciples who sought to shape their lives by those principles. Then, some time after his arrival at Harvard, he hosted a group of reporters and photographers from *Life* magazine; while he lectured them about the wonders of operant conditioning, his students, in the background, were photogenically training pigeons to play Ping-Pong.

Skinner's students were even more fervent than their master. They produced beautiful, world-class data on the progress of their pigeons' conditioning. But if Lick, or Miller, or any of the other Memorial Hall natives tried to suggest a mathematical theory of how that conditioning occurred, they reacted violently. Quoting Skinner, they insisted that mathematical models required the introduction of variables and concepts that could not be directly observed. And that, they decreed, was "unscientific."

Not surprisingly, this stance elicited some thoroughly mixed feelings down at Smitty Stevens's end of the basement. On the one hand, everyone there was in awe of Skinner and his very obvious abilities. Moreover, they could feel his considerable charm. "He read widely and used to send me clippings he thought I would be interested in," recalls Miller. "He was thoughtful that way." On the other hand, with that charisma came ego. Louise Licklider, Kitty Miller, and the other spouses soon learned to avoid Skinner at parties, since he responded to the stimulus of a pretty young woman by becoming what was then called a wolf; he seems to have made passes at every one of them. What was worse, he did so in front of his wife, a large woman who clearly did not feel herself to be smart or attractive enough in such company. Louise, already horrified by the idea of raising a baby in a Skinner box, was not impressed.

And then there was Skinner's fanatical behaviorism. It is hard to know precisely what Lick thought of it because he never said much about Skinner one way or the other. But that silence was in itself probably significant, since he tended to be very vocal about people he liked and respected. "Behaviorism and that business seemed like old hat to Lick," says Karl Kryter. "I think he found it passé and a little boring." The systems engineer wanted to understand what was going *on* inside the skull, not just what went in and came out. "What we couldn't understand," explained McGill, "and what I think offended Lick, was that Skinner had gathered around him this group of followers who were almost cultlike in their resistance to math. Lick thought their data were beautiful—but [also] that they needed systems analysis to explain [them]! Nobody got into open fights. We just went our own way and thought it was too bad that they were missing such an opportunity to understand what was going on."

Miller had much the same reaction, though his opinions characteristically were (and are) more biting. "I tried to be a behaviorist," he says. But he was finding it harder and harder to suppress his common sense. The mind didn't *exist?*

Then what about his first and deepest love, language? He couldn't imagine anything more obviously linked to conscious thought, to feeling, to expectations, to humans' whole inner life. "I used to say that behaviorism is to psychology as meter reading is to physics," Miller declares, even as he concedes that he and Skinner shared a wary lifelong friendship and a grudging mutual respect. "Fred Skinner's notion of science was control, not understanding. He wanted to turn a knob and see the behavior change—you know, turn up the reinforcement and see the behavior increase, turn down the reinforcement and see the behavior decrease. Most of his innovations were devices that would enable him to do that. Skinner boxes, baby cribs, and God knows what. He had one of the finest minds of the nineteenth century."

Still, the younger men's reaction was as nothing compared to the tension between Skinner and Smitty Stevens himself. There seemed to be no rapport between them whatsoever, even though it was Stevens who had taken the initiative in bringing Skinner to Harvard in the first place. If nothing else, the two men had profoundly different views of human nature. Skinner was the ultimate blank-slate man, arguing that all behavior in an organism was imposed by the environment through stimulus-response conditioning. Stevens had an equally strong conviction that nature was predominant over nurture, that intelligence was inherited, and that mental disorders such as schizophrenia resulted not from improper conditioning in early childhood but from some sort of genetic flaw. And most of all, as Miller points out, Stevens and Skinner were two alpha males trapped in too small a cage. "Whatever group they were in," he says, "they each had to lead it."

The resulting polarization along the Stevens-Skinner axis made a deep impression on one frequent visitor to the Memorial Hall basement, a precocious kid from New York City who had entered Harvard as an undergraduate in 1946, after a brief stint learning electronics in the navy, and soon found himself fascinated with the workings of the mind. "It was a whole universe in that basement," Marvin Minsky recalled for a 1981 profile in *The New Yorker.* "On the west were the behaviorists, who were trying to understand behavior without a theory. [And] on the east were the physiological psychologists [Stevens and company], who were trying to understand some little bit of the nervous system without any picture of the rest." However, Minsky added, in the middle of that basement he also found some young assistant professors who were "new kinds of people." He specifically mentioned two: George A. Miller and J. C. R. Licklider. "I worked with Miller on theories of problem-solving and learning," he said, "and with Licklider on theories of perception and brain models."[1]

Indeed they *were* new kinds of people, though they barely knew it themselves. In effect, Lick, Miller, and their fellow Young Turks were groping toward a third way of thinking about the human condition, one that was neither blind mysticism nor cold reductionism. In the intellectual turmoil of Cambridge in 1948, they were beginning to glimpse a way to be completely rigorous and scientific and yet still believe in the existence of mind and the reality of our inner lives. Skinner and the other behaviorists had tried to make psychology "scientific" by reducing us all to the status of machines—blind, rigid, nineteenth-century-style machines. We were nothing more than walking, talking bundles of stimulus and response, organic robots to be controlled and manipulated at will. But Lick and Miller had begun to sense the subversive possibilities of twentieth-century machines. Feedback systems were the stuff of everyday engineering, and yet they could embody "mental states" such as *purpose* and *expectation*. Electronic computing systems were little more than arrays of very fast switches, and yet they, too, could embody "mental states," in the form of data stored in (where else?) a *memory*. Computer programs were nothing more than lists of precisely defined rules that the machine followed to the letter, and yet as Turing had shown, they could generate results that were inherently surprising and unpredictable.

"These ideas had a freeing effect," says Miller. "It seemed to me that there were all these forbidden words like *expectation* and *memory* that the behaviorists hadn't allowed us to use but that had found instantiation in engineering." Obviously, the brain was far more complex than any computer or feedback circuit, but that wasn't the point. The question was, if the engineers could use those concepts—and, more important, demonstrate them in working hardware—then precisely what was so "unscientific" about them?

As always, unfortunately, the answer—that there is *nothing* unscientific about those concepts—is a lot clearer in hindsight than it was at the time. But then again, notes Miller, he can now look back with that same hindsight and see exactly when the crucial spark was struck—certainly in his own mind, and quite probably in Lick's and others' as well. It must have been sometime in the late spring or early summer of 1948, he recalls, because Smitty Stevens had just gotten the July issue of the *Bell Systems Technical Journal* in the mail. The publication was exactly what it sounded like: a magazine devoted to the gritty details of communications technology. But no matter. Stevens would go through all the journals as they came in, and whenever there was an article he thought the staff ought to read, he would clip on a routing slip and circulate it. And that was what he did with this issue, flagging a long article entitled "A Mathematical Theory of Communication." It looked at first like just one more dense and incredibly boring paper full of equations and diagrams. Yet when Miller actually started to read it, the author's mathematics turned out to be astonishingly elegant. Moreover, his analysis of the communication process was remarkably broad, applying equally well to the transmission of signals through a telephone wire, the ebb and flow of binary digits inside a computer, the exchange of words in a conversation, or even the surging of neural impulses in the brain: the author had somehow found a way to describe the fundamental unity of communication in all its forms. And on almost every page he used yet another psychological term: "information."

Miller kept on reading. And by the time he was finished, he now says, he knew that his life had changed.

THE CONJURER

Legend has it that Claude Shannon published "A Mathematical Theory of Communication" in 1948 only because his boss at Bell Labs finally badgered him into it. And whatever the truth of that story, the point is that no one who knew Shannon has any trouble believing it.

"He wrote beautiful papers—when he wrote," says Robert Fano, who became a leader of MIT's information-theory group in the 1950s and still has a reverential photograph of Shannon hanging in his office. "And he gave beautiful talks—when he gave a talk. But he hated to do it."

It wasn't a matter of Shannon's being lazy, Fano says; he was constantly filling up notebooks with ideas, theorems, and calculations. He just wouldn't publish—or not very often, anyway. No, Shannon's reticence seems to have been more a matter of extraordinary self-sufficiency. Most people thought of him as being very shy, and in many ways he was. He was the kind of guy who was friendly and open enough once he got to know you but who kept to himself and rarely sought people out. Nor did he seem to care about credit or fame. As he told *Omni* magazine in 1987, "I was more motivated by curiosity. . . . I just wondered how things were put together. Or what laws or rules govern a situation, or if there are theorems about what one can't or can do. Mainly because I wanted to know myself. After I had found the answers it was always painful to write them up or to publish them (which is how you get the acclaim)."

In short, mathematics for Shannon was more like a game, something that he did for the pure intellectual joy of it. And indeed, that's what his friends tend to remember best about him: his extraordinary playfulness. Once you got past that shyness and reserve, you found a little boy who still loved to tinker and explore—as well as a grown man with the soul of a conjurer, an illusionist, a stage magician. "Claude loved to laugh and to dream up things that were offbeat," says David Slepian, a mathematician who worked closely with Shannon at Bell Labs in the 1950s. Shannon had a large repertoire of card tricks, for example. He taught himself to juggle and in later years even published a paper on the mathematical theory of juggling. He likewise taught himself to ride a unicycle (his wife, Betty, formerly a "computer" at Bell Labs, had given it to him as a Christmas present) and soon became notorious for riding the thing up and down the hallways at night—while juggling. And he filled up his home with all manner of bizarre machines he'd built. There was the THROBAC, a calculator that performed all its arithmetic operations with Roman numerals, and Theseus, a life-size mechanical mouse that could find its way through a maze. And perhaps most famously, there was the Universal Machine, which was described by the science-fiction writer Arthur C. Clarke in *Voices across the Sea:* "Nothing could be simpler. It is merely a small wooden casket, the size and shape of a cigar box, with a single switch on one face. When you throw the switch, there is an angry, purposeful buzzing. The lid slowly rises, and from beneath it emerges a hand. The hand reaches down, turns the switch off and retreats into the box. With the finality of a closing coffin, the lid snaps shut, the buzzing ceases and peace reigns once more."

Some of that same sleight-of-hand quality came through in Shannon's mathematics. "Claude had a very particular kind of cleverness," says Slepian. "He wasn't deeply learned, like the kind of mathematician who knows a lot of theorems and can attack a problem head on. Claude never did that. He could just see a field whole and get it right. And he had a quick, cunning mind. He invented whatever he needed. He would circle around and attack the problem from a direction you never would have thought of." Then, when he was done, adds Slepian, he would amaze you with an answer that you had never seen but that had been right there in front of your face the entire time.

That was certainly how Shannon had astonished the electrical-engineering world in 1938, with his MIT master's thesis on switching circuits and logic. And all through the decade since then, that was how he'd approached his Mathematical Theory of Communication, better known as information theory.

It's hard to say precisely what first piqued Shannon's interest in communication. It may have started during his teenage years in Gaylord, Michigan, when he and a friend connected their houses via a makeshift telegraph line that made use of a handy stretch of barbed-wire fence. Or it may have dated back to a summer job in 1937, when he spent several months at Bell Labs in New York City while still working on his master's. But wherever the spark had come from, communication was very much on his mind by the autumn of 1938, when he was newly enrolled in the Ph.D. program in mathematics at MIT and casting about for a suitable thesis topic. "Off and on," he wrote to his mentor Vannevar Bush on February 16, 1939, "I have been working on an analysis of some of the fundamental properties of general systems for the transmission of intelligence, including telephony, radio, television, telegraphy, etc."

It's telling that Shannon felt no need to explain why; in 1939 Bush would have known that communication engineering was still mostly a matter of trial and error, lore, and rules of thumb. Shannon's tacit hope was that the field could be transformed from an art into a science, that a rigorous mathematical theory of communication would provide engineers with the tools to design their systems with assurance. And indeed, the approach he outlined in the remainder of his letter would prove remarkably prescient, comprising many of the ideas he would publish in 1948 and giving a hint that he had already thought about most of the rest. (Toward the end of the letter, which is now preserved in the Library of Congress archives, Shannon alluded to "several other theorems at the foundations of communication engineering which have not been thoroughly investigated.")

In the end, Shannon settled on another thesis topic suggested by Bush. But he returned to the communication problem as soon as he'd received his Ph.D., in the spring of 1940. True, the war was a distraction. In the summer of 1941, after spending the previous academic year at the Institute for Advanced Study in Princeton, Shannon returned to Bell Labs in New York City as a full-time employee of its mathematics department and immediately went to work on the anti-aircraft fire-control problem. (The control system developed at Bell Labs was the one that beat out the Wiener-Bigelow system from MIT for eventual deployment in the field.) Shannon also helped out on the highly classified "X-system," the first radiotelephone to encrypt voice communications in real time.

Nonetheless, he still made time to plug away on his fundamental theory of communication. The key word here is *fundamental:* Shannon was determined to find a theory that could be applied not just to telephones but to *any* form of communication, from telegraphs and televisions to casual conversation or even nerve impulses. To achieve that ambitious goal, he first imagined the communication process as being divided into five parts:

1. An *information source:* the person or thing generating the original message.
2. A *transmitter:* the instrument that transforms the message into a signal suitable for transmission (the voice that produces a sound wave, the telephone that produces an electrical signal, etc.).
3. A communication *channel:* the medium that conducts the signal (air, a telephone wire, a coaxial cable, a beam of light, etc.).
4. A *receiver:* the instrument that takes the signal and tries to reconstruct the message (the ear, the telephone on the receiving end, etc.).
5. A *destination:* the person or thing the message is intended for.

This deceptively simple five-part framework had the great virtue of clarity: just as John von Neumann's abstract functional design for EDVAC would later do for computer engineering, Shannon's outline gave him a way to think about the *architecture* of communication and what a given system was supposed to accomplish, preventing him from getting bogged down in the vagaries of vacuum tubes and cable connections. It provided the generality he needed to devise a truly fundamental theory of communications, in much the same way that Sir Isaac Newton had derived a fundamental theory of physics from just three general laws of motion and one universal law of gravitation. And yet it simultaneously gave him a framework that he could tailor to any given problem, much as Newton's laws could be applied to the fall of an apple or the motions of the Moon. So when Shannon wanted to analyze a telegraph system, for example, he could take the English text generated by the information source and model it as a simple string of symbols having the right statistical properties (such-and-such a proportion of *a*s, such-and-such a proportion of *b*s, etc.). He could model the transmitter as a simple computer that translated each letter into dots and dashes. He could model the communication channel as a wire that injected a certain amount of random electrical noise into the message. And so on.

To do a full mathematical analysis, however, Shannon needed a way to quantify what was being transmitted through those five stages of communication. Happily, he didn't have to look far. The term "information" had been common parlance around Bell Labs since 1928, when an engineer there named Ralph Hartley had first used it to describe the amount of message flowing through a telephone wire, as opposed to static. Building on earlier work by Bell Labs mathematician Harry Nyquist—who had called it intelligence—Hartley had defined "information" as the useful part of the signal, the part that people wanted to listen to, the part that the engineers were always trying to maximize in the face of noise and distortion. More precisely, he said, "information" ought to measure how much you actually learned from a given message. For instance, if you already knew what the message was going to say, then you learned nothing new by receiving it, and the information content should work out to be zero. But if you didn't have a clue what the message was going to say, then you learned a lot by receiving it, so its information content ought to be high.

Hartley had then gone on to supply a mathematical expression of this idea, which Shannon liked very much. Giving Hartley full credit, he accordingly took it over and made the definition rigorous—and to modern eyes very familiar. At least in the simplest cases, said Shannon, the information content of a message was just the number of binary *1*s and *0*s required to encode it. For example, if you knew in advance that a message would convey a simple binary choice—yes or no, true or false, black or white—then one binary digit would suffice: a single *1* or a single *0* would tell you all you needed to know. That message would be said to contain one unit of information. However, if the message was more complicated, it would require more digits to encode and would contain that much more information (think of the thousands of electronic *1*s and *0*s that make up a modern word-processing file).

Of course, this definition did have its perverse aspects, as Shannon knew full well. If the two alternatives were *yes* and *no*, say, then the actual message might carry a world of meaning—as in "*Yes,* I will marry you"—but only one binary unit of information. The content and significance of the message, being unquantifiable, would just be ignored. Conversely, a long and turgid legal document might convey no meaning whatsoever to the recipient while, because of its very length, comprising a huge amount of information.

Nonetheless, as Shannon would point out in his 1948 paper, the separation of information and meaning did have the virtue of putting the interpretation of meaning where it belonged: in the brains of the people sending and receiving the message. The engineers' job was merely to get the message from here to there with a minimum of distortion, whatever it might say. And for that purpose, the digital definition of information was ideal because it allowed for a precise mathematical analysis via questions such as, What are the fundamental limits of a given communication channel's carrying capacity? How much of that capacity can be used in practice? How much is it degraded by the inevitable presence of noise in the line? What are the best and most efficient ways to encode the information for transmittal in the presence of noise?

Judging by what he would tell Fano and others many years later, Shannon had outlined his answer to these and other questions as early as 1942 or 1943. Suppose you're trying to send a birthday greeting down a telegraph line, say, or through a wireless link. The communication channel has an information capacity of so many binary digits per second. And the message likewise carries an average information content of so many binary digits per letter. But taken together, Shannon realized, these two quantities determine a fundamental speed limit, measured in binary digits per second. Above that speed limit, perfect fidelity is impossible: however cleverly you encode your message and compress it, you simply cannot make it go any faster unless you first throw some information away. *Below* that speed limit, however, the transmission is potentially perfect. Shannon was able to show that there must exist codes that will get you right up to the limit without losing any information at all. Moreover, such perfection is possible even in the presence of noise. Shannon demonstrated that no matter how much static and distortion there may be in a given communications channel, and no matter how faint the signal, messages can still get through with perfect fidelity. Of course, if the signal is very, very faint, you may have to assign a huge number of *1*s and *0*s to each letter or pixel so that some of them have a chance of getting through. And you may have to devise all kinds of elaborate error-correcting codes so that corrupted parts of the message can be reconstructed at the other end. In practice, diminishing returns will set in: the codes will eventually get so long and the communication so slow that you'll have to give up and let the noise win. But in principle, you can make the probability of error as close to zero as you want.

This "fundamental theorem" of information theory, as Shannon later called it, surprised even him: the conquest of noise seemed to violate common sense. And yet it was true. Just a few years earlier, with his master's thesis on logic and switching, Shannon had laid the foundation for all of modern computer circuitry; now, working on his own in the middle of the war, he had quietly done the same for most of the rest of the modern digital world. Ultimately his fundamental theorem explains how, for example, we can casually toss around compact discs in a way that no one would have ever dared do with long-playing vinyl records: error-correcting codes inspired by Shannon's work allow the CD player to eliminate noise due to scratches and fingerprints. Shannon's theorem likewise explains how error-correcting computer modems can transmit data at the rate of tens of thousands of bits per second over ordinary (and relatively noisy) telephone lines. It explains how NASA scientists were able to get the *Voyager* spacecraft's imagery of the planet Neptune back to Earth across two billion miles of interplanetary space. And in general, it goes a long way toward explaining why today we live in an increasingly digital world—and why the very word *digital* has become synonymous with the highest possible standard in data quality and reliability.

For the most part, Shannon seems to have kept very quiet about all this; some of his closest colleagues from this period swear they knew nothing about information theory until 1948. Nonetheless, even a shy and reticent young man occasionally has to talk to *some*body.

Between January and March 1943, for example, Shannon struck up a brief friendship with Alan Turing, who was visiting Bell Labs to make a report on the X-system for the British government. They certainly couldn't talk about Turing's wartime work back in England, where he was a leading member of the now-famous code-breaking team at Bletchley Park; in 1943 the very existence of the Bletchley Park effort had to be kept secret, lest the Germans realize that Turing and company could crack the messages being encrypted by the Wehrmacht's "Enigma" machines, and move to another system. But the two young men could and did start meeting in the laboratory cafeteria around "teatime," as Turing put it, to chat about their common interest in information and computation.

Indeed, as Turing's biographer Andrew Hodges points out, there had been not only a parallel between Shannon's work and Turing's, but a kind of reciprocity. Turing had started from the concept of computability and come to realize that computation was inextricably bound up with information and encryption; the very fact that his abstract machine had to read and write symbols encoded on a data tape guaranteed it. (At Bletchley Park, in fact, Turing had invented the "deciban," a measure of information content that was conceptually identical to Shannon's.) Shannon, conversely, had started from the communication and information end and come to much the same conclusion. In his five-part architecture of communication, after all, there was always a *transmitter* to encode the message and a *receiver* to decode it at the other end (his example in the 1948 paper would be a transmitter that scanned through an English text letter by letter and turned it into Morse code). But whatever the physical structure of these devices—whether you were talking about the vocal chords and the ear, say, or a microphone and a radio receiver—they each took an input stream of information and converted it into an output stream using a well-defined procedure. And that process, broadly defined, was computation.

Another of Shannon's confidants was John von Neumann. It was that word *information,* Shannon explained to the great mathematician at one point: he had never liked it. The technical distinction between *information* and *meaning* was too much a violation of common usage, he felt, and would just end up confusing people. Could von Neumann suggest anything better?

Von Neumann's answer was immediate, as Shannon later recounted the story: "You should call it entropy, and for two reasons." First, von Neumann told the younger man, his formula for the information content of a message was mathematically identical to the physicists' formula for entropy, a mathematical variable related to the flow of heat.* (Shannon was astounded to learn this; he had derived his formulation totally on his own.) But second, and much more important, said von Neumann, "most people don't know what entropy really is, and if you use the word *entropy* in an argument, you will win every time!"

In the end Shannon elected to stick with *information,* since engineers had been using the word since Hartley's day and weren't about to abandon it. But in the meantime he had considerably better luck with another word. Over lunch one day in late 1946, a group of Bell Labs researchers were grousing about the awkwardness of the term "binary digit" and deploring the lack of any good substitute (existing proposals included hybrids such as *binit* and *bigit,* both considered loathsome). But then the statistician John Tukey joined the discussion. "Well," he asked with a grin, "isn't the word obviously *bit?*" And it was.[2] Shannon liked the new word so much that he started using it himself and gave Tukey the credit in his 1948 paper—which, according to the *Oxford English Dictionary,* marked the first time *bit* was ever used in that sense in print.

Yet another intimate of Shannon's was MIT's Jerry Wiesner, who had been a good friend since their undergraduate days together at the University of Michigan. Shannon apparently told Wiesner about his information-theory work sometime in late 1946 or early 1947. And that, of course, was the same as telling all of MIT, Norbert Wiener most definitely included.

Wiener was not amused. More than once in his long career, the portly mathematician had been known to snore his way through a lecture until the speaker was just reaching his main conclusion: "I did that," he would snap, suddenly wide awake. "I did that *years* ago." And that appears to have been his position on information theory as well.

He had a point, sort of. It all depended on how you interpreted "information theory." If you meant the specific set of mathematical tools that Shannon was

---

* This was no accident; the links between information and entropy turn out to go very deep. Entropy was originally defined in the mid–nineteenth century as a mysterious variable that played a starring role in the equally mysterious second law of thermodynamics: heat always flows from hot to cold, said the second law, and in such a way that it ends up *increasing* the total entropy of the universe. A *decrease* in the total entropy seemed to be impossible, though the reasons for that fact were entirely obscure. It was only at the end of the nineteenth century that physicists began to realize that entropy was really just a measure of randomness at the molecular level: how thoroughly two chemicals are mixed in a beaker, for example, or how chaotically the atoms are moving within a hot gas. That fact, in turn, implied that the second law was just the statement that molecules will never unmix themselves spontaneously. The randomness of the universe always increases, never decreases. Thus the link to information: the more random something is at the molecular level, the less information you have about how its molecules are arranged. Entropy is, in effect, "missing information."

inventing for communication engineers, then no, Wiener had not had much, if anything, to do with it. However, if you meant the broader concept of *message* in all its aspects, then yes, Wiener had been immersed in the field since the 1920s. His inspiration had come from Vannevar Bush, who'd explained to him that electrical engineering actually embraced two separate disciplines: power engineering, where the goal was to optimize the flow of energy through a line, and communication engineering, where the goal was to modulate a comparatively trivial energy flow into a meaningful signal. Wiener had been enchanted with this insight and deeply influenced by it. It had led directly to some of his most important technical work in the 1920s and 1930s, work that was certainly similar in spirit to Shannon's theory and that ultimately resulted in a definition of information flow that was mathematically identical to his. And during the war, of course, it had proved central to Wiener's work on feedback and purpose. Think of a telegraph key tapping out a signal to open the floodgates of a multi-megawatt dam: *communication,* Wiener realized, was just the flip side of *control.* The concept of information lay at the very heart of his new science.

Shannon, perhaps fortunately for everyone's peace of mind, was perfectly willing to acknowledge Wiener's claim; in part 3 of his paper, which was published separately in October 1948, he followed Jerry Wiesner's advice and included two effusive testimonials to Wiener's contributions. Wiener himself, meanwhile, seemed perfectly willing to be mollified; he would later write of Shannon's "profound originality" and how he had "gone from triumph to triumph." Still, Wiener's boundless vanity had been pricked, and he never quite got over the feeling that he had not received his due credit for information theory—which may have been one reason he was so anxious to lay out his ideas before the public under his own name.

AN IMAGINATIVE FORWARD GLANCE

It had all begun in the spring of 1947, as Wiener later told the story, on one of those happy trips that start out well and get better. His eventual destination was a mathematics conference in Nancy, France. But his more immediate goal was England, where he was eager to see how his "new science" was faring amid the postwar reconstruction.

Certainly it was doing well in the United States. At MIT his ideas on communication and control had found a natural home in the ramshackle warren of the Rad Lab, which had reconstituted itself after the war as RLE, the Research Laboratory for Electronics. Thanks to Wiener, as Jerry Wiesner later recounted, he and his companions at RLE had taken their charter to be "the universal role of communication processes in man's universe. . . . Our interests ranged from man-made communication and computing systems to the sciences of man, to inquiries into the structure and development of his unique nervous system, the phenomena of his inner life, and finally his behavior and relation to other men."[3]

Nationally, meanwhile, Wiener's new science had inspired a kind of traveling road show known as the Macy meetings. Organized by Warren McCulloch, with funding from the Macy Foundation, and inaugurated with a gathering in New York City on March 8 and 9, 1946, the meetings were intended to pick up where the 1945 conference at Princeton had left off. The idea was for Wiener, von Neumann, McCulloch, Pitts, and the rest of the usual suspects to get together every six months or so to discuss their new science in the company of like-minded neurophysiologists such as Rosenblueth, plus a contingent of equally like-minded social scientists led by Gregory Bateson and Margaret Mead, the famed husband-and-wife team of anthropologists. Wiener, not surprisingly, had liked the idea very much—especially since he got to be the star. He was the "brilliant originator of ideas and enfant terrible," according to his biographer Steve Heims. "Sometimes he got up from his chair and in his ducklike fashion walked around and around the circle of tables, holding forth exuberantly, cigar in hand, apparently unstoppable."[4]

And now, as he discovered on his 1947 trip, his new science seemed to be taking root in England as well. In addition to seeing many colleagues in the neuroscience and mathematics communities, he later recounted, "I had an excellent chance to meet most of those doing work on ultra-rapid computing machines, especially at Manchester and at the National Physical Laboratories at Teddington"—the latter being roughly the British equivalent of the U.S. National Bureau of Standards. Above all, he continued, his visit to Teddington gave him an opportunity to discuss the fundamentals of his new science with a "Mr. Turing," who was now trying to make use of the wartime advances in electronics to build his abstract machine for real. (The NPL's mathematics division had given top priority to catching up with the Americans in computers and had hired Turing to lead the charge; his plans for the Automatic Computing Engine, or ACE, were already well along by 1947.) So Wiener left for France feeling very satisfied: interest in his new science was "about as great and well-informed in England as in the United States," he wrote, "and the engineering work excellent."[5]

However, there was even better yet to come. At the mathematics conference in Nancy, he was introduced to a Monsieur Freyman, a Mexican-born mathematician who was representing the French publishing house Hermann et Cie. M. Freyman had a proposal: might Professor Wiener be interested in writing a short book to explain his ideas about communication and control?

Professor Wiener would be fascinated; at the moment the state of his personal finances was rather, um, *fragile*. So, he later recalled, "over a cup of cocoa at a neighboring patisserie," he happily signed a contract. He went to work almost immediately, writing his book in Mexico City during the summer and fall of 1947 while spending a sabbatical semester with Arturo Rosenblueth at the Instituto Nacional de Cardiologia. (In fact, he would dedicate the volume to Rosenblueth, "for many years my companion in science.")

In the process, moreover, Wiener finally figured out what to call his new science.

Wiener's assistant Oliver Selfridge, who was also down in Mexico City that semester, along with Walter Pitts, remembers Wiener's going through his reasoning for them one evening over coffee as they sat in the lovely rooftop garden of his apartment house. Given the central role of communication in his new science, Wiener explained, his first thought had been to derive a name from the Greek word for "messenger." Unfortunately, that word was *angelos*, which in English had long since taken on the specific meaning of "a messenger from God." Somehow, a new science of *angelics* wasn't quite what he was looking for. So instead, said Wiener, he had decided to focus on the theme of control. In Latin, he knew, the word for "steersman" was *gubernator*, from which we get the English word *governor*. That was better, since the word *governor* could sometimes refer to a device used to control the speed of an engine. More often, though, it referred to a human "governor," or steersman for policy: the wrong connotation again. However, the Latin *gubernator* turned out to be a corruption of the Greek word for "steersman," *kybernetes*. And that, Wiener felt, could be transmuted into English very nicely, as *cybernetics*.

Selfridge and Pitts agreed that cybernetics was indeed an excellent name for the new science. And so, blissfully ignorant that he had just given later generations the means to coin an endless string of buzzwords—*cyberspace, cybercash, cyberpunk, cybersex*, ad infinitum—Wiener continued writing.

Cybernetics was not an easy book to write, says Selfridge, who watched Wiener struggle through many revisions.

At first, Wiener had planned a rather straightforward book that would appeal mainly to his colleagues. In it he would provide a mathematical account of the new information theory "developed by Shannon and myself," and a similarly technical explanation of the prediction theory he'd developed for antiaircraft fire control. (The latter work had been classified until well after the war, much to Wiener's annoyance; this would be his first chance to make it public.) He would write about computation, feedback, and purpose, with particular emphasis on his neurology work with Rosenblueth. And that would be about it.

Even as he wrote, however, Wiener found the scope of his book expanding. The new technology of machines guided by the feedback principle had the power to remake the world, he believed. And the public deserved to know that—especially since, as he put it, "this new development has unbounded possibilities for good and for evil."

Those possibilities were much on Wiener's mind in 1947. During the war, it was true, he had devoted himself wholeheartedly to military research. But the obliteration of Hiroshima and Nagasaki left him horrified by what science and technology had done. As the physicist Daniel Q. Posten wrote to Einstein in October 1945, "Wiener stands aghast—as though a man in a confused dream—and wonders what we must do, and he protests at scientific meetings the 'Massacre of Nagasaki' which makes it easier, for some, to contemplate other massacres." When Wiener dropped into people's offices around MIT during this period, his first comment was often "Do you think there will be another war?" He talked darkly of the "tragic insolence of the military mind" and brooded over the prospect of his own work's being used for destructive ends. On October 18, 1945, he even drafted a letter to MIT president Karl Compton submitting his resignation: "[I intend] to leave scientific work completely and finally. . . . I see no other course which accords with my conscience."[6]

Wiener's immediate sense of despair evidently lifted, for his resignation letter was almost certainly never sent; his gusto for work shortly returned, as fervent as ever. Nevertheless, the autumn of 1945 had been a turning point. "I do not expect to publish any future work of mine . . . which may do damage in the hands of irresponsible militarists," he declared in "A Scientist Rebels," a one-page letter that appeared simultaneously in the January 1947 issues of the *Atlantic Monthly* and the *Bulletin of the Atomic Scientists.* "It is perfectly clear . . . that to disseminate information about a weapon in the present state of our civilization is to make it practically certain that that weapon will be used."[7]

His letter struck a chord with many other scientists who were haunted by that same question of responsibility in the post-Hiroshima world. The *Bulletin of the Atomic Scientists,* after all, had been launched by Manhattan Project veterans at the University of Chicago expressly as a forum for debate on the implications of nuclear energy. "I greatly admire and approve the attitude of Professor Wiener," said Albert Einstein when a reporter asked him about the letter. "I believe that a similar attitude on the part of all the prominent scientists in this country would contribute much toward solving the urgent problems of national security."[8]

Nonetheless, Wiener also paid a price for that letter. For one thing, it isolated him from some of the most exciting research of the day, including computer research. Almost immediately, for example, Wiener felt obliged to withdraw from an international computer conference that Howard Aiken was convening at Harvard to celebrate the inauguration of his Mark II relay calculator—which was, of course, being paid for by the navy.

For another, his going public with his concerns may well have deepened his growing estrangement from his good friend John von Neumann.

In truth, von Neumann was an exceedingly busy man in 1947. If nothing else, the war had made him something of a scientific superstar, the very Hollywood image of what a scientist ought to be, up to and including that faint, delicious touch of a Middle European accent. He was one of the chief wizards of the Bomb at a time when many still considered it a technological triumph. He was the inventor of game theory* and the coauthor of *Theory of Games and Economic Behavior,*[9] a 1944 volume that held out the hope that economics—and perhaps even human behavior in general—might be made into a mathematical science just as rigorous as physics. Indeed, von Neumann seemed to be the living embodiment of rationality, objectivity, and the power of reason in human affairs. "Johnny von Neumann . . . is one of the two or three men whom I knew contemporarily who were my ideals," recalls the neurophysiologist Ralph Gerard, who had made his acquaintance during the Macy meetings. "He was always gentle, always kind, always penetrating and always magnificently lucid. . . . He managed to make extremely complicated things crystal clear. . . . He was very eager to work things through intellectually, extremely able to do so, and would defuse a confused situation."[10]

That calm, dispassionate clarity had made von Neumann a valued adviser on scientific and nuclear matters in Washington, where he was the confidant of generals, congressmen, agency heads, and cabinet officers. His deeply informed scientific judgment was likewise prized on the now half-deserted mesa at Los Alamos, where he continued to spend his summers helping his fellow Hungarian Edward Teller develop what was then considered a long shot: the hydrogen bomb. (In late 1945, when the still-secret ENIAC was finally ready to carry out its first real calculation, von Neumann had arranged for the shakedown run to be a three-week-long simulation of a hydrogen-bomb detonation.)

And of course, von Neumann's enthusiasm for computers probably did as much as any other single thing to make computing respectable.

That was another reason he was busy: amid all that high-level advising and politicking, von Neumann was also, along with his colleagues at the Institute for Advanced Study, inventing most of computer science as we know it today. He *had* to: in late 1945, after he talked the IAS into letting him build a stored-program computer right there on campus, he'd had nothing to go on except his own incomplete and highly abstract "First Draft" on the EDVAC. In order to turn that abstraction into a working computer, he and his team had been obliged to make up the details as they went along. For example:

Hardware. Von Neumann had wasted no time putting his team in place. By November 1945 he had recruited a chief engineer who came highly recommended by Wiener—Julian Bigelow—plus his two strongest supporters during the EDVAC controversy at the Moore School: Herman Goldstine and the mathematician-

---

* Von Neumann's initial paper on the subject had appeared in 1928. For simplicity's sake, imagine a game with two players, A and B. It is A's move. Now imagine that player A first evaluates all his possible moves and all his opponent's possible countermoves, estimating his numerical payoff for each combination (if the result is actually a setback for A, then the "payoff" is a negative number). Von Neumann showed that under very general conditions, player A can always choose a move that maximizes his payoff and minimizes B's. This is obviously the optimum strategy, and a rational player will always choose it.

turned-engineer Arthur W. Burks. (Von Neumann had also made an offer to Presper Eckert, but bitterness over the stored-program patents was escalating so rapidly at that point that he soon withdrew it.) And by 1946, von Neumann, Goldstine, and Burks had laid out a more refined concept of machine architecture in "Preliminary Discussion of the Logical Design of an Electronic Computing Instrument," a report that would prove to be even more influential than "First Draft." Among many other things, the 1946 report pointed out how much more efficient a computer would be if it could get access to each memory address at "random"—that is, instantaneously, without having to wait until the correct address came around on a circulating tape or a mercury delay line. Naturally, such a storage scheme became known as Random-Access Memory, or RAM. (Of course, the report also concluded that the memory unit should store the data as charged spots on the face of a cathode-ray tube—a cutting-edge technology in 1946, but now so thoroughly obsolete that almost no one remembers it.)

Bigelow, meanwhile, was turning these ideas into working hardware, with far-reaching results. Because it was the first true stored-program computer—and because the Pentagon wanted replicas of its own for nuclear-weapons calculations—the Institute for Advanced Study's machine would serve as the model for first-generation computers constructed during the 1950s at the University of Illinois, the RAND Corporation, IBM, and the national laboratories at Los Alamos, Oak Ridge, and Argonne.

Software. By 1947, von Neumann and Goldstine had laid the foundations for software engineering with the first installment of "Planning and Coding Problems for an Electronic Computing Instrument," a report that they would publish in three parts over the next year and that by default would become the standard textbook for the whole first generation of programmers. Among its innovations were conceptual aids such as the "flow chart," a graphical technique for keeping track of how data flow from one program step to the next. The report also included numerous examples of how complex programs could be built up from simpler blocks of code, which today would be known as subroutines.

Simulation. All during this period, von Neumann had been pioneering a new branch of applied mathematics known as numerical analysis, the art of taking the real physical equations that govern a given process, such as the weather, and approximating them with digital algorithms. The challenge was to make the algorithms at once simple enough to run efficiently on the computer and sophisticated enough to keep the approximations within an acceptable range of error (otherwise, the results would quickly become meaningless). Von Neumann was not only one of the first mathematicians to address this challenge systematically, but also a leader in showing a younger generation of scientists what numerical simulation could accomplish. Very early on in the project, for example, he pulled some of his many strings to give outside scientists a chance to do simulations on ENIAC, which was still the only general-purpose electronic computer that was actually up and running. He also arranged separate funding for a research group at the IAS focused entirely on numerical meteorology, which he regarded as the most promising computer application of all. (He was right: in 1950, with von Neumann's help, the IAS group would use ENIAC to run the first reasonably accurate numerical weather forecast. In 1954, its increasing success—augmented by von Neumann's behind-the-scenes lobbying—would lead directly to the founding of the Joint Numerical Weather Prediction Unit, which continues even now to make routine daily weather forecasts under the auspices of the National Weather Service.) And when the IAS computer itself finally became operational, in 1952, von Neumann would make *it* available to many outside scientists, who would use it for problems ranging from algebraic number theory and linear algebra to special functions, gravitational theory, fluid dynamics, and stellar evolution.

Theory. Since 1946, meanwhile, von Neumann had been attempting to create a "General and Logical Theory of Automata" that would integrate his own ideas about computing with the work of Wiener, Turing, Shannon, McCulloch, Pitts, and many others. Unfortunately, von Neumann was too preoccupied with his other duties ever to take his theory very far. Nonetheless, remarks he made at various times suggested that he intended the word *automaton* to encompass not just computers but brains, radar systems, the telephone system, homeostatic systems within biology, and anything else that processed information and regulated itself. In at least one instance, moreover, his theory proved to be spectacularly prescient.

In the case of, say, a factory machine tool, von Neumann observed, you have an automaton that can turn out very complex parts but *not* another machine tool. Likewise, a universal Turing machine can output an arbitrarily complex tape but *not* another Turing machine. However, in almost any biological organism, you have an automaton that can not only reproduce identical copies of itself but also (through evolution) give rise to organisms that are *more* complex than itself. So von Neumann asked, What are the essential features required for an automaton to reproduce itself and to evolve?

To give his readers a feel for the issues involved, von Neumann started out with a thought experiment. Imagine a machine that floats around on the surface of a pond, he said, along with lots of machine parts. Furthermore, imagine that this machine is a *universal constructor:* given a description of any other machine, it will paddle around the pond until it locates the proper parts, which it will then use to assemble that machine. In particular, given a description of itself, it will construct a copy of itself.

Now, that is almost self-reproduction, said von Neumann—but not quite. The newly created copy of the first machine will have all the right parts, but it won't have a description of itself, which means that it won't be able to make any further copies of itself. So von Neumann also postulated that the original machine must have a *description copier,* a device that will take the original description, duplicate it, and then attach the duplicate description to the offspring machine. Then, he said, the offspring will have everything it needs to continue reproducing indefinitely. And that *will* be self-reproduction.*

As a thought experiment, von Neumann's analysis was simplicity itself. He was saying that the genetic material of any self-reproducing system, whether natural or artificial, must function very much like a stored program in a computer: on the one hand, it had to serve as live, executable machine code, a kind of algorithm that could be carried out to guide the construction of the system's offspring; on the other hand, it had to serve as passive data, a description that could be duplicated and passed along to the offspring.

As a scientific prediction, that same analysis was breathtaking: in 1953, when James Watson and Francis Crick finally determined the molecular structure of DNA, it would fulfill von Neumann's two requirements exactly. As a genetic program, DNA encodes the instructions for making all the enzymes and structural proteins that the cell needs in order to function. And as a repository of genetic data, the DNA double helix unwinds and makes a copy of itself every time the cell divides in two. Nature thus built the dual role of the genetic material into the structure of the DNA molecule itself.

In any case, John von Neumann was a very busy man indeed in the late 1940s–to the point where attendees at the Macy meetings began to talk of his "meteoric appearances," when he would show up for no more than half a day, or even just for lunch. His General and Logical Theory of Automata was clearly in

---

* Actually, von Neumann wasn't completely satisfied with his thought experiment; the image of a self-reproducing machine on a pond was still too concrete, too tied to the material details of the process. As a mathematician, he wanted something completely formal and abstract, something that he could use to prove theorems. The solution, a formalism that would eventually become known as the cellular automaton, was suggested by his good friend Stanislaw Ulam, a full-time resident of Los Alamos who had pondered many of these same issues himself. Imagine a programmable universe, said Ulam. "Time" in this universe would be defined by the ticking of a cosmic clock, and "space" would be defined as a discrete lattice of cells, with each cell occupied by a very simple, abstractly defined computer—a *finite automaton*. At any given time and in any given cell, the automaton could be in only one of a finite number of states, which could be thought of as *red, white, blue, green,* and *yellow,* or *1, 2, 3,* and *4,* or *living* and *dead,* or whatever. At each tick of the clock, the automaton would make a transition to a new state, which would be determined by its own current state and that of its neighbors. Complex objects in this universe would exist as patterns of states in the lattice. And the "physical laws" that governed those patterns would be encoded in a table of rules connecting the new state of each cell to its current state.

  Von Neumann loved the cellular-automaton idea and soon sketched out a proof that there existed at least one cellular-automaton pattern that could indeed reproduce itself. The pattern was immensely complicated, requiring a huge lattice and twenty-nine different states per cell. It was far beyond the simulation capacity of any existing computer. But for von Neumann, the very fact of its existence settled the essential question of principle: self-reproduction, once considered to be an exclusive characteristic of living things, could indeed be achieved by machines.

the cybernetic spirit. But the "cybernetical circle," as Norbert Wiener and his acolytes came to be called, didn't rank very high among his priorities.

Once again, Wiener was not amused, and it wasn't long before the strains in their friendship became noticeable. When von Neumann did address the Macy meetings, for example, Wiener would ostentatiously take up a pencil and doodle— or even do his snoring trick, much to von Neumann's annoyance. Conversely, when Wiener once spoke at a mathematics conference, von Neumann sat in the front row and read the *New York Times* as noisily as possible. Granted, the friction never escalated into open hostilities, and the two men would continue their correspondence on scientific issues until the mid-1950s. Yet the old camaraderie was gone.

Perhaps, however, they should have been commended for keeping their relationship as civil as they did. As their joint biographer Steve Heims points out, World War II had been a watershed for science (as for everything else in the twentieth century): after Hiroshima and Nagasaki, no one could pretend any longer that science was apolitical. And the fact was that this watershed had left Wiener and von Neumann on opposite sides.

Von Neumann, for his part, was a figure of the sort that many in the post-Vietnam generation find all but incomprehensible: a genuinely kind, outgoing, and supremely rational man who was also a grimly determined Cold Warrior. In fairness, his attitudes were quite comprehensible in context. John von Neumann had reason to be profoundly afraid of the Soviet Union. Not only had he just seen his native Hungary forcibly made into a Soviet satellite state, but he had experienced Soviet-style communism firsthand: in 1919, in the turmoil following the collapse of the Austro-Hungarian Empire, a short-lived Bolshevik regime in Hungary had filled the streets of Budapest with gangs of homicidal thugs known as Lenin Boys and driven the wealthy Neumann family into temporary exile.

That said, though, he did throw himself into the arms race with an undeniable zeal. He had no compunction about continuing to work on the hydrogen bomb—which would be thousands of times more powerful than the uranium and plutonium bombs that had destroyed Hiroshima and Nagasaki—even as many Manhattan Project veterans were openly questioning the morality of such a device, not to mention its technical feasibility. "I believe there is no such thing as saturation," he once told Robert Oppenheimer. "I don't think any weapon can be too large."[11] And after August 1949, when the Soviets stunned the world by testing an atomic weapon of their own, von Neumann was reportedly willing to contemplate a preemptive war: "If you say why not bomb them tomorrow," he was later quoted as having suggested, "I say why not today? If you say today at five o'clock, I say why not one o'clock?"[12]

Wiener, not surprisingly, was appalled by that attitude. Not only did he see it as the kind of adolescent saber-rattling that was going to get us all incinerated, but he felt that his old friend had come to personify a dangerously seductive brand of intellectual hubris. Through the use of innovative analytical tools such as von Neumann's game theory, went the argument—an argument that was already being embraced by strategic thinkers in the government and in newly formed think tanks such as the RAND Corporation—the nuclear-arms race could be rationalized, mathematized, reasoned about, and *managed.*

Wiener begged to differ. He certainly didn't favor *ir*rationality in human affairs; the world, he felt, had already heard entirely too much about the "triumph of the will" from Hitler and his ilk. But he did want to see this rising generation of mathematical Cold Warriors be a little less naive about the uncertainties of the world. And in this regard, von Neumann's game theory seemed to him particularly insidious. Underlying its vaunted objectivity were built-in assumptions about the nature of human beings and human society that Wiener judged to be "an abstraction and a perversion of the facts."[13]

For one thing, Wiener noted, game theory presumed that the rules of the game were fixed, that competition was inevitable, that all players were perfectly ruthless, and that they would always choose the strategy that furthered their own self-interest, period. And yet, he said, while there was a depressing element of truth to that assumption, it did sometimes happen that player A genuinely cared what happened to player B. And it also sometimes happened that player A's fate was intertwined with that of player B—as when they were two nuclear-armed nations inhabiting the same planet, for example. Indeed, he wrote (echoing such luminaries as Niels Bohr and Albert Einstein), when war becomes a form of mutual suicide, you don't want to win the game. You want to get *out* of the game.

A second problem was that game theory depicted every player as being perfectly rational and capable of foreseeing the full consequences of every possible action. To be fair, it must be said that von Neumann wasn't alone in this assumption; it was (and remains) a standard axiom of mathematical economics. But with the possible exception of von Neumann himself, such flawless rationality and foresight eluded everyone Wiener knew. "Where knaves assemble," he wrote in *Cybernetics,* "there will always be fools." Furthermore, he maintained, in the real world it was rarely just a matter of two players' meeting one-on-one. Society, for him, was more closely akin to what would now be called a complex adaptive system—a constantly evolving, endlessly surprising web of interacting players and overlapping feedback loops. "In the overwhelming majority of cases," Wiener insisted, "when the number of players is large, the result is one of extreme indeterminacy and instability."

In that kind of environment, he continued, it was foolhardy to assume that technological "progress" would always be benign. Consider the Industrial Revolution of the nineteenth century, for example, which had vastly increased the wealth and economic productivity of England, the United States, and many other countries but had also created the fetid slums and hellish factories described by Dickens. To Wiener, there seemed every possibility that computers and other such technologies of the cybernetic age—he would later coin the phrase "the Second Industrial Revolution"—would have consequences just as dire. Inevitably, he felt, the rich and the powerful would seek to use these new technologies of communication and control to cement their power even further. (Wiener tended toward the utopian socialism of his father, to the extent that he was interested in politics at all, and could be every bit as shrill about "these merchants of lies, these exploiters of gullibility" as von Neumann was about the Soviets; he would have been right at home in the radical antiwar movement of the 1960s.) "The first industrial revolution, the revolution of the 'dark, satanic mills,' was the devaluation of the human arm by the competition of machinery," he wrote. "The modern industrial revolution is similarly bound to devalue the human brain, at least in its simpler and more routine decisions."[14]

To Wiener, this was an argument not for suppressing the new technology—which he felt would be impossible anyway—but for embracing it cautiously. Indeed, he argued that for all the danger the technology posed, its real message was one of hope, for "a new interpretation of man, of man's knowledge of the universe, and of society."[15]

In effect, though Wiener didn't quite express it this way, cybernetics was offering an alternative to the Skinnerian worldview, in which human beings were just stimulus-response machines to be manipulated and conditioned for their own good. It was likewise offering an alternative to von Neumann's worldview, wherein human beings were unrealistically rational technocrats capable of anticipating, controlling, and managing their society with perfect confidence. Instead, cybernetics held out a vision of humans as neither gods nor clay but rather "machines" of the new kind, embodying purpose—and thus, autonomy. No, we were not the absolute masters of our universe; we lived in a world that was complex, confusing, and largely uncontrollable. But neither were we helpless. We were embedded in our world, in constant communication with our environment and one another. We had the power to act, to observe, to learn from our mistakes, and to grow. "From the point of view of cybernetics, the world is an organism," Wiener declared in his autobiography. "In such a world, knowledge is in its essence the process of knowing. . . . Knowledge is an aspect of life which must be interpreted while we are living, if it is to be interpreted at all. Life is the continual interplay between the individual and his environment rather than a way of existing under the form of eternity."[16]

The question, said Wiener, was how to keep this message of hope from being perverted. And the only answer, for him, was for scientists to engage in that same process of knowing, through constant vigilance, constant thought, and constant, well, *feedback*. "Even when the individual believes that science contributes to the human ends which he has at heart," he would write in 1960, "his belief needs a continual scanning and re-evaluation which is only partly possible. For the individual scientist, even the partial appraisal of this liaison between the man and the [historical] process requires an imaginative forward glance at history which is difficult, exacting, and only limitedly achievable. . . . We must always exert the full strength of our imagination."[17]

And that, of course, was exactly what Wiener planned to do with *Cybernetics*.

THE INFORMATION BOMB

The spring and summer of 1948 did not lack for headlines. In May, the Jewish Agency's executive committee created the independent state of Israel, which was promptly attacked by its Arab neighbors on all sides; the war for independence would continue for another year. Then, in June, the Soviets blockaded the British, French, and American sectors of Berlin, which lay deep within the Russian-occupied zone of conquered Germany; a military airlift was immediately organized to take in food and supplies. And on June 30, Bell Labs announced a strange new amplifying device based on the cutting-edge physics of crystals and quantum mechanics—something called a transistor.

And yet even in the midst of all that, Claude Shannon's long-delayed opus on information theory exploded like a bomb. His analysis of communication was breathtaking in scope, masterful in execution—and, for most people, totally unexpected. "It was like a bolt out of the blue, a really unique thing," recalls his Bell Labs colleague John Pierce. "I don't know of any other theory that came in a complete form like that, with very few antecedents or history."

"It was a revelation," agrees Oliver Selfridge. "Around MIT the reaction was 'Brilliant! Why didn't I think of that?' Information theory gave us a whole conceptual vocabulary, as well as a technical vocabulary."

Indeed, there was something about this notion of quantifying information that fired the imagination, much as the notion of quantifying "chaos" would do in the 1980s. *Fortune* magazine declared Shannon's work to be one of man's "proudest and rarest creations, a great scientific theory which could profoundly and rapidly alter man's view of the world." *Scientific American* agreed wholeheartedly: "[Information theory encompasses] all of the procedures by which one mind may affect another," held an article* written by the director of the Rockefeller Foundation's Natural Sciences Division, Warren Weaver. "[It] involves not only written and oral speech, but also music, the pictorial arts, the theatre, the ballet, and in fact all human behavior."

Shannon's fellow mathematicians were enthralled. "A Mathematical Theory of Communication" had created a whole new domain of applied mathematics at a stroke, and suddenly there were a million questions to play with. What was the information capacity of a two-way channel? What did information theory have to say about cryptography? What were the most efficient encodings for English text as opposed to numerical data? Academic bulletin boards soon

---

*Weaver would subsequently arrange to have a collection of Shannon's papers published as a book, with his own article serving as the introduction and with himself billed as full coauthor. Published in 1949, *The Mathematical Theory of Communication* by Shannon and Weaver exposed the theory to a much wider audience and became such a standard reference in the field that many people to this day refer to its subject matter (incorrectly) as "Shannon-Weaver" information theory.

began to display announcements for seminars, conferences, and courses on information theory.* Bell Labs, MIT, and many other places quickly formed whole research groups devoted to the new field. In short order there was even a professional society for information theory.

And then there were the communications engineers, Shannon's original audience, who would happily have voted him a Nobel Prize in applied mathematics if such a thing had existed. Since it didn't, they simply showered him with any number of lesser awards. Nor were they alone: the walls of Shannon's house soon began to fill up with citations, plaques, and other testimonials. And Shannon, being human, was not displeased.

He was exceedingly disturbed, however, by a related development. Within a year or two of his initial publication, much to his horror, information theory started to become . . . *popular*. Anybody and everybody seemed to be plunging into the field, many without a clue as to what they were talking about. Scientists were beginning to submit grant applications that included references to information theory, whether or not their proposals actually had anything to do with it. It was becoming a scientific buzzword, much as *chaos, artificial intelligence,* and *complexity* would become buzzwords in the 1980s and 1990s. And Shannon hated all of it—the hype, the attention, the constant badgering by reporters. Most of all, he hated the way his ideas were being stretched beyond all recognition by idiots. Information theory was being greatly oversold, he asserted in a 1956 editorial entitled "The Bandwagon," and some moderation was in order. "It has perhaps ballooned to an importance beyond its actual accomplishments," he wrote. "Seldom do more than a few of nature's secrets give way at one time."[18]

Yet Shannon was bailing against the tide, and he knew it. So he essentially dropped out. He did continue to do cutting-edge research on information theory, at least for a while. But he turned down almost all the endless invitations to lecture and give newspaper interviews; he didn't *want* to be a celebrity. Instead he devoted more and more time to his gadgets, and to getting rich. (The first of the modern "high-technology" companies were just getting under way in the 1950s, and Shannon was asked to sit on several boards of directors; he did quite well for himself.)

In 1957, after Robert Fano and the information-theory group at MIT finally

---

* Occasionally with startling results. When Robert Fano taught MIT's first graduate course on information theory in 1950, for example, he suggested several topics for the students' term papers, including one based on a challenge implicit in Shannon's fundamental theorem: find an algorithm that compressed any given message into the smallest possible number of bits, taking into account the fact that some letters are more probable than others. One of his students, David Huffman, took on that challenge. His basic idea was to assign one-bit codes to the most frequent symbols, two-bit codes to the next-most-frequent, and so on. This was essentially what Samuel Morse had intuitively done in the 1840s with his own telegraph code, but Huffman took it a step further by devising a rigorous algorithm that could be applied to any set of symbols. Indeed, the Huffman encoding scheme is still the foundation for optimal data compression in everything from compact discs to interplanetary spacecraft.

lured Shannon back home for good, some of his colleagues were cleaning out his office at Bell Labs when they found a file he'd left behind, labeled "Letters I've Procrastinated Too Long On." It contained exactly what it said: letters, including some from major figures in science and government.

Shannon hadn't bothered to answer any of them.

For all Shannon's unwanted celebrity, it's hard to say whether he or Norbert Wiener bore more responsibility for the Information Bomb. After all, *Cybernetics* hit the bookstores only a few months after Shannon's paper appeared, and greatly reinforced its message. Together, the two theories became like a self-exciting system, inseparable in their effects.

There's no question at all, however, about which of the two men had more fun.

In the beginning, it's true, most people found *Cybernetics* almost as hard to read as it had been for Wiener to write. Browsers who flipped it open at random were likely to find page after page of mathematical equations and diagrams. Even to the engineering types at MIT, it seemed intimidating. And it didn't help that the book was full of typographical errors. The publisher had sent Wiener two copies of the printed galleys for proofreading, and when they arrived, it was painfully obvious that the manuscript had been edited and typeset by French employees who understood neither English nor mathematics. "Walter Pitts and I carefully proofread one galley," says Oliver Selfridge, "and we gave it back to Wiener—who promptly mixed them up and sent back the unedited version to the publisher." Every mistake was thus dutifully printed.

In the end, however, *Cybernetics* became a best-seller anyway. After reviewers finished grousing about the math and the typos, they typically went on to praise the author's "clear and direct" style *(Annals of the American Academy of Science)* and his "provocative analysis of some of the most exciting developments in modern science" *(New York Times).*[19] Von Neumann himself reviewed the book for the journal *Physics Today,* commending its "brilliancy" and "deeply original character," though he couldn't resist getting in a dig: "The freshness of the approach excuses some exaggeration of emphasis."[20] The *Saturday Review of Literature* was downright giddy: "It appears impossible for anyone seriously interested in our civilization to ignore this book."[21] And in later years, *Cybernetics* would be hailed as one of the "seminal books . . . comparable in ultimate importance to   Galileo or Malthus or Rousseau or Mill" *(New York Times* again).[22]

Meanwhile, readers were making Wiener into a celebrity. Unlike Shannon, Wiener reveled in it, rapidly carving out a niche for himself as a public scientist-philosopher—and in the process, as Heims notes, pioneering a role that would later be taken up by figures such as Carl Sagan, Lewis Thomas, and E. O. Wilson. (In 1950 Wiener would amplify the themes of *Cybernetics* in a second and far more accessible book, *The Human Use of Human Beings.*) He supplied the newspapers and magazines with colorful quotes. He posed for the photographers who tracked him down on his wanderings through the MIT campus, gazing quizzically at the camera, cigar in hand. And as Heims describes, he gave many public lectures, usually to overflow audiences: "He spoke in a fluent, discursive manner (without notes), throwing out a multitude of ideas for an hour or two, until somebody signaled him that it was time to stop. The talk, amplified over a loudspeaker system, would be punctuated by the sound he made as he drew on his ever-present cigar stub."[23]

While many members of Wiener's audiences undoubtedly came for the entertainment value, others were eager to hear what this odd little man had to say. After all, these were people who had just seen World War II bring forth atomic bombs, "electronic brains," radar, guided missiles, antibiotics, and jet engines—a wave of technologies so astonishing that it was hard to tell sober news stories from pulp science fiction. And now here was Norbert Wiener with his vision of a new age in history, helping them make sense of it all. He didn't call it the Information Age; that term would be invented later, by others. But he made it clear that this magical stuff called information lay at its heart. Information was a substance as old as the first living cell and as new as the latest technology. It was the stuff that flowed through communication channels; indeed, it was the stuff that messages were made of. But it was also the stuff that concepts and images and stored programs were made of. It was the stuff that entered the eyes and the ears, that flowed through the brain, that provided the feedback for purposeful action. Information was what computers and brains were *about*. It was the one central concept that unified *communication, computation,* and *control* and made them all seem like different facets of one underlying reality.

Information was at once the stuff of a new world and a whole new way of understanding that world.

Certainly that was how it seemed down in the basement of Memorial Hall. George Miller, J. C. R. Licklider, and the other Young Turks had grasped the implications of the Information Bomb immediately: information was the concept that promised to make the mind and the brain comprehensible.

"There is no question in my mind," said William McGill, that "what we're looking at here is the origin of cognitive science." Cognitive science is our modern view of the brain as an organ designed to acquire, communicate, and transmute information, he explained; the discipline would rout behaviorism almost completely during the "cognitive revolution" of the 1960s and 1970s and eventually cut broad swaths through psychology, neuroscience, artificial intelligence, linguistics, and even anthropology. "But you can pin down [its birth] at MIT and Harvard in the late nineteen-forties, when people were sensing that these new mathematical methods would transform everything."

George Miller, in particular, would go on to become one of the commanding generals of the cognitive revolution, doing as much as or more than any other single person to break behaviorism's stranglehold. But even in 1948, said

McGill, it was clear that Miller was on fire with this idea of information. He saw Shannon's theory as a pathway to the deep mysteries of human discourse, human language, and human thought.

True enough, concedes Miller, who smiles when he thinks back to the excitement of those days. Take the perception of words, he says. Imagine that you're at a loud party, or in the cockpit of a fighter plane, or anyplace else where you're trying to hear over a lot of background noise. It turns out that you can understand what's being said much more easily if the words are selected from a very limited vocabulary, so there can be no confusion (this is why fighter jocks go in for that clipped *alpha-foxtrot-bravo* jargon). Shortly after Shannon's paper came out, Miller says, he and two of his students decided to measure this effect in terms of information theory. They found that their experimental subjects could detect the difference between, say, "boy" and "heel" at quite high noise levels: it was just an either-or choice, requiring the perception of only one bit of information. But as they added more words to the list of possibilities—that is, as they demanded the perception of more and more bits of information—it rapidly became impossible for their subjects to detect the differences. They literally could no longer hear "boy" and "heel" in the noise, even though the stimulus was exactly the same.

"Now, if you believe that," Miller says, "and God knows I had to, then what a person *expects* to hear is critical to what he *does* hear. That doesn't sound like a great idea now—that expectation should influence behavior. But it was anathema to the behaviorists. I remember writing up some of this research and showing it to one psychologist who was a rabid Skinnerian, and he said, 'Well, that's very interesting, but you don't dare publish it. If you write a paper about expectations, then your scientific reputation will be destroyed.' "

To hell with it: Miller published anyway. And oddly enough, his reputation did just fine. Maybe the wall wasn't quite as impervious as it seemed. In any case, Miller says, by 1949 he had written a paper with the Harvard psychologist Frederick Frick advertising the many different ways in which information theory could be used in psychology. And he had arranged to take a sabbatical at the Institute for Advanced Study in Princeton, so that he could learn enough math to understand Shannon's work thoroughly. "I took a course from von Neumann, but I couldn't understand him," Miller recalls with a laugh, "so I had to go take other courses. I ended up spending most of my time there in the math department!"

Lick, for his part, was embracing his systems-engineering approach to the brain more enthusiastically than ever. To his mind, cybernetics in general and information theory in particular were lighting the way toward a "hard-science" psychology, one that would be far richer, more satisfying, and more productive than anything that had ever come out of behaviorism. Indeed, said McGill, if Lick's career had turned out differently, he might well have gone on to become a pioneer in cognitive neuroscience, the branch of cognitive science that explores precisely how the brain processes information and gives rise to conscious experience. It's certainly true that by the time McGill entered Harvard, in 1948, just a few months after Shannon's paper appeared, Lick had already started incorporating notions such as channel capacity and signal-to-noise ratios into his work—even as he was spreading the word to anyone who would listen. Within a year, moreover, he was also talking up the work of the Canadian psychologist Donald Hebb, who suggested in 1949—correctly, as it later turned out—that the connections between neurons grew stronger with use. Lick the cyberneticist found this idea electrifying: McCulloch and Pitts had shown that a network of neurons could compute anything computable, and now Hebb had shown how such a network could learn. Lick became one of Hebb's earliest and most ardent advocates.

Meanwhile, when it came to the larger meaning of cybernetics for history and culture—well, there's no way to know if Lick was directly influenced by Wiener's views on the subject, since he never discussed the matter one way or another. But he definitely resonated with those views. When Lick later spoke about the power of computers to transform human society on an epic scale, about their potential to create a future that would be "intellectually the most creative and exciting [period] in the history of mankind," he sounded very much like Norbert Wiener in prophet mode. Lick likewise echoed Wiener as he worried about technology's potential for harm: "If all the Industrial Revolution accomplished was to turn people into drones in a factory," he was sometimes heard to say, "then what was the point?" Indeed, Lick's entire later career in computers can be seen as a thirty-year exercise in the human use of human beings, an effort to eliminate mind-numbing drudgery so that we could be free to use our full creative powers. Even in the 1940s, moreover, much of Lick's work as a psychologist still revolved around the notion of the human-machine interface and the question of how to use technology not to replace people but to help them. He had learned the power of that deceptively simple idea during the war, when he and his colleagues at the Psycho-Acoustics Lab pioneered technologies that helped pilots function more effectively in their inhumanly noisy aircraft. And now he was reinforcing that lesson every day through his own experience. "Patchcords in one hand and potentiometer knob in another," he rhapsodized in 1967, looking back on the wonderful electronic devices he used to play with in his laboratory, "the modeler observes through the screen of an oscilloscope selected aspects of the model's behavior and adjusts the model's parameters . . . until its behavior satisfies his criteria. To anyone who has had the pleasure of close interaction with a good, fast, responsive analog simulation, a mathematical model consisting of mere pencil marks on paper is likely to seem a static, lifeless thing."[24]

Of course, Lick was still dealing here with analog devices—tunable electronic circuits. Digital computers weren't even an option for him, since the very few machines that were actually operational in the 1940s were largely reserved for military work. Nevertheless, he was certainly aware of digital computers, and fascinated by them: Howard Aiken's Mark II was right there on the Harvard campus, after all, and Lick would later tell the story of how he'd met Aiken on an airplane trip in the late 1940s and gotten himself invited to his lab to see it.

But the time was fast approaching when Lick would get an even closer acquaintance with computers, courtesy of the rapidly chilling Cold War.

CHAPTER 4

THE FREEDOM TO MAKE MISTAKES

The ultra-top-secret news hit Washington like an electric shock: on September 3, 1949, a U.S. reconnaissance plane patrolling off the Kamchatka Peninsula had detected a cloud of radioactive debris drifting eastward with the prevailing winds out of Siberia. The immediate reaction in the U.S. intelligence community was confusion and denial, as certain high-level officials continued to insist that the Soviet Union was at least five years away from testing an atomic bomb of its own. Half the country was still in ruins from World War II, they pointed out. The radioactivity *had* to be from a reactor accident.

Wrong. Detailed radiochemical measurements soon forced the wise men to admit just how badly they had underestimated Joseph Stalin's determination to catch up in the nuclear-arms race. "Joe One," as intelligence analysts now code-named the Soviets' first atomic bomb, had in fact been detonated on August 29, near the town of Semipalatinsk, in the high, grassy steppes of Kazakhstan.

President Truman's public announcement of the Soviet test, on September 23, made banner headlines around the world. Coming as it did only three months after the end of the Berlin Blockade, at a time when most of Eastern Europe had already fallen under Soviet domination, when China was falling to the

---

The birth of interactive computing: an operator communes with the SAGE system

forces of Chairman Mao, when Communist guerrilla wars were raging in Greece and Turkey, and when Communist North Korea was making threats against the southern half of that country, Joe One so unnerved official Washington that calls could be heard for a "preventative war": *Do it to the Russians before they can do it to us.* Cooler heads prevailed, fortunately. But the arms race had begun in earnest. In January 1950, after intense debate, Truman was convinced that the new Soviet atomic threat had to be countered with an even more powerful weapon: the *hydrogen* bomb, which thus far had been little more than a long-shot research project. The half-deserted mesa-top laboratory at Los Alamos was soon bustling again as a new generation of scientists launched a crash program to meet the challenge.

Meanwhile, the Pentagon was also rethinking the suddenly vital issue of air defense. Not only did the Soviets now have the Bomb, but the latest intelligence suggested that they were developing long-range bombers that could deliver the weapon to targets in the United States. In December 1949, the Pentagon deputized a panel of technical experts headed by the MIT physicist George E. Valley to investigate U.S. air-defense capabilities. Valley and his colleagues were appalled by what they discovered. True, the air force had already reactivated the first-generation radar network left over from the war, in one of its initial acts after being separated from the army in 1947. But Valley's committee considered that network disastrously inadequate to meet the new threat. The installations were concentrated along the coasts, leaving no coverage for the polar approaches over Canada. The big radars left gaps at low altitude, where an attack could slip through unnoticed. And the system had virtually no automation whatsoever. Human operators were supposed to track aircraft by making grease-pencil marks on a Plexiglas map, and then they were to place a phone call to the nearest air force base if any of those aircraft seemed suspicious, so that interceptors could go up and take a look.

The Valley committee's preliminary report, in January 1950, accordingly made two key recommendations: first, the existing system should be upgraded as quickly as possible by "a competent technical organization"; and second, a long-term solution should be devised that would encompass a multitude of smaller radars for greater coverage nationwide plus an advanced communications system connecting the radar stations with one another and with the air bases. In particular, advised the committee, the new radar system should coordinate surveillance, target tracking, and all other operations using electronic digital computers.

This last proposal was definitely a gamble. Half a decade after ENIAC and John von Neumann's draft report on EDVAC, the first generation of stored-program computers was still pretty much stuck in the laboratory. The only such machines actually running in January 1950 were the prototype MARK I computer at Manchester University and the relatively modest EDSAC at Cambridge University. In the United States, EDVAC itself was still two years from completion at the Aberdeen Proving Ground in Maryland, having been badly delayed by the exodus of so many essential people from the Moore School. At the Eckert-Mauchly Corporation, the two former leaders of the Moore School team were still a year away from finishing a business-oriented data processor they called the UNIVAC. And in Princeton, von Neumann's team was likewise at least a year away from completing the Institute for Advanced Study's computer. When it came to digital computers that could actually compute, the United States had ENIAC, now relocated to Aberdeen and hard at work on H-bomb calculations, and Howard Aiken's electromechanical machines at Harvard—and that was about it.

Gamble or not, however, Washington was in a mood to listen. On June 24, 1950, Communist troops from North Korea attacked southward across the Thirty-eighth Parallel, drawing the United States and its allies into a "police action" to contain them. By the time the Valley committee submitted its final report, in October, the Cold War seemed on the verge of becoming a very hot war indeed; the report's recommendations were accepted without question. The air force immediately contracted with Bell Laboratories and General Electric to upgrade the existing radars. And in December 1950 the air force asked MIT, home of the wartime Radiation Laboratory, to turn the Valley committee's vision of computer-based air defense into working hardware.

MIT was dubious; its leaders weren't at all sure that the school ought to tackle such a huge project, especially given the real possibility of failure. But the air force insisted, and patriotism prevailed. After all, who was better qualified? The Rad Lab's successor organization, the Research Laboratory for Electronics, was still the finest collection of electronics wizards in the world, not to mention one of the Defense Department's biggest research contractors. RLE was ready to roll; the people working on the existing defense projects could simply transfer to the new effort en masse.

And besides, where else could you find a computer like Whirlwind?

REAL-TIME

The irony of it was that Whirlwind had started out as an analog computer. Back in December 1944, the U.S. Navy had asked the MIT Servomechanism Laboratory to do a feasibility study on an all-electronic flight simulator, a dual-purpose system that could be used both to evaluate new aircraft designs and to train student pilots. The navy's problem was that the existing trainer-analyzers were mechanical monstrosities, each requiring an elaborate array of linkages and gears to model the behavior of just one specific type of aircraft. The navy's question was whether it might not be cheaper and easier to do such modeling with analog electronics, since in principle the circuitry could be rewired to simulate *any* type of aircraft.

The job of finding out fell to Jay Forrester, a University of Nebraska graduate who was still working on his master's degree at the Servomechanism Lab. His conclusion was "Maybe": an all-electronic trainer-analyzer would be pushing the state of the art but might be just barely doable. This being wartime, however, that was good enough for the navy. A new contract was in place by the summer of 1945, and young Mr. Forrester suddenly found himself in charge of a team that was supposed to build the device for real.

Now, that might have been a disaster right there; no one ever claimed that Jay Forrester was an easy man to work for. Even at age twenty-six, he was cool, formal, and remote, always the chief, never one of the boys. Happily, however, Forrester seemed to recognize that quality in himself, and he had the good sense to choose the affable, outgoing Robert Everett as his deputy. Together they formed an exceptionally effective leadership team. And that was a lucky thing, too, because building a trainer-analyzer in the lab quickly proved to be even more difficult than it had appeared on paper. The cockpit simulator—the movable box where the pilot would sit—was tricky enough in itself. But far worse was the electronic controller, which had to turn and tilt the cockpit in response to the pilot's every touch on the controls; update the instrument panel; mimic the wind resistance; record every move for later analysis; *and* be adaptable enough to simulate multiple aircraft.

For a few months the group struggled along, making very slow progress. But then in October 1945, as Forrester was to recount in a 1983 retrospective on the project, a chance encounter showed him a different way: "Perry Crawford, who . . . at that time was in the Special Devices Center of the U.S. Navy [the office that was funding Forrester's project], was standing with me on the front steps of MIT at 77 Massachusetts Avenue late one afternoon. He called my attention for the first time to digital computation, to the mechanical Harvard Mark I computer, to the electronic ENIAC computer." Crawford pointed out that Forrester and Everett had been beating their brains out trying to make their analog system work. Why not try the digital approach instead?[1]

Being an MIT man, Forrester was steeped in analog thinking. But once he started looking into digital, stored-program computing—von Neumann's "First Draft" was just then making the rounds—he immediately grasped the advantages for the trainer-analyzer. Rather than trying to hard-wire the system's behavior with electric circuits, they could abstract the behavior, turn it into a set of algorithmic procedures, and encode it in software, a far more flexible medium. "We are no longer building an analog computer," Forrester informed his team shortly thereafter. "We are building a digital computer."

Um, sure, Jay. "Things were different in those days," Everett would recall in his own memoir. "We didn't have a big study group, and when Jay decided to build a digital computer, we all thought that was great."[2]

Actually, Everett adds, it was probably just as well that they didn't think about it too hard. There they were, a pack of kids with no experience in digital computing, blithely proposing to build a machine of unprecedented speed and reliability. And worse, they were just as blithely proposing to take the technology in a radically new direction. So far, from Babbage's Analytical Engine in the 1830s through ENIAC in the 1940s, digital computers had always been conceived of as calculating machines. Whatever else they might offer in the way of electronic speed and/or universal programmability, they were still designed to take in data for one specific problem at a time, grind away until they spit out an answer, and then stop and wait for new input.

But *this* digital computer was supposed to act as a flight simulator, a machine for which there was never any "answer," just a constantly changing sequence of pilot actions and simulated aircraft responses. So Forrester and his team would have to create not a calculator but a computer that could monitor its inputs constantly, staying ready for whatever might come along; that could respond to events as fast as they occurred, without ever falling behind when things got hectic; and that could keep going until the simulation was over, however long that took. In short, they would have to create the world's first *real-time* computer.

Fortunately, writes Everett, they did have one advantage that more than made up for their lack of experience: they were right next door to the Rad Lab, with all the wonderful technology developed for radar. Indeed, that advantage allowed them to settle on a logical design for the new machine as early as 1947, by which point they had also begun to realize that their "trainer-analyzer" contained the germ of a much more interesting idea: a *general-purpose* real-time computer. In that same year of 1947, Forrester and Everett wrote a paper describing how such a computer could be used to coordinate the activities of a naval task force, including submarines under the surface, ships on the surface, and aircraft overhead.[3] A year later that notion was reinforced when MIT president Karl Compton, who also happened to be the head of the Pentagon's R&D board, asked Forrester, Everett, and their team to examine the future of computers in the military. For their report they produced a fifteen-year timetable for real-time computer use in ten application areas, including logistics, antiballistic missile defense, air-traffic control, the coordination of naval task forces—and an air-defense system. Their estimated fifteen-year price tag for the research to create these machines was $1 billion; adding in development and production costs would bring it to an estimated grand total of $2 billion.

In 1948 those were the kinds of numbers that still made people swallow hard, according to Forrester: "The report created a communications gap when we went into a meeting with the Office of Naval Research, where they thought the agenda was whether or not we were going to get our next hundred thousand dollars—and we came in with a forecast ten thousand times that for the next fifteen years."[4] The agency eventually agreed to let Forrester and Everett abandon the trainer-analyzer idea and instead refocus the project on real-time computing in general. It also agreed to expand their budget to $1 million per year, enough to pay for seventy engineers and technicians plus another one hundred support staff. True, they weren't talking billions. But even so, the Whirlwind project, as it was now known, had suddenly emerged as by far the largest and most expensive computer effort of its day.

The machine had a bulk to match. It's impossible to say when Whirlwind was completed, since it continually evolved as Forrester and his team experimented with new components. But when it first became operational, in 1951, Whirlwind took up more than twice as much floor space as ENIAC: twenty-five hundred square feet versus about a thousand. In fact, Forrester and company could actually walk around inside the machine; for ease of maintenance, they had distributed the electronics into eight tall racks extending on either side of a central corridor. By 1951 standards, moreover, Whirlwind's performance was impressive, being roughly the equivalent of a 1980-vintage personal computer such as the TRS-80.

And most important, Whirlwind *worked*. In fact, it was a 1950 visit to the nearly completed Whirlwind that had persuaded the Valley committee to endorse a computer-based air-defense system in the first place. ("What [they] needed was a digital computer, and what we needed was air defense," Everett wrote: the navy had quickly gotten tired of paying that $1 million per year and was on the verge of cutting them off entirely.)[5] Not long afterward, moreover, it was Whirlwind that had demonstrated the real-world feasibility of such a system: on April 20, 1951, the computer successfully tracked two prop-driven fighter planes in the skies over Massachusetts, repeatedly computing interception trajectories that allowed the controller to steer the "defending" fighter by radio to within a thousand yards of the "attacker." That was good enough to silence the skeptics at MIT. In August 1951, with full funding from the air force, the institute formally launched "Project Lincoln" in the warren of ramshackle structures that had housed the Rad Lab a decade earlier. Forrester and his team joined the project wholesale the following October, becoming Division Number 6. And Whirlwind became the test bed for the new, computerized air-defense system.

Of course, Project Lincoln wasn't just RLE plus Whirlwind. As good as those two groups were at building hardware, the project's planners had recognized that the success or failure of this air-defense system would also depend upon human beings. The radar operators, in particular, would have to take in huge amounts of incoming data, coordinate with other operators who were coping with their own floods of data, and then make split-second decisions that were *right*. A mistake could spell catastrophe for the whole nation. So somehow Project Lincoln would have to make all of this computerized hardware work together with human beings as a system. And to accomplish *that*, the project leaders realized, they would have to recruit a team of top-rank psychologists who thoroughly understood human-machine interactions.

Fortunately, they had one such psychologist ready at hand. When they gave out the carefully numbered keys to the Project Lincoln enclave—the program was classified, after all—key number 12 went to MIT's J. C. R. Licklider.

THE DREAM TEAM

It had only been a matter of time, really. By the late 1940s, most of Lick's friends were at MIT. His interests were at MIT. And he had increasingly come to feel that his future was at MIT.

It certainly wasn't at Harvard. Quite aside from anything else, Harvard's abysmally low salary scale for nontenured instructors was beginning to pinch; with the birth of his son, Tracy, in 1947, and his daughter, Linda, in 1949, Lick had a growing family to support. Indeed, it was partly in an effort to make ends meet that he spent the summers of 1948 and 1949 as a consultant to the Navy Electronics Laboratory in San Diego. Louise Licklider had long since lost count of the number of times they'd moved apartments. And the reality was that no tenure slots seemed likely to open up in Harvard's psychology department for many, many years.

Fortunately, however, as Lick approached his thirty-fifth birthday, his star quality was as bright as ever. In late 1988, in the one extensive interview he ever gave about his own life and work,* Lick characteristically painted himself as an ordinary guy who just happened to fall into things by luck: "There came a time that I thought that I had better go pay attention to my career. . . . I am not a good looker for jobs; [so] I just came down to the nearest place I could, which was in our city."

In fact, MIT came looking for *him*. Several years earlier, the institute had lured the Harvard physicist Leo Beranek across town to help organize a new acoustics lab, which was codirected by the MIT physicist Richard Bolt. And by 1949, says Beranek, who had been the wartime head of Harvard's Electro-Acoustics Laboratory and who therefore knew his counterparts at the Psycho-Acoustics Laboratory very well, "I started agitating for MIT to bring Lick down from Harvard, too. 'This guy is brilliant,' I told them. 'We need this kind of addition to the lab.'"

MIT didn't need much persuading, apparently. Nor did Lick. Bolt and Beranek's lab sounded like wonderful fun: acoustical work, visual work, the theory of hearing, and just plain playing around with human communication and all that went on in the brain. MIT was also promising him tenure after only a few years' probation—and then sweetening the deal even more with an offer he *really*

---

* The three-hour interview was conducted on October 28, 1988, by the historians of science William Aspray and Arthur Norberg on behalf of the University of Minnesota's Charles Babbage Institute, as part of a larger study of ARPA's role in computing and networking. The interview is available from the Babbage Institute as oral history number OH 150. Unless otherwise indicated, all direct quotations from Lick are taken from this interview.

couldn't refuse. In those days the university had no psychology department as such, just a small group of researchers based in the economics department, where they focused on labor relations, social psychology, and the like. So the idea was to have Lick take over (when he wasn't working in the acoustics lab) and build the group into a full-fledged department—*his* kind of department, based on rigorous, experimental, "hard-science" psychology.

Lick lost no time in saying yes. He followed Beranek across town on February 1, 1950, and never looked back.

Given the timing, however, he also walked right into the Cold War.

Lick's wartime experience at the Psycho-Acoustics Lab had not gone unnoticed at MIT. In the spring of 1950, for example, even as he was settling into Bolt and Beranek's acoustics lab, he was invited to take part in Project Hartwell, a navy-sponsored summer study that focused on the broad issues of undersea warfare and the security of overseas transport. Then, in February 1951, having made a favorable impression during the Hartwell deliberations, Lick was asked to join another summer study, Project Charles, which was supposed to hammer out a technical blueprint for the Valley committee's air-defense idea. As the only psychologist among twenty physicists, moreover, he was the one member who pushed harder than anyone else to include human factors in the project—which, of course, was how he got key number 12: having opened his mouth, Lick was promptly made codirector of Project Lincoln's radar-display development group.

That was fine by him. Project Lincoln's goal of understanding how machines and humans could work together as a system was just an extension of Lick's own inner quest: understanding how the human brain *itself* worked as a system. Besides, it had already occurred to him that in building up the Project Lincoln effort, he would also be putting together an absolutely first-rate psychology group—using the Pentagon's money. After all, he told the MIT hierarchy with blue-eyed sincerity, he would never be able to attract the best young psychologists to Project Lincoln unless he could also offer them some sort of joint appointment in his psychology section; they would be too worried about leaving the academic track and not being able to get back on.

They bought it. "[So] I went to what I thought were the ten best graduate schools in psychology," Lick explained in the 1988 interview. "I tried to get at least one and sometimes two people from each place, people who were just then getting their Ph.D.s. I was looking for an orientation to theory and experiment [as opposed to social or clinical psychology]. But I really did not care very much what they knew or what they were interested in. My bias was toward very bright people."

Few of Lick's chosen refused. Although he wasn't quite the all-American boy anymore—his sweet tooth had long since gotten the better of his waistline—those enthusiastic blue eyes and that lopsided grin made them feel that working at

Project Lincoln would be incredibly exciting. Besides, in an appeal that carried a great deal of weight in those pre-Vietnam days, he offered them a chance to do something vital for their country: "Everybody was kind of excited about the Russians' getting the bomb," he said. "One kid was saying, 'Well, I don't want to write the last journal paper.' There was some feeling of imminence."

The upshot was that Lick had no trouble sweeping up about a dozen young researchers into what he would always remember as "one of the best groups of psychologists there ever was," made up of people such as Bert Green, Herb Jenkins, Joe Bennet, Bill Harris, George Miller, and Bill McGill.

McGill, for one, was thrilled about the move, "What you have to imagine is this seedy backstreet environment, with all these various groups stashed cheek by jowl in the temporary buildings," he says. "And they were all working on utterly fascinating things." RLE, in particular, still had labs scattered everywhere—and not just for Project Lincoln. Thanks to Jerry Wiesner, who was already a power in RLE and would become its director in 1952, that warren of temporary structures was alive with new ideas about the nature of mind and brain, and their relation to technology. Step through one door, for example, and you might find yourself in the neurophysiology group, greeting Warren McCulloch and Walter Pitts, who were continuing to develop their neural-network ideas, or Jerome Lettvin, who was making the first measurements of electrical activity in single neurons. Step through another door and you might encounter Walter Rosenblith and his crew analyzing neural activity in their "communications biophysics" laboratory. Step through yet another door and you might see Peter Elias, one of the brightest of the theorists working on communications problems and information theory, or Robert Fano, leader of the MIT information-theory group. And of course, you might run into Norbert Wiener wandering the halls almost anywhere—at RLE if not at Project Lincoln, of which he heartily disapproved.

George Miller, meanwhile, had been somewhat ambivalent about the move. "I was always surprised that I had the gumption to do it," he says. "I had such a bad case of Harvarditis." Nonetheless, he adds, Lick didn't have to do much to persuade him; his tenure prospects at Harvard were just as bleak as everyone else's. He followed his once and future colleague across town before 1951 was out. And Lick was so overjoyed to have him there that he made him his de facto joint team leader. "Lick and I sort of time-shared," says Miller. "I would run the psychology section sometimes while he was in charge at Lincoln, then we would switch and he would worry about the academic program for a while."

In truth, Lick needed all the help he could get. Convinced that psychologists should work with the engineers from the very beginning of the design process as opposed to coming in at the end, when it was too late to change anything, Lick had organized the Project Lincoln radar-display group jointly with the engineer Herbert Weiss, with a team membership that was half and half. It did not start out well. "The engineers never took to it at all," says Bert Green, who had been just finishing up his Ph.D. at Princeton when Lick recruited him. Like most technical people of that era (and many today), the engineers tended to view human-factors issues as a fuzzy-minded obsession with trivia such as the shapes of knobs and the colors of dials, things that were a distraction from the serious business of designing hardware. Moreover, it must be said that Lick's band of greenhorns gave them reason for that view. "The engineers would come ask, 'How bright should this display be?' " recalls Green. "Well, we'd all been trained as academic psychologists. So we'd say, 'Come back in three months, after we do the experiments.' But they wanted the answer right *now*."

It was a culture clash that took quite a while to bridge. But Lick was determined that his people do so, said McGill—and in time, they did. Once Project Lincoln started working with real radar signals, for example, the aircraft blips turned out to be surrounded by all kinds of extraneous noise and ground clutter. The engineers knew that they could filter out a lot of that stuff electronically, according to McGill, but the question was, How much? Filter too little and the radar operator might be confused at a critical moment; filter too much and a real attacker might be filtered right off the screen. It wasn't really a technical question: finding the best balance required a deep understanding of human perception and our ability to detect patterns in the midst of chaos.

And then there was the question of how the operator was supposed to communicate with the computer and designate this or that blip for special attention. "A great moment was when Robert Everett invented the light gun," said McGill—a device that Lick's team helped refine into the light pen. "The idea was that you should give the operator a pen with a light-sensitive device on it, so that he could put it over the target blip, press a button on the pen, and acquire the location of the target in the computer." It worked—and indeed, it would go on to be used widely in nonmilitary applications. "Lick was always immensely proud of the light pen," said McGill. It seemed to prove that his kind of psychology really could make a difference.

No doubt about it, says Louise Licklider: she was married to a hopeless romantic. On summer evenings Lick liked to take her dancing at the roof garden of the Ritz Hotel in Boston. Or he might turn on the radio at home and waltz her around the kitchen. And every Saturday, of course, she was still getting a fresh gardenia. "People said, 'You two don't act like old married people. You *court* each other,' " she says with a laugh. "And I told them, 'We have to—we don't have time for anything else!' "

True enough: what with doing his own research at MIT's acoustics lab, institution-building in the psychology section, and defending the free world through Project Lincoln, Lick was seriously overcommitted in those years. "So he would come home for dinner with me and the kids," she says, "and afterward he would put them to bed. He was a great storyteller. But then he would go back to work at the lab and wouldn't come home until about eleven or so. That's when we'd tell each other about our day and I'd learn what was uppermost on his mind. Early on, he started calling this the nightly core dump."

If it was warm out, she adds, they might sit on the back-porch swing and sip iced tea as they talked. (This would have been after they moved into their wonderful old Victorian house in suburban Arlington, Massachusetts; Lick's job at MIT had finally given them enough financial security to settle down a bit.) Of course, she didn't pay too much attention to the technical aspects of his work, which she wasn't particularly interested in anyway. But it was hard not to get caught up in his infectious sense of excitement. Take the saga of Lick's "hard-science" psychology section, for example: it appealed to her sense of drama. Not only did it have a hero—this guy sitting next to her in the swing—but like any good story, it also had a villain.

That it did, agrees George Miller. The problem, he explains, was that psychology had always been a part of economics and social science at MIT—which meant that Lick now found himself reporting to the dean of Humanities, John Burchard, one of the very few people around the institute who appeared to be immune to his enthusiasm. Indeed, Burchard seemed to be one of the very few people in the world who could make Lick visibly lose his temper. In fairness, Dean Burchard was a distinguished architectural historian who had ably served in Vannevar Bush's NDRC during the war and done yeoman's work afterward in reorganizing and modernizing MIT's library system. Miller remembers him as a pleasant, friendly sort of fellow, if a little on the dim side. Yet he drove Lick and his crew to distraction through his utter inability to comprehend what they were trying to do. To Burchard, psychology still meant industrial relations, group dynamics, and organizational behavior. So when Lick and his young Turks showed up talking about cybernetics, information theory, neural networks, and the systems analysis of the brain—and generally sounding like refugees from RLE, where they probably should have been in the first place—Burchard just couldn't see the point.

Of course, adds Miller, that wouldn't have been so serious if Burchard hadn't also had control of the group's purse strings. "MIT was enormously conservative in those days," he explains. "They had a very small endowment, and they spent it like it was their personal money. They'd be wild with government money. They'd do all kinds of crazy, wonderful things [such as Project Lincoln or Whirlwind]. But with their endowment, they were very, very tight." In that environment, Dean Burchard wasn't about to grant departmental status or tenure slots to a wild-eyed group that had come in under Project Lincoln. If he made any such commitment to them now, then MIT would have to support them on its own once the government project was over.

The upshot was that almost every encounter he had with the dean left Lick steaming. He would return to the group proclaiming that things were "all Burchard up"—*again*.

It was a situation that brought out all his instincts for subverting officious stupidity. If Burchard and company weren't going to help them build a modern psychology department, they would have to do it themselves. "We formed a little cabal in which the MIT administrators, particularly Dean Burchard, were the enemy," says Miller. A little checking revealed that under previous leaders, the psychology group had had the authority to take on graduate students and grant them Ph.D. degrees independently. So Lick and Miller just decided to proceed as if they still had that authority—without asking permission from Burchard. In retrospect, this was incredibly naive, Lick admitted in his 1988 interview. But being young and cocky, they did it. And in fact, he said, they soon had quite a show going: "We got some Ph.D. students—five of them. We had outside research contracts, so that we really had lots of money. And we had the whole basement of the Sloan Building, which is where the Faculty Club is now. We were doing pretty well."

Meanwhile, Lick was also bringing graduate students into the acoustics lab and setting them to work on understanding behavior in terms of cybernetics-style control and communications. "This was nineteen fifty-one," recalls Jerry Elkind, who was one of those recruits. "I was looking around for a master's-thesis topic, and since I was interested in control systems, one of the faculty members suggested that I talk with Lick at the acoustics lab. Well, Lick was looking for somebody to run this machine he had [an analog, tape-based computer he'd just inherited from a friend in another lab]. So I walked through his door and he said, 'Great! How about a job?' I was delighted. I did my master's degree with him, looking at how people perform in control tasks—in fact, I think my project grew out of a note that Norbert Wiener had written to Lick—and then I continued down the same direction for my Ph.D."

Indeed, Lick was already honing the leadership style that he would use to such effect a decade later with the nationwide computer community. Call it rigorous laissez-faire. On the one hand, like his mentor Smitty Stevens, Lick expected his students to work very, very hard; he had nothing but contempt for laziness and no time to waste on sloppy work or sloppy thinking. Moreover, he insisted that his students master the tools of their craft, whether they be experimental technique or mathematical analysis. On the other hand, Lick almost never told his students what to do in the lab, figuring that it was far better to let them make their own mistakes and find their own way. And imagination, of course, was always welcome; the point here was to have *fun*.

The Licklider style wasn't for everyone, and not everyone stayed. But for self-starters who had a clear sense of where they were going, it was heaven. Good people liked to be with Lick; he seemed to be surrounded by an atmosphere of ideas and excitement. "He communicated the feeling that you could understand any field you wanted to," explains Jerry Elkind. "He loved gadgets and putting things together. He loved to apply information and new ideas. So any area of science was interesting to him; he pulled in ideas from all kinds of domains. And he was always looking for novel ways of challenging your understanding of the domain, by constructing problems or puzzles that would require insight into the theory to solve."

McGill had much the same reaction: "There was nothing like listening to Lick give a talk," he said. "He'd go on like some hayseed with this homespun Missouri accent, speculating and scratching his head. And he would formulate the problem right there in front of you. Once he was talking about measuring the intensity of the croak of a bullfrog. And he suddenly wondered out loud, 'Why doesn't the frog deafen himself?' And he immediately came up with a hypothesis about how the frog must have some sort of protective mechanism, because otherwise the croak was so loud that it would damage the inner ear. It was beautiful to watch—like a child at play."

Of course, it also has to be acknowledged that Lick wasn't much of a manager in the conventional sense. He hated talking on the telephone, for some reason, and always seemed to figure that the administrative stuff would be easier if he left it for later. He would rather be back in the laboratory having fun—which meant that people on his team were forever digging paperwork out of his in-box and doing it for him. Miller never got over Lick's attempts to hire a secretary: "Lick gave all the applicants the Miller Analogies Test [created by a different Miller, no relation] and hired the one with the highest score. We had a string of brilliant young women who lasted about three months before resigning in violent boredom. I finally took over and hired an ordinary, quiet young woman of slightly-better-than-average intelligence. . . . She soon learned her job, and when she had nothing to do she sat waiting at her desk. Everyone adored her—she was working there years after I left."[6]

But never mind the paperwork. Lick was, well, not the patriarch. He had far too sunny a personality to indulge in Smitty Stevens's thunderstorm tactics. But like his mentor, he clearly treated his group as his family. "They were his sons," says Louise Licklider. He and Louise welcomed each of the team members personally when they arrived in town, and helped them get settled. He made sure they were invited to Faculty Club receptions. He opened his home to them at all hours. And he was there for them in emergencies. When Bill McGill's wife was stricken with a pulmonary embolism shortly after the birth of their second child, McGill's first anguished call was to his boss, who got out of bed in the middle of the night, threw on his ratty old trench coat, and came straight to the hospital. "He actually paid the bill so that I wouldn't be hassled," recalled McGill. "I didn't even know about that until days afterward. And then he sat with me and held my hand until she was out of danger."

Everyone in the group seems to have felt it to one degree or another. Ask Miller: "We were a tribe." Or Elkind: "Lick was very warm and supportive with his graduate students. I look back in amazement, but when I was getting into the serious part of my thesis work, I'd pop into his house at ten o'clock at night and we'd spend a few hours spreading data over his kitchen table. He was always willing to provide insights, without taking over the project at all—and Louise was most tolerant about these intrusions at every hour of the day and night."

Or ask Thomas Marill, who also started working on his Ph.D. under Lick in 1952: "Lick very rarely got angry. But he got furious with me one time because I moved some cables around in a laboratory setup he had. Lick had spent a great deal of time and effort putting it together, so he even called me some kind of name. But he was immediately overcome by guilt and that night invited me to his house for dinner. He was very upset at himself for the outburst."

Or McGill again: "Rather than having formal colloquia, which is the way departments usually do it, we'd all sit around at someone's house with beer and pretzels, we'd look at a tough problem, and we'd brainstorm it. In the end, nobody could figure out who was responsible for which ideas. But the wonderful thing about Lick was that he didn't give a damn. If you wrote a paper using an idea he'd come up with, he'd give it to you! He much preferred this kind of brainstorming to the task of knocking out a long chapter himself." Indeed, McGill added, talking in the 1990s like a man in his seventies who remembered all too clearly what might have been, "it was a much warmer, more interactive style of doing science than anything I've ever experienced since then. I think none of us really understood at the time what a golden moment it was."

A STATE OF CONTROLLED PANIC

It was a golden time for all of them, really. When veterans reminisce about Project Lincoln—or Lincoln Laboratory, as it was renamed in 1952—they sound a lot like veterans of the Manhattan Project, or the Radiation Lab, or even the Apollo moon program of the 1960s. They were young. They were blazing new trails, thinking new thoughts, and doing what had never been done before. They were in the midst of a war—or close enough, anyway—doing work that was vital to the nation's survival. And they had the Pentagon behind them all the way, so that money was no object, and the petty bureaucratic obstacles seemed to vanish before them like mist.

All of which was probably just as well, since the Lincoln veterans also admit that they had plunged into the air-defense project with only the vaguest idea of what they were doing. It wasn't just that the whole air-defense system was being based on a technology—real-time computing—that was still so new it was beyond the cutting edge. *Everything* was beyond the cutting edge: the input devices, the display screens, the communications systems, even the radars themselves. Moreover, those individual pieces not only had to be mastered separately but then had to be integrated into a unified whole. And it all had to be done *now*.

The atmosphere was "a state of controlled panic," recalled Albert Shiely, the air force general who administered the overall air-defense effort from an office in New York City. In effect, he said in a 1983 retrospective on the project, the MIT team improvised the whole air-defense system on the fly, "building and designing and doing everything simultaneously."[7]

Take communications, for example. The idea was to have computers at the various radar centers automatically exchange data over the existing telephone lines. But when the Lincoln Lab researchers tried it, they found that the digital signals came through the telephone system hopelessly corrupted by noise and frequency shifts. The problem was that the phone lines were analog, designed to carry the electronic equivalent of the human voice. In frustration, the engineers developed a desk-sized box full of vacuum tubes that would take the digital bits on one end, modulate them into something the analog connections could handle, and then fire them down the line. An identical box on the other end would then demodulate the signal and retrieve the bits unscathed. The engineers were able to achieve transmission speeds of up to 1,300 bits per second with the device, which they called a modulator-demodulator—a mouthful that they quickly shortened to "modem."

Meanwhile, there was Whirlwind. In the early days its greatest claim to fame around MIT was not its status as the world's first real-time computer but its spectacular unreliability. Even with an average of four hours' maintenance per day, Whirlwind still crashed so often that the local wits took to calling it Headwind. This was no joke when you were talking about an air-defense system that had to operate twenty-four hours a day and could not be allowed to fail. "Everything having to do with the reliability and long-term performance had to be explored from the ground up," noted Forrester.[8]

The vacuum tubes were a pain, of course. But the really worrisome components were the thirty-two cathode ray tubes that provided Whirlwind's memory (think of them as TV picture tubes that didn't display pictures, but instead stored little spots of electric charge on their surface corresponding to binary *1*s and *0*s). They had to be made by hand in the Whirlwind workshops, at a cost of about a thousand dollars apiece. They were clumsy and temperamental in the extreme. And they usually burned out within a few weeks. The project was thus spending some thirty-two thousand dollars per month on data storage alone.

No one was more aware of this than Forrester, who had started looking for a substitute for the tubes long before Project Lincoln. And in fact he had found one, one night back in the spring of 1949. He was browsing through the journal *Electrical Engineering,* he says, when he happened to notice an advertisement featuring a new magnetic material called Deltamax, which had been developed for use in amplifier circuits. Something clicked: maybe, just maybe, you could use this stuff in a computer memory. "The idea immediately began to dominate my thinking," he later recalled, "and for the next two evenings I went out after dinner and walked the streets in the dark thinking about it, turning over various configurations and control methods in my mind."[9]

Forrester's basic insight was that a piece of material magnetized in one direction—say, *north*—could represent a binary *1*, while a piece magnetized the other way—*south*—could represent a binary *0*. So he ordered some Deltamax and began to experiment. He found that by shaping the material into a little ring and then running a wire through the hole, he could flip the ring's magnetization whichever way he wanted by changing the current in the wire. Moreover, he discovered that the magnetization in the ring would stay put until he flipped it again.

Encouraged, Forrester took the concept back to the Whirlwind workshops for development. The next step was to make a grid of wires, with each intersection encircled by one of the "cores," as the little magnetic rings came to be called. The idea was that if you energized, say, the third wire down and the fifth wire across, a computer could detect the magnetization of the core at the intersection of those particular wires and, if need be, flip it—that is, the computer could read or write that particular bit of data. What was more, it could read or write to any given intersection just as easily as any other. It was truly "random access."

It was an elegant concept, and it worked beautifully—in the lab. Putting the concept into practice was considerably harder. (Among other things, Deltamax proved too unstable for long-term use; Forrester eventually switched to ceramic ferrite, the same form of iron oxide that is now used in magnetic tapes and floppy disks.) Nonetheless, by the time Project Lincoln came along in the summer of 1951, the technology of magnetic-core memory was well in hand.

However—and this is where the controlled panic came in—the advent of Project Lincoln had also placed the Whirlwind group in what would now be known as a classic catch-22. Their machine was suddenly in constant demand as a test bed for radar tracking experiments and the like, which meant that they desperately needed to finish development on the more reliable core memory; the endless maintenance and constant crashes had become intolerable. But now they couldn't finish development on the memory, because they had no way to try it out in the computer—Whirlwind was in constant demand as a test bed.

Forrester and his colleagues could see only one solution, albeit a fabulously extravagant one: spend $1 million on a whole new computer just to test core memory. So that was exactly what they did, with the Pentagon's signing the checks. Work on the Memory Test Computer started in May 1952. And a little over a year later, the first eight-thousand-word bank of fully tested core memory was wired into Whirlwind itself. The results were dramatic: the operating speed doubled, the data-input rate quadrupled, and maintenance time was reduced from four hours per day to two hours per week. Memory had gone from being the least reliable component of the computer to being the *most* reliable.

Looking back on it, says Norman Taylor, then the chief engineer of Lincoln Lab's Division Number 6, this story is a prime example of why this enormous and half-panicked project worked as well as it did. For whatever reason—the perceived urgency of the task, perhaps, or the good sense of General Shiely and his oversight team—the researchers had remarkable freedom to make decisions without being second-guessed from the top. They simply paid for the Memory Test Computer out of Division 6's "advanced research" budget, which they could dip into for whatever they considered needful—with no committee meetings, no studying the question to death, and nobody's pointing out a thousand ways they ought to do it differently. "As long as [our decisions] were plausible and could be explained," agreed Forrester, "we could carry other people with us."[10]

Just as important, Forrester adds, they also had the freedom to make mistakes and learn from them: project managers wasted very little time on finger-pointing. "[Mistakes were] admitted and fixed rather than evaded or denied," he says. Perhaps the most spectacular example was recognized one weekend in the autumn of 1953, when the Whirlwind staffers finally had to admit to themselves that they would never be able to deliver a computer as reliable as they had promised the air force. They had actually made phenomenal progress with the vacuum tubes by that point, with the average tube's lifetime now being measured in years instead of days. But even that wasn't good enough, not for a machine whose tubes would number in the tens of thousands: the odds of one's failing at a critical moment were still just too great. So after working and reworking the numbers, the group decided that the only answer was redundancy. Every computer in the air-defense system would have to be paired with an identical backup computer that would be ready to take over at a moment's notice. The implications of this were horrendous: the Pentagon had already penciled in billions for the air-defense system—the exact figure was still unclear—and now the Lincoln Lab computer team would have to tell them to take that figure, whatever it was, and double it.

And so it was that in November 1953, chief engineer Norman Taylor found himself giving a talk on reliability to the Air Defense Command in Colorado Springs and feeling extremely young: "There was what seemed like a whole roomful of generals sitting in front of me—and I hadn't even met a general before that." But they listened politely, he says, and at the cocktail party afterward, "one of the generals whose name I don't remember came up to me and said, 'Norm, that was a very convincing speech. We're going to go with this thing, this duplex.' Just right there, that afternoon. It would take a year to get that decision nowadays."[11]

Meanwhile, say the veterans, still another reason the air-defense effort worked as well as it did was that Lincoln Lab didn't just build a fancy system in the laboratory. The engineers tried to tackle real-world problems from the beginning. Early on in the project, for example, they established a data link with a cluster of long- and short-range radars at South Truro on Cape Cod, so that they could confront Whirlwind with signals from an actual environment. This "Cape Cod System" then became the test bed for the ultimate design. (It was this system that Whirlwind used in December 1953 to track forty-eight aircraft at once—the most impressive demonstration of its life.) As early as July 1952, moreover, long before the final air-defense computers were even designed, Forrester, Everett, Taylor, and Cape Cod System designer Robert Weiser started looking for an industrial partner with hands-on experience in high-tech manufacturing. Time was of the essence, they realized. Once the design work was completed, those assembly lines had to be ready to *go*.

Their choice, announced in October 1952, was IBM—though this wasn't quite the foregone conclusion that it might have been a few years later. Forrester and his team had also seriously considered the office-machine manufacturer Remington Rand, a company that had foreseen the commercial possibilities in computing very early in the game, and then promptly made itself into the market leader by buying out the Eckert-Mauchly Corporation. The company's first UNIVAC 1 had passed a rigorous acceptance test and been delivered to the U.S.

Census Bureau in March 1951. At the time Lincoln Lab went looking for partners, in fact, Remington Rand was still the only commercial computer maker in the world.

Nonetheless, IBM was the biggest data-processing firm in the country, thanks to its sophisticated punch-card tabulators and its office machines. It had the resources, the engineers, and the management to tackle a project of this scale. Its main plant, in Poughkeepsie, New York, had relatively good train connections to the Boston area, a significant consideration at a time when air travel was still something of a luxury. And IBM, however belatedly, had begun its own move into the computer arena.

That move was another reaction to the Cold War, as it happened. Soon after the Korean conflict broke out, in June 1950, IBM sent two scientists on a tour of defense contractors, research institutes, and the military services, asking what the company could do to help the defense effort. James Birkenstock, executive assistant to Thomas J. Watson, Jr., the son and heir apparent of the company's chairman, and mathematician Cuthbert C. Hurd, head of IBM's applied-science department, reported back that the answers kept adding up to the same thing: build computers. The result was IBM's first fully electronic stored-program computer, the Defense Calculator, later renamed the IBM 701. Designed along the lines of von Neumann's machine at the Institute for Advanced Study, the 701 was still under construction when Lincoln Lab came calling in 1952. But interest in the machine was already comparatively brisk, even at the stiff rental price of fifteen thousand dollars per month. (The first one would be installed at Los Alamos in March 1953; a total of nineteen would eventually be produced, with all of them going to aircraft companies, universities, and government laboratories.) That response, in turn, had already encouraged Hurd to start drawing up plans for a midsize business computer that would rent for three to four thousand dollars a month; he figured that he could place at least fifty of them with the military's paper-choked supply services alone.*

So when Forrester and his team made their offer, IBM accepted, though not without some serious reservations. For one thing, managers worried that the air-defense project would drain off much-needed talent and money from other, more lucrative efforts. For another, IBM's commitment to computers was still fairly shaky at that point. Internal opposition to the 701 had been fierce—it had been rammed through the system largely by the boss's son, Tom junior—and resistance to Hurd's business machine was even tougher. The visionaries in the product planning department couldn't see a significant market for computers and wanted to put the company's resources where the real money was: developing better punch-card tabulators.

---

* He was right. By December 1955, a year after the first of these machines was delivered, IBM had installed 120 of them and taken orders for 750 more; the IBM 650, as it came to be called, would eventually become known as the Model T of the computer industry, the first mass-market computer.

Nevertheless, there was a certain element of patriotism involved here, as quaint as that now seems. Even granting that this was a classic chance for the company to do well by doing good—collaborating with Lincoln Lab, after all, meant getting the first crack at that new technology from Whirlwind—there is little doubt that IBM's higher-ups felt the same sense of national peril as everyone else. So in the end, as Forrester would note in 1983, "IBM management really threw their resources into the program without restraint." Indeed, the company actually used its own money to build a factory for the air-defense computers in Kingston, New York—even before the air force had officially signed the contract. IBM likewise committed dozens of its best engineers, setting them up in a former necktie factory on High Street in Poughkeepsie and then allowing them to operate as "Project High," without any of the usual commercial constraints.

After a rocky start, moreover, the IBM engineers and their Lincoln Lab counterparts found themselves sharing a remarkable spirit of camaraderie. Henry Tropp, then a young engineer at Lincoln, remembers the first joint meetings of the two groups in Hartford, Connecticut, starting on January 20, 1953: "The Lincoln people, filled with the hubris of young engineers and fresh from Whirlwind, had the idea that they would design the machine and that IBM would do the production engineering, whatever that was, and build the necessary quantity. The IBM people, also proud and capable, fresh from the 701, and much more knowledgeable about what it took to produce equipment, had the idea that a page or two of specifications was all that Lincoln need supply. The first meetings of these two groups were loud and rancorous. As I look back on them, they were social rather than technical. We argued about everything. IBM used square steel tubing for racks, MIT used L-shaped aluminum. The amount of time spent on this subject was remarkable unless one sees it (as I do now but didn't then) as a process of getting acquainted. After a while, as the two groups began to know and respect each other, the arguments became more cogent and took place between individuals, instead of between organizations. . . . From my point of view it was a fine relationship."[12]

One source of the camaraderie may well have been a shared sense of awe at what they were trying to accomplish. The Semi-Automated Ground Environment, or SAGE, as the overall air-defense architecture came to be known, was immense even in its preliminary version, which was formulated in 1954. And it only got bigger as Lincoln and IBM filled in the details. The plans called for twenty-two "direction centers" in the United States and a twenty-third at North Bay, Ontario, each housed in a windowless, concrete, and hopefully atomic-bomb-proof fortress that would loom over its surroundings like some grim, modern version of the Great Pyramid. Each direction center, in turn, would include communications equipment, air conditioners, electrical generators, battle stations, and two identical copies of probably the largest computer ever built. The SAGE computer design that evolved out of Whirlwind—it would eventually receive the poetic military designation AN/FSQ-7—called for a machine weighing 250 tons. It would contain fifty-five thousand vacuum tubes and be capable of storing up to one million bits of data in its internal and external memory. It would be able to integrate information from as many as one hundred radars and observation stations. It would be able to respond simultaneously to fifty human operators sitting at fifty separate monitors, while tracking up to four hundred airplanes at once. And it would do all this in real time, while the aircraft were moving.

In February 1954, with the design work completed, IBM was officially awarded the first contract to start producing these behemoths. The company's gamble had paid off: its factory would be building SAGE computers for another ten years.

Down at the oversight office in New York City, General Shiely was watching all this in a state of chronic anxiety, certain that the SAGE project was going to fall flat on its face. He tried not to let it show, he would say in 1983. But whenever he was told that Lincoln's Robert Everett or IBM's Robert Crago or any of the other key engineers was coming down, he knew he had to brace himself. "I didn't know what hand grenade someone was going to roll out on the table," he said, "but I knew there was going to be one." The only blessing, he added, was that the big problems tended to come one at a time instead of all at once, so that they had a chance to grapple with each crisis in turn before the next one came along to overwhelm them.

Take software, for example. Lincoln Lab's initial guess for the programming requirements on SAGE—that it would require perhaps a few thousand lines of computer code to run the entire air-defense system—was turning out to be the most laughable underestimate of the whole project. True, the Lincoln Lab team was hardly alone in that regard. Many computer engineers still regarded programming as an afterthought: what could be so hard about writing down a logical sequence of commands? Nonetheless, the Lincoln Lab programmers were being asked to create what would now be called a real-time operating system for the most complex computer/communications system in the world, and they had no modern tools to help them—no Fortran, no Cobol, no Algol; no computer languages, period. All they had was the most basic, hardware-level computerese, alphanumeric codes that corresponded to operations like "Add the contents of register A to the contents of register B and place the results in register C." Anyone who tried to program with such codes quickly discovered that it was terribly easy to make mistakes in even the simplest algorithms. And the SAGE system was anything but simple.

Indeed, it soon became all too clear that the air-defense software was going to need something like two thousand programmers, which presented a problem, to put it mildly. First, MIT had no desire to put so many people on the Lincoln Lab payroll for a single project. One day the air-defense programming would be finished, and *then* what would they do? IBM had much the same reaction, as did Bell Labs, another subcontractor. The upshot was that the programming responsibility, along with many of the original Lincoln Lab programmers, were eventually transferred to Santa Monica and the RAND Corporation's system development division, which in December 1956 would break away and become the independent Systems Development Corporation.

Second, in the early 1950s there probably were no more than a few thousand programmers in the whole country. So the SAGE project soon found itself in the business of mass education. Special programming courses were set up at MIT, IBM, and RAND, and people from every walk of life were invited to enroll. The trainers quickly discovered that it was impossible to predict who their best pupils would be—not even professional mathematicians were a sure bet; they often lost patience with the details—but it was very easy to spot the talented ones once they got started. As a general rule of thumb, for example, music teachers proved to be particularly adept. And much to the project leaders' astonishment (this being the 1950s) women often turned out to be more proficient than men at worrying about the details while simultaneously keeping the big picture in mind. One of the project's best programming groups was 80 percent female.

So month after month, the grenades kept rolling across Shiely's desk, and the problems kept coming. But so did the solutions. In July 1958, after numerous shakedown tests—and some major slips in the software schedule—the first SAGE direction center became operational at McGuire Air Force Base in New Jersey. The rest soon followed; by 1963 all twenty-three centers were up and running, and the system was being tied in to new generations of weapons systems. It would continue to scan the skies for another two decades, until the last six remaining stations were decommissioned in January 1984. And right up to the end, almost a decade into the age of microcomputers, the old vacuum-tube monsters were winning praise for their reliability. Thanks to the Whirlwind-derived fault-checking system, the SAGE computers were out of commission an average of only 3.77 hours per year—or just 0.0433 percent of the time.

Of course, it's hard to say how effective the SAGE system really was in military terms, since it was (fortunately) never used in combat. Arguably, in fact, the SAGE system was obsolete almost from the day it was commissioned, since by that point the United States and the Soviet Union were hard at work on intercontinental ballistic missiles that could deliver a nuclear warhead across the North Pole in under an hour. But in hindsight there is no uncertainty whatsoever about the project's enormous impact on the history of computing. First, it helped catalyze the formation of the Silicon Valley of the East. In the summer of 1953, Lincoln Laboratory moved from the MIT campus into a shiny new complex in suburban Lexington, Massachusetts, not too far from a major ring road around Boston, Route 128. In July 1958, when SAGE was about to come on line and it was time to start integrating new jet fighters and missiles into the system, Lincoln Lab spun off its Division Number 6 as a whole new independent consulting firm just to deal with that task, renaming it the MITRE Corporation. And so it went.

Second, SAGE came along just as computers were starting to move out of the laboratory and into the marketplace, so that all those billions of Pentagon dollars came pouring into the infant industry right at the most critical moment of its development. If nothing else, those cascading billions helped the nation's leading punch-card-tabulator manufacturer transform itself into the world's leading computer company—not least because SAGE left IBM with a cadre of engineers who understood how to tackle big, interconnected, real-time data-processing projects.

Third, SAGE was the pipeline that transported Whirlwind's real-time computing technology into the commercial world. In 1955, for example, long before the SAGE computers themselves were deployed, IBM introduced its model 705, a business-oriented data-processing machine that was the first commercial computer to use magnetic-core memory. It would not be the last: core memory would dominate the industry for another generation, to the point where phrases such as "core dumping" (i.e., printing out the entire contents of a computer's memory for debugging purposes) would become common slang. Indeed, magnetic technology wouldn't give way to semiconductor memory chips until the mid-1970s, by which time the cores themselves would be miniaturized to the size of pepper grains and produced by the billions every month.

Meanwhile, IBM had taken some of its technicians off SAGE and put them to work on a big, interconnected, real-time reservations system for American Airlines. The basic idea was to create a simplified commercial version of SAGE, with a duplex computer at the airline's main computing center north of New York City linked by phone lines to twelve hundred Teletypes all over the country. Ten years and $300 million later, when the reservations system finally became operational, late in 1964, the Semi-Automated Ground Environment had become the Semi-Automated Business-Related Environment, or SABRE. It was the largest real-time commercial data-processing network in the world—but once again, not the last. Just as the Whirlwind group foresaw in 1948, real-time computer applications have since expanded to include air-traffic control, automatic factory management, and a host of other uses.

Finally—and in the long run, perhaps most significantly—SAGE planted the seeds of a truly powerful idea, the notion that humans and computers working together could be far more effective than either working separately. Of course, SAGE by itself didn't get us all the way to the modern idea of personal computers' being used for personal empowerment; the SAGE computers were definitely not "personal," and the controllers could use them only for that one, tightly constrained task of air defense. Nonetheless, it's no coincidence that the basic setup still seems so eerily familiar. An operator watching his CRT display screen, giving commands to a computer via a keyboard and a handheld light gun, and sending data to other computers via a digital communications link: SAGE may not have been the technological ancestor of the modern PC, mouse, and network, but it was definitely their conceptual and spiritual ancestor.

MORTALITY

One thing that SAGE did not do, unfortunately, was halt the sudden, sharp decline of the cybernetics movement.

Of course, it does sound a little odd to talk about the "decline" of a movement whose concepts have become so integrated into our culture that we take them for granted. Norbert Wiener could already feel it happening in 1961, when he fretted in the second edition of *Cybernetics* that the book might seem "trite and commonplace."[13] And these days, some four decades later, it's getting harder and harder for anyone to remember when we *didn't* live in an Information Age. Thanks to microelectronics, we've gotten used to being surrounded by computational devices that can sense their environments and actively respond through feedback, instead of just proceeding blindly. Thanks to digital telecommunications, we're beginning to learn what it means to be immersed in a sea of information—also known as cyberspace, courtesy of Wiener. On the scientific front, meanwhile, it now seems natural to hear molecular geneticists describe DNA as a molecule that encodes information—namely, the blueprint for building a new organism. It seems just as natural to hear neuroscientists, cognitive psychologists, and artificial-intelligence researchers define the mind and brain in terms of information processing (and in some cases, in terms of neural networks) or to hear computer scientists talking about information flow, Turing machines, and chip-level logic gates.

So, yes: as a set of concepts, the cybernetics movement is still very much alive, and we are all members. We hear the voices of Wiener, Shannon, Turing, von Neumann, McCulloch, and Pitts from every side. But as a single, unified new science that encompasses both "the animal and the machine"? No. That dream of unity was already failing by the early 1950s, when Norbert Wiener's supper seminars and the Macy meetings both came to an end. And by mid-decade, with the movement increasingly left to third-raters and crackpots, it was effectively dead.

What happened? The SAGE project, for one thing: it landed right at the movement's nerve center, RLE, and ended up diverting many of the young engineers who might otherwise have been eager participants in cybernetics. After all, noted Bill McGill, "these man-machine ideas didn't seem so important when you were talking about the annihilation of the human race." Even worse, perhaps, SAGE helped deflect the attention of lab director Jerry Wiesner, the impresario who had done so much to *make* RLE the nerve center. Convinced that our ever-escalating arms race with the Soviet Union was fast becoming a mutual suicide pact, Wiesner was now refocusing his energies on the quest for disarmament and/or arms control (he would later serve in the White House as President Kennedy's science adviser).

Another factor in the decline, sadly, was Norbert Wiener himself. "He had a tragic falling-out with many people in the fifties," says his former assistant Oliver Selfridge. "Norbert could take offense at imaginary things. I don't know what drove him. But once the offense was there, it could not be erased. He fell out with Warren McCulloch, I'm really not sure why. Then he broke with Walter Pitts because Walter was working with Warren. Then he feuded with me because I was working at Lincoln Lab, supported by the air force." In his 1956 autobiography, *I Am a Mathematician*, Wiener managed to write at length about the origins of cybernetics without ever mentioning McCulloch and Pitts—or Selfridge, for that matter. His fellowship of the new science had long since become a shambles.

Added to that, Wiener had largely ceased doing original mathematics research. By modern standards, he wasn't really an old man: he celebrated his sixtieth birthday in 1954. But he was almost blind after a battle with cataracts. And he was increasingly preoccupied with the human impact of modern technology, as well as the responsibility that he felt scientists had to assume for that impact. He would continue to speak out on these issues. In *God and Golem, Inc.*, published in 1964, the year of his death, he would even discuss the philosophical linkage between cybernetics and religion. But he would never again be the unifying, catalytic figure he had once been.

And that, even more sadly, was yet another factor in the decline of cybernetics: mortality. The giants of the heroic age were passing from the scene, including two of the greatest, Alan Turing and John von Neumann.

Turing, of course, had spent much of his time since the war trying to create his abstract machine "in the metal," first with his designs for the ACE computer at the National Physical Laboratory and then, starting in 1948, with his work as chief programmer for the Mark I computer project at Manchester University. In parallel, however—and to Turing's way of thinking, far more important—he was also continuing his struggle to understand the fundamental nature of intelligence.

That effort culminated in 1950 with his paper "Computing Machinery and Intelligence,"[14] in which he addressed the fundamental question: Can a machine think? Instead of trying to answer that directly, however—an exercise that had already generated entirely too much philosophical hot air for his taste—he parsed it into two questions that were even more elemental. First, What do we mean by a "machine"? And second, What do we mean by "think"?

Turing's answer to the first question was not too surprising: by "machine" he meant "a digital computer." This in itself was no real restriction, he argued, since a digital computer was *universal* and could simulate any other machine—including, presumably, the human mind. His answer to the second question, however, was vintage Turing: idiosyncratic, startling, and yet utterly logical. "Thinking," he declared, could be defined via his now-famous Turing test, a kind of party-game affair in which a human and a computer are hidden from view and answer queries posed by an interrogator who is trying to determine which is which. Thus

> Q: Please write me a sonnet on the subject of the Forth Bridge.
> A: Count me out on this one. I never could write poetry.
> Q: Add 34957 to 70764.
> A: (Pause about 30 seconds and then give as answer) 105621.*
> Q: Do you play chess?
> A: Yes.
> Q: I have K at my K1, and no other pieces. You have only K at K6
>   and R at R1. It is your move. What do you play?
> A: (After a pause of 15 seconds) R–R8 mate.[15]

and so on. Turing argued that if the interrogator *could not* tell, no matter how many questions he or she asked, then one had to admit that the machine was really thinking. After all, he noted, at that point the interrogator would have precisely as much evidence for the computer's thinking ability as for the human's.

Viewed in retrospect, Turing's test for machine intelligence has to rank as one of the most provocative assertions in all of modern science. To this day, people are still talking about it, writing commentaries on it, and voicing outraged objections to it (most of which he anticipated in his original paper, by the way).[†] Of course, like so much of Turing's work, the 1950 paper wasn't widely read at the time, and it had essentially no impact on the artificial-intelligence research that was just beginning in the United States. But then, Turing didn't really seem to care about any of that. By 1951, having said all he wanted to say about machine intelligence, he was already immersed in a brand-new interest: biological growth and form. How did an undifferentiated blob of cells turn itself into an intricately structured sea urchin, say, or a butterfly, or a human being? Being Turing, he once again took his own, highly original approach to the problem. Focusing first on the formation of patterns, such as the spots on a butterfly's wing or the scrollwork on a seashell, he showed that those patterns were exactly what one would expect if cell development was directed by "morphogens," a hypothetical

---

* The correct answer, of course, is *105721*. Presumably the number given here is either a typographical error or Turing's little joke.

† Turing listed nine objections: *theological* (computers have no souls); *head-in-the-sand* (the consequences of thinking computers are too dreadful to think about); *mathematical* (computers are subject to Gödel's theorem, but humans aren't); *consciousness* (computers don't have it); *computers-can't-do-X* (where X = learn from experience, fall in love, laugh, enjoy strawberries in cream, etc.); *computers only do what you tell them to do* (humans can take initiative); *computers are digital* (humans are analog and continuous); *computers are bound by definite sets of rules* (humans aren't); and *computers don't have ESP* (humans do—maybe). He answered each of these objections, with greater or lesser success. But it's interesting that he seemed genuinely troubled only by the last one. Purported laboratory evidence for extrasensory perception was taken more seriously in 1950 than it is today—at least by mainstream scientists.

family of compounds whose reaction rates varied radically with concentration. After a full-scale commercial version of the Mark I computer started operation in Manchester in July 1951, moreover, Turing demonstrated that such "nonlinear" patterns could be simulated on a computer screen, in one of the first visual computer simulations in history. Of course, it's still not completely clear whether Turing's morphogen model holds true for real embryos (though in some ways it seems pretty close to the mark). In mathematical terms, however, his idea was remarkably prescient: his analysis of nonlinear chemical equations, first published in December 1951, anticipated by more than a decade the modern field of nonlinear systems dynamics, better known as chaos theory.

Unfortunately, Turing would have little opportunity to pursue these ideas in peace. Just one month later, in January 1952, he was arrested in Manchester and charged with three acts of consensual sex with a teenage boy. His trial that spring resulted in a quick conviction; homosexual acts were still a crime in the United Kingdom, and in any case, the boy was underage. Because of the importance of his work with the Manchester computer project, though, Turing was offered an alternative to jail: a one-year course of estrogen treatments, which, according to the medical thinking of the day, would reduce his supposedly perverted libido. Turing agreed to the treatments. Back at the computing laboratory, he did his best to make a joke of the whole experience, including the fact that the estrogen was making his breasts develop. Moreover, he refused to show any remorse or shame over his actions; he had become increasingly open about and even proud of his homosexuality since the late 1940s and maintained that it was the law that was wrong. Nonetheless, he seems secretly to have been deeply humiliated by the experience. He was also bitter over the resulting termination of his security clearance, which cut him off from any further research on cryptography (he had continued to do part-time government work in this field ever since the war). And he was further outraged when, in 1953, British security forces hounded a Norwegian friend who was simply trying to visit him.

Then, on June 8, 1954, at five o'clock in the afternoon, Turing's cleaning lady entered his house near Manchester, England, and found him lying neatly on his bed, stone cold. The police determined that he had been dead for about a day. There was no note, but since the autopsy revealed that his body contained cyanide, and since there was a bottle of potassium cyanide in his home chemical laboratory—which he'd used in electroplating experiments—Turing's death was ruled a suicide. A half-eaten apple that was found on the floor by the bed was not tested, but the presumption was that Turing had dipped it in the poison before taking a bite.

Whether it *was* suicide, however, we will never know. His mother would always stoutly maintain that her son's death had been accidental, the tragic outcome of an unnoticed spill during a chemical experiment, combined with Turing's pensive habit of putting his fingers in his mouth. And she had a point. From all accounts, Turing had been in good spirits during the preceding weeks.

His estrogen treatments had ended more than a year earlier. He had found acceptance among his colleagues on the computer project, and his position there was secure. And he was enthusiastic about his research: at the time of his death, he was extending his biological models to cover spherical organisms, such as the radiolarians, and cylindrical objects, such as plant stems.

In short, Alan Turing had every reason to live in June 1954, and in any event, he gave no indication that he was contemplating suicide. About the only thing that will ever be clear about his death is that it silenced one of the most original minds of the twentieth century, at age forty-one.

In the latter part of that same year, as it happened, John von Neumann's scientific work also came to an abrupt end, albeit in a seemingly less drastic fashion. In late 1954 he was offered a seat on the Atomic Energy Commission—at that time the highest official position in the U.S. government available to a scientist.

Stanislaw Ulam remembers his friend's agonizing over whether to accept. Like virtually everyone else in the U.S. scientific community, von Neumann had been repulsed by the AEC hearings in April and May of 1954, when the rabidly anti-Communist AEC chairman Lewis Strauss had had J. Robert Oppenheimer's security clearance revoked to punish him for advocating a go-slow approach to the development of the hydrogen bomb. As von Neumann and many others had vigorously testified at the hearings, a political and technical disagreement was hardly the same thing as treason. On a personal level, too, von Neumann's loyalty lay with Oppenheimer, who had been director of the Institute for Advanced Study since 1947 and had strongly supported his computer project. Nevertheless, wrote Ulam, von Neumann "was flattered and proud that although foreign born he would be entrusted with a high government position of great potential influence in directing large areas of technology and science. He knew this could be an activity of great national importance."[16]

So after many sleepless nights, he accepted the AEC's offer. He was confirmed by the Senate on January 10, 1955, and moved with his family to Washington, D.C., that spring, setting aside all his research and consulting activities. For the duration of his five-year term, he planned to focus his prodigious energy almost exclusively on the business of the AEC.

But even that was not to be. Just a few months later, in the summer of 1955, von Neumann slipped in the corridor of an office building and bumped his left shoulder painfully. *Very* painfully; the soreness refused to go away. That August, the diagnosis came back that this was no bruise: von Neumann had bone cancer. Metastasized. Inoperable. Terminal.

His initial reaction was denial. He threw himself into his AEC duties more energetically than ever. He likewise started drafting what would be his last sustained intellectual effort: the Silliman Lectures, a weeklong series of talks that he'd been invited to give at Yale University the following year, during the spring term of 1956. "The Computer and the Brain," as he called the series, was to be a synthesis and an extension of all he had done so far in his General and Logical Theory of Automata—or, as he described it, "an approach toward an understanding of the nervous system from the mathematician's point of view." He was even able to write parts of two lectures. And yet there inevitably came a time when even John von Neumann couldn't keep up the pace. His customary four to five hours of sleep per night no longer sufficed. His energy flagged. He underwent surgery, but to no avail: the cancer continued its spread. He dropped all his other research efforts to concentrate on the Silliman Lectures, even after he was confined to a wheelchair in the spring of 1956. This was to be his legacy. Yet he found it increasingly impossible to concentrate. Assaulted by a steady drumbeat of pain, even his incomparable mind was starting to fail him. In April 1956 von Neumann entered Walter Reed Army Hospital in Washington for the last time.

He seemed inordinately terrified of death. "When von Neumann realized that he was incurably ill," wrote his childhood friend and lifelong colleague Eugene Wigner, "his logic forced him to realize also that he would cease to exist, and hence cease to have thoughts. Yet this is a conclusion the full content of which is incomprehensible to the human intellect and which, therefore, horrified him."[17] To the astonishment of everyone who knew him, von Neumann the lifelong agnostic now sought spiritual counsel—not from a rabbi but from Father Anselm Strittmatter, who began visiting him regularly in the spring of 1956 to instruct him in the Catholic faith. Perhaps it helped. But even if it did, it was not enough to stave off the panic attacks or the screams of terror in the night. "I think that von Neumann suffered more when his mind would no longer function, than I have ever seen any human being suffer," said Edward Teller.[18]

For all of that, however, von Neumann remained a member of the AEC until the last. Indeed, the dying man was so central to the nation's nuclear-weapons program that he could be attended only by air force orderlies with top-secret security clearance, as there was considerable concern that his pain and mental distraction might lead him to babble classified information. Years later AEC chairman Lewis Strauss told of one final meeting at Walter Reed Hospital near the end: "Gathered around his bedside and attentive to his last words of advice and wisdom were the Secretary of Defense and his Deputies, the Secretaries of the Army, Navy and Air Force, and all the military Chiefs of Staff. . . . I have never witnessed a more dramatic scene or a more moving tribute to a great intelligence."[19]

John von Neumann died on February 8, 1957. He was fifty-three years old.

WHAT MACHINES DO, AND WHAT HUMANS DO

One other thing that the SAGE project did not accomplish, unfortunately, was to save J. C. R. Licklider's psychology group. Indeed, the mortal blow had been struck all the way back in 1953. Lick, as promised, had been granted tenure at MIT. But as far as Dean John Burchard was concerned, that was that. It was bad enough making a lifetime commitment to just one of those soft-money Lincoln Lab types; he wasn't about to do it for the lot of them. "Thank God," Burchard remarked with at least one person in earshot. "That's the last appointment we'll have to make in *that* field!"

When this comment got back to Lick, he was aghast. He felt humiliated and betrayed, not to mention enraged. He and his group had all the money they needed, having gone out and raised it on their own. They had the students, some of whom were already finishing up their dissertations and expecting Ph.D. degrees. They had the enthusiastic support of people such as Jerry Wiesner. They had the endorsement of Smitty Stevens and many other top rank experimental psychologists, who had been telling Burchard in person that in them he had the makings of a great psychology department. And Burchard just couldn't *see* it.

Lick was still furious when he broke the news to Miller. And Miller, equally appalled, immediately understood the significance of Burchard's remark: their attempt to build a department on the sly had backfired, and badly. "It signaled to me that there wasn't going to *be* a department," says Miller. And without a department, much less a chance at tenure, neither he nor anyone else in the group would have a long-term future at MIT.

That bleak assessment proved to be dead on target, especially after Burchard found out that the psychology section was on the verge of granting Ph.D.s to several students on its own authority. No more students, he decreed, and once the current crop was through, no more Ph.D.s. Lick was helpless; he had to just watch the whole thing fall apart.

Miller was the first to leave: about a year later, when Harvard offered him a tenured position that had newly opened up, he took it. He had long since gotten over his Harvarditis, he explains: "I loved MIT. I thought it was a marvelous place." But if he stayed there, he realized, his career would be confined to Lincoln Lab. And by that point, with the SAGE project moving inexorably from research toward production, it was clear that a career at Lincoln meant a career increasingly devoted to applied, defense-related work, with very little time left over in which to study the fundamental nature of language and cognition. So, looking back wistfully, Miller returned to the Memorial Hall basement in February 1955.

Then it was Bill McGill's turn, when he got an offer from Columbia University about a year after Miller left. "When I said I was going," McGill later recalled, "Julius Stratton, who was the MIT provost then, called me into his office and asked me, 'Why are all the psychologists leaving?' I told him about Burchard. Stratton told me that the real problem was that Lick was a lousy administrator, that the institute couldn't figure out where the psychology group was going, and so forth. I told him that that wasn't the problem. If the group had been allowed to grow, we would have stayed."

Stratton was unconvinced, however, and so McGill went on his way, eventually carving out quite a career for himself. He became chairman of the Columbia psychology department in 1961; cofounder of the psychology department at the University of California's newly established San Diego campus in 1965; chancellor of UC San Diego in 1968; president of Columbia in 1970; and finally, in 1982, cofounder of UC San Diego's cognitive science department, where he continued to work out of his office in William J. McGill Hall until he was well into his seventies. And yet for all of that, he said in an interview shortly before his death, in 1997, he still grieved for that golden moment at MIT. "Without any doubt," he explained, "this group, although not so labeled, was the first department of cognitive science in the country. And it was so original in concept—because, of course, Lick was its dynamism and its leader—that if MIT had had the wisdom to stick with this man, and to understand the qualities of mind that he had, they would have had a twenty-year lead on the development of cognitive science. As it was, all these people disseminated to other parts of the country. And in each instance, so help me God—it was true of Miller, and it's certainly true of me—they tried to re-create, in the other places they were, the fervor and excitement of this group."

Miller chuckles a bit when told of that comment. "I always thought I was trying to re-create the Psycho-Acoustics Laboratory!" he says. But he concedes the point: "I guess whatever laboratory it is that first gets under your skin is special, like a first love."

Miller, of course, would go on to make quite a career of his own. Unlike Lick, who tended to ignore the behaviorists who then dominated psychology (when he wasn't laughing at them), Miller had long since crossed the line into open rebellion. And he was winning.

He wasn't alone. That introductory article he'd published with Fred Frick in 1949 had made information theory into something of a fad in psychology, especially among the rising generation. And as these young researchers applied the theory to more and more aspects of human perception, they found more and more evidence for the same kind of "channel capacity" that Miller had found in the perception of words. Their experimental subjects could distinguish very well between different musical pitches, say, or different positions of points on a line, or even different levels of saltiness in a taste of water—*if* there were only two alternatives. Salty-not salty and so on were really just yes-no choices, meaning that the subjects had to perceive only one bit of information. But as the number of alternatives increased, the subjects inevitably began to falter and make mistakes at the level of roughly seven choices, or slightly less than three bits of information.*

The clear implication was that perception didn't just *happen*, as the behaviorists would have it. Perception was a real physical process with capacities and limits, in precisely the same way a telephone line had limits. Why this limit should be seven alternatives as opposed to, say, three, or nineteen, was a mys-

---

* Strictly speaking, this limit applies only to high-level, conscious perception. Unconscious neural processes such as perception handle vastly more information, with the retina alone taking in visual information at a rate measured in billions of bits per second.

tery (and still is). But the number turned up so consistently that in 1956, when Miller reviewed the evidence for human information-processing limits, he would entitle his article "The Magical Number Seven, Plus or Minus Two"—and begin it with one of the most memorable laments in the scientific literature: "My problem is that I have been persecuted by an integer. For seven years this number has followed me around, has intruded in my most private data, and has assaulted me from the pages of our most public journals. . . . There is, to quote a famous senator [the rabidly anti-Communist Joseph McCarthy], a design behind [the persistence of this number], some pattern governing its appearances."[20]

Or was there? He'd had something of a scare in 1953 or 1954, says Miller. "One of our graduate students, Dick Hayes, who is now at Carnegie Mellon University, was working with Lick to apply information theory to memory. One day he went in to see Lick in total frustration and said, 'Information theory is no damn good! It doesn't work!' Well, that challenged me. Another student and I began playing with it—and we decided that Dick was right!"

The type of memory in question was "short-term" or "working" memory, the mental scratch pad where we keep the images and concepts that we're focusing on at any given moment. As the name suggests, says Miller, the contents of working memory are volatile in the extreme: look up a telephone number, for example, and your memory of it evaporates almost as soon as you've finished dialing (if not before). But much more interesting for his purposes, he says, was that the capacity of working memory is very limited: like the perceptual parts of the brain, it can handle no more than about seven pieces of data at a time.* That's why a seven-digit telephone number is easy enough to remember, while longer sequences are much harder.

Miller had tacitly assumed that this was just another example of the magical number 7. As he investigated, however, he quickly verified what Hayes had found—namely, that what limits the capacity of short-term memory is not just the *amount* of information but the *kind* of information. For example, if you try to remember a list of random words, you do indeed begin to falter at about seven. But if those words form a meaningful sentence, then you can repeat them verbatim out to about sixteen words. "So information wasn't a constant in memory," says Miller. "Instead, we cooked up this notion that the amount of information you can hold is measured in 'chunks,' " or meaningful clusters of items that you can remember or units. That's why sentences are easier to remember than words listed at random: you can remember them as phrases, or chunks of meaning. And it's why a twelve-digit sequence such as 149217761066 is very dif-

---

* The brain also has a totally separate, *long*-term memory, which is what we usually mean by the word *memory*. This type of storage is effectively infinite in capacity, easily capable of holding the experiences of a lifetime and retrieving any one of them in an instant. The brain mechanisms underlying these two forms of memory are still something of a mystery and are among the most active research areas in modern neuroscience.

ficult to remember until you see it as three famous dates—1492 + 1776 + 1066—at which point it suddenly becomes trivial.

In short, says Miller, "the scare" led to an even deeper confirmation of the phenomenon: the magical number was 7, all right, but seven *chunks*, not just seven *items*. Moreover, he says, this recognition of chunking was what finally led him to make an open break with Skinner and company. Not only did the data prove the existence of mental states—namely, concepts in memory—but they showed that these mental states have *structure*. Indeed, says Miller, chunking implied that our minds are capable of organizing whole hierarchies of data: each chunk in short-term memory can hold several pieces of information and perhaps several other chunks; these in turn can point to yet more information and yet more chunks, and so on. Suddenly, he says, the "black box" of the mind seemed filled with light.

Still, admits Miller, for all of that, he retained a funny blind spot: he'd been so intent on measuring information *flow* that it hadn't yet occurred to him to think about information *processing*—the notion that perception, problem solving, memory retrieval, and every other mental function could be understood as types of computation. However, says Miller, he can definitely remember when and how the computational idea at last began to sink in. It was sometime in the summer of 1956, he recalls, and a former colleague, Walter Rosenblith, was insisting that there was this wonderfully bright guy over in MIT's linguistics group whom he should know about—one of the people Jerry Wiesner had just brought in for an RLE project on machine translation. His name was Noam Chomsky.

Never heard of him. "So I kept saying, 'Sure, sure,' and putting it off," Miller says with a laugh. But Rosenblith was persistent. So eventually, remembers Miller, he and his Harvard colleagues invited Chomsky over to give a talk.

They didn't regret it.

Chomsky, it turned out, had been staging his own revolt against behaviorism, and doing so with all the intellectual ferocity that he would later make famous. Brilliant, intense, and blessed with extraordinary abilities in mathematics and symbolic logic, the twenty-eight-year-old Philadelphia native was a near unknown in 1956. But even then he relished the chance to take on conventional wisdom and demolish it. In this particular case, Chomsky was arguing that the behaviorist account of human language wasn't just wrong, it was ludicrous. Moreover, he could prove that assertion mathematically.

In truth, language had always been the ultimate hurdle for behaviorism. How could you avoid talking about "mental states" such as ideas, images, feelings, and intentions when the whole point of language was to *communicate* those states? The behaviorist answer—naturally—was that such apparent mental states were illusory. We produced certain sequences of words only because we'd learned associations between individual words: *John* had a certain probability of being followed by *kissed*, which had a certain probability of being followed by *Mary*, and so on. Thus *John kissed Mary*. Indeed, B. F. Skinner himself was just finishing up a major book on the subject and had declared that a behaviorist account of language would be the culmination of his life's work.

Chomsky was contemptuous. "I thought that [Skinner's theory] was completely wrong from every point of view," he would say many years later. "You could ask a question that for some reason investigators rarely asked: namely, can the proposed model *even in principle* account for the known facts? As soon as you ask that it becomes clear that [the behaviorist model] can't possibly account for the known facts about language."[21]

First, Chomsky pointed out, human language is not something that can be learned word-for-word· assuming ordinary intelligence, any one of us can produce and understand vastly more sentences than there are seconds in a human lifetime.* Instead, Chomsky argued, we must be *generating* our utterances, using some simple set of rules analogous to the rules of algebra, or to a computer program. (One of his more provocative claims was that some generalized form of these rules must be genetically hard-wired into our brains; otherwise, he asserted, babies would never be able to learn to talk.) Chomsky called such a set of rules a grammar, and declared that the task of linguistics was to discover those rules for each language that can generate all the grammatical sentences in that language and *only* those sentences.

Second, said Chomsky, sentences aren't just arbitrary strings of words. They have an obvious structure, consisting of noun phrases, verb phrases, and prepositional phrases—all the components that high school students study when they learn how to "diagram" sentences. And the behaviorist model couldn't even begin to account for that structure.

Why not? Well, consider a given set of grammatical rules, said Chomsky. Viewing them in the abstract, you can think of them as collectively forming a "machine"—an automaton that generates word strings in much the same way that an abstract Turing machine computes numbers. And once you do that, he said, you can ask essentially the same question that Alan Turing asked of *his* imaginary machine: What can this grammatical machine actually compute? What kinds of word strings can it produce, and how complex can those word strings be, even in principle?

The answer to that last question was, Not very complex at all, if the grammatical rules correspond to a behaviorist word-association model. In purely computational terms, such a language generator is about as feeble as a grammar machine can get. Indeed, Chomsky showed that grammar machines fall into a

---

* To learn how to produce all possible grammatical sentences and *only* those sentences via behaviorist-style word associations, went the argument, you would have to listen to each one at least once. But even if you confined yourself to grammatical sentences less than twenty words long, or about the average for *Reader's Digest* articles, and even if you listened nonstop at the rate of one sentence per second, it would take you roughly thirty-two trillion years to hear them all.

hierarchy, with each level of the hierarchy's being more computationally powerful than the levels below it—and with simple word-association machines at the
bottom. These latter can produce word strings, but not much else. To generate
the phrase structure of standard English (or of any other human language),
Chomsky demonstrated, you need a formalized grammar from one of the more
powerful levels. And to enable transformations between sentence structures—the
sort of thing we do all the time when we go from, say, active voice *(John kissed
Mary)* to passive *(Mary was kissed by John)* or to a question *(Whom did John kiss?)*—
you need a grammar from the most powerful class of all, the one that is mathematically equivalent to a Turing machine.

To put it another way, the very fact that we human beings use language in the
way we do is proof that, in some sense, our *brains* have the computational power
of a Turing machine. Or to express it still another way, the pinnacle of all possible mathematical machines—the Turing machine—is also the baseline, the minimum needed for human cognition. Anybody who seriously wants to understand
the workings of the mind had better start from there, because nothing less will do.

Today, of course, Chomsky's proof that computational machines form a hierarchy is considered a major refinement of Turing's original insight, and one of
the foundation stones of modern computer science. But to Chomsky himself
that was just a pleasant side effect. His own goal was to overturn a linguistics establishment that he regarded as stuffy and unimaginative. And he did: starting
in 1957, when some of his lecture notes were published as the book *Syntactic
Structures,* his notions of transformational grammar and mathematical linguistics
swept the field, to the point where Chomsky himself would soon become the
new establishment. Along the way, moreover, he also did a great deal to help undermine behaviorism. When Skinner's book on language came out in that same
year of 1957, for example, Chomsky wrote a review that turned out to be more
influential than the book itself, pointing out the inadequacies of Skinner's account in withering detail.

Certainly the implications of Chomsky's work were clear enough to George
Miller at their first meeting, in 1956. Miller's own work on chunking had already
shown him that the mind organized concepts into a hierarchy. "But Chomsky's
model seemed to make that idea much more general," he says. Indeed, his meeting with Chomsky was as pivotal as the moment eight years earlier when he'd
picked up that July 1948 issue of the *Bell Systems Technical Journal* containing
Claude Shannon's article—and all the more so because Chomsky's message was
to be strongly reinforced just a short time later.

Miller remembers the day very clearly: Tuesday, September 11, 1956, the second
day of the second international conference on information theory. Actually, the
whole conference was good. Held in MIT's Sloan Building, right on the river-
front, it included talks by Jerry Wiesner, Bob Fano, Peter Elias, Oliver Selfridge,
Walter Rosenblith, and even Shannon himself. But that second day was what

Miller had really been looking forward to. He was scheduled to give a talk about his work on human memory limits, and Chomsky was likewise slated to make one of the first public presentations of his abstract-grammar work. Nathaniel Rochester from IBM was going to describe a simulation of neural networks that he and his colleagues had done on the big new IBM 704 computer. And in the lead-off slot, first thing in the morning, the program listed "The Logic Theory Machine," a Ph.D. thesis project by a young fellow from the Carnegie Institute of Technology in Pittsburgh: Allen Newell.

Miller knew Newell slightly, as it happened. He had been a mathematics grad student at Princeton back in the fall of 1949, when Miller had spent a sabbatical year there trying to learn enough math to understand information theory. Newell was a tall, beefy man, a San Francisco native with a beaming, exuberant grin and a rapidly thinning head of pale, pale blond hair. He had gotten a physics degree from Stanford after the war and had then come east to get his Ph.D. A heck of a nice guy, Miller had thought. But then, after a semester or so, Newell had just . . . disappeared.

Miller didn't have a clue what had happened to him until several years later, when his own SAGE work led him to visit Santa Monica, where the RAND Corporation had built a simulated air-defense center to study the social and psychological aspects of the operation. "And when I arrived," he says, "there was Allen with his usual muscling enthusiasm."

Math, it appeared, had been too arcane for Newell's taste; he wanted problems he could get his hands on, problems he could *solve*. And after dropping out of Princeton, he'd found exactly what he had in mind at RAND. Not only had he been a prime mover in planning the air-defense simulator, itself a multi-million-dollar effort, but he had become the project's resident computer wizard—the computer in this case being an old card-programmed calculator used to print out the simulated radar maps on big sheets of fan-fold paper. This task, in turn, had led Newell to another discovery, in the person of the Carnegie Tech economist Herbert A. Simon, who regularly came out to Santa Monica as a summer consultant on that same air defense study and who would eventually become Newell's major professor and collaborator on the Logic Theory Machine. Simon had been captivated by Newell's computer setup the first time he saw it. "What was remarkable about this application," he would later write, "was that the computer was being used not to generate numbers, but locations—points—on a two-dimensional map. Computers, then, were not merely number-crunchers, they were general symbol manipulators, capable of processing symbols of any kind—numerical or not!" Newell, conversely, had already begun to notice some striking similarities between Simon's specialty—the way humans organized their work in the center—and the way data were processed in his computers: both activities had a critical dependence on information flow, and on how that information was used to make decisions. Personally, too, the two men had hit it off immediately, so much so that Newell's wife, Noël, listening to their endless happy arguments about human information processing, was reminded of two dogs playing in the park: wrestling, snarling, chasing each other around, and generally having a wonderful time.

Simon himself was only forty years old in 1956, but he already had a formidable reputation in the social and behavioral sciences. At the University of Chicago, right after the war, he had been one of a group of young rebels who were reinventing the whole field of economics, trying to bring mathematical rigor to a discipline that had until then been based largely on anecdotes and history. In 1947, moreover, Simon's book *Administrative Behavior* had been hailed as a classic in the newly emerging field of management and organizational decision making. And in 1949, he had helped found Carnegie Tech's new Graduate School of Industrial Administration, which was fast building a national name for science-based professionalism in business education.

Simon was also renowned as a man who liked nothing better than a good intellectual tussle—and who was so fiercely intelligent that he usually won. Born into a German-Jewish family in Milwaukee, he was short, rather stout, and supremely self-confident, feeling no compunction whatsoever about critiquing the emperor's new clothes. When it came to human behavior in particular, Simon could be just as scathing about behaviorist ideology as Miller or Chomsky. After all, he pointed out, it was hard to see how human beings could build steel mills or manufacture TV sets unless somebody somewhere had a mental image of what he or she was trying to accomplish.

Simon was even more scornful about the ideology in his own field. Whereas Skinner and company had been trying to deny the existence of the mind entirely, the postwar economists had been swinging to the other extreme, describing "economic man" as a godlike intellect blessed with perfect knowledge of the world around him, perfect knowledge of what he wanted, and perfect knowledge of how to achieve his desires in any conceivable situation.

Bull. Simon had spent enough time in real organizations to know that the world was a chaotic, unpredictable place and that human beings were a very long way from being perfectly rational. Our "bounded rationality," as he put it in *Administrative Behavior,* constrains and shapes everything we do. When you go out shopping for, say, a new pair of shoes, do you systematically visit every shoe store in town to find the lowest possible price? Probably not; the savings just aren't worth it. The majority of us simply visit one or two convenient stores and settle for the most satisfactory buy we can find. Simon coined a new word for this kind of behavior—*satisficing*—and suggested that it was ubiquitous in human decision making, even at the highest levels of government and business. Indeed, this is why very smart people will sometimes arrive at disastrously wrong decisions (for a later generation, the Vietnam war comes to mind). But trying to avoid every conceivable error is a recipe for paralysis: having the freedom to make mistakes is necessary for achieving any progress at all.

This kind of analysis was almost unheard of in 1947, which is why *Administrative Behavior* was considered so incisive when it was published, and why it would be among the works cited by the Swedish Academy of Sciences when Simon was awarded the 1978 Nobel Prize in economics. Viewed in retrospect, moreover, Simon's thinking on bounded rationality and "satisficing" may be seen as a pioneering effort to grapple with the same kind of cognitive limitations that George Miller would later find in his work on memory.

Still, Simon was somewhat frustrated by the early 1950s. Bounded rationality was a negative definition, he knew. It told us what human behavior is *not*—namely, perfectly rational—but not what human behavior *is*. If you really want to understand what people do, Simon reasoned, you need a language you can use to formulate rigorous theories about behavior, in much the same way Galileo and Newton used mathematics to analyze the motion of physical objects. Obviously, mathematics itself wasn't the right language for *this* application: people are far too complicated to be described by mathematical variables (Let X = motivation?). But as for what to put in its place, well, that was the significance of the computer program that Simon and Newell called the Logic Theory Machine, or Logic Theorist. They hadn't created it to prove that a computer could "think" (though it would later be hailed as one of the first and most influential examples of artificial intelligence); rather, they had created it to show that the proper analytical language for describing human thought and human behavior was computation.

As the name indicated, Newell and Simon's program proved theorems in symbolic logic, a task they chose partly because the material was ready at hand—Simon happened to have in his home library a copy of *Principia Mathematica*, Bertrand Russell and Alfred North Whitehead's classic treatise on logic and the foundations of mathematics—but mostly because it seemed to be such a perfect model of bounded rationality in general. Basically, they knew, a proof in logic was just like the proofs that every new generation of high school students had to struggle through in algebra or geometry. The starting point was a set of symbolic statements known as the premises, and the goal was another symbolic statement known as the theorem. To get from one to the other, you just had to combine and modify the premises by using the rules of inference (of which there were about half a dozen). In time, if the rules were applied in the right order, the manipulations would produce the theorem. The trick was to figure out that right order. Although the proof in its final form was obviously "logical," the process of discovering it was not at all logical. Ask mathematicians how they did it, and they would likely describe a process of befuddled groping as they went through days, weeks, even years of trial and error—followed (sometimes) by flashes of insight and intuition. In short, they would describe bounded rationality in action.

Indeed, Newell and Simon realized, in a well-defined domain such as logic-theorem proving, you could even show why rationality *had* to be bounded, quite aside from the limits on short-term memory or anything else. Imagine for a moment that you wanted Logic Theorist to prove its theorems by brute force. The method sounds easy enough: just start from the given premises and have the program systematically apply every inference rule until the correct solution is found. And in fact, that approach is guaranteed to produce the theorem—

eventually. It's an example of what is known in the trade as an algorithm, a well-defined procedure that takes you step by step from the input data to the final answer. (The classic analogy is a cookbook recipe: take so many cups of this and so many tablespoons of that, mix it all together, shake it, bake it, and voilà!) Algorithms, of course, are usually seen as the essence of computation. The whole point of a computer is to execute such step-by-step procedures very, very fast—which is precisely why the machines are so useful for grinding out scientific calculations, generating corporate payrolls, and doing anything else that can be formulated as an algorithm.

The problem in logic-theorem proving, however, is that "eventually" may not come for a very, very, very long time. Newell and Simon estimated how many proofs the algorithm would have to generate before it found all sixty-odd theorems contained in chapter 2 of the *Principia*, and got an answer—"almost certainly greater than 10^{1000}"—that was a number so vast as to defy all metaphor. There is no computer that could generate so many proofs. There is no way even to *conceive* of a computer that could do it. There haven't been that many microseconds since the Big Bang. There aren't that many atoms in the observable universe. Whatever mathematicians (and high school students) were doing when they proved theorems, Newell and Simon concluded, it obviously wasn't this.[22]

This "combinatoric explosion" of insanely multiplying possibilities was what threw logic-theorem proving into the realm of what Newell and Simon called complex information processing. There were any number of alternatives that you *could* choose at every step, but there was no practical way to anticipate what you *should* choose until you got there: the proper choice depended on context and on all the choices that had been made up until that point. Moreover, the two men pointed out, this same explosion crops up everywhere, not just in logic but in games such as chess and in everyday life. As an illustration, try calculating how many conceivable outfits you could put on tomorrow morning, counting every possible combination of shirts, slacks, belts, underwear, socks, and so forth that you own. It doesn't take a very big closet to generate a number in the billions. So here's the problem in a nutshell: how does anyone even manage to get dressed in the morning?

The answer—which is also the key to understanding our human mode of boundedly rational problem solving—is that we don't go through all our billion-plus options one by one. Instead, we half consciously apply Simon's "satisficing" strategy, thinking things like Those two look good together, or Do these shoes match? or I can't wear that to work! And very quickly, using a few rules of thumb, we whittle the choices down to something manageable.

Such rules of thumb are known as heuristics, from the Greek word *heuriskein*, meaning "to invent" or "to discover." Newell and Simon had accordingly made heuristic reasoning the central strategy of Logic Theorist: at each decision point, instead of blindly trying to test all possible paths, the program would apply its own rules of thumb. Its reasoning process could be paraphrased as "First try approach A; if that doesn't work, try approach B, and so on." For example, Logic

Theorist might start with substitution, simply replacing one symbol with another symbol or perhaps with another whole expression. If that did not produce the desired theorem right away, the program might then try such things as working backward from the theorem to find intermediate expressions that might be easier to prove. In effect, Logic Theorist would go out searching for a proof—not blindly, as the brute-force algorithm would, and not with an instantaneous leap to the right answer, as the economists' perfect rationality called for, but by a process of intelligent exploration. In effect, Logic Theorist would apply Simon's "satisficing" strategy and proceed with bounded rationality.

Clearly, Logic Theorist's dependence on heuristic search moved it a very long way from standard algorithmic programming. Unlike algorithms, heuristic procedures are not guaranteed to work; they are guaranteed only to be worth trying. Their outcome is not predictable because there is no way to know in advance what kind of situations they will encounter. And they are not precise and tidy—in fact, there is something fundamentally messy and ad hoc about heuristics. But then the world is a messy, ad hoc place. For any organism hoping to cope with that world, "complex information processing" must be the norm rather than the exception. The unique power of heuristic reasoning lies in its ability to cope with the complex and the unexpected, to make acceptable choices when there isn't time enough to make the ideal choice, to hunker down and keep on going when a precisely defined algorithm would be overwhelmed by the combinatoric explosion. In effect, heuristic reasoning is what allows us to go through life in a chronic state of controlled panic.

That said, however, Newell and Simon faced the inevitable question: How do you actually get a computer to do that?

With some difficulty, it turned out. Unlike algorithmic programming, which was well understood millennia before the computer came along—Euclid worked out systematic calculational methods some twenty-three hundred years ago—heuristic programming was something new, and Newell and Simon pretty much had to make it up as they went along. As they would later explain to the writer Pamela McCorduck, they had begun to rough out the design of Logic Theorist by October 1955. "My method of working," Simon told her, "was to take theorems in the *Principia* and work out proofs while trying to dissect as minutely as possible, not only the proof steps, but the cues that led me to each one. Then we tried to incorporate what I had learned into a flow diagram. We repeated this day after day, with the flow diagram steadily approaching a description that could be programmed on the machine. On December 15, 1955, I simulated by hand a proof of Theorem 2.15 of *Principia* in such detail that we agreed the scheme was programmable. I have always celebrated that day as the birthday of heuristic problem-solving by computer."[23] Newell recalled their elation: "Kind of crude, but it works, boy, does it work."[24] A few weeks later, when classes had resumed after Christmas break, Simon greeted the students in his course "Mathematical Models for the Social Sciences" by announcing, "Over Christmas Allen Newell and I invented a thinking machine."

Meanwhile, however, Newell had been running up what were then horrendously large telephone bills (two hundred dollars per month!) by exchanging incessant Teletype messages with J. Clifford Shaw, who was doing most of the actual programming of Logic Theorist on the JOHNNIAC computer back at RAND. (This was another clone of John von Neumann's computer at the IAS—which had received its name, apparently, over von Neumann's strenuous objections.) Not surprisingly, considering that Newell and Shaw were inventing everything from scratch, they had a bear of a time getting it all to work. The programming and refinement of Logic Theorist consumed most of the spring. By June 1956, though, the program was working well enough that Newell and Simon could exhibit preliminary runs at Dartmouth College during a summer conference on "artificial intelligence"—a title invented for the conference by its organizer, a young Dartmouth mathematics instructor named John McCarthy. On August 9, 1956, Logic Theorist produced its first complete proof, of *Principia*'s Theorem 2.01. And a month later, on Tuesday, September 11, 1956, as the first speaker on the second day of the MIT Symposium on Information Theory, Allen Newell stepped up to the podium to explain how at least one kind of intelligent behavior—theorem proving—could be understood in terms of simple computational mechanisms.

Miller, who had been hearing rumors about Logic Theorist all summer, listened closely to everything Newell had to say. Then he listened even more carefully to Chomsky's talk, which came next. And for Miller, it was one of those rare moments when everything begins to click into place. Chomsky's grammars, Newell and Simon's heuristic programming, his own work on chunking in short-term memory—they were all very different, "but in the enthusiasm of the moment I felt that we were all doing the same thing. The notion that there *were* rules that could generate behavior, that behavior wasn't just the accumulation of reinforced responses—this was a thread that ran through all of the presentations."

Once again, says Miller, these ideas had a freeing effect, just as the whole cybernetics movement had a decade earlier. After Wiener, Rosenblueth, and Bigelow, we could begin to understand how an ordinary physical mechanism could embody a mental state—purpose—through the simple mechanism of feedback. Purpose wasn't a kind of stuff, it was a kind of organization.

After McCulloch and Pitts, likewise, we could begin to understand what it meant to say that the brain was a processor of information. Again, the magic was not in the individual neurons but in their organization—how they turned one another on and off through a web of interconnections.

And now, after Chomsky and Newell and Simon (and Miller himself), we could begin to understand how deliberative, purposeful reasoning—indeed, the mind itself—might arise from information processing. Obviously, there was a great deal that was left out of this picture—emotion, for example, not to mention sensory input, motor control, and learning. Nonetheless, says Miller, "Logic Theorist was a demonstration that you could have artifacts that would behave intelligently. Even if you didn't believe the further assumption that the way the computer did it was the same way we do it, this in itself was enough to free the psychological imagination. If you talk about the computer's having a memory, then certainly the behavioristic ban on concepts like memory was no longer necessary. And if a computer could be prepared to anticipate any one of *n* alternatives, then certainly the ban on expectations was no longer valid. The behaviorists kept saying 'metaphysics,' and we kept saying, 'Not at all: here is a perfectly nonmetaphysical way to do it.' So for those of us who were crippled by the antiseptic approach of behaviorism, it gave us license to think about the mind again."

Indeed, Miller has always looked back on that Tuesday morning in 1956 as the real birthday of cognitive science. He remembers walking out of that meeting like a man transformed. He had found his mission: the computational, information-processing view of human cognition would guide his career from that day forward. In 1958, for example, he joined with Simon and the psychologist Carl Hovland to organize a summer school on computer simulation at RAND, where he himself spent much of the session learning the basics of computation from Newell. The experience soon paid off. "The next year I spent at the Stanford Center for Advanced Study in the Behavioral Sciences," Miller recalls, "and Eugene Galanter and Karl Pribram were there. And I'd come along with all this material from this summer seminar. We began meeting together, and our discussions got rather interesting, so we decided we should record them; and the first thing we knew we'd written a book."[25] In that book, published in 1960 as *Plans and the Structure of Behavior,* Miller and his coauthors started from the antibehaviorist (but commonsense) notion that all behavior arises from internal plans. Then they went on to provide a general model for such plans using cybernetic feedback, in a form that was strongly reminiscent of Newell and Simon's heuristic search model and that a programmer would instantly recognize as a loop:

TEST (Has the goal been achieved? If not, OPERATE.)
   OPERATE (Take steps to reach the goal.)
   TEST (Has the goal been achieved? If so, EXIT. If not, OPERATE
      again.)
EXIT (We're there.)

During the 1960s, as the computational approach to psychology increasingly began to erode psychologists' allegiance to behaviorism, *Plans* would come to be seen as a landmark in the shift (as would Miller's "Magical Number Seven," which by 1975 would rank as the most-often-cited paper in experimental psychology). Yet another landmark came in 1960, when Miller joined with his colleague Jerome Bruner at Harvard to found a Center for Cognitive Studies and thereby endow the emerging movement with a name. Indeed, as Bernard Baars has pointed out, "there is little doubt that George A. Miller has been the single most effective leader in the emergence of cognitive psychology. . . . His career demonstrates, throughout, the importance of being in the right place thinking the right thoughts at the right time. [But] many of his personal characteristics were no doubt helpful: his notably lucid writing style, his capacity for presenting novel arguments with humor and drama, his consummate lecture style, his personal charm and ability to maintain relationships with other actors in the 'invisible college' that was creating cognitive psychology in the '50s and '60s."[26] George Miller, in short, became one of the commanding generals of the cognitive revolution. And by the 1970s, he and his colleagues would have routed behaviorism almost completely.

But of course, none of that happened at MIT.

And Lick?

On the surface, at least, he hadn't let the disintegration of his psychology section bother him too much. Like another Missouri boy, Harry S Truman, he had been brought up to keep a cheerful face on things and never complain. So to his students and colleagues, he simply said, Gee, that's too bad about the psychology group—and then renewed his efforts at Bolt and Beranek's acoustics lab and at Lincoln Laboratory, as if that were all he really cared about.

Beneath the surface, however, the fiasco with Dean Burchard seems to have affected Lick more deeply than most people realized. "He was very unhappy during that time," says Louise Licklider, who listened to her husband voice his frustrations every evening. It's telling that Lick didn't even bother to go to the 1956 information-theory symposium that so enraptured Miller (or if he did, that he was so uncharacteristically quiet that no one can remember his being there). It's possible that he was just too busy: he was an active member of the air force's scientific advisory board in those days, and he spent quite a bit of time that fall flying around the country to evaluate new weapons systems. But the fact remains that he didn't make time for a major meeting that was being held right upstairs from his office, in the same building. It was as if the gradual loss of his Dream Team had leached the joy out of Lick's work, out of MIT, out of his very profession. Lick, too, had once been a man with a mission: to build a great "hard-science" psychology department. And now he was left with—what?

Well, he still had his irrepressible imagination. He still had the vision he had been exploring since his World War II days in Harvard's Psycho-Acoustics Lab— the vision of humans and machines working together as a system. He still had the warm afterglow of the cybernetics movement and Norbert Wiener's dream of understanding both "the animal and the machine" through the same deep principles of control, communication, and information processing. He still had Wiener's conviction that the new technology was carrying us into a new age of history. And of course, he still had the cognitive revolution being waged by

Miller, Chomsky, Newell and Simon, and many others, with its portrait of the mind as an information processor.

In short, Lick still had the whole rich ferment of ideas that had been swirling around him for more than a decade. What he needed now was a catalyst, something that could reshape this inchoate brew of ideas into a new sense of purpose. Something that could excite him.

Something that he could rededicate his life to.

As near as Wesley Clark can recall after forty five years, it was in late 1956 or early 1957—sometime around then, anyway—out at the Lincoln Laboratory site in Lexington, Massachusetts, where Clark and his colleagues in the advanced computer development group were set up at one end of a long basement hallway. What Clark does remember about that particular day is that he had gone down to the stockroom, which was at the far end of the hall. And he also remembers that for some reason his eye was drawn to a door he'd never paid much attention to. "Off to the side," he says, "way near the end of the hall, was this very dark laboratory. So I went in, and after probing around in the dark for a while I found this man sitting in front of some displays. He was doing some kind of—a piece of psychometrics, maybe, or preparing an experiment for such a study. We began to chat, and he was clearly an interesting fellow. So I told him about my room." Clark explained that he and his group were developing a new machine called the TX-2. Very cutting-edge stuff: not only was it one of the first computers to be built with transistors instead of vacuum tubes, but it had a display screen that you could program and interact with in real time. The whole idea was to make the computer interactive, exciting, and *fun,* said Clark. In fact, he told the man, you could use a computer like that to do everything you're trying to do with these analog electronics—and do it a whole lot easier and faster. So why don't you come on down sometime and give it a try?

Sure, said J. C. R. Licklider. Sure. That does sound like fun

CHAPTER 5

THE TALE OF THE FIG TREE
AND THE WASP

Interactive. Exciting. *Fun.*

That wasn't how most people thought about computers, not in 1956. Within the industry, noted Kenneth Olsen, who was then one of the leaders of Lincoln Lab's advanced computing group, the concept of having fun with a computer was, well, *strange.* "Some people thought it was wrong," Olsen later told an interviewer. "They almost spoke in ethical terms. Computers are serious, you shouldn't treat them lightly. You shouldn't have fun with them. They shouldn't be exciting. They should be formal and distant with red tape involved."[1]

Indeed, recalled Olsen, to most people in the 1950s, that was what computers *were:* big, impersonal oracles sitting off in air-conditioned rooms somewhere, crunching data for big, impersonal institutions. Back then, after all, only the largest institutions could afford them. Except for a few special applications such as SAGE, the only real market for the machines lay in automating those huge data-processing engines called bureaucracies, which had been growing in government and business for more than a century. That was how IBM had been able so quickly to overcome its late start in computers and achieve near-hegemony in the field: a great many corporations and government agencies had already invested

---

Interactivity for the masses: MIT hackers play Spacewar on the PDP-1

heavily in IBM office equipment, and its sales force was already in place. Starting in 1954 with the IBM 650, moreover, the company had designed its computers as upgrades to its ubiquitous punch-card tabulators, which meant that customers could just slide the new machines into their preexisting operations with little pain or fuss.

Then, too, this perception of bigness and remoteness was a reflection of engineering reality: from the start of modern computing in the 1940s, all the way through the 1950s, the prevailing vacuum-tube technology had meant that the machines *had* to be big and expensive. According to "Grosch's law," a bit of industry lore first formulated by IBM executive Herbert Grosch in the late 1940s, spending twice the money would get you roughly four times the processing power—which meant that customers could always get much more bang for their buck by buying the biggest machines they could afford. So that was what they did.

On a day-to-day basis, however, what most powerfully reinforced the standard perception of computers was batch processing, an assembly-line style of operation inherited from the punch-card-tabulator era. In batch processing, the users weren't allowed anywhere near the machine; instead, they were supposed to hand in their decks of IBM punch cards to the computer-room technicians, who would run a group of jobs through the computer as a "batch" for efficiency's sake and then deliver the resulting stacks of fan-fold printout a few hours later. Or maybe eight hours, twelve hours, or even twenty-four hours later. And if the only result was an error message about a comma that had been omitted on card 43—well, next time be more careful.

This kind of regimentation undoubtedly appealed to the gray-flannel corporate mentality of the 1950s. And in fairness, if you thought about computers purely in terms of routine jobs like payroll and billing, in which the programs were already written and the task was simply to churn through masses of data again and again, then batch processing was a perfectly efficient way to operate. True, it was "efficiency" for the machines, not for the users. But on a strict dollar-per-hour basis, so what? Machine time was far more valuable than human time.

However, if you thought about computers in Lincoln Lab terms, with the focus on *creating* programs and exploring the frontiers of what the machines could do, batch processing was an outrage. Olsen, Clark, and their companions in the advanced computing group had a crazy idea that these "ultrafast calculating machines" could become a new, creative medium of expression. And because of that, they had an even crazier idea that computers ought to be helping humans, not vice versa.

It was an inspiration born of luck, recalls Clark, who joined the Whirlwind project as a programmer in 1952. Most members of his group had been around MIT long enough to remember what life had been like before SAGE really got going—before Whirlwind, the Memory Test Computer, and every other machine in sight had been totally monopolized by national security. Back in those days, says Clark, Whirlwind had actually functioned as a very large personal computer, complete with an interactive display screen. "What now sits comfortably on a small desktop, in those days required an entire room for the control [console] alone," he wrote in a memoir. "But its early users did indeed walk into the control room and, for their assigned block of time, typically fifteen minutes or so depending on the time of day, the entire machine was theirs."[2] For that whole glorious fifteen minutes they could play around, type in commands, try things out, see what happened, get new ideas and try those out—all the things we now take for granted. But then it was an experience unique in all the world.

What really drove the point home, added Clark, was the MTC: the Memory Test Computer, which Ken Olsen and Harlan Anderson had completed in 1953—when they were still graduate students—to test core memory for Whirlwind. As an exercise, Clark remembered, he and his colleague Belmont Farley once used the MTC to simulate a series of McCulloch-and-Pitts-style neural networks. It was a joy. "Blocks of time measured in hours were available and the entire machine, quite similar in architecture to Whirlwind, now occupied only a single large room," he wrote. "Bristling with toggle-switches, push buttons, and indicator lights, and provided with audio output as well as versatile CRT displays, it made interactive use a quite lively and memorable experience." Indeed, Olsen and Anderson had designed the MTC that way deliberately: "We put a loudspeaker on every computer we built because you always wanted to be able to play music or make it do things," Olsen said. "We believed computers should be fun. They were exciting. They could do so many things. The opportunities were just without bounds."[3]

That vision really began to flower after 1955, when Olsen, Clark, Anderson, and a number of other free spirits first organized the advanced computer research group. By then the stakes were clear. It was one thing to have personal control of a Whirlwind or a Memory Test Computer in the laboratory, but out in the real world, the batch-processing mentality was becoming more deeply entrenched than ever. The people paying the bills wanted to see their million-dollar IBMs and UNIVACs crunching numbers every possible second; they weren't about to let you play around with one of their machines just to see what happened. So if interactive computing was ever going to be anything more than a laboratory pastime, the group realized, computers were going to have to become a lot smaller, faster, and cheaper. And the only way to ensure that was to replace those big, power-hungry vacuum tubes with transistors.

The problem, of course, was that transistor technology was still on the cutting edge in 1955, which was why industry engineers generally preferred to stick with the proven, reliable, and familiar vacuum-tube technology. "The commercial world just smiled at us and said we were 'academic,' " recalls Olsen, who was not a man to take such condescension lightly. So there was nothing for it but to build a fully transistorized computer themselves, just to show it could be done.

The overall design came from Clark, who sketched out a very small, "primitively simple" machine that was dubbed the TX-0. Responsibility for its actual construction fell to Olsen—who soon discovered that the establishment engineers weren't crazy for being skeptical about transistors. The only versions that were commercially available in the mid-1950s were still incredibly delicate. "If you combed your hair and touched one, you burned it out," says Olsen. To make the devices work at all, the team had to design special circuitry to protect them from power spikes, sparks, or discharges of any kind.

Then there were all the *non*technical problems. "The rules were, I could hire nobody and have no space," Olsen explains. "So I studied the rules carefully and found all the loopholes. . . . We discovered that the hallway was not 'space.' So we moved my office into the hall and put walls around it. We then traded that space for a space in the basement which was less desirable but bigger. . . . When [the lab managers] discovered what we had done they said, 'Never again.' Then we talked people into twice the light level of any place else in the laboratory. And when [the management] found that out they said, 'Never again.' And then we had the walls painted with a different color. The walls there were just a normal military-type beige or green, but we had a bright color."

Still, says Olsen, the real purpose of this exercise was to make the TX-0 *exciting*. And that the group did: "There was a loudspeaker and amplifier underneath the control table for playing music or anything else you wanted with the computer. We automatically built in a cathode-ray tube. It had four thousand lines, I think, because we focused on one spot at a time instead of [scanning back and forth in] a raster like we do today. Then we had a light pen, which was what we used in the air-defense system and which was the equivalent of the mouse or joystick we use today. With that you could draw, play games, be creative—it was very close to being the modern personal computer."

Indeed, Olsen was proving to be a born showman. "We'd discovered with the MTC that blah-looking computers never really attracted attention," he says. "So we made the TX-0 as modern a design as we could. Now it looks quite naive. But it had rakish lines like a race car. And we picked a color [that was] just the opposite of the traditional black wrinkle finish, which was World War II. It looked so military and blah. So we picked brown and beige, which seemed like a dramatic change. Then we set the computer back from the door for good pictures and to show it off with a little bit of flair. The result was [that] when the head of Lincoln Laboratory had visitors, he of course brought them to our laboratory."

In short, the TX-0 was a hit—so successful at both computing and attention grabbing that in 1956, when it was finally finished, Olsen, Clark, and their colleagues decided to spread the gospel by sending it over to the MIT campus. (They needed the space in any case, since they had already started construction on the TX-2, a new, larger, and ambitious machine designed by Clark to experiment with computer-driven graphical displays.)* "By that time both of the Cam-

---

* And the TX-1? That was the name Olsen and Clark had used for the *very* ambitious machine outlined in their original proposal. It was after the Lincoln Lab higher-ups vetoed that project, suggesting that the members of the advanced computing group might want to walk before they tried to run, that Clark came up with the TX-0 design. Olsen and Clark later joked that they'd decided not to build the TX-1 because they didn't like the color scheme.

bridge machines, Whirlwind and MTC, had been completely committed to the air defense effort and were no longer available for general use," Clark would later write. "The only surviving computing system paradigm seen by MIT students and faculty was that of a very large International Business Machine in a tightly sealed Computation Center: the computer not as *tool*, but as *demigod*. Although we were not happy about giving up the TX-0, it was clear that making this small part of Lincoln's advanced technology available to a larger MIT community would be an important corrective step."[4]

Indeed it would—and far more so than anyone could have imagined, as things turned out. But in the meantime, Ken Olsen was getting impatient. He wasn't satisfied with telling the MIT campus about interactive computing; he wanted to tell the *world*. While he was at it, moreover, he wanted to wipe that condescending smile off the industry's face once and for all. "We had published what we had done with the TX-0," the still-exasperated Olsen would recall nearly forty years later. "We had demonstrated all the ideas of high-speed transistor computers and shown that you could make computers much better than anything done with vacuum tubes by far. And we thought the world would be waiting with open arms for this. But nobody cared! It turns out that it takes more than ideas. You've got to *sell* your idea."

So that was exactly what he and Harlan Anderson decided to do. Taking what was then a radical, almost unheard-of step, they decided to go out and form their own company—a company that would sell interactive, transistor-based computers to the world. A company that would beat the industry at its own game.

All they needed was money. Unfortunately, Olsen and Anderson soon discovered that Wall Street had little interest in high technology. "A number of companies had started during the Korean War," says Olsen. "But in nineteen fifty-seven a recession was starting, many of them were in trouble, and some were no longer in existence. So this was not a popular idea."

By and by, however, he and Anderson heard about the American Research and Development Corporation, one of the nation's first venture-capital firms. "So we went to see them," recounts Olsen. "It turned out that they were worried, too, because some of their investments hadn't paid off very well. But the staff was fascinated enough to listen to our proposal. Then they told us we could go present it to their board of directors and see what happened—but gave us three bits of advice before we went. One was, Don't use the word *computer* in the proposal, because *Fortune* magazine said that no one was making money in computers and no one was about to. So we took it out." Instead, Olsen and Anderson promised not to market a computer until they were making a profit by selling modular, transistor-based circuit boards—which would be the components of the computer in any case. The second bit of advice was "Don't promise five percent profit." That was about average for a well-run company in those days, says Olsen. "But the staff said that if you're asking someone to give you money, you've got to promise better results than that. So we promised ten percent. Then the third thing they said was, 'Most of the board is over eighty [years old], so promise fast results.' So we promised to make a profit in a year."

Olsen and Anderson gave their pitch on May 27, 1957, and apparently made a good impression: the ARDC board decided that they were worth a risk of seventy thousand dollars. It was enough to give them a chance. That July, after a long search for affordable space, the two men found what they were looking for in a renovated nineteenth-century mill out in rural Maynard, Massachusetts, where military blankets and uniforms had been manufactured during the Civil War. The second floor of Building 12, containing precisely 8,680 square feet, rented for only three hundred dollars per month. Olsen and Anderson took it. And on August 27, 1957, having settled on a name that said "computer" without actually *saying* "computer," Olsen, Anderson, and Olsen's brother Stanley opened for business as DEC, the Digital Equipment Corporation.

EPIPHANY

Interactive. Exciting. *Fun*–J. C. R. Licklider loved it. He absolutely loved it.

After that first encounter in his basement psychology lab at Lincoln, back in late 1956 or so, Lick and Wes Clark had quickly recognized their mutual fascination with computers, neuroscience, neural networks, and all the rest. They had just as quickly become friends. Clark found Lick to be a delightful companion (he remembers one particularly lively discussion over dinner at Joyce Chen's Chinese restaurant on the outskirts of Cambridge. Since that establishment had no liquor license at the time, Lick helped the conversation along by handing Mrs. Chen a brown bag as they came in and asking her for "the usual tea").

Lick, for his part, was captivated by Clark's ideas on interactive computing. True, the concept wasn't completely new to him: Lick had been working with Whirlwind and its successors ever since he joined Project Lincoln, and he'd long since turned the basement of the Sloan Building into his personal toy shop of analog machines. But sitting at the console of a digital computer that he could program and reprogram all by himself without changing a single wire, watching its responses appear instantly on the display screen, having all that power under his own two hands—now *that* was something. Lick was instantly bursting with ideas for using such a computer in the laboratory: he could run his experiments automatically, process his data in a flash, reprogram whole new experiments with a few taps on the keyboard. All his life he had been longing for a magical device that would let him implement ideas as fast as he could think of them. Now he'd found it.

But even more compelling was the sense of freedom and autonomy he felt at the console. The TX-2 was the first digital computer Lick had ever seen that didn't just execute a preplanned algorithm; instead, it invited him in and helped him to create new programs of his own. It was likewise the first computer he'd ever seen that didn't just serve the needs of some large institution; it was a *personal* computer, serving the needs of his own imagination. The TX-2 put him in charge. Using it, Lick later wrote, "[was] like sitting at the controls of a 707 jet aircraft after having been merely an airline passenger for years."[5]

Most of all, however, Lick's encounter with the TX-2 began to spark an idea. Unfortunately, there is no way to reconstruct his exact train of thought in this period, assuming he could even follow it himself. But it's clear that at some point he began to reflect on a curious dichotomy. When it came to things that computers did well, Lick realized, thinking of rote, algorithmic tasks such as arithmetic and data sorting, humans were pathetically slow and mistake-prone. Most of us could be beaten several times over by the simplest desktop calculator. And no wonder: all our mental processing had to be done through laborious, conscious deliberation, using biological information processors that could execute no more than a few operations per second and handle only about seven chunks of data at a time.

And yet, Lick further reasoned, when it came to things that computers did poorly or not at all—intuitive processes such as perception, goal setting, judgment, insight, and all the rest—our human capabilities surpassed the most powerful machines on the planet (and still do). Having spent much of his career trying to untangle the relatively straightforward mechanisms of hearing, Lick could testify to that point personally: these "heuristic" mental functions, as he called them, actually involved a vast amount of information processing. They seemed effortless and intuitive only because that processing was carried out deep within the brain, by massively parallel networks of neurons operating well below the level of consciousness. It was (and is) a deep mystery why our conscious and unconscious mental functions should behave so differently. But there it was: the human-computer complementarity was almost perfect.

So, Lick wondered, what would happen if you put humans and computers together as a system? What would happen if you let the machines take over all those dreary, algorithmic chores they were so good at? How much of your time would be opened up for real creativity?

Ever the experimenter, Lick decided to answer that question with a measurement. "In the spring and summer of 1957," he wrote a few years later, "I tried to keep track of what one moderately technical person actually did during the hours he regarded as devoted to work. Although I was aware of the inadequacy of the sampling, I served as my own subject. . . . I obtained a picture of my activities that gave me pause. Perhaps my spectrum is not typical—I hope it is not, but I fear it is. About 85 per cent of my 'thinking' time was spent getting into a position to think, to make a decision, to learn something I needed to know. . . . [These getting-into-position activities] were essentially clerical or mechanical: searching, calculating, plotting, transforming, determining the logical or dynamic consequences of a set of assumptions or hypotheses, preparing the way for a decision or an insight. Moreover, my choices of what to attempt and what not to attempt were determined to an embarrassingly great extent by considerations of clerical feasibility, not intellectual capability."[6]

Eighty-five percent!? That figure did more than give Lick pause. It seems to have hit him with the force of a religious epiphany: our minds were slaves to mundane detail, and computers would be our salvation. We and they were destined to unite in an almost mystical partnership: thinking together, sharing, dividing the load. Each half would be preeminent in its own sphere—rote algorithms for computers, creative heuristics for humans. But together we would become a greater whole, a *symbiosis*, a man-machine partnership unique in the history of the world:

"The hope is that, in not too many years, human brains and computing machines will be coupled together very tightly, and that the resulting partnership will think as no human brain has ever thought and process data in a way not approached by the information-handling machines we know today."[7]

However he got there, this was the watershed of Lick's life, the Great Idea that was the culmination of all that had gone before and the vision that would guide him from then on. Granted, it would be quite a while before he could articulate this notion as clearly as he could feel it. But what he did know, and very quickly, was that the TX-2 was *it*. Once he'd had a few sessions at the console with Wes Clark to learn how the machine worked, he said, "I saw that I had really *got* to do that."

But sadly, he added, "I also saw that I was not going to do that trying to build a psychology department at MIT."

This is the part that still has Lick's colleagues from the psychology years shaking their heads in bewilderment. Frustration with MIT? Sure—who wouldn't be frustrated with Dean Burchard? But to give up a career he'd spent twenty years building, right at the moment when cognitive psychology was taking off? To give up tenure? For *computers*? It was beyond belief.

Yet that was exactly what he did. Not immediately; nobody can completely redesign his life overnight. Lick would retain a lively interest in acoustics research well into the 1960s. Nonetheless, it was around this time that Louise Licklider began to notice the word *computer* cropping up more and more often in their late-evening mutual core dumps. It was likewise around this time that her husband's associates began to note a certain evangelism on Lick's part. "I was computer-resistant," said Bill McGill, who wouldn't become a convert until many years later. "And Lick used to lecture me by the hour about the error of my ways!" He could be heard to say things like, "Any psychologist is crazy to keep on working with people if he has access to a computer!" And it wasn't entirely clear that he was joking.[8]

Before very long, Lick's fellow psychologists began to feel, well, not rejected, exactly, but aware somehow that he was inexorably drifting away. "I saw him a couple of times after I went back to Harvard," says George Miller. "But his interests were already elsewhere. The things that we had once talked about with enthusiasm, he was just vaguely interested in." Once again, it would be only a matter of time: Lick was ready for a change.

And once again, change—in the person of Leo Beranek—came looking for him.

DREAM TEAM REDUX

When Leo Beranek thinks back on that period of the 1950s, he mainly remembers being worried. Oh, not about anything specific, he says. His company was doing well, and had been from the beginning. In fact, it had been doing well since *before* the beginning.

Back in May 1948, Beranek explains, not long after he and Richard Bolt started the MIT acoustics lab, a movie-house owner in Hoboken, New Jersey, had wanted to hire him part-time to improve the acoustics of his theaters—all fifty of them. "I was already using each Friday, my one day a week permitted by MIT, to consult for the General Radio Company," Beranek says. "So I agreed that I would find time on the weekends to do the job. But then Bolt came across the hall one day and said, 'A while back I put in a bid to do the acoustics for the new United Nations Building in New York City. Now I've won it, and it's too big for me alone. Let's go in together.' And I said, 'I've got too much work already!' "

In an effort to cope, the two physicists brought in several graduate students to help out, and then formed a legal partnership in November 1948. They started out calling themselves Bolt Beranek, since Bolt was the elder of the two. And they followed the same formula when the young architect Robert B. Newman became the third partner in 1950, arriving at the name Bolt Beranek and Newman.

For a time, says Beranek, they continued to operate BBN as a sideline; their main focus was still the acoustics lab. But not for long. The firm's work at the United Nations soon led to any number of new commissions, and by the time it incorporated in 1953 and moved to its new headquarters on the outskirts of Cambridge, BBN had twenty-six full-time employees and thirteen part-timers. And from there it continued to grow at the rate of about 20 percent a year, branching out into concert-hall acoustics, noise control for the new jet airliners, criteria for speech privacy in office buildings, and anything else having to do with the applied science of sound.

So no, says Beranek, there was nothing obvious to worry about. But he worried anyway: "I said to myself, 'We're consulting on auditoriums, concert halls, and all that. But there's a limited amount of work in that area. We ought to be a company that relates to other senses besides hearing—eyes, touch, and so forth. Why don't we set ourselves up as a company that deals with man-machine systems across the board?' For example, I thought we might find better ways of controlling a car or a motorboat at high speeds—ways to do a more efficient job

nvironment. So I got to thinking, Who would be
d up this kind of activity?"

of the troubles Lick was having with Burchard,"
r lunch three times during the spring of nineteen
rsuaded to come to BBN." Lick seemed willing to
w other irons in the fire. "In August [of the same
cohorts at MIT that he was thinking of a possible
he had just left for a week's secret conference at
s to talk about the future of space research in the
t on a plane and went after him. "I remember
sign and meeting Lick in the glass-walled lobby.
made him a very good offer, and he was consider-

might have been very different if he had, says
istory of BBN. But as it was, the two men agreed
to be a joint one, made with Louise Licklider. So
Santa Monica beach the next day, Sunday, August
ket from the hotel, and we brought some chips
. "I would say we talked for about two hours on
convince Louise that she was so well liked in
ur drama activities at MIT were so successful that
staying there. My offer to Lick was that he would
all psycho-acoustics research at BBN, and that we
option at a greatly undervalued basis of a dollar
public in 1961 at $12 per share).
argument was more persuasive, "but by golly, a
would come."
T was June 30, 1957.

BBN, it developed, was everything Beranek had promised. Beranek himself was
preoccupied with concert-hall acoustics, including his design for a hall being
built for New York City's brand-new Lincoln Center. And Bolt was increasingly
preoccupied with the policy aspects of science; he would soon go off to Washington to serve at the National Science Foundation. So within reason, Lick had a
free hand to do what he wanted, to hire whomever he liked, and, most especially,
to explore whatever new directions he wished. He plunged in immediately, setting up a new engineering psychology branch and a new psychoacoustics lab.
(Overcommitted as ever, he also served as president of the American Acoustical
Society in 1958.) And while he was at it, he reassembled as much of his Dream
Team as he could.

"Lick collected people," says his former student Tom Marill, who was struck
by the way his mentor always tried to bring his favorites along as he moved from

place to place. "He was very bright, he was very articulate, and because of that he was able to get very good people. They *liked* being collected."

Marill himself apparently enjoyed it well enough. Having spent the previous two years at Lincoln Lab, working on a formal evaluation of the SAGE system, he allowed himself to be collected in 1958. So did another of Lick's former students, Jerry Elkind, who came to BBN that same year from RCA. And so did Lick's old friend Karl Kryter. "Karl had spent his lifetime, really, on measuring the effects of noise on humans," Lick explained in his 1988 interview. "And it was pretty big stuff: at that time there were many manufacturing situations that made a terrible clatter. People were just beginning to realize that that was all much harder on hearing than anybody had suspected. And there were six or seven billion dollars' worth of lawsuits stacked up at that point. So Karl came and started up that work."

And all along the way, of course, Lick was trying to figure out how he could develop his transcendent idea of a human-computer partnership.

"Leo," he said to Beranek within a year of his arrival, "I want to buy a computer."

Beranek had had a feeling this was coming; Lick had already dragged him out to Lincoln Lab twice to see the TX-2, as well as a new series of machines that Wes Clark was designing to control laboratory experiments.

"And how much will it cost?" he asked, thinking of all the analog computers he'd already bought for Lick's new laboratories, not to mention that very expensive IBM card machine in the BBN accounting department.

"About thirty thousand dollars."

Thirty thou . . . ?! Beranek was aghast. "Lick, we've never bought anything that expensive! What are you going to do with it?"

"Leo," said Lick, his blue eyes shining, "I don't know. But if BBN wants to go on being a leading company in the future, it's got to get into computers. They are here to stay."

Four decades later, Leo Beranek can testify that J. C. R. Licklider was a very persuasive man. "I consulted with my executive vice president and our treasurer, and we agreed," he says. "So we bought him the computer. He didn't use it very efficiently, as it turned out; basically, he just wanted it so he could learn how to program."

Well, yes, that was basically true. But then, the machine *had* come fully equipped with a teacher: Edward Fredkin, an arrogant, abrasive, insubordinate, and totally brilliant young man who had dropped out of college to fly fighter jets for the air force—and who had then decided to find himself some real excitement by programming computers.

"I had been part of the air force team that went up to Lincoln Lab to test the SAGE system," recalls Fredkin, who has mellowed considerably in the decades since then. "But we discovered that SAGE was going to be a year late, so the air force just decided to assign me to Lincoln Lab. One day this guy came in and said, 'Hi, Lieutenant, I'm supposed to interface with you.' He described the most mundane things that I was supposed to do: make sure the reports came in, that sort of stuff. I said, 'That's totally inappropriate.' So he went away, and nobody else ever told me anything to do. So I worked very hard doing whatever I wanted. I taught all kinds of people to program. When *Sputnik* went up, I wrote programs to calculate orbits. It was a wonderful environment. And after I got out of the air force [in July 1958], I just continued there. I didn't have a degree, so they didn't have any slot for me except 'clerk.' But I said, 'I don't care what you call me, I just want to keep on doing what I've been doing.' "

Eventually, after everyone else at Lincoln Lab had despaired of figuring out what to do with him, Fredkin came to rest in a group headed by Tom Marill, with whom he actually got along. They even kept in touch after Marill departed for BBN. "Sometime after Tom left, I decided that I'd start a company," says Fredkin. "I ordered a computer—a Librascope LGP-30—even though I had no money. I figured that I would pay for it with the money that I would earn. And then Tom suggested I might get business from BBN. So I went and I met Lick. We had a wide-ranging conversation about computation. We just talked about everything, from digital logic to programming to hardware."

Indeed, the two men hit it off immediately. For years, says Fredkin, he'd been trying to tell people that computers mattered. And now, finally, here was someone who understood.

Lick, meanwhile, was just as enthusiastic: "It was obvious from the beginning that here's a young genius," he said in his 1988 interview. And since Lick's idea of a personnel policy hadn't changed a bit since the Project Lincoln days—Hire the most brilliant people you can find, and stand back—it was equally obvious that Fredkin was the kind of guy he wanted to add to his collection. So Lick made him an offer on the spot: Don't start your company, he urged Fredkin. Come to work for BBN. We'll buy the computer you're trying to buy, and you can use it to teach us all about computing.

Done. "So we got Fredkin working at BBN," said Lick. And he didn't disappoint: difficult though the young man might be, he was undeniably a virtuoso with computer code. As promised, he taught Lick to program the LGP-30. (Since the code had to be written in a "hexadecimal" notation, in which the numbers *10* through *15* were abbreviated by the letters *F, G, J, K, Q,* and *W,* Lick soon came up with a mnemonic: "For God and Jesus Christ, Quit Worrying.") Fredkin also taught many others around BBN, including the twelve-year-old Tracy Licklider.

Unfortunately, it soon became all too apparent that the LGP-30 was a hunk of junk—"just totally inadequate," according to Lick. The machine's sole virtue was its ability to respond to the user interactively, in real time; it wasn't just another batch processor (indeed, it was a very early attempt to market what would later be called a minicomputer). But it was still a vacuum-tube machine. It had no core memory, relying instead on a kind of primitive hard drive known as a magnetic drum. And its internal architecture was a kludge [a patched-together mess]. "It was ridiculous!" recalled Lick. "I programmed one of the models I had running on an analog computer. The analog computer solved the problem thirty times a second and displayed it [on an oscilloscope screen] so I could twiddle my knobs and see what was happening. This darn thing [the LPG-30] took two and a half days to run the program. In the process it made many mistakes, so you had to checkpoint yourself [on] paper tape—and the paper tape was unreliable. So the LGP-30 was fun to play with, and I learned to write programs and stuff on it. But we were still an order of magnitude from what it was going to take [for real interactive computing]."

Disappointing though the LGP-30 may have been, however, it was the first computer that Lick ever sat at for hours on end. It helped him clarify in his own mind what a human-computer partnership might actually entail. And it seems to have confirmed him in his instinct that this partnership would require a fundamental shift in the direction of computer technology. Instead of simply making the machines bigger and faster, which was what IBM and the other big manufacturers were constantly trying to do, the industry would somehow have to start making them more intuitive, easier to program, easier to understand, and easier to communicate with—in short, more closely adapted to the workings of the human mind.

THE UN-COMPUTER

About a year after the LGP-30 was delivered, says Leo Beranek, probably around August 1959, DEC's Ken Olsen stopped by to, um, take a look at the thing.

Actually, says Beranek, Olsen was more interested in taking a look at Lick. And after a short chat convinced him that the older man did indeed understand digital computers, Olsen revealed his real purpose: would BBN be interested in a little project?

Out in Maynard, it developed, DEC had finally taken the plunge and built a real computer. It was quite a plunge. The company was still less than three years old. And while Olsen and his colleagues had fulfilled their promise to make a profit on their high-speed transistor modules starting in year one, the margins were paper-thin. Moreover, getting even that far had required some real sweat equity on the part of the Olsen family. "We had no problems on weekends," Ken Olsen's Finnish-born wife, Aulikki, later recalled. "We knew exactly what we were going to do. We were going to the plant to clean. Everybody in the family. Our kids looked forward to it because the watchman would give them a nickel. They would clean the coffee cups, I would clean the bathrooms, Ken would clean the work area. It was regular physical work that we did for the first three years."[9]

Still more daunting, however, was the fact that DEC's vision of interactive computing now lay even further outside the mainstream than when they'd started. IBM, UNIVAC, the new Control Data Corporation up in Minneapolis—by 1959, *everybody* was committed to batch processing.

But DEC went ahead anyway. Ted Johnson, who had been hired the year before as the company's first and only salesman, remembered being in Olsen's office one day in April 1959 when a letter came in from a naval laboratory in California. It was a formal request for a quotation on a new computer. "When Ken opened the letter," said Johnson, "his face lit up, and he said, 'That's the machine I had in mind! Go sell a computer!' "[10]

Of course, that *did* require having a computer to sell. To create one, Olsen and Anderson hired the best engineer they knew: Lincoln Lab's Ben Gurley, who had worked with them on the Memory Test Computer and the TX-0 and had completed the construction of the TX-2 after they left. And that, in turn, proved to be one of the best moves in DEC's early history. In what Ed Fredkin would later call a tour de force of computer design, Gurley had the new machine ready to go in just three and a half months. True, it was only a smaller prototype version of the computer specified in the navy's letter, which never did get built. But it worked beautifully. "Ben Gurley, perhaps better than any other engineer of his time, had the right combination of technical brilliance and engineering conservatism to bring these ideas into reality," noted Fredkin, who later came to know Gurley well. "When he drew something on paper, it worked exactly as drawn. There were no bugs in his logic."[11]

What Gurley had created might be best described as a commercial version of the TX-0, especially on the outside. Their refrigerator-sized cabinets were similar, as were their operator consoles, loudspeakers, light pens, and big CRTs housed in distinctive hexagonal cases. But even more important was what Gurley had done on the inside. Instead of just replacing vacuum tubes with transistors one for one—as IBM was planning to do with *its* first transistorized computer, the just-announced IBM 7090—Gurley had tried to take full advantage of the new technology from the ground up. His design, building on the TX-0, owed nothing to vacuum-tube architectures. And as a result, his new machine was capable of doing a hundred thousand additions per second—which made it not the fastest in the world, even then, but far, far faster than anything else in its price range.

If Gurley's prototype computer was everything Olsen wanted, however, there remained the surprisingly delicate question of what to name it. "When we were almost finished with the computer," explains Olsen, "we had a request from the government to build a machine to collect seismographic information, to help them look for earthquakes.' We didn't quite believe that they were all that interested in earthquakes, but we were willing to let it go at that. The problem was that the Congress, in their wisdom, had said, 'No more computers will be bought until all the computers in Washington are used one hundred percent of the time.' Now, it seemed unlikely that the Russians would wait to try out their next bomb until all the accounting machines were fully utilized. So we said [to our potential customers], You could call it something else. Tell Congress it's really not a computer, it's a Programmed Data Processor."

Thus the name of the new machine, the PDP-1. And thus Olsen's visit to

BBN that August: would the company like to serve as the PDP-1 test site for a month, with DEC paying for the time?

Did he even have to ask? Serial No. 0 was delivered to BBN headquarters in the early fall of 1959, says Beranek. They had to set it up in the main reception lobby; the PDP-1 was at least ten feet long, and there was no other space for it. But with some Japanese screens around the outside to keep visitors at a distance, it operated there for a month, while Ben Gurley carted in the machine's core memory unit every morning and took it back home to Maynard with him every night. It was the only memory unit DEC owned.

Today it would be called a beta test of a new machine, says Beranek, with Lick and Fredkin the main testers. Lick was ecstatic. "[Now] *that* was a serious computer," he said in his 1988 interview; his main suggestion was that the PDP-1 should have a "thin skin" so that experimenters could get inside it. And Fredkin was in his element, critiquing the design, tinkering with the machine, and offering endless suggestions for improvements. "The PDP-1 was revolutionary," Fredkin declares, still marveling four decades later. "Today such things don't happen. Today a machine comes along and is slightly faster than its competitors. But here was a machine that was off the charts. Its price performance was spectacularly better than anything that had come before. Compared to the LGP-30, it cost a little more than twice as much but was more than a thousand times faster. Or to put it another way, it was as fast as a computer that cost a hundred times as much. There had never before been a machine that was this much in front of the competition. And never since. It was a singular event."

Indeed, there was no holding Lick and Fredkin back. Long before the prototype went back home to Maynard, BBN had put in a purchase order for a PDP-1 of its own, at a cost of $150,000, with extras. In December 1959, DEC shipped the prototype to the Eastern Joint Computer Conference in Boston for the PDP-1's official world debut. And in early 1960 the company shipped Serial No. 1 to BBN, where it was installed on the first floor of a newly constructed building.

Fortunately for BBN's bottom line and Beranek's peace of mind, Lick was happy to go out and solicit research contracts that would help pay for his new toy. Once Beranek had taken him down to Washington and introduced him around at the major research agencies, he was off and running. "Finally, this gave him a machine with which he could do something," says Beranek. "Lick's mind was working, he was buzzing back and forth to MIT, he was putting out proposals for contracts with NASA, the air force, this new thing at the Pentagon called ARPA. He was everywhere, bouncing around like a kid with too much candy."

"We kept that thing running day and night," Lick agreed in his 1988 interview. "Everybody connected with it just sat at the console and did on-line interactive programming—and since I was one of the first ones to be involved, I got most of the time!" Indeed, the forty-five-year-old Lick was happily transforming himself into what would now be called a hacker. Every evening after having dinner with his family, he would be back at the console writing programs interactively. "He would spend all night developing a program and never even save it!" marvels Karl Kryter, who watched any number of his friend's efforts disappear into the midnight ether. By some standards, it's true, very few of those efforts were worth saving: "It was like Norbert Wiener's trying to be a chess player," says Fredkin. "Lick always had better ideas than anyone, but rotten execution." But who cared? Lick was playing with the PDP-1, kicking it around, seeing what it could do. "We were living out there in the future," says Fredkin. "And Lick was one of the few people with the vision to understand."

Early on, for example, Lick wrote some of the world's first educational software, using Tracy and Linda Licklider as willing guinea pigs. "In a small and preliminary way," he explained in 1961, "with only a small computer, a computer typewriter, and a few nights of programming, some of us have already created 'motivational traps' for our children, and we are sure that a computer teaching machine can be made more attractive than television. The youngsters love real-time interaction with a thing like a computer. It can tell them immediately, 'No, that was wrong'; it can calculate and post a score as it goes along; with the aid of a simple random process, it can look up in a table a suitable compliment or a suitable sarcastic remark such as 'Oh, oh, you're slipping.' The youngsters will sit there and punch the keys for hours learning spelling and language vocabulary."

In the future, Lick added prophetically, "any student from grade school through graduate school who doesn't get two hours a day at the console will be considered intellectually deprived—and will not like it."[12]

And yet, Lick knew, as compelling as the PDP-1 was, it still wasn't enough. For interactive computing to make a real difference in the world, it would have to be available to *everybody*, not just a chosen few. And that wasn't going to happen with the PDP-1. "You have to remember the time frame," he explained during a panel discussion in October 1988. "[A bare-bones PDP-1] was a hundred twenty thousand dollars in nineteen-sixty. That was way beyond anyone's budget. Now, it's true that if that had been a tenth or a hundredth of that amount, people would have begun to think of personal computers right then and there. It would have been an interesting machine. But [as it was,] it was just way out of the ballpark of what you could afford."[13]

If the time frame had been the 1990s, of course, he could have just waited: the price would have plummeted within a year or so, assuming that DEC had survived until then. But the dynamic of ever-decreasing prices is unique to the microchip age. In 1960 there was every reason to believe that the economies of scale would keep on pushing computers to grow larger and larger. And that, in turn, seems to have convinced Lick that while the Lincoln/DEC approach of stand-alone interactivity was marvelous fun, and a great way to explore the possibilities of the human-computer partnership, it was not the ultimate answer. There had to be a better way, he reasoned.

And of course, it just so happened that he knew what it might be.

MEETING THE STANDARD

It must have been sometime around June or July 1960, says Leo Beranek. He was walking down the hall at BBN when he caught sight of two strangers sitting in a conference room. No surprise there: Lick had been bringing in all sorts of people to work with the PDP-1 as part-time consultants. And indeed, one of the pair turned out to be Marvin Minsky, the kid who had spent so much time hanging around the Harvard Psycho-Acoustics Lab right after the war. But the other one—well, there was something about the wild shock of hair and the maniacal, fixated stare that gave him the look of a nineteenth-century anarchist getting ready to throw a bomb. This, Beranek sensed, was not just another consultant.

"Excuse me," he asked the two younger men. "Who are you?"

The second stranger looked him up and down with those disturbingly direct eyes. "Who are *you?*"

"And that," says Beranek with a laugh, "is how I met John McCarthy!"

Beranek was right: John McCarthy was not just another consultant. At age thirty-three, this aloof, wild-eyed mathematician was already among the most innovative individuals in the short history of computing. In the process of achieving that stature, moreover, he had single-handedly reinvented the whole concept of interactive computing—albeit in his own highly idiosyncratic way. He was the founder of artificial intelligence, which for him meant (among other things) a computer that could respond to you in real time, with humanlike common sense. He was likewise the creator of Lisp, an interactive symbol-processing programming language that not only had a compelling mathematical beauty but would let you grow your programs in a much more open-ended, organic manner than batch processing ever could. And he was the inventor of general-purpose computer "time-sharing," a technique that let individual users interact with batch-processing behemoths in a way that looked very much like present-day personal computing.

Of course, McCarthy had also become famous for being decidedly strange, even by Cambridge standards. If a topic of conversation interested him, for example, he might very well seize on the idea and start to analyze it in a kind of thinking-out-loud monologue—and then analyze it and analyze it and analyze it, until the sound of his robotically precise voice made his listeners want to scream. But if he didn't think he had anything worthwhile to contribute, he was perfectly capable of sitting there in dead silence for minutes at a time, leaving his companions to wonder if they'd offended him or said something unspeakably stupid. He was legendary for answering questions hours or even days after they had been asked, walking up to his questioner in the hallway and launching into a brilliantly reasoned response as if no time had passed.

Actually, to a lot of people this sort of behavior wasn't just strange; it was offensive. It tended to come across as breathtaking arrogance, as if John McCarthy considered himself possessed of powers that had been denied to mere mortals. In particular, as even McCarthy himself now admits, he didn't fit in well with Lick's easygoing style of give-and-take bull sessions: "I tended to hold back on ideas," he says, "hoping to develop them further."

In fairness, however, McCarthy's "arrogance" grew out of his own standards of intellectual quality, which were so obsessive that even he himself rarely met them—especially when it came to the field he'd help invent. "I had certain standards of mathematical precision for what a scientific paper ought to be like," he explained in an interview conducted by the writer Pamela McCorduck in 1974, adding that by "precision" he meant mathematical abstraction, formal definitions, powerful theorems, and rigorous proofs. "But finally it became clear that I wasn't going to solve the artificial intelligence problem in a mathematically rigorous way in reasonable time, so I simply decided to start publishing what I had."

The impatience was understandable: McCarthy had been dreaming about machine intelligence for most of his life. Born in Boston in 1927, the son of a Lithuanian Jewish mother and an Irish Catholic father who had converted to Marxism and then fought for his beliefs as a union leader, young John had decided to become a scientist by the age of ten. "Science seemed to me the only serious search for an explanation of the world," he recalled to McCorduck. "Since I was brought up to be antireligious, the notion of a nonscientific, religious explanation never had any attraction." By no coincidence, he had also become an avid reader of science fiction, and an increasingly skeptical analyst of the sci-fi writers' accounts of how their robots worked. "Every speculation about the mechanism was wrong," McCarthy says, "especially Asimov's Three Laws of Robotics, which are kind of Talmudic laws rather than scientific laws." His fascination with the true nature of intelligence continued throughout his high school, college, and army years and only grew stronger after the war, when the G.I. Bill allowed him to return to Caltech as a graduate student in mathematics. (He had started there as a sixteen-year-old freshman in 1943, and had soon been expelled for refusing to submit to the required physical-education classes.) "I heard the talks [on automata theory] by von Neumann and other people," he told McCorduck, "and immediately arrived at the conclusion that, well, I could do better than that—or anyway, that their approaches didn't seem to me to be correct."

For the moment, unfortunately, McCarthy lacked any alternative approach that met his standards for rigor. So, being unwilling to speak up without having anything worthwhile to say, he spent most of his time on more conventional mathematics, first at Princeton, where he got his Ph.D. in 1951; then briefly at Stanford; and finally, starting in February 1955, at Dartmouth College in New Hampshire.

It was at Dartmouth that McCarthy soon found a whole new realm of possibilities opening up, courtesy of IBM. As luck would have it, the company was just then putting the finishing touches on its newest scientific computer, the IBM 704. And as luck would also have it, the company had just agreed to donate one of the new machines to MIT, where it would be housed in a brand-new

"Computation Center." (MIT was after a replacement for Whirlwind: the vener- able machine was already obsolete and far too expensive to operate, though a number of researchers had come to depend on it. And IBM was happy to oblige; the company had been planning to place 704s at a few strategic universi- ties all along, on the theory that students would then become loyal users and de- mand IBM machines in their future workplaces.) Eight hours of the machine's time per day would be allotted to MIT researchers; another eight hours would go to IBM's own local scientific office; and the final eight hours would be shared by a consortium of smaller New England colleges, including Dartmouth.

The upshot was that in the spring of 1955, as part of the planning for the gift, Dartmouth received a visit from a group of IBMers based at the company's lab- oratory in Poughkeepsie, New York. "I remember that they took us out to din- ner," says McCarthy. "And after we bought our own drinks, they made it clear that they would pay for the dinner. So the idea of my spending the summer at IBM was first broached then."

Fortunately for him, McCarthy said yes. If nothing else, the summer at IBM gave him another crack at machine intelligence; the information research group in Poughkeepsie had been a hotbed of speculation on the subject ever since the company built its first computer, the IBM 701. For example, an engineer named Arthur Samuel had used the prototype 701 to test an early version of his checkers- playing program, which would later be hailed as one of the landmarks of artifi- cial intelligence. And group leader Nathaniel Rochester, along with several of his younger team members, had created a series of biologically realistic neural- network simulations in an effort to test Donald Hebb's theories about learning in the brain. (As it happened, they had been inspired to start this work by one J. C. R. Licklider, who visited the lab in 1951 and gave a compelling lecture on Hebb's ideas.)

More important, however, that summer at IBM gave McCarthy a chance to learn computer programming. Because once he had started to do that, he was al- most immediately struck by a beautifully simple realization: the way to create machine intelligence was to program it on a general-purpose computer. Whether it be using language, forming abstractions, recognizing new concepts, solving problems, or learning from experience, he later declared, "every aspect of learn- ing or any other feature of intelligence can in principle be so precisely described that a machine can be made to simulate it."[14]

Viewed from the perspective of the new millennium, of course, the idea of simulation is so blazingly obvious that it hardly seems to be an insight at all. But in 1955, says McCarthy, "it was just not a common idea. Among the first gener- ation to be interested in artificial intelligence, most thought about building some kind of machine for the purpose." This group included essentially all the cyberneticists, ranging from McCulloch and Pitts, with their neural-network models, to Gray Walter in England with his "turtles," little wheeled devices that used photoelectric cells and feedback circuits to seek out light sources with creepily lifelike determination. For that matter, it included J. C. R. Licklider and his analog circuit models of the brain. Intelligent behavior resided in the hardware, went the cybernetic line. In fact, says McCarthy, the first person to make a reasonably explicit case for the software approach was Alan Turing, in the 1950 paper that introduced his Turing test. And even there, says McCarthy, it was not very prominent; the first time he read that paper, the software idea didn't even sink in.

But it was sinking in now, and bringing with it a new resolve to launch yet another frontal assault on machine intelligence. "I had this idea that if only we could avoid all these distractions and devote some time to it," says McCarthy, "if we could just get everyone who was interested in the subject together, then we might make some real progress." McCarthy got three other people to go in with him on the idea: IBM's Nat Rochester, Bell Labs' Claude Shannon,* and the Harvard mathematician Marvin Minsky, whom McCarthy had met when they were both mathematics students at Princeton. He wrote up a memo to the Rockefeller Foundation asking for money to fund a six-week conference to be held the following summer at Dartmouth. And on August 31, 1955, he sent it off, under the title "A Proposal for the Dartmouth Summer Research Project on Artificial Intelligence."

The phrase was definitely an attention-grabber; even *The New Yorker* got wind of it and made the obligatory joke about the need for "natural" intelligence. But McCarthy maintains that he wasn't trying to be provocative. In fact, he isn't even sure anymore where the phrase "artificial intelligence" came from; he just remembers that it seemed to express precisely what he wanted the conference to be about. And apparently it did just that. Lasting through most of the summer of 1956, the Dartmouth conference has subsequently become famous as the event that finally established artificial intelligence as a field in its own right and gave it the name that has stuck.

Characteristically, however, McCarthy once again felt disappointed. "I simply measured the distance between what I had hoped to accomplish and what we did accomplish, and it was pretty large," he says. He had hoped to see von Neumann there, but the great mathematician already lay dying. Of the ten people who did attend, few stayed for the full six weeks, and almost no one altered his thinking because of anything he learned there. Worse, only two participants had impressive new results to report—Newell and Simon came to Dartmouth bearing printouts from the preliminary runs of Logic Theorist—and *they* had a noticeable chip on their shoulder. "They didn't stay very long," says McCarthy, "and they had a tendency to feel, perhaps quite correctly, that the conference

---

* In one of his few prior forays into artificial intelligence, McCarthy had spent the summer of 1952 working with Shannon at Bell Labs. The main result was a volume of invited papers on the subject, which they jointly edited. It would appear in 1956 as *Automata Studies,* a title that McCarthy hated. But with contributions from von Neumann and many others it added up to present a greatly enriched understanding of what Turing machine–like automata–that is, computers–can and cannot do. The book is now considered a landmark in computer science.

was being run by people who hadn't actually done anything." Besides, neither Newell nor Simon liked the term "artificial intelligence"; the two would continue to use their own phrase, "complex information processing," for years afterward.

It probably didn't help that the feeling was mutual. McCarthy found Newell and Simon's heuristic-reasoning approach sloppy and ad hoc; it didn't have anything like the kind of clarity and precision he was looking for in AI. His own thinking centered around a concept he would eventually call Advice Taker. From the user's point of view, Advice Taker would be a fully interactive machine intelligence, capable of responding to everyday questions with reason and common sense. If you asked it, for example, "How can I get to the airport?" it would reply, "Take your car; it's in the garage." It would also be able to incorporate suggestions from the programmer about how to do better next time—thus the name. Under the hood, however—and this was where McCarthy differed radically from Newell and Simon—Advice Taker would not base its reasoning on heuristic rules of thumb. Instead, it would be supplied with a vast database of propositions expressed in some clear and precise formal notation such as symbolic logic, with each proposition representing a piece of commonsense knowledge about the world. (Example: "If I am at my desk and I walk from my desk to my car, then I am at my car.") Then, when the computer had to make a response, it would automatically draw upon that database and generate all the appropriate logical inferences it needed to reach a decision.*

During the year that followed the AI conference, alas, McCarthy was once again forced to put such matters aside; life as a junior math professor at Dartmouth left him very little time for anything but some fairly standard mathematics research and teaching.

Still, that didn't stop him from thinking about artificial intelligence, or from refining his ideas about how to get there. Not only was there the fundamental problem of commonsense reasoning, for example; there was also the very practical problem of interactivity. After all, McCarthy figured, no matter how you chose to define "intelligence"—and it was a remarkably slippery term—it surely

---

* Formalizing common sense is actually much harder than it sounds, as McCarthy had already realized. For instance, ordinary formal logic is "monotonic," in the sense that learning something new can never cause you to retract a conclusion; it can only allow you to generate more conclusions (or reach a total contradiction). But everyday common sense isn't monotonic at all. Say you've concluded that you must drive your car to the airport and then you discover that your battery is dead. You instantly reverse yourself and conclude that you *cannot* drive your car to the airport.

Such problems implied that ordinary formal logic would have to be drastically modified in order to handle common sense. McCarthy would never be able to achieve that objective to his own satisfaction. Today, after four decades of research, he's still trying: Advice Taker was never implemented. Nonetheless, the logical-deduction philosophy behind Advice Taker gave rise to what is now known as the logic-based approach to artificial intelligence—as opposed to the heuristic approach, the neural-network approach, and so on.

included an ability to perceive the environment, to draw inferences about the situation at hand, to make appropriate responses, and to learn from experience. So at a minimum, an intelligent computer ought to be responsive to its programmer. If you inputted a question, it should at least take a shot at passing the Turing test, and give you an answer. And it should do so *now*, not six hours from now, with the answer buried somewhere in a stack of fan-fold printout.

This last was a point that McCarthy could speak to with some feeling, especially after 1955 and his summer at IBM's Poughkeepsie lab, when he'd been traumatized by his introduction to programming via batch processing. Those bureaucrats wanted *him* to wait for the computer? They considered *his* time to be less valuable than the computer's? Not hardly. Quite aside from any concerns about machine intelligence, McCarthy felt, a truly responsive computer would tell you about those damned missing commas right away, so you could fix them and keep going without breaking your train of thought. Nor would it force you to think through every detail of your programming code beforehand, as a batch-processing operation tacitly expected. That was fine for fields such as physics or data processing, where the equations were already in the textbooks and the only real problem was converting them into bug-free algorithms. But it was a disaster in an arena such as artificial intelligence, where there were no textbooks and you had to grope your way toward the solution every step of the way. What you needed in that case was a programming language and a computer system that would let you explore, try things out, see what happened, and *learn*. In effect—though McCarthy certainly wouldn't have said it this way—what you needed was a system that would help you carry out Newell-and-Simon-style heuristic search. And most especially, what you needed was interactivity day or night, at work or at home, anytime you had an idea, anytime you wanted.

Unfortunately, McCarthy knew, real computers weren't interactive at all—or at least none of the ones *he* knew about were. And that triply shared IBM 704 down at MIT wasn't much of a substitute, not when he had to travel the 135 miles from Dartmouth to Cambridge just to submit a deck of punch cards. But no matter: he had already come up with a solution. And now, thanks to a Sloan Foundation fellowship that would support him for a full twelve-month sabbatical at the MIT Computation Center, it looked as though he would finally get a chance to try his solution out.

He headed south in September 1957, a man with a mission.

McCarthy has always maintained that the idea simply popped into his head during that same IBM summer of 1955, almost as soon as he'd recognized the problem. If you wanted to create your programs interactively, he'd reasoned, and if nobody was going to give you your very own 704 to do it with, then the obvious answer was to get together with a bunch of other users to share a machine. And not just share it in the lockstep, line-forms-at-the-rear manner of batch processing, either. Really *share* it. Give everybody a remote terminal so they could all tap in to the big computer through telephone lines whenever they liked, from wherever they liked. Once they were in, moreover, assign each of them a securely walled-off piece of the computer's memory where they could store data and programming code without anybody else's horning in. And finally, when the users needed some actual processing power, dole it out to them via an artful trick.

You couldn't literally divide a computer's central processing unit, McCarthy knew; the standard von Neumann architecture allowed for only one such unit, which could carry out only one operation at a time. However, even the slowest electronic computer was very, very fast on any human time scale. So, McCarthy wondered, why not let the CPU skip from one user's memory area to the next user's in sequence, executing a few steps of each task as it went? If that cycle was repeated rapidly enough, the users would never notice the gaps (think of a kindergarten teacher holding simultaneous conversations with a dozen insistent five-year-olds). Each of them would perceive his or her program to be executing continuously. And more to the point, each would be able to create and modify and execute programs interactively, as if he or she had sole control of the computer.

Since the users would be sharing the computer's processing time as well as its storage space, McCarthy took to calling his scheme time-sharing. And characteristically, he wasn't too impressed with himself for having thought it up. "Time-sharing to me was one of these ideas that seemed quite inevitable," he says. "When I was first learning about computers, I [thought] that even if [time-sharing] wasn't the way it was already done, surely it must be what everybody had in mind to do."

Wrong. Nobody at IBM had even imagined such a thing, not in 1955. It's true that the company was the prime contractor for the SAGE air-defense system, which called for "time-sharing" in the sense that one central computer would control many radar terminals. But the SAGE computer would be running just one overall program to manage everything, with the individual radar operators' being limited to a few standard queries and responses. McCarthy's proposal was far more radical. He wanted to give the users free rein inside the computer so they could play, experiment, meditate, run programs, modify programs, crash programs, and waste time as they pleased. In effect, he was proposing to optimize *human* time instead of *machine* time. But in 1955, at IBM, that kind of proposal sounded both naive and self-indulgent. For one thing, the technology wasn't up to it. The computer memory that McCarthy was so blithely planning to carve up was in fact very expensive and very small (core memory was just making its commercial debut). Likewise, the central processing unit that he wanted to send skipping so merrily from user to user was made of vacuum tubes, which were giving the engineers enough trouble as they tried to get their machines to work on straightforward arithmetic. And besides, went the argument, even if you could get time-sharing to work technologically, why would you *want* to? If people were too lazy to write their programs correctly in the first place, why was that the computer's problem?

OK. Fine. McCarthy set the idea to one side while he organized the Dartmouth conference. But he picked it up again in September 1957, as soon as he'd started his sabbatical at MIT: What would the Comp Center say to a little time-sharing demonstration on the 704? Given the state of the technology, McCarthy remembers, "I was rather modest in terms of what I thought we could do [on the 704]." But better technology was coming, he argued, and they ought to be ready for it.

The Computation Center would say yes, actually. In principle. With due care. That was what the place was for, after all; its director, the physicist Philip Morse, had founded it to be a center for doing research on computers, not just for chewing on punch cards. That was why he brought in people such as McCarthy in the first place. And in any case, the Computation Center staff had already discovered that administering a batch-processing system was almost as annoying as using one. The center had started out operating the 704 in the "personal computer" style that so many employees there remembered from the Whirlwind days: you just signed up for an hour, and the machine was yours for the duration. Indeed, the 704 had been way underutilized in the beginning, to the point where staffers considered it part of their job to go out and drum up business. Aside from the small cadre of Whirlwind veterans, very few people had the slightest idea of what to do with computers, even at MIT. And those who did were often intimidated by them.

But then came April 1957, and Fortran, the new Formula Translation language created for the IBM 704 by the mathematician John Backus and his team at IBM headquarters in New York City. Suddenly, instead of struggling through the dreary, error-prone process of writing out numeric codes that only the computer could understand, programmers using Fortran could deal with humanly comprehensible variables such as density and velocity. They could manage comparatively transparent commands such as "IF Temperature <273 THEN GOTO Step 55." They could even write their science and engineering programs in a reasonable approximation of standard mathematical notation. Only at the end would they have to worry about translating it all into hard-core computerese—and at that point a special piece of software called a compiler would take over and do the translation for them.

"Fortran made a big splash," says Fernando Corbato, who was then a young Ph.D. physicist serving as Morse's deputy. "It opened the door to people who were intimidated by machine language, [and] enlarged the user base by a factor of ten, at least. And gradually we, and almost everybody else in the country, switched to batch processing." It was the only way to handle the exploding demand. Of course, he adds, that wasn't much of a problem for the large banks or for national laboratories swimming in Cold War defense money. If they found themselves with too many users waiting for the computer, they could just buy another computer. But universities didn't have that luxury, not even MIT. The upshot was that Corbató, as the senior full-time staffer at the Computation Center, constantly had to listen to irate users complain about twenty-four-hour turn-around times—even as he was nagging, cajoling, and threatening other users to run their big jobs at night instead of during peak hours. "It was driving everybody crazy," he says.

So if McCarthy had a better idea, Corbató was all for letting him try it. Still, McCarthy's little time-sharing experiment wasn't going to be easy. If nothing else, the 704 couldn't be modified for time-sharing with a mere change in software. Alterations would have to be made in its hardware as well, with the most obvious being the addition of an "interrupt" system that would allow the machine to put one task on hold and switch to another. "I was very shy of proposing hardware modifications," McCarthy later wrote, "especially as I did not understand electronics well enough to read the logic diagrams."[15]

But he forged ahead anyway. McCarthy learned to read the logic diagrams. He obtained permission from both MIT and IBM to make the required modifications. He got a promise of financial support from the National Science Foundation. He discovered that IBM had already created a gadget that would make a very nice interrupt system—a "real-time package" that had been developed at the request of Boeing, to allow the 704 to take in data from wind-tunnel experiments—and he talked the company into forgoing the package's $2,500-per-month rental charge and letting MIT use one for free. He got the Computation Center engineers to design the hardware for patching in a Flexowriter, a kind of stand-alone typewriter that was more typically used for punching out paper tapes. And then . . .

Well, then McCarthy had to do a lot of marching in place, as IBM delayed delivery of the real-time package for more than a year. Happily, however, he had plenty of other things to keep him occupied in the meantime.

McCarthy's year at MIT had turned out to be a pretty good one, actually, not least because of the presence of Marvin Minsky, who had just relocated to Lincoln Lab. True, the two men were something of an odd couple, and had been ever since their graduate-student days together at Princeton. When it came to the fundamentals of the field, to their most basic assumptions about the nature of intelligence and how to create it in a machine, they didn't agree on anything. The aloof, cerebral McCarthy was still in search of a mathematician's explanation of intelligence; he couldn't be satisfied with anything less than a theory that was deep, precise, logical, and beautiful. And he was more convinced than ever that his Advice Taker notion was the right way to get there.

Minsky, by contrast, was even then a balding gnome who went at AI with the playfulness of a child. Not for nothing was he known as the world's oldest three-year-old: he liked to build things, to see how they behaved, and to learn by doing. He saw no particular requirement for elegance or rigor in a theory of intelligence. After all, he argued, both the mind and the brain were products of evolution, and natural selection was a consummate tinkerer, slapping parts together with the abandon of a Rube Goldberg. Indeed, Minsky would later become famous for declaring that intelligence was a kludge. So he was sympathetic to Newell and Simon's heuristic-reasoning approach. But he was also sympathetic to the neural-network approach, to the robotics approach, and to any other approach that seemed useful; for him, whatever worked, worked.

Odd couple or not, however, Minsky and McCarthy did share a bedrock conviction that AI was important, exciting, and worth pursuing. With the approach of September 1958, moreover, as McCarthy's sabbatical at MIT was coming to an end and he really should have been packing for his return to New Hampshire, the two men also found themselves sharing a conviction that Dartmouth was just too far away for regular schmoozing. So instead, McCarthy and Minsky went to see RLE director Jerry Wiesner and asked, would it be possible to start up a new artificial-intelligence lab there, at RLE?

"Sure," enthused Wiesner, with all the expansiveness of an age when funding agencies were content to let managers manage without a lot of second-guessing. "What do you need?"

So everything started instantly, says McCarthy. "They gave us a room in the basement and a secretary and a keypunch, plus six graduate students from the math department, and they let us hire Steve Russell, who had been working with me at Dartmouth."

It must have been a magical moment: for the first time in his life, McCarthy was finally free to focus totally on AI, with a whole community of like-minded people around him. Indeed, by all reports the AI group jelled almost immediately, becoming yet another of those wonderfully exhilarating cabals that seemed to spring up so often in those days, especially around RLE. "The group was amazingly social," recalls Daniel Bobrow, a graduate student who joined the lab shortly after the first six. "We were over at Minsky's house and McCarthy's house a lot. And there was this span of interests from the frog's brain to the human brain to the mathematics of it all. I remember one discussion among Marvin, Claude Shannon, and me about how hands work. Well, the hand is marvelously complicated. I remember Shannon's going home—he had a great machine shop in his basement—and when he came back several days later, he had made a model of a finger. One finger. I tried to write some software in Fortran that would control the finger. And I couldn't because there was a bug in the compiler!"

McCarthy, for his part, had made it a top priority that first autumn to codify all his thinking on Advice Taker into a formal proposal, which he would eventually present that December at a conference in Teddington, England. But he made it an even higher priority to forge the tools he needed to implement Advice Taker, starting with a computer language suitable for interactive AI.

Of course, the computer world was hardly suffering from a shortage of new languages, even then. Thanks to the proliferation of commercial machines such as the IBM 650 and 704, plus the success of Fortran, the years 1956 and 1957 had already seen new dialects emerging by the dozens (two of the languages produced during this era, Fortran itself and the business-oriented Cobol, are still in widespread use today). However, none of these efforts came close to meeting McCarthy's standards. In all of the new languages, for example, each symbolic variable had to be assigned to a fixed block of storage in the computer's memory, so that one block might be allocated to the variable "name," another block to "salary," and so on. In fact, those assignments had to be made by the programmer up front, before the program could even begin to execute (in many modern languages, such as C++, this is still the case). But in artificial-intelligence research, such a demand would be sure death. The "variables" in an AI program somehow had to represent the quicksilver fluidity of mental states in human working memory—the images, concepts, possibilities, goals, and alternatives that the problem solver focuses on at any given instant. And there was no way for the programmer to know in advance how big or how complex these variables should be, because they would constantly be changing as the problem solving advanced.

Fortunately for McCarthy, however, a solution was already at hand, courtesy of Carnegie Tech's Allen Newell. Some three years earlier, in the course of creating his Logic Theorist program, Newell had come up with something he called a list, a data structure in which chunks of information scattered throughout the computer's memory were connected in a loose chain. In Newell's conception, the list worked a bit like one of those old-fashioned treasure hunts in which each hiding place contained a piece of the treasure, plus a clue to the location of the next hiding place. (Or to use a more up-to-date analogy, a list was like a chain of World Wide Web pages, with each page containing a certain amount of data plus a hyperlink to the next page.) Thus, to go down the list, the computer would just have to follow the links in order, jumping freely through the memory banks as it did so. Moreover, by adding and dropping links or shifting them around, the computer could freely expand or contract the various lists without having to move the data themselves. It could start with, say, a simple linear chain of symbols—*(a b c d)*—and then replace the third element with a whole new list: *(a b (g h) d)*. Indeed, the computer could keep on nesting one list inside another like this until it had built up matrices, tables, trees, forms, or any other kind of data structures. Or conversely, it could lop off whole branches at will. The result would be a style of computation in which the data structures weren't fixed at all but could form, grow, interact, change, and evaporate, much as human ideas do.

Newell had accordingly put list-processing operations at the very heart of Logic Theorist, which was written in a list-based "Information Processing Language" that he had developed with J. Clifford Shaw, the former RAND colleague who did the actual programming on the JOHNNIAC computer in Santa Monica. Newell had also presented a detailed description of the list-processing idea at the Dartmouth conference in 1956, to John McCarthy's intense interest.

McCarthy, not surprisingly, had a low opinion of the Information Processing Language as it stood; Newell and Shaw, being in a hurry, had created a spartan, utilitarian framework in which the commands were abbreviated by arbitrary numeric codes, the program statements had to be rigidly formatted to fit on IBM punch cards, and many of the syntactical features existed only because they were required by JOHNNIAC, the particular computer they were using at RAND. But list processing itself—now *that* was an elegant idea. So in the autumn of 1958, with the new AI lab finally giving him the opportunity and the wherewithal to focus on the problem, McCarthy decided that it was time to take that idea and turn it into a whole new language for AI.* He and his students soon took to calling it List Processor, or Lisp.

Its initial development took the better part of two years, recalls the computer scientist Paul Abrahams of the Courant Institute, who was one of the first and most enthusiastic of those original students. At this late date, the memory of the process has mostly become a happy blur, almost impossible to reconstruct in any detail. Fortran and most of the other early languages had been created by formal teams and committees, Abrahams points out, so there were lots of related documents and memos circulated. "[But] Lisp really was the work of one mind," he says. "The rest of us mostly added things. We implemented it, we added trimmings, but basically Lisp was John McCarthy's invention."[16]

Certainly it was a giant step beyond Newell's original conception of list processing. McCarthy's most striking addition was his insistence that every list define what mathematicians call a function, a kind of abstract machine that takes in raw materials at the input hopper, grinds away, and produces a finished prod-

---

* Actually, before deciding to create a new language from the ground up, McCarthy had already spent a fair amount of time working with Nat Rochester's group at IBM Poughkeepsie in an effort to graft list processing into Fortran. Among McCarthy's closest collaborators there was an energetic young physicist named Herbert Gelernter, who had taken up a suggestion made back at the Dartmouth conference by Marvin Minsky: devise a program that could prove theorems in high school geometry. (Gelernter was also the father of a son named David, who would grow up to become a well-known computer scientist in his own right and who would achieve unwanted fame in 1993 by becoming a victim of the Unabomber.) By the fall of 1958, Gelernter, McCarthy, and their coworkers had gotten their Fortran List Processing Language working well enough for Gelernter to use it; his Geometry Theorem Prover is now considered yet another of the early landmarks of artificial intelligence.

In one sense, however, that particular landmark was a little *too* prominent. By the fall of 1958, the reputation of the AI work in Poughkeepsie had spread to the point of being featured in the *New York Times* and *Scientific American*. Shortly thereafter, IBM president Thomas J. Watson, Jr., having been personally badgered by stockholders demanding to know why the company was wasting money on frivolous junk like this—and having heard far too many IBM salesmen complain that customers were feeling threatened enough by computers as it was, without *artificial intelligence*—told Rochester and his colleagues to cease and desist. Thus IBM essentially abandoned artificial-intelligence research, and for another generation, its salesmen could soothingly tell customers that IBM computers would do only what you told them to do.

uct at the output pipe. Thus, a Lisp function *plus* would be given two or more numbers as input, and the computer would produce their sum as its output: (*plus* 2 2) → 4, and so on.

Now, the point of all this was that a Lisp programmer could start with a handful of basic functions and then systematically create new functions by hooking up the output pipe of one to feed the input hopper of the next. Syntactically speaking, it was simply a matter of nesting one function inside another (and keeping track of all the parentheses). For example, a programmer might combine multiplication, addition, and subtraction in a certain way–(*times* (*plus* w x) (*minus* y z))–and then feed in four numbers such as *2, 2, 5* and *3,* so: (*times* (*plus* 2 2) (*minus* 5 3)) → (*times* 4 2) → 8.

Of course, the process of building new functions out of old needn't stop there. A Lisp programmer could happily go on indefinitely this way, hooking up simple functions into more complicated ones, then hooking up *those* functions to create still *more* complicated functions, until things got as sophisticated as anyone could want. Indeed, this functional structure ultimately turned out to be Lisp's most powerful and compelling feature. First, it fitted in perfectly with McCarthy's goal of interactive, question-and-answer-style computing: all you had to do was think of each Lisp function as a question, and its value as the computer's answer. That was one big reason McCarthy and his students decided not to have Lisp compiled directly into machine code, like Fortran and most other languages. Instead, they created a kind of monitor program called the Lisp interpreter, which would scan each function as the user typed it in, convert it to machine code on the fly, compute the results, and then respond instantly.

Second, Lisp's functional structure gave the programs themselves a striking clarity. Instead of wallowing in undifferentiated masses of numeric machine-language codes (or, for that matter, undifferentiated masses of Fortran statements), McCarthy and his students could build their programs out of meaningful, well-defined building blocks–namely, Lisp functions. And that, in turn, allowed various members of the group to single-handedly tackle what were then considered very large and complex problems–carrying out the symbolic manipulations of ordinary algebra, for example, or symbolically checking a mathematical proof for correctness.

The use of software building blocks was hardly a new concept even then, of course. Starting in the 1940s, long before Lisp, or Fortran, or any of the other languages appeared, programmers had learned that they could save themselves a lot of time and confusion by breaking up programs into *subroutines*–reusable, self-contained procedures that could each do one specific task. One subroutine might calculate the cube root of any given number, for instance, while another might sort any given list of names alphabetically. These procedures could in turn be broken up into still simpler subroutines, and so on, all the way down to the level of individual commands, if need be. Indeed, as programmers tackled tougher and tougher challenges in the 1950s–the SAGE project, for example–

this kind of decomposition had become increasingly critical.* Even today, under the rubric of "structured programming," it is still considered the essence of good programming practice.

Similarities notwithstanding, however, the subroutine approach and Lisp's functional approach did tend to encourage different styles of programming. After all, any language that emphasized *breaking up* a program into subroutines was almost automatically going to emphasize a top-down management philosophy: first decide the large issues, then delegate the details to subordinates, and so on down the line. And indeed, just as in the corporate world, this could be a very effective way to manage complex undertakings *if* you already knew what problems you were trying to solve. The top-down approach also fitted in very well with compiled languages, such as Fortran, which required that all the code be ready and waiting before the compiler could start its work. And that, in turn, meant that the approach lent itself well to batch processing.

Lisp, by contrast, was better suited to the kind of bottom-up, exploratory style of programming that McCarthy had been after all along. Instead of breaking things up, Lisp programmers were constantly putting things together, using simple functions to build more complex functions, trying things out, seeing what worked, seeing what didn't work, and learning as they went. True, Lisp did pay a heavy price for its flexibility: being interpreted line by line made Lisp programs agonizingly slow and inefficient compared to, say, a Fortran program that

---

* It was no coincidence that these same few years at the end of the 1950s saw the emergence of Algorithmic Language, or Algol, which was the first to make this kind of decomposition into subroutines part of its very structure.

Algol has been called the most beautiful computer language ever written by a committee; certainly it was the most influential. The Algol committee, which had its first official meeting in January 1958, was a joint U.S.-European effort to create a new "universal" programming language, a kind of digital lingua franca that would replace Fortran, FLOW-MATIC, and all the rest with a second-generation language that everybody could use, so that different computer groups could communicate with one another and run one another's programs. The explosive proliferation of new languages had gotten out of hand, went the argument, and the infant industry desperately needed standards. More than a dozen leading language mavens agreed to serve, including John Backus of IBM, Alan Perlis of Carnegie Tech, and one John McCarthy, still officially of Dartmouth. Indeed, the challenge of creating a whole new programming language seems to have set McCarthy's mind aboil with ideas about what a beautiful and rigorously designed language ought to look like, and led him to clarify many of the principles he would later incorporate into Lisp.

In one sense, at least, the Algol committee failed: the language community is still Babel. By the time the final version was defined in 1960, older languages such as Fortran were too well entrenched to be abandoned. Algol itself was never widely used except in parts of Europe. Nonetheless, it foreshadowed the structured programming movement of the 1970s and was the direct ancestor of the modern language Pascal. Moreover, it was in developing Algol that John Backus and Denmark's Peter Naur invented the now-universal Backus-Naur form notation for giving the formal definition of a programming language.

had been compiled in the conventional manner. It would be the better part of a decade before more efficient versions of Lisp were developed, and years more before the language became widely used even for AI. Nonetheless, it's fair to say that one of Lisp's two greatest legacies to the art of programming was a certain style, a certain exploratory approach to pushing back the software frontiers.

And the other legacy? An undeniable grace, beauty, and power. As a Lisp programmer continued to link simpler functions into more complex ones, he or she would eventually reach a point where the *whole program* was a function—which, of course, would also be just another list. So to execute that program, the programmer would simply give a command for the list to evaluate itself in the context of all the definitions that had gone before. And in a truly spectacular exercise in self-reference, it would do precisely that. In effect, such a list provided the purest possible embodiment of John von Neumann's original conception of a stored program: it was both data and executable code, at one and the same time.*

In mathematics, the technical name for this sort of thing is recursive function theory, which was why McCarthy called his first public description of Lisp "Recursive Functions of Symbolic Expressions and Their Computation by Machine." Today ranked as one of the most influential documents in the history of computer languages, that paper established that a language could have a rigorous mathematical foundation. And it signified that John McCarthy had finally come up with a framework that was precise enough, rigorous enough, and compelling enough to satisfy even him. Certainly it changed how McCarthy's AI students perceived their creation, says Abrahams: "Now, all of a sudden, Lisp was not merely a language you used to *do* things. It was now something you looked at, an object of beauty. It was something to be studied as an object in and of itself."

So there it was: by the beginning of 1959, just eighteen months after his arrival at MIT, John McCarthy had a secure position as codirector of the AI lab. He had a fully fleshed-out proposal for Advice Taker, his vision of interactive artificial intelligence. He had the outlines of Lisp, an interactive computer language that he could use to create that artificial intelligence. He had a beautiful scheme for time-sharing, which would effectively give him an interactive computer to

---

* Actually, in an even more spectacular exercise in self-reference, the Lisp interpreter itself could be expressed as a Lisp function. In other words, Lisp could be written in Lisp, with only a very small kernel of machine code to keep it moored to the physical operation of the computer. That Lisp kernel, known as *apply*, thus provided a particularly elegant example of a universal Turing machine: it was the universal function that took the definition of any other function as input and then executed that function. By no coincidence, McCarthy implemented this kind of functional programming in Lisp using the notation of the "lambda calculus," which Alonzo Church had created twenty years earlier to solve the decidability problem, and which had allowed him to beat Alan Turing to the punch.

run it all, and he even had everything in place to run the world's first demonstration of that scheme—everything, that is, except the blasted interrupt handler, which IBM still hadn't delivered.

McCarthy was sick of waiting around, to put it mildly, which may account for the urgent tone of the memo he wrote to Comp Center director Philip Morse on January 1, 1959. His basic message was, Seize the moment! Over the next eighteen months the IBM 704 was to be upgraded twice, McCarthy pointed out, first to an IBM 709 and then to a transistor-based 7090 (IBM had promised in the original deal to replace the center's machine with new models as they came out). In anticipation of that, he said, MIT should start work *now* on a brand-new time-shared operating system for the machines. "If [this] proposal is to be considered seriously, it should be considered immediately," McCarthy declared. "I think the proposal points to the way all computers will be operated in the future, and we have a chance to pioneer a big step forward."[17]

Indeed, McCarthy himself seemed ready to seize the moment, with or without Morse. "McCarthy is calling a series of meetings to consider operator and compiler programs for the 7090," wrote the MIT mathematician and computer-aided-design pioneer Douglas T. Ross in his diary for the week of January 12, 1959, less than two weeks after McCarthy's memo.[18]

By that point, of course, McCarthy had had ample opportunity to experience the Lincoln Lab style of interactivity: Wes Clark and his crew had sent the TX-0 down to campus only a short time after he'd arrived, and it was now ensconced just one flight up from the 704, in the very same building. But it just didn't meet his standards. Yes, you could have complete, interactive control of the TX-0 for an hour or more at a time. But with some twenty-five to forty people wanting access in a typical week, he noted, "the user has to sign up well in advance and cannot go on whenever he is ready."[19] Moreover, he wrote, with the clock always ticking and the next person always impatiently waiting his turn, a TX-0 user rarely had the time to think about his results as he got them. No, what McCarthy wanted was time-shared access on demand, whenever and wherever he had an idea.

By that point, too, McCarthy had also heard from time-sharing's many critics. Jay Forrester, for one, was a charter member of the think-it-through-and-get-it-right-ahead-of-time school of programming; like many others on the MIT faculty (and in the mainstream computer industry) he continued to view time-sharing as a horrendous waste of computational resources. But then, McCarthy still thought of authority as something to be challenged, not deferred to. So he simply kept on talking and talking and talking about time-sharing. What ultimately had value was not computer time, he kept insisting, but human time.

It was a message that more and more people were ready to hear. "John was an evangelist, and he was a little cocky about how easy it was to do," remembers Corbató with a laugh.[20] What saved McCarthy, and gave him real credibility among the engineering types around MIT, was his willingness to get down and dirty with the technical details. A few years later, says Corbató, after time-sharing had become a well-funded and well-regarded strategy, "a number of people came out of the woodwork and said, 'Oh, I invented time-sharing,' and 'Did you read my paper?' and this and that. The problem was that everyone had kind of dreamy visions of people interacting with equipment. . . . The thing that John did was to spell out the particulars of how you go about [it]."

Eventually, in fact, he even got to demonstrate those particulars, once IBM had finally gotten around to delivering that interrupt controller. The first public demonstration of time-sharing consisted of a nervous John McCarthy's standing next to the Flexowriter in the 704's first-floor control room and speaking via closed-circuit TV to a fourth-floor lecture hall filled with MIT's prestigious Industrial Affiliates. He was acutely aware that this would also be the first public demonstration of interactive Lisp, with disaster lurking at every turn. And he was even more acutely aware that the system was a kludge, an awkward patchwork of hardware and software that would simply let the 704 execute a bit of Lisp code between every batch job.

Still, as McCarthy later told the story, things were going well enough, if rather slowly—until the program unexpectedly ran out of memory and automatically called a housekeeping routine. THE GARBAGE COLLECTOR HAS BEEN CALLED, typed the Flexowriter at a stately ten characters per second. SOME INTERESTING STATISTICS ARE AS FOLLOWS . . . and on and on and on, with McCarthy unable to stop it. At the other end of the TV link, the audience was helpless with laughter; this being MIT, they were sure he was the victim of a practical joke.[21]

He wasn't. But by that point even McCarthy was laughing, because he knew his point had been made: the 704, however crudely, was sharing its time. This could work.

By mid-1960, in fact, McCarthy had gotten so many people so agitated about time-sharing that the MIT higher-ups couldn't ignore it any longer. So in classic academic style, they formed a Long Range Planning Committee to study the future of computing at MIT.

Actually, MIT did need a plan. The IBM 7090 was now scheduled for delivery in 1962. But after that? The technology was evolving rapidly, and the proliferation of commercially available mainframes was nearly as explosive as the proliferation of personal computers would be some two decades later. So the choices were far from obvious. Should MIT just abandon the overcrowded Computation Center and let its individual departments acquire smaller machines independently, as they were already beginning to do? Or should the center try to acquire a truly massive machine that could accommodate everybody? And if so, what kind of fundamental architecture should that next machine have?

With questions like that on the table, the official Long Range Planning Committee had to include such academic barons as Philip Morse, head of the Computation Center, and Robert Fano, head of the information-theory group— even though none of these official members had any hands-on experience with computers. But no matter: the real work was to be done by a subcommittee of about a dozen young agitators that included Fernando Corbató, Doug Ross, and of course, John McCarthy—a group that McCarthy later called "the actual computer scientists, [who] were persuaded that a revolution in the way computers were used was called for."

Not surprisingly, given the number of true believers involved, this subcommittee was of almost one mind about its ultimate goal: a time-sharing computer. And while the members were far from unanimous about how to reach that goal, that didn't seem so important for the time being. What mattered was that at last they were on their way.

THE SYNTHESIZER

None of this, of course, had been lost on J. C. R. Licklider; the time-sharing idea was one of the main reasons he'd invited Minsky, McCarthy, and their students to consult at BBN in the first place. "Most other people thought we were nuts," recalls Fredkin. "We'd tell them, 'You can do this or that with a computer,' and they'd say, 'This is too expensive! You guys are a bunch of crackpots or charlatans!' But not Lick. What he had was an ability to understand what we were dreaming of, and why."

True, Lick wasn't just another one of the boys. He was the boss, for one thing, and considerably older: on March 11, 1960, he celebrated his forty-fifth birthday. He was also the only psychologist in a crowd of engineers, physicists, and mathematicians—albeit a psychologist who knew just as much about modern electronics as any of the others.

But more than anything else, J. C. R. Licklider was still the guy who collected ideas as avidly as he collected people, who enjoyed nothing more than a good bull session, who loved to turn people's assumptions inside out on them, and who was constantly pushing notions as far as they could go. Within that small, contentious community of interactive-computing pioneers, that made him—well, not the leader; nobody could have gotten very far trying to tell John McCarthy what to do, or Wes Clark, or Ken Olsen, or Ed Fredkin. But it did make him the integrator and synthesizer, the one who was doing as much as or more than any of the others to envision what a fully computerized world might be like, to imagine what interactivity might mean in human terms, to articulate where computers were going and what researchers would have to do to get there.

In short, Lick was the one who provided the road map.

The paper was basically written as a favor, remembers Jerry Elkind. In the fall of 1959, about the same time DEC was setting up its prototype PDP-1 in the lobby, Elkind had agreed to edit a new journal called *IRE Transactions on Human Factors in Electronics* (IRE was the Institute for Radio Engineers, a professional association that has since been renamed the Institute for Electrical and Electronics Engineers, or IEEE). "So for the first issue I wanted a lead article that would speak to the future of human factors, not to the past," he explains. "The future was with computing. And since Lick clearly was the person who could articulate that future better than anyone else, that's what I asked him for."

Lick, who hated to write but hated even more to disappoint anyone in his professional family, agreed to help out. He handed Elkind the completed manuscript of "Man-Computer Symbiosis" on January 13, 1960.

"It was . . . beyond expectations," says Elkind, who still marvels at what his mentor produced. Indeed, he says, when you look back at that paper from the perspective of today, knowing what happened later, you can see that it essentially laid out the vision and the agenda that would animate U.S. computer research for most of the next quarter century, and arguably down to the present day.

Lick began the article with a metaphor: "The fig tree is pollinated only by the insect *Blastophaga grossorum* [the fig wasp]. The larva of the insect lives in the ovary of the fig tree, and there it gets its food. The tree and the insect are thus heavily interdependent: the tree cannot reproduce without the insect; the insect cannot eat without the tree; together, they constitute not only a viable but a productive and thriving partnership. This cooperative 'living together in intimate association, or even close union, of two dissimilar organisms' is called symbiosis. . . . The purposes of this paper are to present the concept [of] and, hopefully, to foster the development of man-computer symbioses."[22]

Such a symbiosis would involve much more than merely programming the computer, Lick emphasized. That kind of relationship reduced the machine to a tool, a gadget for executing an algorithm that the human had thought through in advance. What Lick had in mind was a relationship more like that with "a colleague whose competence supplements your own"—that is, a friend who could help you out when the problems got too *hard* to think through in advance. "[Such problems] would be easier to solve," he wrote, "and they could be solved faster, through an intuitively guided trial-and-error procedure in which the computer cooperated, turning up flaws in the reasoning or revealing unexpected turns in the solution."

A second, and closely related, meaning of symbiosis was real-time command and control, Lick noted: "Imagine trying, for example, to direct a battle with the aid of a computer on [a batch-processing schedule]. You formulate your problem today. Tomorrow you spend with a programmer. Next week the computer devotes five minutes to assembling your program and forty-seven seconds to calculating the answer to your problem. You get a sheet of paper twenty feet long, full of numbers that, instead of providing a final solution, only suggest a tactic that should be explored by simulation. Obviously, the battle would be over before the second step in its planning was begun."

In either case, he said, symbiosis meant humans and computers working together in a partnership, with each side doing what it did best: "[Humans] will set the goals and supply the motivations. . . . They will formulate hypotheses. They will ask questions. They will think of mechanisms, procedures, and models. . . . They will define criteria and serve as evaluators, judging the contributions of the equipment and guiding the general line of thought. . . . The information-processing equipment, for its part, will convert hypotheses into testable models and then test the models against data. . . . The equipment will answer questions. It will simulate the mechanisms and models, carry out the procedures, and display the results to the operator. It will transform data, plot graphs. . . . [It] will interpolate, extrapolate, and transform. It will convert static equations or logical statements into dynamic models so that the human operator can examine their behavior. In general, it will carry out the routinizable, clerical operations that fill the intervals between decisions."

For anyone who still wondered whether the effort of creating such a symbiotic relationship was worthwhile, Lick included the story of his self-administered time study and the infamous 85 percent drudge-work factor. He then devoted the rest of his paper to outlining the technical challenges, most of which involved bridging the immense gap between what humans do and what computers do.

The first and foremost challenge, said Lick, was what he called the speed mismatch: "Any present-day computer is too fast and too costly for real-time cooperative thinking with one man." His phrasing suggests that he had already rejected as impractical the idea of everyone's having a computer of his or her own, as Wes Clark or Ken Olsen might have advocated. "Clearly," he wrote, "for the sake of efficiency and economy, the computer must divide its time among many users." He was deliberately noncommittal about how this would be accomplished, noting only that time-sharing systems were currently under active development. Never one to hold his imagination in check, however, he couldn't resist following the notion to its logical conclusion—namely, an on-line "thinking center" not unlike today's World Wide Web. "[It] will incorporate the functions of present-day libraries together with anticipated advances in information storage and retrieval and the symbiotic functions suggested earlier in this paper. The picture readily enlarges itself into a network of such centers, connected to one another by wide-band communications lines and to individual users by leased-wire services."

In writing that last sentence, Lick may have been influenced by McCarthy, who maintains that he had been thinking about time-shared "computer utilities" from the first. But that's not very likely, given McCarthy's penchant for keeping ideas secret until he felt they were ready. And in any case, Lick had described essentially the same vision of networked centers more than two years earlier, in an unpublished essay[23] he wrote in August 1957 for a special study on the air force's long-range research needs. Both the body of that essay and its title—"The Truly SAGE System, or Toward a Man-Machine System for Thinking"—make it clear that Lick drew his inspiration from the SAGE deployment, in which the twenty-three individual combat centers were linked by telephone lines. (It was in a later essay[24] for that same special study, in November 1958, that Lick first used the term *symbiosis* in writing.) But wherever he got the idea, the essential fact was that by the fall of 1959, Lick was already thinking about computer networks that connected individual users on a continental scale—the essence of today's Internet.

Another challenge was input and output, he wrote, since computer communication via punch cards and printouts was hopelessly impoverished relative to human communication via sight, sound, touch, and even body language. His proposed solution was a desk-sized console that would function much like a turn-of-the-millennium personal computer equipped with voice and handwriting recognition: "[a display surface] approaching the flexibility and convenience of the pencil and doodle pad or the chalk and blackboard used by men in technical discussion."

Yet another set of challenges comprised memory, which was still a limited and precious commodity in 1960-era computers, and computers' inherent literal-mindedness—their need to be told precisely what to do at every step. Lick's proposed solutions included some form of cheap, mass-produced "published memory" for distributing reference works (think CD-ROM); data storage that could access items by content and not just by names or keywords (still a difficult process); and languages that would allow the user to instruct the computer by giving it goals instead of step-by-step procedures (even more difficult).

Finally, Lick's article reveals considerable ambivalence about artificial intelligence. He knew that AI was important to symbiosis, since smart assistants are presumably better than dumb ones. And he was clearly fascinated by the subject; as he said on another occasion, AI seemed to offer "the most direct path toward the understanding of intellectual processes."[25] Yet he also knew far too much about the brain and its complexities to believe the hype about AI, which was abundant even then. Assertions by Minsky and others to the contrary, computers were nowhere near replicating basic human abilities such as judgment or common sense. So for a long time to come, Lick wrote, humans and computers would have to work together. With tongue only partly in cheek, moreover, he referred to a recent study he'd done for the air force predicting that artificial intelligence wouldn't be of much use for another twenty years: "That would leave, say, five years to develop man-computer symbiosis and fifteen years to use it. The fifteen may be ten or five hundred, but those years should be intellectually the most creative and exciting in the history of mankind."

When Lick's article appeared in March 1960, it was quite a hit around BBN, says Tom Marill, who can remember many a conversation about it in the hallways. True, Lick's plan for a personal desk console did cause a certain amount of hilarity. "Everyone thought his idea of a horizontal desktop display that you could write on was harebrained," recalls Fredkin. "Ben Gurley pointed out that if you dropped something on it, it would explode." McCarthy, for his part, thought the whole human-computer-symbiosis notion was obvious; he claims to have found "no surprises" in the paper.

For most of the others, however, symbiosis was "obvious" only in the sense of slapping your forehead and saying, "Of course! Why didn't I see that?" Technical quibbles aside, Lick's metaphor seemed to capture the essence of what they'd all been striving for: DEC, Wes Clark, BBN, the Young Turks on campus, everyone. And then to see all their ideas laid out like that as a coherent *system*, complete with a technological road map of how to get there, was breathtaking. "One of us *had* to write that," Marvin Minsky told Lick.[26]

Of course, someone would also have to make it happen. But then in the Go-for-it! atmosphere that Lick had cultivated around BBN, why wait around? Certainly that wasn't Ed Fredkin's style. As soon as he heard what the company's new consultants were up to, Fredkin was eager to try it himself. "John," he said to McCarthy sometime in the early summer of 1960, "let's do time-sharing on the PDP-1!"

McCarthy stared back as if he thought Fredkin was crazy. It had been hard enough to demonstrate time-sharing on the 704, which was a big machine. And doing it right was going to take bigger machines still. The PDP-1 was a great little computer, if you liked the stand-alone approach to interactivity. But in terms of essentials such as memory and processing power, it was, well, *little*.

Fredkin was not to be denied, however. When it came to ingenuity, not to mention a visceral loathing for batch processing, he was one of the few people in the world who could outdo even John McCarthy. So he took it on himself to transcribe the time-sharing idea into hardware for the PDP-1.

Lick, of course, thought Fredkin's time-sharing experiment was wonderful fun. "Just to prove it could be done," he recalled in the 1988 interview, "we divided the scope into four quadrants and let each person have a quadrant of the scope." That was one each for Fredkin, Lick, Minsky, and McCarthy—a format that quite possibly represented the first "windows" ever to appear on a computer screen.

But the author of "Man-Computer Symbiosis" could also see that Fredkin's time-sharing system was far more than just a finger exercise: it was a prototype for his on-line "thinking center." Indeed, says Beranek, who had already caught the fever himself, "Lick immediately saw it as a possible public utility and urged BBN to consider setting up a time-sharing business. We were so excited with the PDP-1 and its possibilities that we ordered two more. They were aboard by year's end in nineteen-sixty. We also let a contract for thirty thousand dollars to DEC to develop software for a PDP-1 specially designed for time-sharing. In early nineteen sixty-two, with that operating system in hand, we were so certain of the success of time-sharing that we submitted a proposal to the National Institutes of Health to design, develop, and implement a time-shared information system for the Massachusetts General Hospital."

By that point, of course, Lick had long since become the world's leading evangelist for human-computer symbiosis—speculating about, elaborating on, expanding, and pushing the idea at every opportunity. Much of that evangelism went unrecorded, unfortunately; once again, there's no way to be sure precisely how Lick's ideas evolved during this period. Given all that happened immediately thereafter, of course, it would be nice to know. But some sense of his thinking does come through in the computer-related papers and speeches that he started publishing in this period, and most especially in the book *Computers and the World of the Future*,[27] which contains the transcribed recordings of a lecture series held at the Sloan School of Business Administration in April 1961 to celebrate MIT's one hundredth birthday.

For example, in delivering his invited commentary after one of the lectures—"The Computer in the University," given by Carnegie Tech's Alan Perlis—Lick first listed the dubious pleasures of batch processing, which seemed to remind him of certain university officials he had known: "[The computer] lives behind a glass wall. It has a tighter appointment schedule, and a more resolute appointment buffer, than a dean." But then he immediately launched into the promise of interactive computing: "No one knows what it would do to a creative brain to think creatively continuously. Perhaps the brain, like the heart, must devote most of its time to rest between beats. But I doubt that that is true. I hope it is not, because [interactive computers] can give us our first look at unfettered thought. It can allow a decision maker to do almost nothing but decision making, instead of processing data to get into position to make the decision."

Ultimately, he continued, such interactive computers "will participate in almost every intellectual transaction that goes on in the university." This obviously included education itself, he said, giving a short description of Tracy and Linda Licklider's experience on the PDP-1. But much more than that, "the computer will explore consequences of assumptions. It will present complex systems of facts and relations from new points of view, 'cutting the cake another way' in a fraction of a second. It will test proposed plans of action through simulation. It will mediate and facilitate communication among human beings. It will revolutionize their access to information."

Interactive computers might even become a new kind of expressive medium. After all, he joked, "I think the first apes who tried to talk with one another decided that learning language was a dreadful bore. They hoped that a few apes would work the thing out so the rest could avoid the bother. But some people write poetry in the language we speak. Perhaps better poetry will be written in the language of digital computers of the future than has ever been written in English."

Perhaps so—and perhaps Lick wasn't really joking. A similar point had already been made during the main lecture of the evening, wherein Alan Perlis described how he had pioneered one of the first undergraduate programming courses in the country. Programming was for *everyone,* he insisted, not just the science and engineering majors. It was a fundamental intellectual skill, like mathematics or English composition. The point of his course was not merely to teach students how to write in Fortran or Algol, Perlis maintained, but also to teach them how to think about *processes* of all kinds—how to describe them, how to analyze them, and how to build up complex processes out of simpler ones.

Lick couldn't have agreed more: programming, like mathematics (or psychology), was one of those disciplines that could simultaneously yield both practical applications and profound insights: "Through its contribution to formulative thinking, which will be, I think, as significant as its solution of formulated problems, the computer will help us understand the structure of ideas, the nature of intellectual processes. [One] can be convinced that information processing, which now connotes to many 'a technology devoted to reducing data and increasing costs,' will one day be the field of a basic and important science. Planning, management communication, mathematics and logic, and perhaps even psychology and philosophy will draw heavily from and contribute heavily to that science."

In hindsight, that's actually a reasonably good description of "cognitive science," the umbrella term that would emerge in the 1970s to encompass most of the disciplines on Lick's list and a few others besides—including, most especially, the cognitive psychology that was then being created by George Miller and company. But even then, though he didn't go into detail about it on that particular occasion, Lick was already taking steps to define this new science experimentally with his PDP-1s. Along the way, moreover, he was coming to understand just how subtle and far-reaching this notion of symbiosis really was. Not only could symbiotic computers clear away the garbage, so to speak, by taking care of the dull, routine stuff that gets in the way of creativity; they could also actively *enhance* our creative powers, through a mechanism that Lick came to call dynamic modeling.

Now, Lick's eyes had always taken on a particular shine when he talked about models. He was a great one for sketching things on the back of an envelope, for example, and a sketch was a kind of model. He was likewise pretty handy with a mathematical equation when he needed to be; that was another kind of model. (Newton's law of gravity, to take a famous instance, can be used to model the fall of an apple, the orbit of the moon, or even the ebb and flow of the tides.) And of course he had been in love with scale models all his life. "I vividly remember, for example, 'flying' a small, tethered model of a Curtiss 'Robin' years ago in the wind of an electric fan in Curtiss-Wright's display room on Washington Street in St. Louis," he wrote in one paper from the 1960s.[28] "The ailerons, elevators, and rudder of the model could be controlled through linkages to a full-sized joy-stick and rudder bar in a crude operator's cockpit behind the model plane. I could make the plane take off, climb, level out, turn (a bit), dive,

glide, and land—over and over again until some other aviation enthusiast succeeded in dislodging me from the seat of twofold privilege. A place in the breeze
was hard to hold on those hot, humid summer days."

A model, in short, is any convenient simulation of reality. However, as Lick
noted in another paper[29] from the 1960s, there are models, and then there are
models: "Ordinary mathematical models are static models. They are representa-
tions in symbols, usually written in pencil or ink on paper. They do not behave
in any way. They do not 'solve themselves.' For any transformation to be made,
for any solution to be achieved, information contained in the model must be
read out of the static form and processed in some active processor, such as a
mathematician's brain or a computing machine. A dynamic model, on the other
hand, exists in its static form only while it is inactive. The dynamic model can
be set into action, by one means or another, and when it is active, it does exhibit
behavior and does 'solve itself.' " The Curtiss Robin was a dynamic model in
this sense, he explained: once the fan was turned on, it flew. It was active. The
same was true of the analog electronic simulations he had once experimented
with at the MIT acoustics lab.

And digital computer models? In principle, said Lick, they combine the best
of both worlds by being both static *and* dynamic—and more. Indeed, he argued,
it is this characteristic that gives software its unique power as a modeling
medium. A model sculpted from software is static when it exists as binary code
on a disk or a tape. As such, it can be stored, transmitted, archived, and re-
trieved, just as ordinary text can be. It in fact *is* a kind of text. Like ordinary text,
moreover—and unlike the balsa wood of a model airplane, or the electrical circuits of an analog computer, or any other physical medium—a computer model
is infinitely malleable. Software can represent a jet fighter one minute and the
girders of a bridge the next. It is the ultimate expressive medium, Lick later
wrote—"the moldable, retentive, yet dynamic medium—the medium within
which one can create and preserve the most complex and subtle patterns and
through which [one] can make those patterns operate (as programs) upon other
patterns (data)."[30]

And that is how a static computer model becomes dynamic: by being executed on a computer. In that context it becomes a process unfolding over time—
not a text but a behavior. Of course, the process isn't a very exciting one if all
you get at the end is a fan-fold printout containing page after page of numbers.
However, said Lick, if you could have a graphics display that allowed you to *see*
the model's behavior—and then if you could somehow grab the model, move it,
change it, and play with it interactively—then the experience of working with
that model would be very exciting indeed. It would be like sitting in the cool
breeze of that fan and flying the Curtiss Robin: you wouldn't just be thinking
about the model, you would be *feeling* it, viscerally. The interactivity would open
up a high-bandwidth channel to our perceptions, to our instincts—to our deepest understanding.

For the psychologist-turned-computer-evangelist, this was *the* great promise of dynamic modeling. After all, as Lick and his coauthor Robert W. Taylor would write in a 1968 article,[31] "By far the most numerous, most sophisticated, and most important models are those that reside in men's minds. In richness, plasticity, facility, and economy, the mental model has no peer." Included among those mental models are images recalled from memory, expectations about the probable course of events, fantasies of what might be, perceptions of other people's motives, unspoken assumptions about human nature, hopes, dreams, fears, paradigms—essentially all conscious thought. Of course, Lick and Taylor would continue, "[the mental model] has shortcomings. It will not stand still for careful study. It cannot be made to repeat a run. No one knows just how it works. It serves its owner's hopes more faithfully than it serves reason. It has access only to the information stored in one man's head. It can be observed and manipulated only by one person." But if you could join mental models to computer models, Lick reasoned, and if you could get the two of them into just the right kind of symbiotic relationship, then you could overcome every one of those limitations.

First and most obviously, he said, the computer would greatly enhance our ability to handle complexity. Partly this is because computers are so good at processing vast quantities of data, but much more important, according to Lick, is their potential to give us a fundamentally new way of representing knowledge. In addition to our classic formats—text, tables, diagrams, equations, and the like—we now have the power to represent knowledge as a *process,* an executable program. Imagine the equations that describe, say, the development of a hurricane. And now imagine a computer simulation that shows us that hurricane wandering across the Caribbean to devastate southern Florida: the equations have been brought to life by the computer in precisely the same way that the score of Beethoven's Fifth Symphony is brought to life by an orchestra. Just as in music, drama, dance, or any other performing art, Lick declared, "information is a dynamic, living thing, not properly to be confined (though we have long been forced to confine it thus) within the passive pages of a printed document. As soon as information is freed from documental bounds and allowed to take on the form of process, the complexity (as distinguished from the mere amount) of knowledge makes itself evident."[32]

In practice, of course, knowledge could "take on the form of process" only through the medium of programming languages such as Fortran, Algol, or Lisp—none of which was for the fainthearted. But those were only crude first attempts, Lick was convinced. Future programming languages would be graphical, incorporating gestures and strokes and images in a way that would be so transparent and so intuitive that people would be able to use them with little or no training. Getting to that point would not be easy, he knew. (Indeed, it's only partially possible today: some of what Lick had in mind can be found in the graphical user interfaces made famous by the Apple Macintosh and Microsoft Windows, and in the drag-and-drop programmability of Web-page editors, drawing packages, and virtual-reality design tools. But no product has yet achieved the full generality and ease of use he was hoping for.) Nonetheless, Lick seems to have given graphical languages a very high priority, to the point of making it a major focus of his own research from then on—that is, in what little time he had to *spare* for research.

Meanwhile, however, Lick was also convinced that interactive, computer-based modeling could remove a second limitation of mental models: with on-line access, a model would no longer be confined to what was in a single head, but could be displayed on many screens at once, where it could be observed and manipulated by many people. As Lick wrote in the article with Taylor, "Modeling, we believe, is basic and central to communication. . . . If the concepts in the mind of one would-be communicator are very different from those in the mind of another, there is no common model and no communication." Conversely, he and Taylor continued, "[a successful communication] we now define concisely as 'cooperative modeling'—cooperation in the construction, maintenance, and use of a model. [Indeed], when people communicate face to face, they external-ize their models so they can be sure they are talking about the same thing. Even such a simple externalized model as a flow diagram or an outline—because it can be seen by all the communicators—serves as a focus for discussion. It changes the nature of communication: When communicators have no such common frame-work, they merely make speeches *at* each other; but when they have a manipula-ble model before them, they utter a few words, point, sketch, nod, or object."

Nowadays this sort of computer-assisted cooperative work comes under the heading of "groupware." Lick's PDP-1 wasn't up to doing many experiments along those lines, unfortunately. But the subject was much on his mind, thanks to one of the first nonacoustical projects he'd started at BBN: a contract with the Air Force Office of Scientific Research, which charged him with figuring out how systems of humans and machines could be organized to perform more ef-fectively. There was ample cause for concern, he knew: modern "military super-systems" such as SAGE and the Strategic Air Command were rapidly growing too large and too technologically complex for human beings to cope with un-aided. "Modern technology is as much politics and sociology as physics and chemistry," Lick wrote in the project's summary report. "The problem is no longer to design a pulley or a gear. It is to find a mission worthy of a million men, to plan a flow of metal and ideas and of flexibility and change. . . . Re-quirements vary, year by year, and detailed plans must follow, day by day, the vagaries of new solutions over which no rigid schedule can prevail. To harmo-nize great projects thus demands an agent, flexible and fast as well as strong and wise."[33]

That agent, not surprisingly, would be a time-shared computer. Or more pre-cisely, it would be a system of humans and computers working together to plan, build, operate, and maintain another system—a satellite-reconnaissance network, say, or a combat-theater operations center. In the planning and design phases, Lick continued, this "system system" would create a computer model that would serve as the blueprint and specification for the target system—much as in a modern computer-aided design/manufacturing setup, in fact. But in later phases, the model would also serve as the basis for planning and evaluating tests of the target system and for training its users. Then, in still later phases, the model would provide a basis for planning upgrades and improvements to the target system. And so it would go, from cradle to grave.

In addition, Lick wrote, this computerized system system would be of enormous help in meeting the challenge of "ordered information": finding and applying the relevant research results, utilizing the expertise of outside consultants, coordinating the efforts of design engineers, analyzing a flood of test data, scheduling routine maintenance, planning for continual evolution and growth— and on and on.

Here Lick may have been thinking of Vannevar Bush's Memex, which he had heard about by the late 1950s, even though he hadn't yet read Bush's original article. Certainly he saw the connection to the notion of an electronic library, which had fascinated him since at least the time of his "Symbiosis" paper. Indeed, he felt this was yet a third way in which symbiotic computing could transcend the limits of mental models: not only would the machines help us exchange our ideas with other people, but they would help us gain access to *any* kind of knowledge, anywhere.

This was an area in which Lick would be able to experiment firsthand. In November 1961, he received a commission from the Council on Library Resources to explore what computers might mean for the "library of the future"—the future, in this case, being the year 2000. He and his BBN colleagues immediately went to work on their PDP-1s to create one of the world's first demonstrations of library automation. By the end of the project, they had done so much "to facilitate the scholarly process" with computers—reading and studying documents, tracing references, and so on—that the council's delighted president, Verner Clapp, talked the editors at the MIT Press into publishing Lick's final report in toto. It would eventually appear in 1965 as *Libraries of the Future,* Lick's only book.

"I have always been pretty happy with that," Lick said in his 1988 interview. It was about as close as he ever came to bragging. *Libraries of the Future* is, in fact, one of the founding documents of what is now called digital library research, which includes (among other things) our efforts to manage information in that sprawling mass of data known as the World Wide Web. In his report, Lick examined all the computer tools that have since become familiar in libraries, such as the on-line card catalog, keyword bibliography searches, and on-line document retrieval. Yet he also emphasized that these tools were just the beginning. Such techniques can be wonderfully helpful when you already know exactly what documents you want, he said. But if you're not even sure where to look, or if you're still in that hazy beginning stage where you're struggling to figure out what questions to ask, then the classic search techniques won't help much at all. In this "negotiation" phase of the search, as Lick called it, what you desperately need is a good reference librarian (a prize that is, alas, all too rare). And it is in this negotiation phase that good, intelligent symbiotic computing could be a real help—if only we knew how to create such a thing.

The same could be said of the final two phases of the search. The "assimilation" phase is when you've successfully negotiated the first steps and have the information you need but still have to get it past "the brain-desk barrier." That is, you have to recognize the significance of what you're reading and seeing. You have to extract key ideas—the real knowledge—from the welter of details. And you have to fit the pieces into a larger whole: your emerging mental model of the new knowledge. Once you've done that, of course, you still have to face the "application" phase, where you try to figure out how you can make use of what you've learned in order to address the task at hand. Only then can you say that your search is truly complete, said Lick. Both of these last two phases are highly nontrivial, obviously, and both could presumably be greatly enhanced by computers. But that, in turn, would require sophisticated symbiosis between computer models and the mental models we build in our heads—which leads right back to the critical need for good modeling languages.

Once again, it would be quite a while before Lick could articulate all these ideas as precisely as he wanted. But by the time of that MIT centennial celebration in April 1961, they were clear enough. Indeed, he had enough confidence in his overall vision for the "new science" of computing that, with an unconscious foreshadowing of the future, he closed his commentary with an academic call to arms: "It will not suffice to wait until the computer industry develops the [symbiotic] computer the university needs; for commercial, industrial, and military requirements are not leading to development of such a computer on anything like the time scale that is feasible. Moreover, having such a computer is much less than half the battle. The task of preparing the programs required to make it 'go critical' is great. But it is a task that universities can and should handle, for it is itself an intellectual enterprise of high order."

DOING OK

By the beginning of the 1960s, then, a decade and a half before the microcomputer revolution began in the garages of Silicon Valley, and a full thirty years before the dawn of the Internet Age, the air around Cambridge was already alive with the essential ideas:

- graphics-rich personal workstations and the notion of human-computer symbiosis;
- time-sharing and the notion of computer-aided collaborative work;
- networks and the notion of an on-line community;

- on-line libraries and the notion of instant, universal access to knowledge; and
- computer languages and the notion of a new, digital medium of expression.

Better still, the technical substrate required to make those ideas a reality—interactive computing—seemed well in hand. In fact, there were two separate ways to achieve it: the stand-alone interactivity of the TX-0, TX-2, and PDP-1, coming out of the Whirlwind tradition, and the centralized interactivity of time-sharing, springing forth from the brain of John McCarthy.

So that was the good news: by the time Lick spoke at the MIT centennial, in April 1961, interactive computing was doing OK. But that was also the bad news: as time went on, interactive computing was still doing . . . OK. It was *not* taking the world by storm.

Consider the stand-alone approach, for example. On the plus side, it *had* been a hit with one admittedly odd little group—with credit for the introduction going to John McCarthy, ironically enough. Back in the spring of 1959, McCarthy and IBM's Nat Rochester had offered MIT's first programming course, a daunting intellectual gauntlet that included exercises in assembly language, Fortran, and Lisp. But it just so happened that among the students in their class were several members of the Tech Model Railroad Club, a band of techno-geek undergraduates who spent their free time creating ever more elaborate train layouts controlled by ever more intricate electrical switching networks, the more ingenious the better. Borrowing an ancient MIT slang word for a practical joke, the railroad club's members had taken to calling any particularly clever bit of controller design a hack. And as the writer Steven Levy described in 1984, the hackers in McCarthy and Rochester's course soon got so caught up in the fiendishly intricate joys of programming that they started hanging around the Computation Center till all hours, the better to gain access to the 704. There they were discovered one day by former railroad-club member Jack Dennis, now the staffer in charge of the TX-0, who asked them if they would like to come upstairs and see *that* machine.

"The TMRC people were awed," wrote Levy. Not only was the TX-0 an interactive machine that let you modify your program on the fly, but you could sign up for blocks of time to use the computer all by yourself! "There was no way in hall that [they] were going to be kept away from that machine."[34]

Actually, no one *tried* to keep them from it. Dennis and his fellow technicians were happy to let the model railroaders hang around the TX-0 till all hours as well. And if the researcher who had officially signed up for the 2:00 A.M. slot happened to oversleep—well, there was no sense in letting all that computer time go to waste. So the "TX-0 hackers," as they were now calling themselves, would crowd around the display screen and explore what it really meant to *play* on a computer. Freshman Peter Samson, for example, wrote a little program that converted ordinary Arabic numbers to Roman numerals and vice versa. Then he started teaching the TX-0's loudspeaker how to play Bach. Another TX-0 hacker devised what was essentially the first word processor, a program that allowed you to type in your class reports and then format the text for output on the Flexowriter. Since it made the three-million-dollar TX-0 behave like a three-hundred-dollar typewriter—much to the outrage of traditionalists who saw this, too, as a ludicrous waste of computer power—the program became known as Expensive Typewriter. In much the same spirit, freshman Bob Wagner wrote an Expensive Desk Calculator program to help him do his homework for a numerical-analysis class. (Wagner's grade: 0. Using the computer was cheating, his professor told him.)[35]

The incredible possibilities expanded even further a year or so later, according to Levy, after DEC, having delivered its very first PDP-1 to BBN, donated another of the early models to MIT for use by the students. The new PDP-1 was duly installed right next door to the TX-0, where the hackers immediately made it their own. New feats of hacking followed, including one of the most spectacular of all, Spacewar—a kind of on-screen sci-fi adventure in which the user could pilot his spaceship through realistic constellations, whip around a star under the influence of realistic gravitational forces, and fire torpedoes at realistic enemies. It was one of the first arcade-style computer games.

Like the adolescents they mostly were, the hackers felt that they were reinventing the world in a way that no adult could possibly understand. But Ken Olsen, for one, understood exactly what they were up to, and he was delighted. Indeed, giving MIT students a PDP-1 of their very own had been his idea in the first place: "[Spacewar] made a great demonstration for the stockholders!" he later said with a laugh—not to mention for potential customers.[36]

Unfortunately for DEC, however, interactive computing wasn't nearly such a hit outside the academic enclave around Cambridge. The business world was just starting its first great wave of computerization in 1960. Sales were rising exponentially, with IBM, in particular, pushing its Big Iron out the door as fast as it could get the machines off the assembly lines. But most of the orders were for batch-processing mainframes—which meant that all too many of Olsen's potential customers *stayed* potential. "The thing that was unique about [the PDP-1] was that it was simple, very fast, and interactive," he explained. "[But] the market wasn't demanding this. People had never seen it, didn't know about it, and didn't ask for it, with very few exceptions."[37] DEC did have a modest success with the PDP-1, eventually selling about fifty of the machines at the rate of two or three per month. But most of them went to research institutions such as Lawrence Livermore Laboratory, Atomic Energy of Canada, and NASA, where they were generally used to control experiments and collect data (a task for which the PDP-1 was very well suited, as it happened). The only real interactivity was between machines, not between humans and machines.

Out at Lincoln Lab, meanwhile, Wes Clark wasn't even doing *that* well—

though things had certainly started out happily enough. After completing the TX-2 in 1958, Clark had continued to push the idea of stand-alone interactivity and by the early 1960s was already talking about giving small computers to each individual researcher (or at least to each individual laboratory), making them personal computers in everything but the name. Moreover, Clark had the enthusiastic support of the Communications Biophysics Laboratory down at RLE, where students and researchers wanted a desktop computer with which to manage experiments, record data, and display real-time results on-screen. What Clark came up with, starting in mid 1961, was the Linc. Named in honor of its birthplace, Lincoln Lab, it was to be a modular machine whose console could indeed be assembled on a single (large) tabletop. True, the Linc would still require a separate cabinet to hold all its digital electronics, but by now that cabinet was down to the size of only *one* refrigerator. And more important, the Linc design looked as if it had a good shot at meeting Clark's arbitrarily imposed cost limit of twenty-five thousand dollars—startlingly cheap in the computer world of 1961.

By April 1962 Clark and his team were able to assign their prototype Linc its first scientific task during a demonstration at the National Institutes of Health in Bethesda, Maryland. The cat who was serving as the subject of the demonstration—a lab mascot named Jasper, who had had an electrode permanently (and painlessly) implanted in his brain—purred happily while the Linc's five-inch CRT displayed, in real time, a plot of his average neuroelectric responses.

Clark's scientific audience was ecstatic. "It was such a triumph that we danced a jig right there around the equipment," said one participant. "No human being had ever been able to see what we had just witnessed. It was as if we had an opportunity to ski down a virgin snow field of a previously undiscovered mountain."[38] The NIH funding officers were almost visibly reaching for their checkbooks.

The upshot was that by the summer of 1962, Clark and his collaborators had every reason to believe that the NIH would fund them in a considerably larger program in biomedical computing. They had the enthusiastic support of Bill Papian, director of Lincoln's advanced computing group. They were refining their designs for the Linc. And they were laying plans for a new neuroscience "wet lab" in a room adjacent to the TX-2. In effect, they were on the verge of launching a major new neuroscience research facility at Lincoln, a facility in which Linc and its ever more sophisticated descendants might very well have evolved into general-purpose personal computers by the mid-1960s.

Except that the Lincoln Lab higher-ups pulled the plug. Declaring that this new neuroscience lab just wouldn't fit in with the rest of their program, they invited Clark and his team to pursue their vision elsewhere. "We were the only nonclassified research group there," explains Clark, still fuming almost forty years later. "The lab director didn't want nonmilitary money and all the shifts of power that would imply. I was told to knuckle under or leave. So I left."

In recounting to his stunned design team what had happened, Clark announced that Linc had just become LINC: the Laboratory INstrument Computer.

Unfortunately, severing that symbolic connection to the computer's birthplace didn't solve the group's now urgent need to find a new home. Lick immediately tried to snap them up for BBN, of course, but without success: Clark, an academic to his soul, was uncomfortable with the idea of working for a commercial firm. MIT seemed out of the question for a variety of bureaucratic reasons. And nothing else seemed to be panning out. So as the summer of 1962 turned into fall, Clark and his handful of visionaries were still figuratively wandering in the wilderness, looking for their Promised Land.

On the surface, at least, the time-sharing approach to interactivity seemed to be doing much better in this period of the early 1960s. Certainly there was no lack of experimentation: with the limitations of batch processing now becoming increasingly apparent to everyone, various forms of time-sharing were springing up all over the country. At Carnegie Tech, for example, the computation center's director, Alan Perlis, was experimenting with a special-purpose form of time-sharing that would give his programming students a fighting chance at real-time access. Up at Dartmouth, McCarthy's former math department colleagues Thomas Kurtz and John Kemeny were starting to plan a campus-wide time-sharing system for much the same reason; the effort would eventually lead them to create an interactive programming language that they would call BASIC. And out at RAND headquarters in Santa Monica, Cliff Shaw and his colleagues were starting work on JOSS, the JOHNNIAC Open Shop System, which would let scientists and engineers tap into the JOHNNIAC computer and work their equations on-line, in the same spirit as later programs such as Mathematica and Macsyma. (JOSS had much the same kind of interactive question-and-answer style as Lisp or BASIC—and indeed, it would be a major influence on the latter.)

But the epicenter was still MIT—where the Computation Center's deputy director, Fernando Corbató, was already at work on what would prove to be the most influential time-sharing system of all.

It was a project born of exasperation, recalls Corbató. A year or so earlier, shortly after that first demonstration for the MIT Industrial Associates, McCarthy had decided that even he couldn't do a good job on AI, Lisp, *and* time-sharing. So he had agreed to hand off the further development of time-sharing to Herbert Teager, a young engineer who had just joined the Computation Center staff and was absolutely in love with the idea. But that was precisely the problem, says Corbató: Teager immediately set out to create the ultimate time-sharing system, drawing up plans for new software, new hardware, and even a brand-new handwriting-recognition system for users who were nontypists. "Herb was trying to engineer it from the ground up," Corbató says. "He wanted to do it just right; he wanted to see the system running efficiently; he wanted to see good user resources. And if he could not get it that way, he was willing to wait."

Corbató, however, was *not*. By the early spring of 1961, he was so frustrated by Teager's achingly slow progress that he and two other staffers at the center started working on a time-sharing system of their own. They desperately wanted a prototype to give people a *feel* for interactive computing, he says: "You could talk about it on a blackboard until you were blue in the face. You could try all these analogies, like describing it in terms of the difference between mailing a letter to your mother and getting on the telephone. And people would say, 'Oh, yes, but why do you need that?' " The prototype would be crude and primitive— basically a series of software patches applied to an IBM batch-processing operating system—but it would run *now*, on the center's existing 709 (and on the 7090 upgrade when it arrived). Moreover, it would run the software that people already had, without forcing them to change over. Thus the name: Compatible Time-Sharing System, or CTSS.

The first, four-terminal demonstration of CTSS wouldn't be given until November 1961 ("Hey, it talks back! Wow! You just type and you get an answer!"), and the system's actual development would continue for another year or more, even as Teager's project continued to disappear into the future. But at least it would proceed with support from the top, says Corbató. Ever since April 1961, in fact, when the Long Range Planning Committee had finally completed its work on the future of the school's computing, John McCarthy's vision of universal time-sharing had gone from being the slightly kooky obsession of one very junior assistant professor to representing something akin to official MIT policy.

Of course, getting to that point hadn't been easy, either, even with McCarthy, Corbató, and their fellow activists on the technical subcommittee in basic agreement on the concept. This being MIT, the subcommittee members were headstrong, to put it politely: after some turmoil, they had managed to produce not one final report, but two.

The majority report had been drafted in all good faith by the subcommittee's chairman, Herb Teager, Corbató explains. After the group finished its deliberations in late 1960 and early 1961, he said in an interview published in 1992, "Herb went out, holed up for a month or two, and wrote the report without interacting much with the committee. Then he came back and presented the draft to the committee."[39]

Bad move: in short order, McCarthy led the others in open revolt, says Corbató. And when a visibly upset Teager declined to change his draft to their liking—though he did offer his resignation, which they refused—they proceeded to issue a "minority report" signed by almost everyone but the chairman.

What they primarily objected to was Teager's conclusion that MIT ought to keep on getting its computers from its longtime friend and ally IBM. Not so fast, declared McCarthy and his cohort, who had some decidedly mixed feelings about the company. "We had some people working within IBM who were rooting as hard as they could and trying like heck to get the company to move in our direction more," said Corbató. But the majority of IBMers seemed to miss the point of time-sharing entirely. "They were humoring us," he recalled. "They kept saying, 'You do not need time-sharing. We are already doing that; we have our airline reservation system [SABRE].' We kept trying to point out to them that that was a transaction system and had nothing to do with general-purpose programming[, but] they were viewing [time-sharing] as just a special-purpose gadget to amuse some academics. They did not see it as affecting the productivity of people in trying to get things done."[40]

So, said Corbató, since none of the rebels necessarily wanted to trust his fate to the IBM crowd, the dissenting report suggested that MIT shop around for a vendor that would be willing to create exactly the system it needed. The report even sketched out some specifications—"not totally unattainable, but thinking big," according to Corbató. Memory, for example: "[The report said] we ought to go for a computer that had a million words [in modern personal-computer terms, about six megabytes], which, in those days, sounded grandiose."[41]

Actually, in some quarters the rebels' specifications sounded almost irresponsible. Their minority report estimated that the total cost of developing and procuring such a system would fall somewhere between a daunting $8 million and an appalling $25 million. Nonetheless, its basic argument—that this approach would in fact be the cheapest way for MIT to meet its future computing needs—prevailed. Both of the subcommittee reports reached the official committee in April 1961. And while neither was ever formally accepted, it was clear enough that the rebel report was the one that found favor.

It didn't hurt that computing was very much on everyone's mind at that particular moment: the submission of the two reports happened to coincide almost exactly with the MIT centennial lecture series. Nor did it hurt that one of the eight lectures in that series was devoted entirely to time-sharing—with the featured speaker being one John McCarthy. Time-sharing was now a certifiably hot topic on campus, bearing the imprimatur of the Sloan School itself: This is where we're going in the *next* hundred years, went the message.

Of course, McCarthy did manage to feel a bit insulted by the invitation; despite his inclusion on the impressively heavyweight roster of participants, which ranged from C. P. Snow to Vannevar Bush, he couldn't help but notice that he'd been asked to fill in only after someone else had taken ill. "I was not considered important by the MIT higher-ups," he declares.

Nonetheless, McCarthy's talk had an impact, not least because at the very end of it he finally stated in public what he'd long been mulling over in private: "If computers of the kind I have advocated become the computers of the future," he said, "then computation may someday be organized as a public utility, just as the telephone system is a public utility. We can envisage computer service companies whose subscribers are connected to them by telephone lines. Each subscriber needs to pay only for the capacity that he actually uses, but he has access to all programming languages characteristic of a very large system. The system could develop commercially in fairly interesting ways. Certain subscribers might offer services to other subscribers. One example is weather prediction[;] other possible services include . . . programming services. Some subscribers might rent the use of their compilers. Other subscribers might furnish economic predictions. The computing utility could become the basis for a new and important industry."[42]

Indeed, he said, given the economies of scale then prevalent in the computer industry, it wasn't too hard to imagine such service centers' growing to the size of municipal power plants: every city would have at least one computer, and the very biggest cities might have several. And yet in the process, paradoxically, computation would be democratized. Instead of being walled off in air-conditioned sanctums, instead of being nothing but the tools of powerful institutions, computers would become available to *everybody*. Like Thomas Edison, the time-sharing pioneers would quite literally be bringing power to the people.

Bob Fano, for one, remembers being deeply impressed by this argument. The Long Range Planning Committee's reports had been his first real introduction to time-sharing; now, just a short time later, McCarthy's notion of an information utility gave him some sense of where the idea was *going*. And the lecture series' organizer, Martin Greenberger of the Sloan School of Management, was so taken with McCarthy's vision that he eventually published a greatly elaborated version of it in the *Atlantic Monthly*.[43] (Lick had mentioned the same idea in his "Symbiosis" paper a year earlier, of course, but only in passing, and rather sketchily, at that.)

In short, April 1961 was a high point for time-sharing at MIT. But then, a high point can also be the beginning of a long downhill slide. And that was the case here. The failure wasn't the result of anything dramatic; at one point, as per the rebel report, MIT did send out a request for proposals on a new time-sharing machine. But after that, well, so far as McCarthy could tell, nothing happened.

By 1962 McCarthy was furious, convinced that he could recognize a stall when he saw one. Not only were the MIT higher-ups taken aback by the cost estimates—an assertion that Fano confirms—but IBM was continuing to whisper in their ears. "Real soon now," the company representatives kept reassuring them, IBM would be coming out with a new, all-purpose "System/360" computer that would meet all MIT's needs, and at little or no cost.

"Unfortunately, the System/360 design took longer than IBM management expected," McCarthy wrote in 1992, "and along about that time, relations between MIT and IBM became very strained because of the patent lawsuit about the invention of magnetic core memory." For McCarthy himself, the last straw came in mid-1962, when MIT's president, Julius Stratton, proposed yet another new study—a kind of market survey, said Stratton, which would conclusively determine what kind of demand there was for this time-sharing stuff among MIT computer users. "I regarded this as analogous to trying to establish the need for steam shovels by [conducting] market surveys among ditch diggers," snorted McCarthy, "and I didn't want to do it."[44]

Actually, he didn't even want to stay at MIT anymore. Disgusted, and having given up on the place as a lost cause, McCarthy was already looking for a new home elsewhere.

And BBN?

Well, J. C. R. Licklider was having fun. By the spring of 1962, Karl Kryter was head of a whole psychoacoustics group; Jerry Elkind was leading a psychology group; Tom Marill was directing a computer group; and Lick himself was a full-fledged vice president in charge of them all. As promised, he had done his best to transform BBN from a pure acoustics consulting firm into a much broader R&D house with a strong presenßce in computers. The company now had three or four PDP-1s and was getting serious in its planning for a time-sharing service. It had the "library of the future" contract, the Massachusetts General Hospital contract, and a host of others. Symbiotic computing was going *great*.

But all that, alas, was before Ed Fredkin left.

At the time, explains Fredkin, it seemed like a matter of survival. In 1962, fearful that a nuclear war was imminent, he and several colleagues laid plans to start a new company of their own—in Brazil. That scheme soon fell through, thanks to a coup in Rio. And while Beranek insists that he and Lick would have been happy to take him back, Fredkin felt that it was time for him to go. "So I consulted for a while for DEC," he says, "then started a company of my own, Information International."

That was fine for Fredkin, but it left Lick in the lurch on the hospital time-sharing system, which NIH had already paid for. So he hired McCarthy to take charge of the project as a one-day-a-week consultant. Yet that in turn meant that progress suddenly slowed considerably; the BBN time-sharing system wasn't publicly demonstrated until September 1962. And even then it really wasn't much more than an exercise. As Lick himself had to admit—and as McCarthy had pointed out early on—the PDP-1 was such a comparatively weak little computer that it didn't have much to time-share. "You can make an argument that, although it had better graphics, in other respects it was almost exactly a Radio Shack TRS-80 computer," said Lick. "It had about that computing power, a little less memory, and it ran a little slower." As the wave of the future, in other words, time-sharing on the PDP-1 was tantalizing—but not, in itself, very impressive.

So there it was: by 1962, the notion of human-computer symbiosis was doing OK—but no better than OK. What it had going for it was a handful of activists and visionaries who were willing to work together—except when they weren't.

What it also had was a handful of interesting experiments that could have completely overthrown the conventional wisdom of computing—except that they were always on the verge of fizzling out. But what the vision *didn't* have was a single champion, someone with enough resources to make things happen in a big way and enough clout to get people marching in the same direction.

Yet.

CHAPTER 6

THE PHENOMENA
SURROUNDING COMPUTERS

In the early hours of October 5, 1957, local time, an R-7 intercontinental ballistic missile lifted off from the Soviet Union's top-secret launch complex at Baikonur, in Kazakhstan, and arced eastward into the predawn darkness over central Asia. The R-7 was an ungainly, utilitarian vehicle, with four outward-angling first-stage engines that made it look a bit like a corn silo wearing a skirt. More to the point, it was still experimental. The engineers who witnessed its ascent were rolling the dice: they had spent the previous spring watching test flights end explosively; their two lone successes had come only that August. Still, the go-ahead for this attempt had been given by the Kremlin itself. And indeed, the engineers were lucky once again. Less than half an hour after launch, at an altitude of some 141 miles, the R-7 released its payload.

The payload didn't look like much, either: just a 184-pound aluminum sphere, twenty-three inches across, trailing four long antennae. But then, humankind's first artificial satellite was fairly simple even by 1957 standards. It did contain instruments to measure cosmic rays, meteoroids, and conditions in the extreme upper reaches of the atmosphere. But mainly it just signaled its continued existence via two radio transmitters that went *beep—beep—beep* ad nauseam.

The hopelessly unfashionable bureaucrat: Lick ca. 1962

To American ears, even its name sounded silly: *Sputnik Zemli,* Russian for "little traveling companion of the world."

Except that no one in the United States was laughing: the launch of *Sputnik* rocked the nation like no event since Pearl Harbor. As one account of the early space race put it, "[*Sputnik*'s] two transmitters would fail twenty-three days after launch—but their arrogant beeping would continue to sound in the American memory for years to come. . . . Gone forever in this country was the myth of American superiority in all things technical and scientific."[1] The Russians weren't just playing catch-up anymore, as they had with Joe One, their first atomic-bomb test, back in August 1949. With *Sputnik* they seemed to have *passed* the United States: their rockets worked, while ours were still blowing up on the launch pad. Moreover, the real import was lost on no one. A rocket that could put an inane silver beachball into orbit was also a rocket that could lob an atomic bomb into New York, Washington, Chicago, Los Angeles—anywhere. "Mankind," intoned a Rockefeller Foundation panel headed by the young Harvard historian Henry Kissinger, "is faced by two somber threats: the Communist thrust to achieve world domination that seeks to exploit all dissatisfactions and to magnify all tensions; and the new weapons technology capable of obliterating civilization."[2]

On November 3, as if to drive the point home, the Soviet missile wizards launched the 1,120-pound biomedical satellite *Sputnik II,* carrying a Border collie named Laika. (After four days she was automatically put to sleep—peacefully, one hopes—by lethal injection.) And on May 15, 1958, the Baikonur team would triumph yet again, with the seven-thousand-pound *Sputnik III.* But by that point its success would be almost redundant: the American public had long since gone into a media-driven orgy of finger pointing and self-flagellation. The banner of Marx and Lenin was on the advance in every corner of the globe, it seemed, while the faltering, ineffectual West was advancing only toward the dustbin. Magazines featured admiring articles about the harshly disciplined Soviet classrooms, so different from our own indulgent schools and so effective at mass-producing Soviet engineers by the tens of thousands for the dawning age of technocracy. Politicians of both parties spoke ominously of a "missile gap"—among them the junior senator from Massachusetts, an affable and ambitious young man named John Fitzgerald Kennedy. And President Dwight David Eisenhower was widely portrayed as the perfect emblem of our decline: the Hero of Normandy and Conqueror of the Third Reich had become a tired, indecisive, confused old man who was totally out of touch with the Space Age.

In the White House, meanwhile, the Hero of Normandy was privately exasperated with all this carrying-on. Eisenhower knew the missile gap was bogus (as did many of the politicians bemoaning it). He had approved the development of ICBMs as far back as 1954; the U.S. rockets now being tested would be far more sophisticated than their clumsy Soviet counterparts. However, he also had to admit that the U.S. military had been entirely too willing to coast along on the technological triumphs of World War II. We hadn't fallen behind yet, public hysteria to the contrary, but unless we acted quickly, we might very well do so in another ten years, or twenty. Moreover, the disarray in the U.S. space program was all too real, with three warring fiefdoms—the army, navy, and air force—developing missiles and applications almost independently of one another.

So on November 7, 1957, in a solemn address to the nation, Eisenhower announced that he was appointing the president of MIT, James Killian, to the newly created post of presidential science adviser, thereby moving science and technology to the center of U.S. policymaking. He likewise committed his administration to massive new investments in education, research, and development. And in the process, he promised, he was going to clean up the research mess in the Pentagon. A short time later, Eisenhower followed up on that pledge by endorsing a plan to consolidate all the Pentagon's space research under a new civilian agency reporting directly to the secretary of defense. It would be called the Advanced Research Projects Agency, or ARPA.

Well, it was a nice try. True, Eisenhower's research and education initiatives marked the beginning of what participants now remember wistfully as the golden age of R&D in the United States—a multibillion-dollar funding boom that would dwarf the Manhattan Project, the Rad Lab, and all the other World War II efforts put together. For the better part of a decade, until the good times were brought to an end by the escalating cost of Vietnam, science education would flourish as never before, from kindergarten to graduate school. Basic research would do likewise, in fields ranging from cell biology to particle physics. And the rising tide of cash would elevate whole new industries, not the least of which was an exotic new electronics technology known as semiconductors.

Back at the Pentagon, however, the services and the Joint Chiefs of Staff regarded this upstart ARPA as a competitor for funds and a rival for the attention of the secretary, the president, and Congress. By the time the agency opened for business, in February 1958, they had already begun their counterattack. And when the dust settled that fall, all of ARPA's nonmilitary space projects had been spun off to a new civilian agency called the National Aeronautics and Space Administration, or NASA. Most of the military space efforts, meanwhile, had reverted to the services. And ARPA found itself reporting not to the secretary of defense but to the director of defense research and engineering (DDR&E), a newly created deputy secretary responsible for coordinating defense R&D across the board. All the fledgling agency had left was a handful of *really* advanced programs such as ballistic missile defense, nuclear-test detection, and materials science—the kind of beyond-the-cutting-edge stuff that the services were willing to do without.

Still, ARPA did survive, and by January 1961, when the junior senator from Massachusetts was inaugurated as Eisenhower's successor, it had even attained a measure of grudging acceptance. When an air force undersecretary named Jack Ruina took over that same month as ARPA's new director, he discovered that he actually rated a government car and driver, plus a Pentagon office suite furnished with a flag and wall-to-wall carpeting. Better still, his suite lay in the coveted E Ring of offices along the outside of the building, which meant that he had the rare privilege of looking out his window into the real world (lesser mortals assigned to the inner A, B, C, and D rings saw nothing but other windows). Indeed, Ruina was on the same third floor as the new secretary of defense himself, Robert McNamara. Nevertheless, Ruina couldn't help but notice that the intensely rank-conscious culture of the Pentagon had calibrated his status to the millimeter. His government car had a dent, and his office was more than a third of the way around the ring from McNamara's, which would make for a long, rank reducing walk should he ever be called into the secretary's presence.

But so what? The Polish-born, Brooklyn-raised electrical engineer was much more concerned about his top priority, which was to remake his agency from the ground up. His predecessors had done their best, but the fact remained that ARPA had survived by being a grab bag of leftover programs. And now that the arms race had most definitely become a technology race, Ruina was convinced that ARPA had to have much more focus, direction, and purpose. Most of all, it had to become a force for technical excellence. It had to fund research of the very highest quality, whether or not that research was what the services thought they needed for immediate military applications. It had to be staffed with program managers who knew the science and the technology thoroughly. And instead of being subjected to endless review and micromanagement, those managers had to be given autonomy, authority, and the freedom to act. Assault the technological frontiers everywhere you can, Ruina urged them. Go out to the university labs, the national labs, the private sector, anywhere. Look for people with ideas that push the envelope. Give them development money. Be generous. Take risks. Cut through the red tape. Do whatever you have to. But do it.

This free-wheeling "ARPA style" of management was radical for 1961, and nothing like the rigidly top-down hierarchy that prevailed elsewhere in the Pentagon. But it worked. Old Washington hands who transferred into the agency could only shake their heads in disbelief: decisions that would take months anywhere else were made at ARPA in a matter of days. You might see Ruina and his program directors brainstorming an idea one morning and then by four the next afternoon hear secretaries typing out letters for contractors to start work.

Not surprisingly, such energy got noticed, especially in a brand-new administration that had pledged to reinvigorate the military. Ruina's approach fitted right in with John F. Kennedy's New Frontier activism. Moreover, ARPA's largest research programs—ballistic-missile defense and nuclear-test detection—had already emerged as what Ruina liked to call presidential issues: the kinds of subjects a secretary of defense might want to talk about when he met face to face with the commander in chief. So in practice, says Ruina, he found himself taking that long, long walk around the Pentagon periphery fairly often. Robert McNamara's appetite for facts and figures was legendary; he wanted to know exactly what was going on in these areas.

Still, says Ruina, astonishing though it may seem in retrospect, there was one technology that the secretary never asked him about: computing. Why should he? ARPA wasn't funding any computer research. And besides, as far as McNamara or almost anyone else on the third-floor E Ring knew, computers were just gadgets to be used for payroll and accounting, plus a few very exotic applications such as the SAGE air-defense system. True, some people down at the working level of the DoD were trying to think more creatively about computers. But they certainly hadn't come up with anything that Ruina found compelling. "People saw that something was there," he says, "but they were not prepared to invest big money without knowing quite what that something was."

All of which explains why ARPA ended up getting into computer research backward, Ruina says, thanks to a kind of bureaucratic triple conjunction. First came the White Elephants: four surplus computers that had originally been built as spares for the air force's SAGE program. These "Q-32" machines were each the size of a small house and were utterly useless for anything other than air defense, unless they underwent a top-to-bottom reprogramming. But because they had been so fabulously expensive, no sane bureaucrat would dare throw them away. So they landed in the office of the DDR&E, along with a directive: Do something with these things.

Second, there was the Orphan: SDC, the Systems Development Corporation of Santa Monica, the RAND spin-off that had written the software and done the operator training for SAGE. The company was facing severe cutbacks and layoffs now that SAGE was nearing completion. And the air force, anxious to keep the team intact, was lobbying hard: Surely, the brass told the DDR&E office, you can find another job for these guys.

Finally, and most critically, there was the growing crisis in command and control, the military art of making timely decisions and then getting those decisions implemented by forces in the field. As venerable as this art was, predating even the Bronze Age, the Nuclear Age had given it an urgency greater by several orders of magnitude. After *Sputnik*, in fact, it was all too easy to imagine a U.S. president's being confronted with an ICBM sneak attack that would oblige him to make apocalyptic decisions on a time scale of minutes—and then have his decisions carried out with split-second timing by forces scattered around the globe. In March 1961, President Kennedy seemed to have this possibility very much in mind when he sent a special message to Congress on defense spending. Command-and-control systems, he said, needed to become "more flexible, more selective, more deliberate, better protected, and under ultimate civilian authority at all times."

Once again, the research component of this issue promptly made its way to the DDR&E office, where it joined the White Elephants and the Orphan. The obvious solution soon presented itself. In June 1961, word came down to Ruina that ARPA should organize a "command-and-control research project"—which was to say, a $6-million-per-year contract with SDC to study command and control, using one of the Q-32s.

Ruina was dubious; he'd never had a high opinion of SDC. Still, the idea itself seemed sensible enough. At every level—maps, weather reports, intelligence reports, logistics, communications, even war games and training—modern command and control was a massive exercise in information processing. Computers could be a godsend to the military. And $6 million per year was negligible by Pentagon standards, even then. So Ruina signed off on the SDC proposal and then, distracted by ARPA's bigger-ticket programs, pretty much forgot about it. From his perspective, in fact, command and control didn't really become an issue again until the following November, when staffers brought him the results of an outside study that ARPA had commissioned on how to apply computers in the field. ARPA had a real opportunity here, said the report: instead of just accepting whatever the computer industry had to offer, the agency could take on computer research directly, expanding it into "potentially fruitful" areas such as better programming techniques and better communication between machines and their users.

Hmm. Do a little research on computers and information processing? Frankly, the thought hadn't occurred to Ruina. But once he heard the idea, he says, "I remember being impressed that this program sounded reasonable."

And that—given the new ARPA style of decisiveness—was that. So the only real question now was who the agency should get to manage its new command-and-control/computing program. Unfortunately, says Ruina, this is where his memory gets hazy; he cannot for the life of him remember how J. C. R. Licklider's name came up. But from somewhere Ruina got the suggestion that Lick should not only head the new command-and-control effort but simultaneously run a small ARPA behavioral-science program that also needed a director.

Behavior *and* computing!? What kind of combination was that? "It seemed strange," recalls Ruina. But then, Lick was one guy who might be able to pull it off. "When the name came up I had known of him, and maybe met him once or twice," Ruina says. And what he knew sounded good. A very distinguished career in psychology. Human-factors work on SAGE. Lots of consulting with the air force, including half a dozen years on its scientific advisory board. Lots of recent work with computers. Definitely a self-starter.

"I went for it right away," says Ruina. J. C. R. Licklider it was.

The only problem was that J. C. R. Licklider wasn't interested. He was as patriotic as anyone, he told Ruina when the ARPA director called. But what with the PDP-1, time-sharing, the "library of the future" project, and all his other activities at BBN, he was having a wonderful time up in Cambridge. So why should he bury himself in the bureaucratic limbo of Washington?

In the end, in fact, Lick agreed to come to Washington only because he'd been talked into it—though it was Lick himself who'd done most of the talking.

As Lick told the story in his 1988 interview, Ruina invited him down to the Pentagon sometime in early 1962 along with a backup candidate, Fred Frick, who had been a colleague of his at Harvard and now headed the communications division at Lincoln Laboratory. Both of them were skeptical, he says, but then "Ruina got Gene Fubini to give us a sales pitch."

Fubini, then a staffer in the DDR&E's office, was a physicist Lick had known professionally. He described him as "a very sharp, incisive, impatient kind of guy; quite eloquent, and really dedicated to this job at the Pentagon. He made Frick and me feel that the world needed at least one of us." And with that, Lick admitted, "I started to wax eloquent on *my* view of the thing: that the problems of command and control were essentially problems of man-computer interaction. I thought it was just ridiculous having command-and-control systems based on batch processing. Who can direct a battle when he's got to write a program in the middle of the battle? Fubini essentially agreed a hundred percent with that, and so did Ruina. We started seeing that this whole military thing was not developed right."

In effect, Lick explained to them, everyone at the Pentagon was still thinking of computers as giant calculators and data processors. For that matter, so was practically everyone at IBM and in the rest of the computer industry. And that, he remembered telling Ruina and Fubini, was where ARPA had its real opportunity: "This kind of [interactive] computing almost did not exist. But up in Cambridge everybody was excited about making it exist. [So] why didn't we really develop an interactive computing? If the Defense Department's need for that was to provide an underpinning for command and control, fine. But it was probably necessary in intelligence and other parts of the military, too." Indeed, he insisted, interactive computing had the potential to transform human life in the civilian sphere as well—everywhere, in fact. "I was just a true believer," he said. "I was one of the very few people, at that time, who had been sitting at a computer console four or five hours a day, or maybe even more. It was very compelling. I was terribly frustrated at the limitations of the equipment we had, but I also saw how fast it [was] getting better. . . . I thought, This is going to revolutionize how people think, how things are done. . . . I thought we were going to double [human productivity,] or triple it, or multiply it by four or ten."

Now then, Lick told Ruina and Fubini, warming to his subject, let's take a look at your behavioral-science program. Maybe it isn't as far afield from command and control as you think. "Notice that man-computer interaction is heavily involved in the skills and capabilities of the people as well as the machines," he recalled telling them. So suppose he did agree to run that program jointly with the computer office—just suppose. "You won't mind if I spend some of the behavioral-science money on the same dream of getting people to work with computing?"[3]

Um—not at all. By this point, it seems, Ruina was sold. In the normal, bureaucratic course of things, he would have been happy enough to hire a manager whose idea of command-and-control research didn't go beyond the mainstream—improvements in database management, say, or fast-turnaround batch-processing systems. He never would have known the difference. Instead, almost completely by accident, he'd stumbled upon a visionary, a man who could take this amorphous idea of command-and-control research and put it into the larger context of human destiny. Unlike most visionaries, moreover, Lick had done some serious thinking about how to get from here to there; in effect, his 1960 "Symbiosis" paper was a ready-made research agenda for this whole ARPA program.

Lick, meanwhile, seems to have convinced himself that he had to seize the opportunity. Improbably, miraculously, through some once-in-a-lifetime alignment of all the right planets, ARPA was offering him a chance to turn his vision into a reality.* He could reinvent the whole field of computing. He could transform those giant calculating machines into full-fledged partners in human creativity. He could create "information utilities" that spanned the continent. He could democratize information by opening up vast libraries of material to instant access by anyone. And he could do it all with the Pentagon's money.

So by the spring of 1962 it was settled: Licklider would take a one-year leave of absence from BBN and come on down to Washington. Including the SDC contract, his budget would start at some $10 million per year, which seemed like a fortune to him. His first official day would be Monday, October 1, 1962.

A VIEW FROM THE HEIGHTS

Actually, as Louise Licklider remembers it, the family moved down to Washington a couple of months early so the kids could start in their new schools at the beginning of the term. Also, they had to get settled into the apartment Lick had found for them in Arlington Towers, right across the Potomac in Rosslyn, Virginia.

It took quite a bit of settling. Compared to their rambling old Victorian back home, the two-bedroom apartment was . . . *tiny*. In fact, since fifteen-year-old Tracy and thirteen-year-old Linda needed a bedroom apiece, their parents had to sleep on a fold-out couch in the living room. But then, Lick hadn't rented this place for practical reasons. He'd rented it because Arlington Towers stood on the

---

* And Fred Frick? On several occasions, including during his 1988 interview, Lick claimed that the two of them had ended up flipping a coin to decide who would take the job—"and I lost!" The problem is, it's hard to believe that even Lick would make such a major life decision so casually. Besides, Ruina remembers clearly that Frick didn't want the job, partly because of family constraints and partly because it would mean his having to take a big salary cut from his post at Lincoln Lab. Furthermore, in his 1988 interview, Lick told precisely the same story about an earlier decision, in 1953, when he and George Miller were trying to decide which of them would move out to Lexington with Lincoln Lab and which would stay on the MIT campus with the psychology section. They tossed a coin, he said, and he lost. However, while Miller did indeed go out to Lexington, he doesn't remember anything about a coin toss. What he does remember is that Lick decided to stay in Cambridge, and that was that.

In short, the coin-flip story should probably be counted not as literal truth but as one of the many jokes that Lick liked to tell on himself.

very banks of the river, so they could walk out on the balcony and see Arlington National Cemetery, the White House, the riverboats, the bridges, the monuments—the whole broad sweep of the nation's capital. Who cared about a fold-out couch when they had a view like that?

Well, his wife cared, frankly. Wasn't it going to be tough enough starting a new life in a new town without them all being crammed into a shoebox? And yet Louise could never bear to dampen that enthusiasm. This was a guy who got choked up at a sunset or a thoughtful gift, and who would come backstage after her theater performances practically in tears: "Sugar, that was beautiful!" If he needed a magnificent vista, they'd make do as best they could.

So it was that on the morning of Monday, October 1, 1962, J. C. R. Licklider arose from the family's new fold-out couch, drove his own car down to the Pentagon, found his way through the endless corridors to his zero-status office on the second floor of the D Ring, and introduced himself to everyone on his new staff—which was to say, his secretary.

And so it was on just about every morning thereafter. Lick would never be counted among the power brokers at the Pentagon. "He was an incredibly impressive person—*if* you were willing to listen," remembers Charles Herzfeld, a Vienna-born, University of Chicago–trained physicist who had come to ARPA in September 1961 to run the Defender ballistic-missile-defense program. "A lot of people weren't willing, though, because Lick came across as being so modest and shy. Also, he had this air of tentativeness about his ideas. He'd go on for hours saying, 'If one could do *this*, then maybe one could do *that*, and then maybe this other thing would be important.' I thought it was charming. But it just wouldn't fly with the military and the political types."

Still, says Herzfeld, that disingenuousness was also Lick's secret weapon: "Lick was totally unfashionable—and didn't give a damn. What he wanted was to intrigue people and get them involved." And every time it really mattered, he succeeded.

Take for example Herzfeld himself, who would soon be sitting in Ruina's chair as the director of ARPA. He vividly remembers a series of lectures on modern computing that Lick gave in the spring of 1962, months before his official start at ARPA. "The setting was a big lecture hall in the Pentagon," he says. "It was a huge place, cold and ugly, maybe half full. But everybody was interested in this guy. Ruina had given him a strong endorsement." And indeed, says Herzfeld, Lick captured everyone's attention inside the first few sentences. Many of the ARPA and DDR&E staffers had used batch-processed computers for number-crunching and such. But here was Lick talking about time-sharing, interactive graphics, networking—concepts an order of magnitude further advanced. "It was a revelation," says Herzfeld. "My first experience with computers had been listening to a talk by von Neumann in Chicago back in nineteen forty-eight. It sounded like science fiction then: a machine that could carry out algorithms automatically. But the next big shock was Lick: not only could we use these machines for massive calculations, but we could make them useful in our everyday lives. I listened. I got very excited. And in a very real sense, I became a disciple from then on." Certainly Herzfeld became one of Lick's closest friends and staunchest supporters within ARPA.

Out on the E Ring, Ruina and the DDR&E staffers were having much the same reaction. "They did start off thinking that I was running a command-and-control program," Lick said. "But I wanted to make it clear that I wasn't doing battle planning. I was doing the technical substrate that would one day support battle planning. So every time I possibly could, I got them to say 'interactive computing.' I kept trying to convince them of my philosophy that what the military needs, is what the businessman needs, is what the scientist needs." And eventually it must have worked, he said, because they let him take interactive computing wherever he wanted, without any real pressure to do "military" applications. Essentially everything he funded was open and unclassified.* Indeed, according to Lick, he hardly even saw Ruina: true to his antimicromanagement principles, the ARPA director met with his newest recruit maybe once a month or so to see how things were going, and then otherwise left him in a state of what Lick laughingly (and gratefully) called benign neglect.

Meanwhile, Lick was also expanding his network of allies into other corridors of the Pentagon, beginning with his counterparts in the uniformed services' own research agencies. This was a matter of some delicacy: until October 1, 1962, their leadership in cutting-edge computer research had gone unchallenged—witness the army with ENIAC and EDVAC, the navy with Whirlwind, or the air force with the TX-0 and TX-2. And yet now here was J. C. R. Licklider, invading their turf with a budget that dwarfed the three of theirs put together. The potential for some nasty bureaucratic infighting was all too apparent.

But it never happened. "The people who might have been our competitors were really our agents and our friends," Lick explained, referring to a deal struck in ARPA's tumultuous first year: instead of building a new bureaucracy of its own, ARPA would make use of the ones that already existed. So when Lick wanted to start a new research project, all he had to do was give the formal go-ahead, and one or another of the service agencies would then write the contract, disburse the funds, keep track of progress reports, and so on. This was great for him, because he was terrible at managing that kind of paperwork anyway. But more important, it opened up the possibility of a wonderful synergy. Since ARPA and the service agencies were already doing the paperwork as a team, it

---

* Lick actively tried to avoid what he called the cloak-and-dagger side of the agency; classified projects made him nervous. But he couldn't avoid them entirely. In his 1988 interview he said that his office was used as a cover for one project that was so secret at the time that even *he* didn't know what it was. Still being discreet a quarter of a century later, he would say only that he had paid for "digging a hole under Lafayette Square"—directly across the street from the White House.

was just one more step to coordinate their plans and do their computer research as a team. And that, said Lick, was where the friendship came in: these were the very same people who had been funding *his* projects all through his MIT and BBN years. They could trust one another.

Quite soon after his arrival, in fact, Lick organized a monthly meeting at which he and his counterparts could bring each other up to date on the research they were funding, eliminate overlaps, and look for new opportunities to collaborate. And very soon after *that*, Lick expanded the meeting to include funding officers from NASA, the National Science Foundation, the National Institutes of Health, the Atomic Energy Commission, and any other agency around Washington that was putting even a little bit of money into computer research.

Their first time together was almost surreal, remembers Robert Taylor, who was then running an information-research program at NASA. "Some of the professional bureaucrats were dumbfounded. They had a history of giving grants to individual people in twenty-thousand-dollar chunks. But Lick was talking about millions of dollars and whole teams of people. It was as though these folks had encountered this alien creature: friendly, but strange." Another participant, Ivan Sutherland, remembers his first meeting in 1964: "These people met, period. The group had no charter, no responsibilities, no budget, no purpose—but it was a great thing. We would discuss what was important, what was current, and what was going on. Precisely because the group had no charter, it was a wonderful way of getting information flow between the agencies."

And the upshot of all this? Simply that this dreamy, talkative, hopelessly unfashionable academic had entered into a bureaucracy as tough and as turf-conscious as any in the world, and with little apparent effort had carved out a niche for himself in which he could pursue his agenda with near-perfect freedom; deal with his natural enemies as a convivial network of friends and allies; and tend to his real constituency, the researchers, with little or no need to watch his back.

It would have been a bureaucratic triumph worth savoring if J. C. R. Licklider had cared about such things—which he didn't. Who had the time? He was already headed for the airport. At MIT, Stanford, Berkeley, Carnegie Tech, and RAND—indeed, throughout the country—Lick knew there were scattered groups of people who shared his dream. His job now was to seek them out, nurture their work with ARPA cash, and forge them into a self-sustaining community that could carry on after he was gone. But with only a year in which to do it all, he had to *move*.

LICKLIDER'S LIST

From the minute he accepted the ARPA job, as Lick later recounted, "it was natural then to think, Well, gee, the first and most obvious and easiest way to get really rolling on this is to get some of this money into that gang at MIT." It was still the epicenter, after all—the densest concentration of interactive-computing fanatics in the nation. "Happily," he added, "I had been away from MIT long enough that I wasn't too obvious in saying that, so I could do it!"[4]

He could, that is, if the gang at MIT didn't tear itself apart first.

The frustration levels were high, Lick knew: a year and a half after the Long Range Planning Committee's endorsement of time-sharing as a campus-wide goal, progress was nonexistent. But Lick could now offer the perfect antidote: money. So he flew up to Cambridge soon after starting at ARPA and asked to meet with the key computer people at MIT so he could lay out his ideas for them and invite a proposal.

Things were worse than he'd thought. The whole gang turned out, all right; in attendance were McCarthy, Minsky, Fernando Corbató, Martin Greenberger, Jack Dennis, Doug Ross, Bob Fano, and several others. But instead of launching into a collegial brainstorming session, as was Lick's intention, they immediately went for one another's throat: "No, no, no—you don't want to do it *that* way! You want to do it *this* way!" As a rueful Corbató would later put it, "We sounded like a pack of dogs going in all directions. It was really disgraceful."

Lick himself seems to have taken the debacle in stride. He knew what MIT had to offer, and he knew he had planted a seed. Somehow or other, he would bring the group into the fold. For now, however, he had a more immediate problem. His $10 million budget may have been a vast amount by all previous standards of computer research, but $6 million of it was already committed to that one White Elephant out in Santa Monica. Like it or not—and Lick hated confrontations—he was going to have to reorder a few priorities.

Actually, some sort of confrontation was inevitable. Lick had been looking into the SDC command-and-control project since at least May 1962, months before he officially started at ARPA. And while he'd found a lot to like, including some pioneering work on how to deal with large databases and how to create large, complex programs, he'd found a lot more *not* to like, starting with the fact that the researchers were using their Q-32 computer as a batch processor. "I hated to see it," Lick later remembered. So in the fall of 1962, on his first official visit to SDC, he very courteously made it clear that he was cutting back on its budget a bit and that the command-and-control project would henceforth be an investigation into time-sharing. Reprogramming of the Q-32 would start immediately. "I was aware that this was cheating," Lick conceded in his 1988 interview, still feeling a little guilty about it more than a quarter of a century later. "I was insisting on my philosophy, my vision of what I wanted to happen here, and these people had every right to have their own vision." But there it was: he was trying to make a revolution, and he couldn't afford to let 60 percent of his budget be squandered on batch processing.

The SDC group did not take it well—with some justification. SDC had always been a results-oriented, get-it-out-the-door kind of place, an air force contractor accustomed to meeting the operational needs of no-nonsense military types. Moreover, the command-and-control project team had been doing exactly what it had promised to do in its original proposal to ARPA. And yet now, with no warning, the researchers found themselves contractually obligated to satisfy an academic-minded missionary pushing dreams of a new human destiny somewhere off in the far, far future. It was a culture clash, to put it mildly. And besides, most of the SDC engineers were like mainstream computer engineers everywhere else: they found the whole idea of time-sharing ridiculous. According to Ed Fredkin, who had come along with Lick as a freelance consultant, SDC chief Herb Bennington put it in no uncertain terms: "Look, Lick, if you persist in this, we're going to have to get you fired. Our contacts in the DoD are higher than yours. If you persist, that's the end."

Fredkin remembers being badly shaken by that comment; he'd never heard a bureaucratic threat being made quite so nakedly—or so plausibly. But Lick wasn't fazed at all, says Fredkin. It later turned out that Bennington did indeed have pull with the air force, but it was easily trumped by Lick's backing from the E Ring itself. "Lick," says Fredkin, "was a pretty good infighter that way."

Nonetheless, Lick needed enthusiasm and creativity from SDC, not sullen acquiescence. So on the flight back to Washington, recalls Fredkin, "Lick asked me to take on the task of making SDC *want* to do time-sharing." The way to do that, the two men agreed, was to show the folks at SDC what time-sharing could really do for them. The result was a little show-and-tell that Fredkin arranged at SDC in November 1962. "We had Marvin Minsky, John McCarthy, Alan Perlis, Ben Gurley, and Fernando Corbató all sitting around this huge boardroom table," he says. "They each gave a short talk. And around the wall we had people from SDC to serve as an audience. Well, it worked. The SDC guys were totally bowled over. This was the largest collection of topflight computer people that anyone had ever seen in one room. And they were unanimous in their feeling that SDC should drop what they were doing and start time-sharing. So SDC capitulated totally. By the time the meeting was over, time-sharing had become a foregone conclusion."

Even Herb Bennington knew he was beaten. "OK, you've convinced me," he said afterward, when Lick, Fredkin, and Minsky were sitting in his office. "But we're not going to make this a big project. You won't get any more than twenty or thirty programmers to start." Fredkin and Minsky looked at each other, trying hard not to explode with laughter. "SDC had done the software for SAGE," Fredkin explains, "so to Herb that sounded like small potatoes. But we were used to having maybe three or four grad students on a project—part-time. To us that sounded huge!"

In fact, SDC ended up committing fewer than a dozen people to the project. But that handful did yeoman work. The new man in charge was Jules Schwartz, whom Lick had pegged early on as an excellent programmer and an able group leader—what he called "a really positive character." He was right. Schwartz and his crew plunged into the time-sharing project with exemplary energy, now that they had been convinced. As requested, they got the first version of their time-sharing system up and running on the Q-32 by June 1963. Admittedly, says Schwartz, it was far more crash-prone than anyone liked, and it was hampered by its reliance on slow tape drives. But it worked. The Q-32 would go on to support a lot of good research in education, psychology, and display technology. It would even support dial-in connections from Stanford, Berkeley, and several other sites around the state—thus serving as a kind of prototype for the far more ambitious long-distance networks to come.

Lick was extremely gratified by this, of course. But in the meantime, he still had a revolution to organize. Factoring in the small amount of money he'd taken out of SDC, his back-of-the-envelope calculations suggested that he could fund maybe ten separate research groups, large and small. "It was going to take more than that to make a movement," he said. "But I felt I could settle for it if I just had ten."

High on his list, for example, was Carnegie Tech in Pittsburgh, where Allen Newell, Herbert Simon, and Alan Perlis had created a movement all by themselves. To bring them into the ARPA orbit, Lick had immediately sent them three hundred thousand dollars out of his behavioral-science budget. Use it however you want, he told them: grad students, machines, programmers—no questions asked. And there would be more to come later, he promised, once they had had a chance to talk. Newell, Simon, and Perlis were startled; Newell, for one, had never even heard of J. C. R. Licklider. But they took the money.

Also high on Licklider's list was the RAND Corporation, SDC's parent company, located conveniently nearby in Santa Monica. Lick knew RAND as the home of JOHNNIAC, the hand-built clone of John von Neumann's original computer at the Institute of Advanced Study. He also knew it as the place where Newell, Simon, and RAND programmer Clifford Shaw had created their path-breaking artificial-intelligence programs, Logic Theorist and General Problem Solver. But when he went calling in the fall of 1962, Lick discovered that RAND had also begun to do some really innovative work in interactive computing. There was JOSS, of course, the mathematician's "helpful assistant"; Cliff Shaw and his colleagues were getting ready for their first debugging runs, which would start the following spring. But there was also a new project known as the RAND Tablet, a kind of high-tech sketch pad that a user could write or draw on with a stylus, with the results then appearing on a CRT display. Now that, said Lick, was *symbiosis*. Indeed, in his 1960 paper he'd envisioned a graphical interface very much like this. Lick made it quite clear that he wanted to fund this work, recalls Keith Uncapher, who was director of the RAND computer group in 1962. And as Lick left, he says, they agreed to keep talking.

On to the San Francisco Bay Area. At Stanford University, nestled against the foothills of the Santa Clara Mountains, some thirty-five miles south of the city, Lick sought out John McCarthy, who was just settling into his new home there. Stanford, it seems, had made McCarthy a job offer that very fall—he'd stopped off for his official interview on the way home from the SDC meeting—and he had accepted immediately. He wasn't angry at MIT, exactly. But why wait around for time-sharing *not* to happen? McCarthy had immediately talked DEC into giving him a PDP-1, more or less for free. He had fired off a proposal asking Lick for money to do artificial-intelligence research at Stanford. And Lick, despite his annoyance at him for abandoning MIT so abruptly, had decided that he wasn't about to leave the new Californian out of the ARPA family. So yes, he now told the astonished McCarthy, who had fully expected to hear "No," ARPA will fund you.

Then it was across the San Francisco Bay to Berkeley, where Lick met with a young redhead named Edward Feigenbaum. ARPA, Lick hinted broadly, would like to put some serious money into AI research at Berkeley. "I was a young assistant professor," marvels Feigenbaum, "barely visible above the surface. Where did he even get hold of my name?" From Newell and Simon, presumably: Feigenbaum had earned his Ph.D. under Simon in 1960, just before going to Berkeley. Still, it was enough to make you believe in fairy godmothers. And then, to top it off, Licklider also hinted to Feigenbaum and to the hardware maven Harry Huskey that he would like to see a proposal for a Berkeley time-sharing effort—a medium-size system, say, as a complement to what SDC was doing.

And so it went. Lick seems to have spent most of that autumn of 1962 on airplanes, trying to entice computing's best and brightest to join in his vision. Of course, he did have to be a little careful about how he made his pitch. Early on, Lick would later remember, some of the old hands at the Pentagon had reminded him sternly that he was now a government employee, which meant that he wasn't supposed to solicit specific proposals from specific people—a practice that smacked of favoritism, elitism, cronyism, and other undemocratic things. In the name of fairness and equal opportunity, they insisted, a government funding officer had to wait for proposals to come to *him*.

Not a chance. Waiting around would spell disaster, Lick realized—so he quickly figured out a way to circumvent the rules without actually disobeying them. "I took advantage of those other three leftover SAGE computers," he later explained. "I couldn't ask for a proposal directly, but I could go around and ask people, 'Do you want one of these things, and what would you do with it if you had it?' Well, people were pretty sensible. Nobody wanted one. But it did lead to a lot of discussions, and the discussions led to proposals, and so forth. So I was able to get proposals out of quite a few places without ever asking for one. They came pretty fast."

By spring, in fact, those proposals would be overflowing his desk. But then, says Lick, he couldn't pat himself on the back too hard. A few proposals did indeed come looking for him, with perhaps the most notable being a long, visionary document that landed on his desk at the Pentagon almost literally the day he arrived. It made for strange reading, he remembered, as unsolicited proposals often do. But in this case, Lick didn't automatically dismiss it as the work of a crackpot, if only because he'd met the author once or twice before. At age thirty-seven, ten years younger than Lick himself, he was a handsome, dark-haired, but rather lonely fellow—the quiet sort who ordinarily might not stick in your memory. But then once you got to know him a bit, you saw that much of this man's quietness came from his habit of *listening*—deeply, profoundly listening to everything that was happening around him, and trying to work out its most fundamental meaning. And then when he did talk, his soft, diffident baritone somehow managed to be hypnotic in its intensity.

His name was Douglas Engelbart.

It was in December 1950, says Doug Engelbart, thinking back to the morning when it all changed for him. He was twenty-five years old, and by every objective measure, life was good. A decade earlier he'd been a semipoor teenager in the semirural outskirts of Portland, Oregon, still getting up every morning to milk the family cow. Now he was several years out of school, with a bachelor's degree in electrical engineering plus two years' experience as a navy radar technician. He was working at a good job at Ames Research in California, where he did electrical engineering for one of NASA's ancestors, the National Advisory Committee on Aeronautics. He had met a girl there. And that very weekend, the two of them had decided to get married.

And yet, as Engelbart explained to Howard Rheingold for the latter's 1985 book, *Tools for Thought,* "the Monday after we got engaged, I was driving to work when I was hit with the shocking realization that *I no longer had any goals.* As a kid who had grown up in the Depression, I was imbued with three goals: get an education, get a steady job, get married. Now I had achieved them. Nothing was left."[5]

Driving onward through the vast prune orchards of what would later be known as Silicon Valley, Engelbart calculated that he had about 5.5 million working minutes remaining in his life. What was he going to do with them? He'd never much cared about getting rich. And changing careers seemed like too much work. So that just left . . . saving the world? Well, he figured, the world could certainly use a little saving, what with overpopulation, the nuclear-arms race, the Korean War, and a vast array of other problems. The question was, How?

Engelbart mulled over that question for several months, considering and then rejecting any number of idealistic crusades. But then suddenly, as he wrote in his own memoir, "up through all this delightful, youthful abstraction bobbed the following clear realization":

- FLASH-1: The difficulty of mankind's problems was increasing at a greater rate than our ability to cope. (We are in trouble.)
- FLASH-2: Boosting mankind's ability to deal with complex, urgent problems would be an attractive candidate as an arena in which a young person might try to "make the most difference." (Yes, but there's that question of what does the young electrical en-

gineer do about it? Retread for a role as educator, research psychologist, legislator . . . ? Is there any handle there that an electrical engineer could . . . ?)

- FLASH-3: Aha—graphic vision surges forth of me sitting at a large CRT console, working in ways that are rapidly evolving in front of my eyes (beginning from memories of the radar-screen consoles I used to service).[6]

The whole idea came together in about half an hour, Engelbart told Rheingold: "I started sketching a system in which a computer draws symbols on the screen for you, and you can steer it through different domains with knobs and levers and transducers. I was designing all kinds of things you might want to do if you had [such] a system . . . how to expand it to a theater-like environment, for example, where you could sit with a colleague and exchange information. God! Think of how that would let you cut loose in solving problems!"[7]

Within a few days, he said, the imagery of FLASH-3 had evolved into a vision of a general-purpose, computer-powered information environment. It would include documents mixing text and graphics on the same CRT display. It would include whole new systems of symbols and methodologies to help users do their heavy thinking. And it would include network-assisted collaborations to allow people to work together in ways that would be more effective than anything anyone had ever seen before.

Within a few weeks he had committed his career to this vision, which he now called "augmenting the human intellect."

Within a few months he had left Ames Research to enroll as a graduate student at Berkeley, where he joined a group that was working on computers.

And within a few years, said Engelbart, he was forced to face the fact that his augmentation idea wasn't going to earn him any applause in academia, let alone a Ph.D. This was still the mid-1950s, and the state of the art was still the IBM 704. When he talked about using computers interactively to help people, or to teach people, his professors and fellow students reacted with an incomprehension that bordered on hostility. So, bowing to reality, Engelbart wrote a dissertation on bistable gaseous plasma digital devices, a worthy topic that was solidly centered in the mainstream. Then, with Ph.D. in hand—along with half a dozen patents on the plasma devices—he went out looking for a more congenial atmosphere in private industry. In October 1957 he accepted an offer from a think tank known as SRI, the Stanford Research Institute, a university spin-off located just north of Palo Alto, in Menlo Park, California.

He very quickly learned to keep a low profile even there. ("Don't tell anybody else," urged one colleague when he heard about Engelbart's ambitions. "It will prejudice people against you.")[8] Engelbart continued to do conventional work at SRI for another year and a half, in the process earning a dozen more patents. Only in 1959 was he able officially to work on augmentation, thanks to a small grant from the air force's office of scientific research, plus some reluctant support wrangled out of the SRI higher-ups. And even then, he later wrote, "it was remarkably slow and sweaty work. I first tried to find close relevance within established disciplines [such as artificial intelligence,] but in each case I found that the people I would talk with would immediately translate my admittedly strange (for the times) statements of purpose and possibility into their own discipline's framework."[9] At the 1960 meeting of the American Documentation Institute, a talk he gave was greeted with yawns, and his proposed augmentation environment was dismissed as just another information-retrieval system.

No, Engelbart realized, if his augmentation ideas were ever going to fly, he would have to create a new discipline from scratch. And to do *that,* he would have to give this new discipline a conceptual framework all its own—a manifesto that would lay out his thinking in the most compelling way possible.

Creating that manifesto took him the better part of two years. "Augmenting the Human Intellect: A Conceptual Framework" wouldn't be completed until October 1962. But Engelbart was nothing if not dogged. "By 'augmenting man's intellect,'" he wrote, struggling to articulate his own gut feelings, "we mean increasing the capability of a man to approach a complex problem situation, to gain comprehension to suit his particular needs, and to derive solutions to problems. . . . We do not speak of isolated clever tricks that help in particular situations. We refer to a way of life in an integrated domain where hunches, cut-and-try, intangibles, and the human 'feel for a situation' usefully coexist with powerful concepts, streamlined technology and notation, sophisticated methods, and high-powered electronic aids."

"Electronic aids" meant computers, of course, along with all the associated technologies for information storage and transmittal. But it was no accident that Engelbart started that sentence with words such as *hunches* and *feel.* Much of what followed was in fact an extended meditation on the human side of the equation: Precisely what was the nature of this system that he was proposing to augment?

Well, Engelbart wrote, let's ask ourselves how humans have managed to cope with complex situations until now. Imagine a native of the Amazonian rain forest who is magically teleported into New York City: confronted with taxicabs, pay phones, and delicatessens—phenomena that natives of the urban jungle can handle with ease—he would be helpless as a baby. But wherein lies the difficulty? Not in the biological "hardware," said Engelbart: at the level of neurons and neurochemistry, all human brains are essentially identical. No, the difference lies in the biological "software," the repertoire of concepts and skills that each of us acquires over a lifetime. The ability to drive a car, the ability to place a telephone call, the very notion of money—these are units of knowledge that can be applied in a wide variety of situations, Engelbart explained, which makes them roughly analogous to data structures and subroutines in a computer program.

However, he emphasized, isolated skills and concepts are useless unless they can be organized for a larger purpose: "Just as the mechanic must know what his tools can do and how to use them, so the intellectual worker must know the capabilities of his tools and have suitable methods, strategies, and rules of thumb for making use of them." Again, this is basically analogous to the way low-level subroutines are assembled into larger subroutines, which in turn are assembled into complete programs. In exactly the same way, wrote Engelbart, human capabilities exhibit a whole hierarchy of levels, ranging from the neural routines that are wired into our brains before birth all the way up to the high-level impulses we absorb from the surrounding culture—the commitment to liberty, equality, and fairness, for instance. Indeed, he suggested, this hierarchy is what we're actually talking about when we use that mystical word *intelligence:* "If there is any one thing upon which this 'intelligence' depends it would seem to be [the hierarchy's] *organization.*" And of course, that elaborately organized hierarchy was what Engelbart proposed to augment. *All* of it—"the system . . . comprising a trained human being together with his artifacts, language, and methodology."

As a concrete example, he offered the office memorandum. To create it, the author has to engage in a variety of standard subtasks such as planning, composing, and dictating, each of which is composed of even simpler subtasks. But by the same token, the memo itself will be just one component of some higher-order process, such as organizing a committee or changing a policy. Like everything else in human life, Engelbart wrote, the simple office memo is embedded in a vast, tangled hierarchy of activities.

Now, he continued, consider the reverberations up and down this hierarchy when a bit of augmentation technology is introduced at a relatively low level: "Suppose you had a new writing machine—think of it as a high-speed electric typewriter with some special features. . . . For instance, trial drafts could rapidly be composed from rearranged excerpts of old drafts, together with new words or passages which you stop to type in. Your first draft could represent a free outpouring of thoughts in any order, with the inspection of foregoing thoughts continuously stimulating new considerations and ideas to be entered. If the tangle of thoughts represented by the draft became too complex, you would compile a reordered draft quickly."

With the possible exception of the Expensive Typewriter program being created by the MIT hackers at about the same time, this remarkable passage is the first written description of a modern word-processing system. And for the life of him, Engelbart can't remember where the idea came from. "It was just a part of all those years of thinking," he says. "Manipulating words seemed like the obvious place to start because it was a way to manipulate your ideas. That's the very essence of your knowledge and thinking: the concepts in your mind that you're converting to words and symbols. And since the computer should be able to manipulate symbols for you, it was 'Well, of course.' It could just help you in so many ways."

Engelbart would later make it his top priority to show exactly how this could happen. In the meantime, however, he was wrestling with a more cosmic theme: his notion that computer-based augmentation represented an entirely new phase in human evolution. Our ancestors had taken their first great step beyond the lower forms of life when they developed the capacity for *concept manipulation,* he asserted: "We speak here of concepts in their raw, unverbalized form. For example, a person letting a door swing shut behind him suddenly visualizes the person who follows him carrying a cup of hot coffee and some sticky pastries . . . a solution comes to mind immediately as an image of a quick stop and an arm stab back toward the door."

The second phase had been *symbol manipulation,* Engelbart continued, when our ancestors evolved the capacity to represent concepts with mental symbols such as words and numbers. For example, this capacity would allow a shepherd to keep track of all twenty-seven sheep in his flock by counting them instead of having to remember what each sheep looked like. Then the third phase was *manual external symbol manipulation,* he wrote, when our forebears invented a variety of ways to represent their mental symbols graphically: "a stick and sand, pencil and paper and eraser, straightedge or compass, and so on." This allowed us to overcome the limitations of working memory and greatly enhanced our ability to visualize things.

But now, wrote Engelbart, thanks to computers that were capable of executing programs on their own, humans were embarking upon a fourth stage: *automated external symbol manipulation.* "In this stage," he noted, "symbols with which the human represents the concepts he is manipulating can be arranged before his eyes, moved, stored, recalled, operated upon according to extremely complex rules. . . . In the limit of what we might now imagine, this could be [done by] a computer . . . with which we could communicate rapidly and easily, coupled to a three-dimensional color display within which it could construct extremely sophisticated images."

Again, both the vision and the proposed technology sound amazingly modern. And again Engelbart has no clear idea of where the notions came from. But it was at roughly this point in his struggles to write his manifesto, he says, that another image struck him—fifteen years after the fact. Back in the late summer of 1945, he suddenly remembered, just after the surrender of Japan, when he was a twenty-year-old radar technician assigned to the Philippines to work on demobilization, he had stopped in one afternoon at the Red Cross library. It was a cool, airy place, built up on stilts like a native hut, and full of polished bamboo. One of the magazines available there was the July 1945 issue of the *Atlantic Monthly.* And on page 101, right after the latest installment of Betty McDonald's comic novel *The Egg and I,* he had come across an article entitled "As We May Think."

Of course! Engelbart hadn't thought about Vannevar Bush's Memex in years. But now that he had, he could see that it was just a matter of replacing Bush's microfilm with modern computers. The essence of the system was all there: a device that could put the entirety of human knowledge at your fingertips, linked and cross-linked into an ever-expanding web of associations, on and on. Bush had been talking about augmentation all along.

Engelbart immediately added a whole section to his manifesto laying out Bush's proposals in detail—most especially the latter's speculations about the

Memex's somehow communicating directly with the brain. Although he hadn't consciously based his own ideas on Bush's, says Engelbart—his own statement was already in close to its final form by the time he remembered the Memex—he still believed that even unconscious debts should be acknowledged.

In the meantime, as Engelbart wrote in his conclusion, the challenge was to get from here to there. His proposal for meeting that challenge—not surprisingly—was an interdisciplinary Knowledge Augmentation Laboratory that could pursue the technology of human augmentation as quickly as possible.

And that, he told J. C. R. Licklider, was what he wanted ARPA to fund.

Engelbart knew exactly who Lick was, of course; in fact, he had led off the "Related Work" section of "Framework" with a discussion of Lick's symbiosis article. Early in 1962, moreover, he had briefly met Lick at the Spring Joint Computer Conference in San Francisco. And when he heard that Lick was going to be starting up a computing office at ARPA, he couldn't help but feel a glimmer of hope.

Hope was actually in short supply by that point, Engelbart remembers. He will never forget the pitch he made to the National Institute of Mental Health in 1961, using an earlier draft of his manifesto: "Since your interesting research would require exceptionally advanced programming support," read the rejection letter, "and since your Palo Alto area is so far from the centers of computer expertise, we don't think that you could staff your project adequately."[10]

Well, the committee had a point; in 1961, the area around Stanford really was the outer boondocks of computing. And for all Engelbart knew, Lick might react the same way. But it was worth a try: Engelbart had a formal proposal and a copy of his manifesto waiting on Lick's desk the day he arrived at the Pentagon. After all, he later wrote, "there the unlucky fellow was, having advertised that 'man-computer symbiosis,' computer time-sharing, and man-computer interfaces were the new directions. How could he in reasonable consistency turn this down, even if it was way out there in Menlo Park?"[11]

He couldn't. Although Lick never publicly described his first response to "Framework," it must have included a strong component of déjà vu. Here was the entire idea of human-computer symbiosis, re-created by a complete unknown out in the middle of nowhere. Lick had to admire that—even though Engelbart had been quite right in anticipating some skepticism on his part ("Later," says Engelbart, "a couple of his friends told me that his reaction was, 'Well, he's way out there'—meaning far from MIT—'in Palo Alto, so we probably can't expect much. But he's using the right words, so we're sort of honor-bound to fund him' "). So by early 1963, Engelbart had funding for his project. Of course, he didn't receive a huge amount of money, thanks to Lick's steadily growing list of academic dependents. But it was a start. And at age thirty-seven, Douglas Engelbart was feeling, well, something this quiet, lonely man hadn't felt very often before, certainly not in relation to his augmentation ideas.

"Lick was the first person to believe in me," says Engelbart. "And he was the first person to stick his neck out and give me a chance. In fact, if he hadn't done that, if he hadn't stuck his neck out and given me money, I don't think anybody ever would have done so. That was why I trusted him. Lick was like my big brother."

So there it was: SDC, RAND, Berkeley, Stanford, SRI, Carnegie Tech—by the late autumn of 1962, Lick had convinced himself that the Cambridge vision could be established on the West Coast, and in the Heartland.

Now if he could only get it established in *Cambridge*. . .

JUST DO IT

Even now, almost four decades after the fact, it's still Robert Fano's favorite story. And Fano, a short, round Italian who bears a notable resemblance to the actor Danny DeVito, and whose letter-perfect English still bears a strong imprint of his native Turin, needs no encouragement to tell it again.

"I was there to chair a session on communication," he says, thinking back to late November 1962 and the Homestead resort in Hot Springs, Virginia, where the MITRE Corporation had organized a meeting for the air force on unclassified command-and-control research. "But I did attend some of the other sessions, and I came out with the feeling that command and control was a mess. Intellectually the speakers were poor. They had no vision. You couldn't see any solution coming from them."

This was not a reassuring thought, says Fano—not when it came less than a month after the Cuban Missile Crisis, and especially not when the hallways of the Homestead meeting were abuzz with rumors of a near-catastrophe. For a full thirteen days in October, the United States had placed nuclear weaponry amounting to some seven thousand megatons—the equivalent of more than a hundred thousand Hiroshimas—on hair-trigger alert. The Soviets had presumably done much the same with their arsenal. And none of it, on either side, had been held in check by anything more than a slapdash, wired-together command-and-control system that virtually invited a fatal mistake. On any number of occasions, apparently, humankind had come *this* close to a conflagration that would have made World War II look like a playground spat—and we had survived largely by luck.*

---

* The rumors were all too true. According to the Pentagon's doleful postmortem, individual people, individual units, and individual technologies had performed very well, for the most part. But the command-and-control system that tied them all together had been a shambles. For example, there was the "missile" that U.S. radars detected rising from Cuba, with a projected impact near Tampa; the signal turned out to be a computer test tape. There was the first U.S. squadron of Minuteman I solid-fueled missiles, which were then undergoing testing

So no, says Fano. This was not a good time to be hearing that the air force didn't have a clue what to do about the command-and-control mess. On the very last morning of the meeting, in fact—Wednesday, November 21, 1962, the day before Thanksgiving—as the participants were settling into the special train that would take them back to Washington, Fano voiced his concerns to his seat-mate, J. C. R. Licklider.

Exactly his point, his old friend declared: this was what he had come to the Pentagon to fix (indeed, Lick had been one of the instigators of the conference). And with that, remembers Fano, they were off. He and Lick talked about the sessions, about Lick's insistence that command and control ought to be done through time-sharing, and about his vision of symbiosis. Especially his vision of symbiosis. By the time the train was well on its way, says Fano, he and Lick had started moving from seat to seat, and their conversation had turned into a free-for-all. "This was a special train for the meeting," he explains, "so everybody there was a colleague"—including Marvin Minsky, Oliver Selfridge, and any number of others belonging to what Lick later called the budding nucleus of his movement.[12] Lick was in his element, making jokes and terrible puns and finding a ready audience for his favorite subject. Along the way, moreover, without ever saying it in so many words, he made it very clear that he still wanted a proposal out of MIT—a big proposal. MIT could make symbiosis real, Lick let it be known. MIT could give the world its first time-shared "information utility," a pilot plant for the continent-spanning utilities to come. And with some of the brightest students in the world, MIT could recruit a whole new generation to the cause.

Fano got the message. He and Lick were both aware that MIT's real problem was lack of leadership. The young Turks who actually knew something about computing had no seniority and no clout and were scattered all over campus. If interactive computing was ever going to become a reality at MIT—and funding from Lick was perhaps the last, best chance to *make* it a reality—then a senior person was going to have to take the lead. "But nobody here was biting," says Fano. Phil Morse, founder and head of the Computation Center, had already spread himself far too thin. Wes Clark and his LINC computer team were still searching for a home after their exile from Lincoln Lab. And that left—well, Fano knew damn well who that left. But he didn't want it. He'd never liked administrative responsibility. Besides, he didn't really know anything about computers: he was an information theorist, a pencil-and-paper man.

---

and certification in Montana; in the rush to get them ready for use, squadron officials ended up overriding such key safety features as the multiple-key system, which meant that the missiles could have been launched by one person, acting on his own initiative. And there was the "intruder" detected in the middle of the night at Volk Field in Wisconsin, resulting in a series of alarms that the crews interpreted as the signal to scramble their nuclear-armed F-106 interceptors. With the air full of B-52s circling on full alert, and the interceptors expecting to find Soviet bombers, the result might well have been nuclear friendly fire over U.S. soil. The F-106s stood down only after an officer drove out onto the airstrip and got their attention by flashing his car lights; the "intruder" had turned out to be a bear.

Nonetheless, says Fano, all the arrows kept pointing his way. Information theory wasn't so very far from computing, after all. He'd served on the Long Range Planning Committee. He'd even taken a computer course or two. "So I was the only senior member of the faculty who had some kind of connection, as feeble as it was, to computation," he says.

And now, of course, here was his old friend J. C. R. Licklider, sweeping him up in that contagious enthusiasm.

It was a *lo-o-ong* train ride, Fano remembers—four or five hours. Still, when the train finally rolled into Washington's Union Station, he said good-bye to Lick without making any promises. Excited though he was, he says, "I was afraid I could not make it work." Instead, he just kept turning the possibilities over and over in his mind during the flight back to Boston. Then the next day, after he and his wife had survived Thanksgiving dinner with their children, who were very small, he found himself pacing around and around the house. No, he didn't want to be an administrator. But Lick was offering MIT a chance to do something very important for the future—which meant, in turn, something very important for MIT.

Important for MIT. That resonated with Fano. He had been at the school almost continuously since 1939, he explains, when he and his brother arrived in the United States as refugees from Mussolini's Nazi-inspired edicts against Jews. During the war, when most of the electrical engineering department had vanished into the Radiation Laboratory and he couldn't follow them—as an Italian citizen, he was technically an enemy alien—he had taken over many of the department's courses and taught them single-handedly. After the war, he had experienced the glory years of RLE, when ideas, people, and energy just seemed to gravitate to it from all directions. And in the mid-1950s he had felt a keen disappointment as the glory years faded, as people got more and more preoccupied with their own disciplines. For a long time now, in fact, he had been wondering how to revive that old spirit. Creating an explosive intellectual environment was a bit like planning a successful party: it had to have an indefinable spark, an unpredictable, near-magical mix of just the right time, just the right place, just the right people, and just the right challenge to get them pumping. No one had a surefire formula. But this new program of Lick's—now maybe, just maybe. . . .

"What got into me I do not know," says Fano. "But at a certain point on Thanksgiving Day I said, 'OK, if nobody else is going to do it, *I* am going to do it.' "

The next morning, he continues, as he was finishing up a previously scheduled meeting with MIT's provost, Charles Townes, he recounted Lick's idea for a prototype information utility, and his own intention to lead the effort to build it at MIT. "Do you want to think about this and let me know whether you think it makes sense?" Fano asked.

"No," said Townes with barely a hesitation. "Go ahead."

Fano was startled; he'd expected at least *some* academic dithering. But then, he wasn't asking Townes for money, was he? He was offering him Lick's money.

And that had always been the problem: the MIT higher-ups weren't *against* time-sharing; they just didn't want to pay for it.

Holing up for the weekend, Fano hammered out a two-page memo summarizing what he had in mind. Then on Monday morning, he says, "I distributed the memo to key people, like my department head, who was Peter Elias at that time, and then to Gordon Brown, who was dean of engineering, and of course, Charlie Townes, and Julius Stratton, who was president of MIT. I found out later that Gordon Brown told his secretary to file it under 'FF,' meaning 'Fano's Folly.' "

No matter. On Tuesday, Fano and Phil Morse went to see Stratton. The MIT president listened carefully, says Fano. And once he, too, was satisfied that he wasn't being asked for money, he had only one real question: "*Where* are you going to do this?"

Oops. This was not a trivial matter, Fano knew. MIT was desperately overcrowded as it was. There was literally no place on campus to put a new project. Fortunately, however, Fano had spent Monday doing his homework. According to rumor, he told Stratton, an out-of-town computer company had fallen on hard times and wanted someone to take over its lease at a brand-new office park just north of campus. The lease covered the top two floors of one of the highrises, 545 Technology Square. "That would be practically ideal," Fano explains, "because we were going to have a big computing installation, and the ninth floor was already planned for that purpose. We could use the eighth floor for offices."

"OK," said Stratton. "Look into it."

And that was that, says Fano: "I did look into it, and the rumor turned out to be true. Well, anyway, that was a Tuesday. Lick had already planned to come to town on Thursday for some other reason. On Thursday we all gathered in Stratton's office—Lick, Stratton, Phil Morse, and I—where the four of us shook hands. And that was the decision!"

Just think, says Fano, still marveling after all this time—just think what a decision like that would involve today. "There would be all kinds of committees, and inquiries, and peer review, and red tape. It would take forever. But we did it all within one week: Thursday to Thursday. Because all the parties concerned just decided that yes, we were going to do this."

Alas, no one can escape red tape quite that easily. Handshakes were a beginning. But Lick couldn't give MIT a penny of real money until ARPA had approved a formal proposal. So Fano and his colleagues had to produce one, fast.

Of course, as Fano discovered very quickly, they wouldn't be working alone. If Lick didn't actually write the proposal for them, he certainly made his wishes known. "I felt bad," Lick admitted in his 1988 interview. "I knew Bob very well, and he was a good writer. But this proposal was just absolutely crucial to me, because it was going to be my first one. A lot of people [at ARPA] were going to see it, and he had to make it a good one. So I talked with Bob at some length on two or three occasions. [I told him that] MIT had a lot of flexibility in just which projects to do. But it was clear that there's got to be artificial intelligence. There's got to be time-sharing. There's got to be interactive stuff like the Teager Tablet [a penlike input device developed by the engineer Herbert Teager], and graphics like the Kludge [an experimental graphics display terminal]. I would have been very unhappy not to have that. . . . I wanted to hear 'Computers are as much for communication as they are for calculation.' Then I wanted assurance there were going to be good people working on the project. I wanted a summer study that would bring people from all over the industry, that would try to shape up this field and make it clear what we were doing. And I also said that I wanted a lot of help, although I didn't want that written in the proposal. I wanted to be able to get an MIT person to visit SDC [and evaluate what was going on there]. I wanted people to take time off from their research to have meetings to think about how the program was going to go."

Nowadays, of course, this sort of involvement in a proposal would land Lick in jail for massive violations of the federal procurement laws, which were greatly tightened in the 1980s. And even though it was legal enough in 1962, it would have been considered far beyond the pale for most agencies. But Jack Ruina's ARPA wasn't most agencies. As Ruina told Fano not long afterward, "I delegated to Lick. He agreed. It goes."

In any case, Fano was willing enough to go along. For one thing, he and Lick were on the same wavelength here, so there was no real conflict. For another, he had more urgent matters to worry about, such as getting this project *organized.*

Happily, he had plenty of help on that job, too. Corbató, Dennis, Greenberger, Minsky, Ross, Selfridge, Teager—once again, the whole gang was there. Only this time they were rallying around instead of tearing one another apart. After all, says Corbató, this was what they had been aching for all along. "I think most of us felt that it was more than just a project," he says. "It was a major milestone. It served as a focal point for all the computer-science activity at MIT. [In fact], it was really the beginning of the computer-science community here, because the theme seemed such a deep one that it would last a long time."

Taking them at their word, a much-relieved Fano formed them into a Tiger Team—"a committee of the computer guys who were about to be parties to the proposal," as he describes it. "They each wrote a little piece. I wrote the introduction." In a nutshell, he says, their proposal was to create the world's first true information utility. "[This information utility] will be by necessity a sort of bootstrap operation," they wrote. "The evolving system will be, simultaneously, the main research tool and the primary object of experimentation, as well as the tangible product of the development effort."[13] Their basic strategy was to start with Corbató's CTSS, which was already running on the Computation Center's IBM 7090. Then, as soon as possible, they would move CTSS to a new computer dedicated to this project alone—namely, an upgraded version of the 7090 known as the IBM 7094 (paid for by ARPA, of course). True, the 7094 would be just about as crummy at time-sharing as the 7090 was. But it would at least give its users something like twenty-four-hour-a-day access right away. This would then provide the project with enough real-world experience to replace CTSS with a new and much-improved time-sharing system. And that system, in turn, would run on a totally new computer, along the lines envisioned in the Long Range Planning Committee's report.

Thus the bootstrap: from kludgy CTSS to a sophisticated new generation. And all along the way, of course, MIT would be fostering computer education, as well as cultivating a wider user community. Indeed, Fano was already out beating the bushes. "From the end of 1962 until the spring of 1963 I was busy giving talks around MIT," he says. "It was important in my mind, and it was important in Lick's mind, to get people hooked on time-sharing. . . . You couldn't develop and improve the system without seeing how a large and diverse community used it in practice."

With innumerable details to get right, and dozens of people to bring on board, Fano and his colleagues had a frazzled holiday season that year. Even so—and perhaps this was a measure of how badly they wanted the project—they managed to finish their proposal just before Christmas, barely three weeks after the handshake in Stratton's office. The week after Christmas gave them a moment to catch their collective breath. Then, in January 1963, they took the proposal to Washington for a formal presentation to ARPA.

Lick was at least as nervous as they were (Louise Licklider remembers his becoming noticeably preoccupied and uncommunicative in the days leading up to the presentation, even during his sacrosanct dinner hour with Tracy and Linda). But in fact it went beautifully. "This was a fantastically good proposal," Lick said in his 1988 interview. "The people who could turn me on or off easily [Ruina, Fubini, and their associates] were very bright people, and they had seen too many bum proposals in their lives. [But] they saw this one and said, 'That's legitimate.' And just like Charlie Townes said, 'Do it,' and Stratton said, 'Do it,' *they* said, 'Do it'—and it was all over."[14] Miraculously, improbably, the planets were still lining up.

But once again, that was only the prelude to a whole new round of headaches. First and foremost, the MIT team was now facing a new deadline: Lick had wanted a summer study, so the proposal had promised him a six-week program starting on July 1, 1963. That did *not* leave a lot of time. Lick had made it very clear that this summer study would be the de facto launch of the ARPA program, the event that would bring his people together and turn them into a real community. So Fano and his new deputy, Oliver Selfridge, started drafting an invitation list that included McCarthy, Perlis, Engelbart, Feigenbaum, Uncapher—everyone who was involved with the ARPA program or even thinking about getting involved. At the same time, the project team swiftly realized that this was an opportunity to look over the latest crop of young hotshots from the computer industry, with an eye toward recruiting some of the best for MIT. At the very least, it might encourage a few people in industry to start thinking about something besides bigger batch processors. And then there were the various government-agency people, the Lincoln Laboratory people, and on and on. Fano and Selfridge eventually compiled a list of more than a hundred invitees.

Meanwhile, since this summer study was also going to be the first extended, real-world demonstration of time-sharing, Corbató and his team now had to prepare for the fact that every know-it-all in the computer world would be standing right there watching for foul-ups. Then, too, the participants gathered at Tech Square would actually have a choice of time-sharing systems. They could use CTSS terminals connected to the Computation Center just a few blocks away, or terminals connected via telephone lines to SDC's new Q-32 time-sharing system out in Santa Monica. This arrangement wasn't intended to be a competition, exactly. Corbató had already spent a fair amount of time advising the SDC group, and he got along very well with the team leader, Jules Schwartz. But still, he was not going to let CTSS be an embarrassment.

"All spring we were furiously getting the system ready for the summer," Corbató says. It was one thing to demonstrate CTSS for an hour or two; they'd been able to do that since November 1961. But now the Computation Center's poor, overworked 7090 was going to have dozens of people pounding on it nonstop for several weeks. "We had to make several key upgrades to make the equipment viable," he says. "One of them was that we attached an IBM 1301 random-access disk file." This was essentially what we'd now call a hard disk, except that in those days it was the size of a washing machine. But it had enough storage capacity to keep all the users' programs and data available for instant retrieval, and it was much, much faster than the tape drives that had been their only storage option up to that point. "The second thing we had to do was get typewriters [the remote terminals] connected in quantity," he says. "They turned out to be Teletypes or IBM [typewriters] initially. There were no lower-case letters, so the output looked like a telegram. Then we ended up getting a huge kludge of a controller, which attached to the IBM 7090, and which allowed us to run with a nominal load of, I think, sixteen active terminals at one time. We also got a telephone switch installed so that people could dial in from many sites with the use of modems." On the administrative front, meanwhile, Corbató arranged to "borrow" huge chunks of the Computation Center's computer time for the summer study to use free of charge, while the higher-ups tacitly agreed to look the other way. So all in all, he says, "that was a very exciting period!"

Perhaps a little *too* exciting at times—especially for Bob Fano, who knew that every bit of this work was being done without a formal contract from ARPA. MIT was actually a part owner of the Tech Square development, Fano explains. So in effect, the school was asking for taxpayer money that it would pay to itself. It took forever to draw up a contract that would allow this to happen legally, he says with a sigh; "the lawyers had a field day." And yet, incredible though it may sound today, everyone continued to operate in a spirit of trust. Long before the contract was officially signed—which finally happened on the last day of June, with the summer study due to start the following morning—Fano and his colleagues were comfortably ensconced on the eighth floor of 545 Technology

Square. MIT had rented and furnished the space on its own initiative, and ARPA had authorized them to start spending money.

Along the way, moreover, the project had acquired a name. "The decision was made in a great hurry," Fano recalls. "This was about March nineteen sixty-three. I had just hired Dick Mills to be my assistant, and he needed a parking sticker. But then the parking office asked him what organization he worked for. Well, the only name that existed until then was FF–'Fano's Folly.' "

Surely they could do better than that. So Fano mulled over a few ideas that day, he says, and then tried them out that night. "McCarthy had come to town, so Minsky had a little party for him at his house. We talked about the name, and I pointed out that you could use the same acronym to stand for Machine-Aided Cognition–the ultimate goal of the project–and for Multiple Access Computer, the tool we were creating. It sounded good to them. So there was never any objection to the name."

Project MAC it was.

GOING GALACTIC

Back in the D Ring of the Pentagon, meanwhile, Lick had at last admitted to himself that it was time for him to hire a deputy. By the spring of 1963 he could see that there were too many turf fights around the Pentagon that he knew nothing about, too many personalities, too many political nuances, and too many land mines that he could step on without ever knowing they were there. What he needed was someone who could keep him out of trouble.

"I talked to a lot of people in the Pentagon," he remembered in his 1988 interview. "They said, 'You want to get somebody who's really credible to the military. If you can, find a war hero. Also, he's got to know something about science, or mathematics.' Well, I wound up with a guy who *was* a war hero. He was a flier who had escaped from captivity by the Germans twice. He also had run the computer center in the Pentagon, but the records did not say that. It turned out he had hated running the computer center so much that he tore up his records before he left. But I finally got him anyway. He was an astronomer by trade. Buck Cleven was his name–a very interesting guy."

Colonel Cleven was indeed a gem. Not only did he skillfully steer his new boss around the land mines, but he proved to be a lifesaver in another sense as well. The proposals that Lick had solicited in the fall from RAND, Berkeley, and elsewhere had started showing up in the spring. But now that the word was out that he really had money to give away, a tidal wave of other proposals was pouring in as well. "Most of it was unbelievably irrelevant or low-grade," Lick recalled. "And there were problems, like this character who came around with a gun that shot little rockets–fifty or a hundred little rockets–and demonstrated it in my office. These things got going and the place was left a shambles. We were lucky not to be blinded. So I just took advantage of Cleven and made him listen to all those things. He was fantastically charming. But his main job was just to make the visitors feel good and not give them any money unless he could spot that they had something—which was one case in thirty or forty."

Lick felt a little guilty about that. Still, this division of labor did leave him free to concentrate on more pressing matters—such as how he was going to hold his movement together. While he had high hopes for the MIT summer study, it was only one event. What would keep that community spirit alive in the fall, after all the participants went back home?

The problem, he knew, was that any given research group might be working on one or another piece of the vision, but none of them, with the possible exception of Project MAC, would be working on *every* piece. The only integration was in J. C. R. Licklider's head. "It was not a clear vision," Lick explained, "certainly not in the sense of saying, 'We'll plug these technologies all together, and that will be the system.' Rather, it was a matter of getting a supply of parts and methods and techniques, so that eventually, different people could put together different systems out of it." To keep it all straight, he added, "I used to draw big sketches on big sheets of paper. Then I would lose the paper. [So] I had pretty well wrapped up in *me* all of the topics that it would take to put interactive computing together."

But what if he got hit by a bus? No, Lick realized, if this vision was ever going to outlast his tenure at ARPA, he would somehow have to forge all these groups into a self-reinforcing, self-sustaining community. Putting MIT to work on the summer study had been one big step in that direction. And by the spring of 1963, he had taken another step by arranging to meet periodically with the leaders of all the groups he was underwriting—people such as Fano, McCarthy, Uncapher, Engelbart, Feigenbaum, and Perlis, who were known as the principal investigators, or PIs. "[The idea was that] we would get our gang together frequently and have special Computer Society meetings," Lick said. "There would be lots of discussion, and we would stay up late at night, and maybe drink a little alcohol and such. So we would have one place interact with another place that way."

But Lick also had the kernel of an even better idea, if they could ever get it working. Early in his tenure—on Thursday, April 25, 1963—he laid it out for the PIs in a long, rambling memorandum that he dictated just before he rushed off to catch an airplane.

"[To the] Members and Affiliates of the Intergalactic Computer Network," he began—making it clear by his phrasing that he meant the word *Network* to refer to his ARPA community in general and to his PIs in particular, not to a collection of wires. Nonetheless, wires were obviously on his mind. "It is evident that we have among us a collection of individual (personal and/or organizational) aspirations, efforts, activities, and projects," Lick continued. The challenge now was to exploit "the possibilities for mutual advantage." He saw a very real danger that those possibilities might never materialize—that his nascent ARPA community might never become anything more than a high-tech Tower of Babel, in which widely scattered enclaves produced incompatible machines, incompatible languages, and incompatible software. Geography almost guaranteed that, he wrote. So they would have to transcend geography. They would have to make it second nature for participants to build on one another's work, by making it trivially easy for them to exchange programs and data no matter where they were.

In short, he said, the various ARPA-funded sites would have to take all their time-sharing computers, once they became operational, and link them into a national system. "If such a network as I envisage nebulously could be brought into operation," Lick wrote, "we would have at least four large computers, perhaps six or eight small computers, and a great assortment of disc files and magnetic tape units—not to mention the remote consoles and Teletype stations—all churning away."

Lick went on to discuss many examples of how people might use such a system, as well as the technical challenges of bringing it into being. At one point he even described something strikingly similar to the migratory Java applets that would appear at the turn of the millennium: "With a sophisticated network-control system, I would not [have to] decide whether to send the data and have them worked on by programs somewhere else, or bring in programs and have them work on my data." The computer could make such decisions automatically, he said—meaning that software could float free of individual machines. Programs and data would actually live on the net.

And so it went for seven pages, in what was arguably the most significant document that Lick would ever write. It's true that his proposal was just an elaboration on the network of "thinking centers" he had envisioned in his 1960 "Symbiosis" paper—enriched, perhaps, by the more recent speculations around MIT about citywide information utilities. But in just a few years, this memorandum to the Intergalactic Network would become the direct inspiration for the Arpanet, which would eventually evolve into today's Internet.

Still, that was later. In 1963, as Lick knew full well, any attempt to build such a system would have been premature. "I deliberately talked about the Intergalactic Network," he would explain in 1988. "But I deliberately did not try to do anything about netting them [the ARPA-funded sites] together, because it was becoming very difficult for them just to get their main projects to run." Besides, the networking technology wasn't even close to being ready. Lick did fund one small experiment at UCLA, which boasted no less than three IBM 7090-series computers on its campus, in hopes that the effort to link those machines would provide valuable experience. He also arranged to have Berkeley, UCLA, and SRI form what he called the California Network, tapping into the SDC time-sharing system via phone line. But that was about as far as he went.

Nevertheless, he knew he had planted another seed. And in the meantime, if he wanted to see what might happen when that seed began to sprout, he could always just drop in at MIT.

The goal, as Project MAC activists never tired of explaining, was nothing less than the democratization of computing. As Martin Greenberger of MIT's Sloan

School of Management declared in the first written manifesto of the information-utility idea, published in the *Atlantic Monthly*'s May 1964 issue, "Computing services and establishments will begin to spread throughout every sector of American life, reaching into homes, offices, classrooms, laboratories, factories, and businesses of all kinds."[15]

Of course, Greenberger intended his analogy to central-station electric power to be taken quite literally. Because Grosch's law (holding that twice the money bought you four times the computing power) was still gospel in 1964, people were convinced that the most efficient computers of the future, like the most efficient power plants, would be very large and expensive. Nonetheless, the MIT vision that Greenberger outlined was remarkably prescient in describing how computers would be used. Assuming that Project MAC and the other time-sharing experiments could master the technical and economic challenges, he wrote, "An on-line interactive computer service, provided commercially by an information utility, may be as commonplace by 2000 A.D. as telephone service is today." One major application would be electronic commerce, with on-line catalogs, on-line ordering and billing, and electronic cash. Others would include the routine use of computer simulation and dynamic modeling, "medical-information systems for hospitals and clinics, centralized traffic control for cities and highways, catalogue shopping from a convenience terminal at home, automatic libraries linked to home and office, integrated management-control systems for companies and factories, teaching consoles in the classroom, research consoles in the laboratory, design consoles in the engineering firm, editing consoles in the publishing office, computerized communities. Different subscribers to the same information utility will be able to use one another's programs and facilities through intersubscriber arrangements worked out with the utility on a fee basis."

That is not a bad description of what computing has actually become in the Internet era. Certainly it was a compelling ideal for the Project MAC team in 1964. The only problem, admits Bob Fano, as he looks back on that frenetic first year, was that he and his colleagues were hopelessly naive about how to get there. They really thought that they could create a computer utility through pure engineering. Just set up the terminals, string the wires, and keep that central power plant—um, keep that *central computer* humming.

They found out differently the instant they got real users involved. "It was a sociological phenomenon," says Fano, who has never quite gotten over his sense of amazement. "All sorts of human things happened, ranging from people destroying keyboards out of frustration, to friendship being born out of using somebody else's program, [to] people communicating through the system and then meeting by accident and saying, 'Oh, that's you.' "[16] True, he adds, this kind of "virtual community" has practically become a cliché in the Internet era. But Project MAC was the first. And there'll never be a first again.

The roots of this sociological explosion reached back to July and August 1963, when a total of fifty-seven people gathered at MIT for various portions of the six-week summer study. "For many of them," Fano later wrote, "this was their first opportunity to use a time-sharing system and to explore the potential of on-line, interactive computation. The reaction was very favorable in spite of the fact that the system was often overloaded and therefore slow in responding to service requests."[17]

"Slow" was a bit of an understatement. As Lick himself remembered it, "There were times in the daytime when this thing was just clunking along and everybody was frustrated. [It was fortunate that] people were around at night enough when it was fairly lightly loaded to get a little insight into what they were really working toward."[18] Still, the very fact that users were willing to turn hacker and spend their summer evenings at a clacking Teletype terminal was a testament to how seductive a responsive computer could be. It was also a testament to the frantic hours Corbató and his team had spent getting CTSS ready. The unofficial "contest" with SDC proved to be no contest at all. Even though CTSS had to run on the overloaded IBM 7090, which effectively ground to a halt anytime there were more than about a dozen users, and even though Jules Schwartz and his team had performed heroic deeds to get the SDC system ready in time, the Q-32 was still using tapes for mass storage. MIT's new hard disk beat it every time; the CTSS terminals were crowded.

Unfortunately, Fano admits, it's hard to say much more than that about what actually happened at the summer study. He and his colleagues were so frazzled after getting ready for the big event that none of them thought to take notes. But Lick clearly felt that he got his money's worth, he says: "A whole community got together, became friends, exchanged ideas, went back home—and the whole movement went on."[19] And of course, that six weeks of near-total immersion in time-sharing had whetted people's appetites for Project MAC's very own IBM 7094, which had been on order since January.

The 7094 arrived and was duly installed on the ninth floor of Tech Square in October 1963. True, this was no ordinary 7094: among other things, IBM had customized it with two extra banks of memory, one for the operating system and the other for the users. (It also had red side panels instead of IBM's traditional blue ones; MAC insiders often spoke of their "red machine" versus the "blue machine" over in the Computation Center.) But its real virtue was its compatibility with that original 7090. Fano remembers one colleague mournfully predicting that it would take months to get CTSS working on the red machine, only to be amazed when Corbató and his team had it running within a week. Very shortly afterward, moreover, they connected the remote Teletype terminals and hooked in the 1,200-bit-per-second modems that would provide dial-up access. And by mid-1964, they were making the system available to users around the clock. The information utility was on-line.

Almost immediately, it seemed, they had their first irate customer.

"One morning I arrived first thing and there was Joe Weizenbaum sitting in my office, mad as a hatter," says Fano. Weizenbaum had attended the summer study as an engineer from General Electric, and he had been so inspired by interactive computing that he had started writing a program that could carry on a real-time, interactive conversation in English. Weizenbaum had joined the Project MAC staff that fall and was now busy expanding his program into what would later become famous (or infamous) as ELIZA, the artificial psychotherapist. But he obviously wasn't making much progress that morning, says Fano: "Dick Mills had come in, too, from his office, and Joe said, 'CTSS was not working last night. I was told that it would be working shortly. It is not working this morning. What the hell is going on?'" Fano and Mills did their best to calm their visitor down, but all the while, says Fano, he was thinking, Victory! "Joe spoke like a public utility customer," he explained "His attitude was just what we wanted!"[19]

But of course, Weizenbaum's complaint was no joke, especially since it was only the first of many. Fano and his colleagues were starting to learn their first great lesson about utilities: what begins as a convenience quickly becomes a necessity. People begin to structure their work, their daily routines, even their entire lives around the utility. And as a consequence they become very, very proprietary about it. "Emotions got very high," says Fano.

In truth, the Project MAC time-sharing system could be irritating at the best of times. If nothing else, there were the terminals: computer-driven typewriters whose sole virtue was that they were comparatively cheap, costing maybe a few hundred dollars apiece. "The problem was that these typewriters were produced for Teletype work," says Fano. "And the Teletype standard was to use just capital letters! It was a real battle with the companies to allow lower-case letters. When we asked, they said 'No, who would need that?' The best we had was an IBM Selectric typewriter. And we had to modify it because the typewriter didn't come with a sprocket drive to keep the pages straight."

Then there was the system's sluggishness. When it came to raw processing power, the IBM 7094 was roughly equivalent to the first IBM PC, marketed in 1981. So when that small amount of power was divided among multiple users, the machine's response time plummeted to what Fano called the Threshold of Impatience: "I could watch a user sitting there giving a command, and then I would see his hand falling down from the keyboard [into his lap]. That took place in about five seconds."[21] And when CTSS crashed—which it did with distressing frequency, being an admitted kludge—well, Project MAC had to replace more than a few smashed keyboards.

Nonetheless, for all its irritations, CTSS was interactive. Make your way up to the public terminal room on the ninth floor of Tech Square, type in a command, and the computer would type out a response (eventually). Type in another command, and the computer would respond again. On and on it would go, in an endless (and rather Lisplike) cycle of command and response. Users found the experience eerily seductive, to the point where Fano had to take some steps: "I was very well aware of how hard the programming staff worked," he says. "They were willing to stay here even in the evening to use the time-sharing system. That inhibited their family life. So I made the decision to provide teletypewriter connections for people at home. I only needed the terminal and a private telephone line to their house. My argument was that the extra work they'd do would very quickly pay back the cost. That was true. But what I didn't anticipate was that I saved a number of marriages. The staffers were still hunched over their machines, but at least their wives could see their backs!" In effect, these terminals were the first "home computers"—a notion that would percolate through the computer community for another decade, to emerge in a new guise from the garages of Silicon Valley.

Meanwhile, Corbató and his team were doing their best to work the bugs out of CTSS as they went along, getting the system to the point where it became downright usable. In fact, they would continue that effort well into 1965, eventually adding some fifteen thousand lines of computer code. And along the way they would produce what was arguably the most influential single operating system in computer history. Simply because they were the pioneers in time-sharing, says Corbató, "we knew we were laying down a pattern that would probably be emulated and imitated. Indeed, that is the way it worked out." CTSS's fundamental model of human-computer interaction—the notion of a user's sitting down at a keyboard and having a set of commands at his or her disposal—can still be found today at the heart of MS-DOS, Windows, Macintosh, Linux, and many other operating systems. So can such innovations as CTSS's "hierarchical" file system, a scheme devised by Corbató and his team to enable individual groups and individual users to organize their personal files into "directories" and "subdirectories" on the 7094's hard disks. (In modern-day Windows and Macintosh systems, of course, such directories are called folders.)

CTSS was also instrumental in alerting outsiders to the fact that something new and strange had come into the world. Interacting with computers? Communicating with computers? Working with computers in your *home?* (Well, with terminals, anyway.) In 1964 this sounded a bit like working with a nuclear reactor in your home. "Very quickly, the lab became like a shrine," says Fano. "We had continuous groups coming through: reporters, technical people, businesspeople, U.S. government officials. The Japanese were all over the place. Officers of their major electronic companies were arriving with gifts, interpreters, and everything."

CTSS was even good enough to convert a few die-hard skeptics, adds Fano. "There were a lot of old-timers who thought it was sacrilege to sit at a computer without thinking out what you were going to do beforehand," he says. "Jay Forrester felt that very strongly. I couldn't get him involved with time-sharing at all. But one day not long after we got started, he called me and said he had a visitor who wanted to see the MAC system. I got Dick Mills, my assistant director, to see them. Now, this turned out to be a little bit of a conspiracy, because Forrester's chief programmer would have loved to use time-sharing. And Dick Mills knew that, because the two knew each other quite well. So when Forrester came here with his visitor, Dick Mills had one of Forrester's own simulations up on the machine, so that he could deal with it interactively. Well, at nine the next morning the telephone rang. It was Jay Forrester: 'Can I get computer time? And by the way, can I get it at a terminal at home?' I said, 'Sure!' "

And yet, says Fano, as gratifying as all that was, it was really nothing more than what they'd expected based on the electric-power analogy. Give people something useful, and they'll use it. What he found truly fascinating, though, was the *second* great lesson of time-sharing: in an information utility, the power flows both ways. Unlike a power utility, which basically just provides a resource that people consume, an information utility lets the users give back to the system. It even lets them create whole new resources *in the* system. "More than anything else," says Fano, "the system became the repository of the knowledge of the community. And that was a totally new thing."

Once again, the credit had to go to Corbató's CTSS. No matter how frustrating the operating system could be, CTSS did have one great thing going for it: Corbató had designed it as an *open* system, in the sense that users could modify it, tailor it, and extend it however they wanted. As Corbató himself said, "This open system quality . . . allowed everyone to make the system be *their thing*, rather than what somebody imposed on them."[22]

To achieve that quality, Corbató had designed CTSS as an inner core of lower-level functions surrounded by an outer periphery of higher-level commands. The core took care of chores such as reading and writing data to the disk, interpreting user commands, and shifting the machine's attention millisecond by millisecond from one user to the next—the time-sharing equivalent of unconscious functions such as heartbeat, breathing, and digestion. Corbató and his team took direct responsibility for that part of the system and worked hard to keep it as simple and as comprehensible as possible. "I was always conscious of how we would explain this to a newcomer in a way where he could understand it quickly, without having to read a manual," he said.[23]

The periphery, meanwhile, was the user's software toolbox, the collection of programs that he or she could invoke to rename a file, say, or print out a list of files in a directory. And it was here that creativity reigned. Users could write whatever new software tools they needed for the task at hand. If enough other users liked it, too, it would be placed in the public library; in effect, it would become a part of CTSS itself. This was a golden opportunity, and the Project MAC community wasted no time before taking advantage of it. Graduate student Jerry Saltzer created the TYPSET and RUNOFF commands to write his thesis proposal; together they constituted the rudiments of a word processor. "Practically from the beginning, we started using them to publish all the Project MAC reports," says Fano.

Even more popular was undergraduate Tom Van Vleck's MAIL command, which allowed users to send text messages to one another—and which thus probably ranks as the world's first implementation of E-mail. Meanwhile, there was Allan Scherr's widely used ARCHIVE utility, which would take a bunch of little files and compress them into one big file, with a substantial saving in disk space. And there was the OPS, or the On-line Programming and Simulation system, created by Martin Greenberger and his students at the Sloan School in the fall of 1963. It offered commands to simulate the stock market, handle accounting, do production scheduling, perform on-line modeling, and do all manner of other things.

And so it went, leaving Fano and his colleagues to shake their heads in wonder at how fast the sociology of software had turned itself inside out. In the batch-processing world, Fano says, programming had generally been a do-it-yourself affair. Sharing a program with someone else would have meant duplicating a massive deck of punch cards, physically carrying the deck around, explaining how to format the input data on the cards, coaxing the program into running on an alien computer, and so on. Who had the time? In fact, says Fano, "in those days programmers never even documented their programs, because it was assumed that nobody else would ever use them." Now, however, time-sharing had made exchanging software trivial: you just stored one copy in the public repository and thereby effectively gave it to the world. "Immediately," says Fano, "people began to document their programs and to think of them as being usable by others. They started to build on each other's work."

Indeed, the very existence of that public data repository on the 7094 quickly transformed Project MAC's central "power plant" into the intellectual center of the community. Through E-mail, the exchange of files, and the sharing of programs, it functioned as the town square, the village market, the Roman forum, and the Athenian agora all in one—the place where citizens gathered to talk, to gossip, to conduct business, to propose ideas, and then to argue until they came up with *better* ideas. Within six months of the system's November 1963 startup, CTSS and the on-line environment it supported had become, at least in embryo, everything that would later be claimed for the on-line world of the Internet.

In fact, it was even more than that. With more than half of its system commands now written by the users themselves, CTSS had proved to be just what Lick had been hoping for in his Intergalactic Network memo, and what Fano had predicted in the Project MAC proposal: a self-guided system. The very fact that CTSS was being (partially) created by its users meant that it was adapting to those users and evolving to meet needs that its founders never could have envisioned. Far more than Corbató himself had originally realized, his open-systems design for CTSS had deep parallels to the notions of free speech in a political democracy, free competition in a market economy, and the free exchange of ideas in scientific research. In each there is a core set of rules that everyone must accept to make the interchange possible at all. But beyond that, creativity reigns. Certainly the open-systems idea would prove crucial a decade or more later, as entrepreneurs struggled to bring computing to a mass market. In the personal-computer boom of the 1980s, for example, computer architectures such as the Apple II and the IBM PC would become wildly successful in large part by being open to anyone who wanted to build a plug-in card. In the 1990s a leaderless, unorganized thing called the Internet would be able to sustain explosive rates of global expansion in large part by being open to any network that wanted to join.

And through it all, standard operating systems such as MS-DOS, Windows, and Macintosh would sustain a burgeoning commercial software industry in large part by being open to any developer who wanted to write an application. Indeed, the beginnings of that industry were already discernible in the new commands being written for CTSS. At the time, of course, no one could have imagined asking to be *paid* for his or her handiwork; the spirit of the enterprise was closer to what would later be known as shareware. Nonetheless, there is a clear line of evolution from Project MAC to the shrink-wrapped boxes now lining the shelves at your local computer superstore.

At the time, however, a more immediate concern for Project MAC was the *third* great lesson of time-sharing: openness requires trust. After all, notes Fano, "the whole goal was to have people store their programs in mass memory, instead of on punch cards. But that meant we had to build faith in the system." The users had to believe that the system would be up and running when they needed it, that a software glitch wouldn't destroy a month's work, and that their files would be safe from damage of any kind. And no one understood that better than Fernando Corbató. "Corby was the key to the success of time-sharing here," Fano declares. "He can't tell you that, but I can. We never had any major disaster in Project MAC, even though people were using an experimental system. We came close a couple of times. But it never happened, because Corby never said, '*If* this goes wrong...'; he always said, '*When* this goes wrong....' He'd thought it through. He was not caught by surprise. His forethought about all the problems, all the tools he built into the system to avoid problems or to provide recovery when problems did occur—he was unique as a systems engineer."

Indeed, Corbató had been obsessive about reliability from the beginning. "A lot of my thinking was formed by being an electronic technician in World War Two," he explains. "I was brought face to face with sonars and radars, which were very new. And it was pretty exciting, keeping those babies running. You got a lot of practical experience debugging."

That background had given Corbató a keen appreciation for Murphy's Law: If anything can go wrong, it will, and at the worst possible moment. But then, in the 1950s, when he first started to get involved with computers, he began to realize that Murphy had been an optimist. Not only did you still have to worry about hardware failures—and before transistors came in, he says, the hardware was *really* flaky—you now also had to worry about software glitches, which could turn perfectly healthy hardware into an inert pile of junk. One bad keystroke could bring a multimillion-dollar system to its knees. And then when he started CTSS, says Corbató, he discovered that time-sharing made simple software glitches seem like the good old days. At least with batch processing you were dealing with one computer and one program at a time, so that everything was more or less under your control. But in a time-shared computer, your computational elbow was constantly being jostled by other people's files and other people's programs, all of which had just as much right to be in there as yours did. It was like moving from a secluded cabin in the woods to Manhattan. Added to that were the newly encountered problems of communicating with the remote terminals: noise on the line, broken connections, modems that suddenly decided to freeze—on and on. And then added to *that,* says Corbató, you had a whole community of users whose behavior was neither controllable nor predictable.

The result, he says, was a system with a certain potential for, um, *spontaneity.* You were never quite sure how it would respond or what was going to crash next. And the only hope for managing in the face of such chaos, Corbató recognized, was to accept that spontaneity. He and his colleagues would have to give up every engineer's first instinct, which was to control things so that problems could not happen, and instead design a system that was guaranteed to fail—but that would keep running anyhow.

Nowadays this is known as a fault-tolerant system, and designing one is still considered a cutting-edge challenge. It means giving the system some of the same quality possessed by a superbly trained military unit, or a talented football team, or, for that matter, any living organism—namely, an ability to react to the unexpected. But in the early 1960s, with CTSS, Corbató and his colleagues had to pioneer fault-tolerant design even as they were pioneering time-sharing itself. For example, among their early innovations were "firewalls," or software barriers that kept each user's area of computer memory isolated from its neighbors, so that a flameout in one program wouldn't necessarily consume the others.

Another deceptively simple innovation was the introduction of a clock, which would (for example) allow the computer to grab control back from a user's program if the program hit a glitch and locked up. This sounds obvious enough now, says Corbató, "but people were very cavalier about time in the early days. Computers didn't have clocks. There was no way to time-stamp files or to do something automatically after so many seconds. So in the very early days of CTSS, we used just an electronic time-of-day clock kludged in through the printer port."

Still another innovation was that most elemental of precautions, the backup. "We were forced into it," Corbató recalls, "because a disk failed fairly frequently, the systems crashed a lot, and so you had to learn to think defensively. Once you start tinkering with a program and you no longer punch out cards [for backup], which people had been used to doing, you are really trapped if the thing vanishes because you can't re-create the details of even a week's worth of work." So in the beginning, he says, the group spent two hours every day copying the entire disk file on tape. Later they automated the process: "We had a little software demon that would run in the background, scavenging the disk to look for newly created files," says Corbató. "We'd back files up within a half-hour of [their] creation, sometimes less."

For the most part, the Project MAC community accepted these precautions with reasonable good cheer, plus a bit of friendly grumbling about the occasional inconveniences involved. However, that was *not* the case for another innovation. With the launching of CTSS, Fernando Corbató instituted what was very probably the world's first computer password system. Everybody in Project

MAC learned the drill: each time you sat down at a terminal, you had to type in your own special code word before the computer would let you do anything. Later on, moreover, after Corbató had refined the CTSS file system, you needed a password even to access the subdirectory containing your personal files.

Users hated this. At a minimum, the notion of passwords in an academic environment seemed vaguely insulting; as Corbató heard it expressed a few years later when the same issue came up at Bell Labs, "We just don't even lock our desk at night. Anyone can walk into anybody else's office, and a gentleman doesn't read anybody else's mail."[24] But to the hackers—that fiendishly clever band of obsessives who had learned to program one on one at the TX-0 and the PDP-1, and who had now found a haven on the ninth floor of Tech Square, in Marvin Minsky's AI Lab—passwords were anathema. As Steven Levy noted in his 1984 book, *Hackers*, "to the hackers, passwords were even more odious than locked doors. What could be worse than someone telling you that you weren't authorized to use his computer?" The whole thing was so . . . *corporate*, like having to wear a nametag and sign in with a guard in the lobby. "To the hackers," wrote Levy, "CTSS represented bureaucracy and IBM-ism."

Corbató understood their grievance but stood his ground. "In order to try to get across the difference," he says, "I used to refer to [hacker-style] systems as 'clubhouse' systems. You had to be trustful of the people using it, and you were basically at the mercy of each other, and that's fine. Everyone knows the clubhouse works fine as long as everyone is friendly. It breaks apart when you begin to get differences."[25] Having been deputy director of the Computation Center, which served hundreds of people, Corbató had a keen sense of just how different those differences could be. Yes, he wanted the on-line community to be an open community, in which all kinds of people would be free to join in. But just as in the outside world, that very openness meant that people needed a zone of privacy. If nothing else, he says, "most individuals are shy about exposing their work prematurely, before they have got it to a point where they are prepared to defend it. Things are still imperfect and flawed and the last thing they want is an attack." Users likewise needed some assurance that they could go about their business without their work's being trashed by the next guy, accidentally or otherwise. "Passwords were just a first cut at security," according to Corbató. "They were a reminder to people that if they tried to trick their way past the passwords, they were doing something wrong." They were also a reminder that you weren't just an anonymous set of fingers typing on an anonymous remote terminal. Once you entered that password, says Corbató, the system knew you as a *person*, a specific human being with a specific identity, responsible for everything you did on-line.

Fano backed Corbató wholeheartedly on the password issue. When users complained, the Project MAC director's standard reply was, "How would you like someone to write a paper before you, using your material?"[26] (This wasn't a hypothetical question; there were several instances of research material's being stolen at Project MAC, both from offices and from the time-sharing system itself.) Nonetheless, with Corbató's attempt to impose a modicum of social order on cyberspace, Project MAC had crossed over from technology into politics, thereby sparking a debate on freedom versus responsibility that was strikingly similar to the one that would roil the Internet in the 1990s. Along the way, prophetically, Project MAC spawned yet another new phenomenon: the hacker as jokester, vandal, and thief.

This called for a considerable change in the definition of the word: "hacker" had originally been a semireputable term for an exceptionally clever programmer. Confronted with CTSS and its passwords, however, the hacker contingent in the AI Lab reacted like the adolescents they mostly were and immediately bent their considerable talents to subverting the system. Usually it was a matter of playing practical jokes just to show they could do it—the digital equivalent of scrawling "Kilroy was here." For example, there was the morning when Bob Fano logged on to his terminal and discovered that the message of the day now talked about a Project HAC. But sometimes the jokes were more serious, as when the "logout" command disappeared. When users finished their work, they couldn't get out of the system to free up space for others; CTSS rapidly ground to a halt. And occasionally a prank would cross the line into criminality. Over at the Sloan School, for instance (not all the subversives were in the AI Lab), some of the young hotshots figured out how to cook the books and get computer time without paying for it.

To the hackers and their allies, it was all a matter of striking a blow for the People against the System. This was the beginning of the sixties, after all, and the information-wants-to-be-free culture of the hackers had a good deal in common with the property-is-theft counterculture that was just starting to emerge in the outside world. To Corbató, such behavior just proved his point: "We really did have in mind to build the prototype of . . . a computer utility," he says—one that really would be as open, as simple, and as easy to work with as possible. But he and his colleagues also knew that if it was going to make a real difference in the world, this new thing they were creating would have to survive being used by real, fallible human beings. And in order for *that* to happen, the technology would have to reinforce such unfashionable values as courtesy, trust, and responsibility—not undermine them.

The passwords stayed.

ART FOR THE MASSES

Whenever her husband was actually in town, says Louise Licklider, thinking back over their stay in Washington, he would come home in the evening to have dinner with the family, and then he'd go back to his office in the Pentagon for another few hours of work—much the same schedule he'd followed back in Cambridge. And just as she had done there, Louise would have the drinks waiting when he got home for good, so they could talk over their day together. But one night, she remembers, Lick didn't return until late, around eleven o'clock. And he was chuckling to himself, as if he'd just heard a great joke.

It was the art prints, he explained.

Ah yes, the art prints; he'd been driving her mad with the art prints. "When Lick was traveling," she says, "he would try to schedule his plane changes to give him a free hour or so in a city that had an interesting art gallery. Or if he hadn't been able to get to an art gallery that trip, he would go to a print shop. Either way, he would come home with an art print or two. Or five." So it wasn't long before cheaply framed prints were taking up all the space on the walls of their tiny apartment, and rolled-up prints were spilling out of the closets.

"Darling," Louise would say as calmly as she could, "what are we going to do with all these prints?"

"Oh, we'll just change them in their frames every week," replied her husband.

Finally, she says, even Lick had to admit that it was getting ridiculous. "Well," he suggested, "I've got lots of wall space in the Pentagon. I'll put some of them up there."

That had brought a certain peace on the subject. But it was not long after that that Lick had come home chuckling. He'd been working, he explained as he and Louise sipped their drinks, and it had gotten to be later than it should be. As he was finishing up, in fact, he could hear the cleaning woman in the hall. So he opened the door.

"Oh, I didn't know you were here!" she apologized, and turned as if to go. No problem, Lick assured her, saying that he was just about to leave anyway, and urging her to go on with her work.

"You know, Dr. Licklider," she said as she was bringing in her brooms and dust cloths, "I always leave your room till last because I like to have time by myself, with nothing pressing, to look at the pictures."

Well, Lick told his wife, he was so touched by that that he stayed for a while. He and the cleaning lady went around to the various pictures and talked about them—the play of light, the colors, how they made you feel.

Which one do you like best, he asked her? And to his enormous delight, she chose one by Cézanne, his favorite artist. So, to *her* enormous delight, he gave it to her.

And that was why he was chuckling, Lick told Louise. The thought had struck him on the way home: Computers for the brasses—art for the masses.

It one of those lame bits of wordplay that he was forever coming up with. But somehow, something about this one seemed to strike him as hilariously apt.

Computers for the brasses—art for the masses.

Leo Beranek was not amused.

Actually, Beranek hadn't been any too happy about Lick's going to Washington in the first place. Lick had gotten BBN into computers. He'd brought in twenty or thirty people, bought several machines, committed the company to fulfilling any number of research contracts. And now he was walking out? Leaving them all in the lurch, with a lot of half-finished projects that somebody else was going to have to deal with?

Well, at least it was only for a year, Beranek had told himself; they could cope for that long. BBN had gotten someone in to take over Lick's programs in the interim—John Swets, who had been one of Lick's postdocs in the old MIT psychology section—so life there could go on.

Except that it didn't—not the way Beranek had hoped, anyway. Once Lick was ensconced in Washington, it soon became all too clear that he was studiously avoiding BBN: even in 1962, and even in the just-do-it culture of ARPA, there was no way he could funnel federal cash straight back to his once and future employer. Beranek was furious. Lick was handing out the biggest chunk of money to come along for computing research since Whirlwind, at least, and BBN wasn't going to see a penny of it. The opportunity cost was horrible.

Still, Beranek decided, there was nothing for it but to grit his teeth; at least the agony would be over in a year.

Except that it wasn't. In mid-1963 Lick came to Beranek and announced that he wasn't coming back: he'd decided to resign from BBN and remain at ARPA.

"Lick," said the appalled Beranek, "are you sure?"

Yes, said Lick, he was sure.

"But Lick, we've really done well by you," Beranek persisted. This was grasping at straws, he knew, but at least they were substantial straws. As a BBN vice president, Lick had received a nice little nest egg of BBN stock; when the company went public in 1961, the shares had jumped from $1 to $12 apiece the first day, and now they were at somewhere around $15.

That had been pleasant, agreed Lick, thereby beginning and ending the one conversation Beranek can remember ever having with him on the subject of money. But even so, Lick said, ARPA was now where the action was in computing. He couldn't let an opportunity like that go by. As Tracy Licklider says, his father was like a monkey—always reaching out for that next piece of fruit. And for now, he'd found it at ARPA.

In truth, his program was going far better than he'd dared hope. And by all reports he was having a wonderful time with it, presiding over his far-flung ARPA community with much the same mix of parental concern, irrepressible enthusiasm, and visionary fervor that he'd brought to his research groups at MIT and BBN. True, Lick's nonstop stream of ideas and suggestions could be exasperating; the recipients sometimes felt as though their sponsor's imagination were voyaging through the stars while they were still struggling to get their first biplane off the ground at Kitty Hawk ("No, Lick, we *still* can't afford to give everybody in Project MAC a display terminal as good as the PDP-1!"). But with rare exceptions—notably that first encounter with SDC—Lick was much more interested in being a mentor than in being a micromanager. As long as people made reasonable progress in the right direction, he would let them find their own way.

At Berkeley, for example, he cheerfully maintained his support even after the senior investigator, Harry Husky, vanished on a sabbatical and left the whole time-sharing project to two unknown quantities: Ed Feigenbaum, who knew nothing about hardware, and David Evans, a very junior professor who had just been hired that semester. "You'll be fine," Lick said soothingly to a very nervous Feigenbaum, who had just informed him that one of the showpiece projects of his whole program would now be headed by an ignoramus and an unknown. And indeed, says Feigenbaum, they *were* fine. Evans, who would later become famous as a cofounder of the graphics-terminal manufacturer Evans and Sutherland, proved to be a great hardware engineer. Together he and Feigenbaum began designing the Berkeley time-sharing system, with Evans tutoring Feigenbaum as they went along. And by the fall of 1964, when Feigenbaum left to join John McCarthy at the brand-new Stanford Artificial Intelligence Laboratory, the Berkeley time-sharing system was almost operational. Project Genie, as it was now known, was well on its way to justifying Lick's faith.

At RAND, meanwhile, Lick was allowing the interactive-computing group to learn from its own mistakes—or so he thought. He was dubious about the RAND group's approach to the electronic "tablet" it was designing for freehand drawing and (eventually) handwriting input. The user was supposed to write with a pencil-like stylus down *here*, on the horizontal tablet, yet the digital "ink" would appear only up *there*, on a vertical CRT screen; the veteran human-factors researcher was convinced that the disconnect would make hand-eye coordination impossible. (Lick would later have the same concerns about the mouse. What he really wanted was a kind of smart paper—a computer you could physically write on.) Nonetheless, Lick left the tablet development team free to try it their way—and within a year was back at RAND admitting his error. The group had shown that their "ink" did indeed provide enough feedback; users could train themselves to write on the CRT display almost instantly. "You were right," he told them. "It works."

Keith Uncapher, who was the leader of RAND's computer group at the time, remembers being so impressed by Lick's brand of intellectual honesty that he essentially bet the farm on him: instead of continuing to seek support for the group through its traditional, safe contracts with the air force, he shifted all its funding to ARPA. "Working with Lick was just so much more fun than dealing with nervous two-star generals who couldn't understand what we were trying to do, or with colonels who had no interest," he says with a laugh.

Finally, Lick started funding Carnegie Tech at a rate of some $1.3 million per year, even though there were only three major computer researchers in the whole school—Allen Newell, Herbert Simon, and Alan Perlis—none of whom had any particular interest in interactive computing as such. To Lick that was far less important than the fact that Newell, Simon, and Perlis were all smart, tough, independent, and devoted to their own standards of excellence—in short, his kind of people. Moreover, the three men embraced the most expansive definition of their discipline imaginable. Just as botany is the study of plants, they declared in a widely quoted essay,[27] computer science is the study of "the phenomena surrounding computers"—*all* the phenomena, from the physics of integrated circuits and the analysis of algorithms to human interface design, the social ferment bubbling up around the ARPA-funded time-sharing systems, and the impact of information technology on human life in general. Newell and Simon even saw a big overlap between computer science and human cognitive psychology. They had based the design of General Problem Solver, their next program after Logic Theorist, on data gathered from human subjects working on puzzles and games in the lab. And when they later came to publish a retrospective of their research with the program—a 920-page tome that is now considered a landmark in both AI *and* cognitive psychology—they would call it *Human Problem Solving*.[28]

Perlis, for his part, had already taken the lead in creating computer science as an organized discipline: computing, he argued, was such a novel and complex endeavor that no existing discipline could cover it all. Indeed, he lived that argument. At one point in the early 1960s Perlis was not only teaching Carnegie Tech's programming course but simultaneously serving as head of the university's Computation Center; chair of its mathematics department; president of the leading professional society, the Association for Computing Machinery; and editor of that organization's widely read magazine, *Communications of the ACM*. In 1965, moreover, when Carnegie Tech became one of the first universities to organize a separate department of computer science, Alan Perlis would be made the chairman of *that*. His energy seemed inexhaustible. Even in his later years, after multiple sclerosis confined him to a wheelchair, Perlis could dominate a room without even trying. Think of a gnome, a gremlin, a leprechaun, a grinning, mischievous, irrepressible Yoda. (Thanks to an otherwise benign condition called *alopecia universalis,* he was completely, utterly bald, without so much as an eyebrow.) He had a talent for deadpan quips that could send a roomful of people into hysterics, even as his words were sliding into his listeners' brains and setting off time bombs. "People in the field would go to conferences and listen to every word he said," Allen Newell recalled in an interview conducted shortly after Perlis's death, in 1990, "because every time he talked about a topic, he was absolutely right about the way it should be. That's why he was the first Turing Award winner.* Not at all for any technical contribution, really, but because he epitomized the nature of computer science."

Certainly this was the kind of thinking that intrigued J. C. R. Licklider, who recognized that Newell, Simon, and Perlis were attempting an intellectual synthesis in computer science that was downright Wienerian in scope. Thus Lick's initial, out-of-the-blue grant of three hundred thousand dollars in October 1962. And thus his decision in mid-1963 to designate Carnegie Tech as an ARPA "Center of Excellence" in the information sciences—the second such, after MIT. While

---

* Established by the ACM in 1966, the Turing Award is the Nobel Prize of computerdom, the most prestigious award that any computer scientist can receive.

Project MAC focused on the practical side of interactive computing, Lick wrote that autumn, Carnegie Tech would devote itself to the "effort to achieve fundamental understanding, to develop the theoretical bases of information processing."[29] By design, moreover, the contract would be open-ended. He already had plenty of people working on time-sharing, he told Newell, Simon, and Perlis. All he wanted from them was excellence. Train some great students. Explore some big ideas. Break the mold. "The evaluation was always, 'If you guys are doing some excellent science, it [doesn't] matter what [it is],' " remembers Newell.

It was an amazingly sweet deal for Carnegie Tech—and one that would pay off handsomely both for ARPA and for the country. By 1967, when the school merged with the neighboring Mellon Institute to become Carnegie Mellon University, the obscure little technical institute in out-of-the-way Pittsburgh would be well on its way toward becoming one of the finest schools for computer science in the world.

If Lick was content to let people find their own way, however, he *did* expect them to perform; he was as impatient as ever with dithering, sloppiness, or mediocre quality. At the end of his first year at ARPA, he in fact pulled the plug on several research projects that hadn't measured up to his standard. And for a while, sadly enough, it seemed as though one of them might be Doug Engelbart's effort at SRI.

Engelbart's project had definitely had its share of growing pains, not the least of which was an ongoing dispute with SRI management about exactly what ARPA was supposed to get for its money. To Engelbart, it felt as if the SRI higher-ups were pressuring him to do something safe and conventional—as if they were so embarrassed by his "Framework," and so terrified that he would cause some debacle in front of a major sponsor, that they didn't dare give him a free hand. Whatever the truth of the matter, though, this was one sponsor that was *not* interested in "safe and conventional." So when Lick came out to SRI toward the end of Engelbart's first year and saw how little he'd been allowed to do, he felt obliged to throw a tantrum with SRI management. "He chewed them out," Engelbart remembers. "He told them, 'Either you put Doug back in and let him do what he wants, or we're through.' "

They acquiesced. In private, moreover, Lick promised Engelbart more time and money to get his project moving. Unfortunately, the extra funding still wasn't enough to pay for both people and a decent computer. As a result, Engelbart later wrote, he and his tiny staff were left trying to augment the human intellect with "the only commercially suitable minicomputer we knew of": a hundred-thousand-dollar Control Data Corporation machine that was painfully slow and memory-limited. It was a pathetic stopgap, he knew; even with Lick's backing, their extinction was only a matter of time.[30]

What saved them was an unexpected offer of support from Bob Taylor at NASA. And no one was more surprised by it than Engelbart himself. Long before, he had visited Taylor at the space agency's headquarters to drop off copies of both "Framework" and his proposal. But that, as far as he was aware, had been the end of it. What Engelbart didn't know was that Taylor had been enchanted by "Framework" and had spent the intervening time drumming up support for the project within NASA. Engelbart might or might not be able to make any of his ideas work, Taylor had argued, but he was definitely headed in the right direction.

It was an argument that had obviously proved effective; for many years afterward, Engelbart's Augmented Human Intellect Research Center would be supported jointly by NASA and ARPA (and later by the air force as well). But the more immediate effect was that he was able to get the equipment he needed, starting with a much more capable computer. Just as important, he was able to get the staff he required, including such notables as Bill English, who joined the project in early 1964 as chief engineer and turned out to have a genius for turning Engelbart's high-flying abstractions into working hardware. Together they were able to start work on the first version of what they came to call NLS, a slightly skewed acronym for oN Line System.

NLS was—of course—an attempt to implement the grand abstractions of Engelbart's "Framework." But its specific starting point was his memorandum example, with its speculations about reordering ideas on the fly. Indeed, to modern eyes NLS looks like nothing so much as a text-based word-processing program from the MS-DOS era. Or make that a word-processing program on steroids: NLS would ultimately offer full-screen text editing, fonts, windows, outlining, hyperlinks, an embedded programming language, integrated E-mail, integrated address books, integrated calendars, and other "groupware" features, and much more besides. Perhaps the nearest contemporary analog would be a flagship package such as Microsoft Office or Lotus OfficeSuite.

But in that initial version of NLS, which the group started working on in early 1964, the big innovation was to have the computer write the text to a CRT display instead of on a standard typewriter terminal. Such displays were fabulously expensive in 1964, but Engelbart figured that by the time his group had really worked out how to use them, their price would have fallen considerably. And there was a surprising amount to learn. If nothing else, there was the 2-D factor—that is, the fact that the computer would now be displaying whole chunks of a file at once, not just one character or one line at a time. So to insert a piece of text, say, you first had to tell the computer exactly where on the screen you wanted it to go. "We needed a screen selection device," Engelbart wrote in his account of the SRI years. "I wanted to find the best thing that would serve us in the context in which we wanted to work—text and structured items and interactive commands."[31] He and his team experimented with every device they knew of—trackballs, light pens, joysticks—but Engelbart wasn't happy with any of them. So they got more adventuresome, experimenting with foot-operated controls, knee-operated controls, even head-operated controls (a nod would make the on-screen cursor move vertically, and a glance to the side would make it move horizontally; users tended to get terrible neck cramps). Engelbart still wasn't satisfied.

Finally, as they were all sitting around brainstorming one day, Engelbart came up with the idea of a little gadget that the user could roll around on the desktop with one hand while the cursor tracked its motion on the screen. Since it didn't seem any sillier than some of the other things they had tried, Bill English went off to the SRI machine shop and made one. It was essentially just a block of wood about the size of a pack of cigarettes, with some rollers set into the bottom and a wire coming out the front end to communicate the motion of the rollers to the computer.

Engelbart wasn't totally satisfied with this contraption, either. And yet when the NLS team hooked up all the selection devices to their computer and gave users a choice, they discovered that people were consistently choosing the little gadget. The preference was so strong, in fact, that they eventually abandoned everything else; the gadget had become their standard. And by that time, of course—though no one on the SRI team can now remember when, or how, or why it started—they had all taken to calling the thing a mouse.

It was more of a joke than a name, really. They would surely find a more dignified term in time. But until then, well, it just seemed to fit.

Of course, there was no mystery about Lick's personal favorite among his projects. Project MAC was his flagship, and he went on board as often as he could. Sometimes he even brought along colleagues to show the place off; Charlie Herzfeld, for one, remembers being particularly impressed by a demonstration of interactive ship design on the Kludge graphics terminal. Not surprisingly, moreover, the enthusiasm was mutual. "I was amazed at how wonderful it was to have a sponsor who believed in what you were doing," says Corbató.

That was probably just as well, since Project MAC was going to need all the support and trust it could get in the coming months. By early 1964, Fano, Corbató, and their colleagues were beginning to take on the computer establishment directly. And in the process, they were getting ready to bet the project's entire future on what was essentially a roll of the dice.

At issue was the next step in their long-range plan: the creation of an industrial-strength time-sharing system to replace CTSS. Granted, says Corbató, they ultimately got CTSS working fairly well, and they could have gone on tweaking it and adding things to it forever. But it was still fundamentally a kludge, with patches on top of patches on top of a batch-processing operating system that ran on a computer never designed for time-sharing. (Anyone who ever struggled to run Windows 3.1 on top of MS-DOS will understand the feeling.) From the beginning, their intention had been to follow it up with a system that would be done *right*, that would be built around the notion of time-sharing from the ground up. Multics, as this hypothetical system came to be known—short for MULTiplexed Information and Computing Service—would have the ability to run without interruption, spreading its workload over multiple processors so that one would always be ready to take charge if another went down. Like a power utility, the information utility would be available to users day and night. Multics would also have a bulletproof, hierarchical file system so that nothing would ever inadvertently disappear. And it would have built-in security: firewalls, passwords, private directories—the works.

Planning for Multics had begun right after the 1963 summer study, with a former Burroughs engineer named Ted Glaser as chief designer. "He was blind, but an amazing systems architect," says Corbató. "We worked together hand and glove." However, the actual programming of Multics wouldn't get under way until 1965. In the interim, the group was absorbed by a much more urgent task: finding a computer for Multics to run on.

The 7094 was out of the question; it had never been anything more than a stopgap machine to begin with. As Corbató and his colleagues had pointed out in the MAC proposal, they needed to replace it as soon as possible with something along the lines laid out by the Long Range Planning Committee. And to do that they would need to forge a partnership, "a working relation with a computer manufacturer which would make it possible to discuss current machine designs, with the objective of acquiring a prototype, perhaps with minor modifications, of a future machine."[32] The manufacturer would then have the right to market a commercial version of whatever they jointly developed. The end result, if all went well, would be the fulfillment of their long-term dream: interactive computing for the masses.

The trick was finding the right partner. Any manufacturer would have been happy to sell them a computer, but what Project MAC needed was a company willing to put some heart and soul into the job. One that would commit time, people, and resources. One that really shared the *vision*.

The job of finding that partner had fallen to a committee of four: Corbató, Glaser, Jack Dennis, and MIT's Robert Graham. Starting in the fall of 1963, with Lick's blessing and with ARPA money in hand, they hit the road.

Their first stop, inevitably, was IBM. If nothing else, it was far and away the largest computer manufacturer on the planet, controlling some 70 percent of the world market. The other seven major producers—Burroughs, Control Data, General Electric, Honeywell, NCR, RCA, and Sperry Rand—had to be content with dividing the remainder among them. Analysts not-so-jokingly described the industry as Snow White and the Seven Dwarfs.

Just as much of a factor, however, was that MIT had been tight with IBM since the old Project Lincoln days, when the company won the contract to build the SAGE computers. IBM had provided the succession of high-end number-crunchers at the Computation Center, not to mention the souped-up 7094 at Tech Square. IBM's director of research, Emmanuel Piore, was a veteran of the Rad Lab and RLE, and Fano had made a point of keeping in close touch with him. "Very early in the planning of Project MAC," Fano would write in a June 29, 1964, report to MIT president Julius Stratton, "before a formal proposal was presented, I went to New York to consult Dr. Piore. One of the first copies of the formal proposal to ARPA was mailed to him and he had been constantly kept informed, either directly or indirectly, of the progress and plans of Project MAC. IBM was strongly represented, more than any other manufacturer, among the invited participants to the 1963 Project MAC Summer Study. There is no question that IBM has had a substantially better opportunity than any other manufacturer to become familiar with our work and to understand our objectives at both the managerial and working levels. If there is a misunderstanding of our objective, I am sure it is not our fault."[33]

Ah, but that was precisely the problem, wasn't it? It was true that IBMers at every level had been friendly and cooperative with the Project MAC team; MIT was a "prestige account," with a public-relations value to the company that far exceeded the actual dollar value of the machines installed there. It was also true that some IBMers seemed genuinely enthusiastic about what ARPA and Project MAC were trying to accomplish. Nat Rochester at the IBM research lab was a notable example, as were some of the engineers who had helped with the modifications on the project's 7094. Nonetheless, Corbató and his colleagues approached the computer giant with decided ambivalence. They couldn't shake the sense that they were being patronized, that the vast majority of IBMers still regarded time-sharing as academic self-indulgence having no relevance to the real job of computing, which was to build better batch processors. The steadily increasing stream of visitors to Project MAC had included not a single top executive from IBM. "I think they were very slow in the company to recognize the need for genuine system research," says Corbató. "The exploration of different kinds of designs was not being done; they thought it was a closed question of how to build machines. They did not recognize that it was an open question."

Still, says Corbató, he and his fellow committee members were expecting that attitude. What they weren't expecting was to find the people at IBM so . . . anxious. So *distracted*. The MAC contingent felt as though no one there could really focus on what they were trying to say.

They were right. In early 1964, after signing a raft of nondisclosure agreements and taking a solemn oath of silence, Corbató, Glaser, Dennis, and Graham were finally let in on the secret: IBM was being convulsed by the biggest gamble in its history. In effect, the company was scrapping every one of its existing computers and replacing them all with a single line of machines that would supposedly accommodate every possible customer. Known as the IBM System/360—for its ability to cover the full circle, so to speak—the new product line was scheduled to debut that very April. It would include smallish machines for small customers, behemoths for big customers, and every size in between; the processing power of the different models would vary by a factor of 50. And yet the upgrade path would be utterly seamless. On the hardware side, every computer in the System/360 line would be assembled from IBM-made solid-state circuits of standardized IBM design. And on the software side, every computer in the line would run the same programs. Indeed, they would all run the same, brand-new operating system, to be known as OS/360. They would likewise offer a brand-new programming language, called PL/1, which would replace Fortran, Cobol, SNOBOL, and the rest for every type of application.

Of course, there was also an obvious potential for disaster with System/360, and the debates within the company had been fierce. After all, dissenters had pointed out, IBM already had two very successful product lines—the 7000-series scientific computers and the 1400-series business computers—both with large cadres of happy customers. Why throw away that advantage by forcing those customers to jump into the unknown, especially when they might just jump to a competitor in the process? Why not simply upgrade and extend the systems that customers had now? If customers ended up liking the System/360 line, then, sure, everyone would be a genius. But if they didn't—well, two years of single-minded devotion to System/360 development had left the company with very little else to fall back on.

And that was the situation that the MIT foursome so innocently walked into. Indeed, they visited at a time when their IBM counterparts were more anxious and preoccupied than ever. Not long before, Honeywell had come out with a machine that was software-compatible with IBM's popular 1400 series of business computers, and had begun to make noticeable inroads into IBM's market share. The resulting agitation in the IBM sales organization had forced the company to move its System/360 announcement up to April 1964, thus giving the engineers roughly a year less than they thought they truly needed to perfect the new system. "They were really in deep trouble," says Corbató. And they had precious little time in which to work their way out of it.

Well, the MIT group could empathize with the IBMers' predicament. But as for the System/360 itself, they were appalled. Yes, System/360 was a beautiful idea in its own way, and a lot of heroic engineering had gone into it. No doubt it would do splendidly in the business market it had been designed for. "But they got fixated on building an engine to support batch processing," says Corbató—and in the process, of course, they went in precisely the opposite direction from what the MIT engineers were looking for. In order to achieve compatibility over such a wide range of machines, the System/360 engineers had designed the smallest computer in the lineup to handle data eight bits—or one "byte," in the IBM jargon—at a time.* (The term "byte" had originated with the IBM 7030

---

* Today, of course, the eight-bit standard is universal. But there was nothing magic about that number. Many of the System/360 designers had in fact argued for a six-bit standard. This would be sufficient to encode 64 different characters, they said—the minimum required for all the upper- and lower-case letters of the alphabet plus the most common punctuation marks. Eight bits would encode 256 characters, far more than anyone would ever need. It would be a waste of space in the computer's core memory, then a very expensive commodity. The eight-bit contingent prevailed, however, on the ground that the extra two bits would allow for much more flexibility in programming.

"Stretch" computer in the late 1950s.) For the more powerful models they simply doubled that, so that the machines would handle sixteen bits at a time. For still more powerful models they doubled that again, to thirty-two bits, and so on up to sixty-four bits at the top of the line.

Now, the MIT team thought this doubling trick was kind of neat, and it certainly simplified the hardware engineering. But once they began to understand what it meant for software, it became all too clear that the rush to announce the new system had taken its toll. "An extremely awkward system to program," they later called it in their report. "The entire system is very poorly meshed together. [All major design decisions have been] subservient to the demand that the machine language be compatible with the language appropriate for the smallest machine. In spite of this demand, even here the struggle was so great that they did not succeed and many subtleties exist which prevent programs from operating identically on all processors. . . . Finally, the design is an extremely difficult one to learn because of the Byzantine-like detail, exceptions, and ad hoc solutions that exist throughout."[34] Building Multics was going to be hard enough in any case; building it for System/360 would be a nightmare.

That wasn't the worst of it, though. "All through their design," they wrote, "both in hardware and in programming, [the IBM engineers] seem to have taken the view that a system is a static thing which is only created once and never modified." In particular, IBM had built in the assumption that each computer would have one and only one processing unit sitting at its center like a spider in a web, with all the memory banks and input-output equipment feeding into it. And indeed, for most business applications, that assumption was perfectly adequate. For the Project MAC representatives, however, it was wrongheaded on two counts. First, because they wanted to create an information utility that would be able to grow and evolve in ways they could not anticipate, they needed a machine that could operate with *many* central processing units at once. Not only would this greatly enhance reliability—since if one processor failed, the others could keep the system running—but it would provide a natural path for expansion: you could just add more processors. With the System/360 design, by contrast, there would be no way to upgrade without replacing the whole computer. Second, because the job of coordinating all those processors would have to be handled by Multics itself, which would reside in the computer's memory banks, Corbató and his colleagues were looking for a "memory-centered" architecture. Trying to make the processor-centered architecture of System/360 work in that fashion seemed like an exercise in futility.

"We were stupefied," says Corbató, "because it was apparent that we were watching a locomotive speeding down the tracks. Nothing we could say or do would cause them to deviate from what they were doing. They had too much at stake. They were not able to change [System/360], or were unwilling to change it. And to some extent, they didn't believe that what we were trying to do mattered that much anyway."

The upshot was that Corbató, Glaser, Dennis, and Graham politely thanked their hosts for the information and quietly made plans to look elsewhere.

By far the most obvious alternative was DEC. Although the company was a niche-market player that didn't even rank as a Dwarf in computer manufacturing, Ken Olsen himself was a strong supporter of time-sharing. Better still, DEC already had a time-sharing computer under development. Gordon Bell, the company's irrepressible chief designer, had taken up the challenge as soon as he heard what MIT was up to. As DEC's Alan Kotok remembers it, "Gordon came running into the lab one day in nineteen sixty-two, as he was wont to do, and said, 'Time to build a big computer, guys!' MIT was looking for a large time-sharing system, and Gordon felt there was no reason that we shouldn't build one for them. So we started designing what would become the PDP-6."[35]

That design process was well advanced by the time Corbató, Glaser, Dennis, and Graham came calling. And not surprisingly, they found that the PDP-6's thirty-six-bit architecture had a lot to recommend it. "It seemed to strike the fancy of every computer hacker who came in contact with it," said Kotok, an MIT graduate who had been one of the original TX-0 and PDP-1 hackers himself.[36] And, of course, the PDP-6 had time-sharing built in from the start; there would be no fighting with the architecture here.

In short, DEC had come up with a beautiful machine, one that might have ended the MIT team's quest right there—except for one little thing. The PDP-6 was the company's first *big* computer, with a size and complexity comparable to those of IBM's heftiest mainframes. And it had become painfully clear that DEC had gotten in way over its head. Certainly the company wasn't yet ready to make the kind of commitment Project MAC needed. As it was, says Kotok, "we built a few [PDP-6s] and sold a few, but none of them ever worked right. There were a lot of flaky electrical signal problems—a static discharge from the line printer would crash the system, the memory would die, things like that. It was hard to get up and running for more than a few hours or so. It was a great embarrassment."[37]

Happily, the PDP-6 would turn out to be a useful learning experience. Its creators would go on to design the very successful PDP-10, which they thought of as "a PDP-6 done right"—and which would ultimately become the time-sharing computer of choice for the ARPA community in the late 1960s and early 1970s. In the meantime, just as happily, DEC donated the PDP-6 prototype to Marvin Minsky's AI Lab. This eased the tension with the rest of Project MAC considerably, since the hackers there fell in love with the new machine and generally left off their depredations on CTSS. Now that they had a computer all to themselves, they could go their own way in splendid isolation, refusing to be part of any community but their own. (When they later wrote a time-sharing system for the PDP-6, they pointedly mocked CTSS, the Compatible Time-Sharing System, by calling their version ITS: the *In*compatible Time-sharing System. As befitting the information-wants-to-be-free hacker culture, ITS was completely open, with no passwords—indeed, no file protection whatsoever.)

However, all that was in the future, and of no help to Corbató and company in early 1964. So it was on to the Seven Dwarfs.

"The responses varied," says Corbató. "Some companies viewed it as an opportunity to bid [the way they would] on a military defense contract. That is, they would build computers if the military thought it needed them—but just as a way to make money on the margin, rather than affecting any of their own product plans." Other companies weren't even that polite. At the Control Data Corporation in Minneapolis, a six-year-old firm that was then making quite a splash in the scientific number-crunching market, chief designer Seymour Cray walked out of the MIT group's presentation in contempt. "We found similar attitudes in most hardware designers, if not quite so extreme," says Corbató. "They all had their own vision of what a computer ought to be, and that didn't include the social environment of [multiple] people trying to use the computer all at once. They really didn't want to be bothered with somebody introducing new ideas."

It was beginning to look like IBM or nothing.

The IBM public-relations operation pulled out all the stops. Press kits, information packets, and advertising campaigns were prepared for all six of the new processor models and all forty-four of the brand-new peripheral devices that would go with them. Announcements were sent to some one hundred thousand IBM customers and prospective customers, inviting them to presentations in 165 branch offices throughout the country. A special train was hired to bring some two hundred writers and editors up to Poughkeepsie for a mass press conference. And on Tuesday, April 7, 1964, IBM System/360 was presented to the world. "Today we are making the most significant product announcement in IBM history," Tom Watson declared. "This is the beginning of a new generation—not only of computers—but of their applications in business, science, and government."[38]

It was the beginning for everyone, that is, except IBM's prospective customers in Tech Square. Other than the private show-and-tell they'd been given several months earlier, Corbató and his companions hadn't heard a word about how (or if) IBM's new equipment could help Project MAC. Indeed, as Bob Fano would write in his June 29 report to Julius Stratton, it wasn't until two weeks after the System/360 announcement that they finally thought they might see some action: "We were visited on April 20, 1964 by a group of IBM people, including Dr. Amdahl, for what I thought was a first presentation of equipment configurations that might be of interest to us." Now, this was the big time: the forty-one-year-old Gene Amdahl had been a chief architect of System/360. His presence at Project MAC could be read as a measure of how prestigious this Prestige Account really was. But alas, wrote Fano, "I was very shocked to realize at that time that nobody of any consequence had yet given any thought to our problems."[39]

In fairness, the frazzled IBMers had only just had a chance to catch their collective breath after getting their new product to the finish line. Still, says Jack Dennis, who had had long private conversations about Project MAC's needs with Amdahl and others, it was as if they couldn't *listen* anymore: "They looked at it as, 'Oh, you want to do time-sharing? You've got all this government money to do time-sharing? Sure, our machine is good for that.' Or maybe to say it more accurately, 'We are determined to prove to you that our machine is good for that, because our machine is the best one that we could possibly build.' "

Fano could see that this was far more than a technical disagreement. "I do know from many sources that the design of the new line of IBM equipment was the subject of very strong controversies within IBM, that left very deep wounds," he later wrote to Stratton. "Some of the designers, and Dr. Amdahl in particular, ended up in assuming very rigid positions, that they are not emotionally free of abandoning."[40] On top of that, he added, the IBM delegation had told him that Project MAC could not expect delivery of any new equipment until the middle of 1966—a slippage of eight months from previous estimates.

Fearing that the situation was shaping up to be a disaster, Fano and Corbató arranged to meet with IBM's chief scientist, Emmanuel Piore, in New York on May 11, in hopes that he could somehow make something *move*. They certainly didn't have any leverage they could use on the company. But they had to try.

Fate, as if on cue, proceeded to hand them a lever.

Right in the middle of this turmoil, Joe Weizenbaum mentioned that he'd just paid a visit to his former colleagues at General Electric. "He'd found out there that they were planning to have a commercial version of a military machine they'd developed," says Fano, "and he thought we ought to look at it, because it had many of the features we wanted." So in late April, shortly after the Amdahl visit, Jack Dennis and Ted Glaser flew out to the GE plant in Phoenix to see what was really going on.

They came back with stars in their eyes: the GE 635, as it was known, was remarkably close to what they were looking for. Just as important, the company was remarkably receptive to a partnership. "When we visited GE," says Jack Dennis, "the responsible engineer [for the GE 635 computer] was John Couleur, who understood the nature of what we expected from the hardware and realized that the 635 was as close as anything in the industry to it. He was willing almost immediately to start thinking about how they could make the machine better for us, which was very different from IBM's approach."

The news put stars in Corbató's eyes as well. "They were naive and we were naive," he says with a laugh, looking back on his younger self. "We all thought we could do more than we could." Nonetheless, he and Fano could now meet with Piore on May 11 feeling confident that IBM was no longer the only game in town. "I made a point to convey this fact to Dr. Piore," Fano wrote, "of course without mentioning the company's name because some of the information I had was still regarded as company-confidential." Piore apparently believed that Fano's alternative was DEC, a company he would have had trouble taking seriously. Nonetheless, he took Fano's point; until that moment, he had assumed that the Project MAC account was in the bag. As Fano would later explain to Stratton, Piore immediately ordered his subordinates "to see to it that an appropriate team would get to work immediately on a specific proposal to meet Project MAC's needs."

And that was what happened. In early June, a team of IBM engineers camped out at Project MAC for a week, talking to everybody. Then they disappeared until June 23, when they submitted their formal proposal.

It was the disaster the MAC group had been braced for all along, says Corbató: more of the same, warmed over. Once the IBMers realized they had a competitor, he says, "their immediate reaction was, 'We have got to save this account.' But that was one of the problems: they kept thinking of us as an *account* rather than as a coparticipant in a new kind of product. That reflected the marketing viewpoint of the company. As a result they ordered all the engineers to do something to make MIT happy. So what they did was build a Rube Goldberg–like offering, which they then priced accordingly—namely, pretty high. But what they were shoving at us was just hopelessly flawed, starting with some deep design decisions that had been made inside the 360. They had been misled by their senior designers to believe that the 360 was good for everything. That just wasn't true, but the company wanted to believe it. Yet the only way to satisfy us would have been to embark on a new product design which was different from the 360. So we got into this impasse."

The result was Fano's letter to Stratton on June 29, which was basically intended to give the MIT president a storm warning: Brace yourself. And on July 15, 1964, Fano and his colleagues announced that their next-generation information utility would be created in partnership with the General Electric Company.

The storm exploded immediately.

In hindsight, it's true, IBM had nothing to worry about. Within the first four weeks after its launch, the company had received well over a thousand orders for System/360, far beyond its most optimistic forecasts; corporate data managers who had been tearing their hair over every machine upgrade absolutely *loved* the idea of software compatibility. Within the first five years, IBM's worldwide inventory of installed computers would more than triple, to an estimated $24 billion worth. And within the first decade or so, it would become clear that the introduction of System/360 had been one of the watersheds of computer history—the beginning of an era that would see computers integrated into every major corporation and every major government operation in the industrialized world. The Dwarfs would prosper as well, of course. But System/360 and its successors—most notably System/370, in 1970—would become de facto standard architecture for mainframes well into the 1990s, in much the same way that the

Macintosh and IBM PC would later become the standard architectures for personal computers. (Indeed, they would even inspire a market for "clones," work-alike mainframes from other manufacturers.) And of course, System/360 would popularize such terms as "byte," along with the convention of encoding data eight bits at a time.

But all that was later. In the summer of 1964, the flood of orders was just one more reason for anxiety: with delivery of the machines promised for 1965, IBM had yet to prove that it could even *build* the things. Its new manufacturing plants were still plagued by start-up problems, particularly when it came to System/360's new "Solid Logic Technology" transistor modules. Worse, the system software was a mess. At the announcement, in April, IBM had promised that the first version of OS/360 would be available by the end of 1965, and that through the selective omission of various bells and whistles, it would run on System/360 processors with as little as sixteen kilobytes of memory. By summer, however, with the programmers falling further and further behind schedule every day, it was clear that the planners had grossly underestimated the magnitude of the task. Indeed, OS/360 would eventually go down in history as one of computerdom's classic horror stories, just as project chief Fred Brooks's rueful meditation on the lessons he'd drawn from that experience, *The Mythical Man-Month*,[41] would go down as one of the classics of the software-engineering literature. The OS/360 that he and his team finally managed to deliver, in 1966, would be late, slow, and a memory hog; much of the work in subsequent releases would be devoted to fixing its deficiencies. In the meantime, moreover, IBM would be forced to renege on its promise of a universal operating system and develop three much simpler systems for its low-end machines: BOS, TOS, and DOS (respectively, the Basic, Tape, and Disk Operating Systems).

And now, right in the middle of this chaos, came Project MAC's defection to General Electric—the one competitor that IBM truly feared. GE was a huge, diversified conglomerate that manufactured everything from jet engines to household appliances. Its pockets were deep; indeed, its annual revenues were several times larger than IBM's. Its research and development organizations were world-class. It had the wherewithal to buy its way into any market it really wanted to enter. And of late it seemed that GE really wanted to enter the computer market. The company had already made considerable inroads in banking applications; now it appeared to be going after the "prestige accounts." The word was that Ohio State University and the University of Michigan were both considering GE machines, as were the computer people at Bell Labs. To IBM management, in the summer of 1964, it seemed very possible that MIT's decision would trigger an academic avalanche toward GE, thereby costing IBM its market leadership in scientific computing.

IBM management did not take kindly to losing market leadership.

Mannie Piore, in a recent interview, insisted that he'd actually breathed a sigh of relief when he heard that MIT was going with General Electric instead of DEC. But that's not the story they heard at Project MAC. "When word got out that the real competitor was GE," says Fano, "Watson called Mannie on the carpet and was real upset about it. 'You're supposed to be the contact with MIT. How could this happen?' So Mannie came to talk to Stratton and really blew his top." After all IBM had done for it, how could MIT be so disloyal?

Stratton, to his credit, refused to second-guess his own people. Nonetheless, Piore was not the only one asking that question around MIT. Corbató was obliged to give up his post as deputy director of the Computation Center, where most of the staff was still quite loyal to its longtime benefactor. And for years thereafter, he says, his relations with his former colleagues remained wary, if friendly: "I became known as one of the people who had let IBM down."

But there was worse to come, at least from IBM's standpoint. "Quickly after our decision to go with GE," says Fano, "Bell Labs decided to join us [in developing Multics]. So it became a three-way partnership. And when that happened, IBM hit the ceiling. Then the next thing that happened was that Lincoln Lab put out a request for proposals on a new computing system with the spec that it had to be time-shared—in exactly the same way Multics was. So IBM decided they *had* to have that contract. They couldn't afford to lose again."

Orders came down from the top: create a time-sharing computer for the System/360 line, starting from the System/360 architecture. This was a measure of just how nervous the top brass really were about GE, notes Corbató. As any engineer could have told them, adapting a 360 for time-sharing was going to require such deep, deep engineering changes that the result would barely be a 360 anymore. But then, no one seems to have objected too strongly; insufficient enthusiasm for System/360 had terminated more than one IBMer's career already.

Thus was born the System/360 Model 67, IBM's answer to General Electric. "They rushed to create a competitive product because they thought *we* would create a product," says Corbató, still shaking his head at the folly of the move. And the irony of it was that most of IBM's work on Model 67 was concentrated at the IBM Cambridge Scientific Center, located right at 545 Technology Square, just a few floors below Project MAC.

THE FATHER OF IT ALL

In the long rebellion against batch processing, and against the authoritarian mind-set that went with it, Project MAC's break with IBM was about as close as things ever got to open warfare. Yet J. C. R. Licklider—the commander in chief, so to speak—kept his distance from the fray. The choice of a next-generation computer was Tech Square's to make, he felt. And besides, why get involved with that battle when he was preparing to infiltrate the enemy lines all by himself? In mid-1964, at about the same time that Fano and company were deciding on the GE 635, IBM chief scientist Mannie Piore had invited Lick to join IBM's research division, headquartered at the new Thomas J. Watson Research Center in Yorktown Heights, New York. And Lick had accepted.

With the correspondence long gone and memories hazy at best, it's hard to know exactly what Lick's intentions were—or Piore's, either, for that matter. Piore himself now maintains that the invitation had nothing to do with Project MAC; he was simply on the lookout for outstanding talent. And indeed, that would have been very much in character. The Russian-born physicist had taken over as IBM's first director of research in September 1956 and had spent the intervening years building an organization with world-class capabilities in solid-state physics, mathematics, superconductivity, and a host of other fields. (In the process, he had also consolidated many of IBM's scattered research activities at the Yorktown Heights site, some twenty-five miles north of Manhattan; the laboratory building, designed by Eero Saarinen, opened its doors in April 1961.) Nevertheless, the coincidence was striking. Given the embarrassment of losing Project MAC to GE, Piore could hardly be blamed if he saw hiring Lick as an effective way to acquire some in-house expertise in interactive computing.

As for Lick, he may simply have felt that it was time for him to move on. The tradition was for ARPA program directors to step down after two or three years in any case, to make room for new people with new ideas. Then, too, with two teenagers who would soon be starting college, Lick can't have been entirely unaware of IBM's lavish salary scale. Nor could he ignore how stressful Washington had been for his family. Louise had been forced to give up an apprenticeship at the Arena Stage, which she loved, and put her theater career on indefinite hold; with two children at home, a husband who was always on the road, and rehearsals at night, scheduling had become a nightmare. Tracy and Linda, meanwhile, had had a rough time adjusting to school. Their parents had innocently enrolled them in private school before they arrived, since a commitment had to be made and Lick had not yet located a place for the family to live. But to their horror, the school had turned out to be a segregation academy: pristine white, very conservative, and run by not-very-bright retired colonels. (Tracy, in particular, loathed the place. In the Cambridge public schools, he had been going through the fast-track science curriculum ever since *Sputnik*. And now *this?* Never mind that his classmates had elected him president of his class for the next year; he refused to go back, announcing that the Arlington public school system was good enough for him, "mixed races" and all.)

While all these considerations undoubtedly influenced Lick, however, what really seems to have persuaded him was the opportunity: at IBM he could preach the gospel of interactive computing in the very high temple of batch processing. If he could convert IBM to the cause, he often rhapsodized to Louise—if he could get even a fraction of that immense marketing power behind the vision of human-computer symbiosis—then he would be one giant step closer to converting the world.

In any case, the plan was for Lick to start at Yorktown Heights in September 1964, which left the problem of what to do about his command-and-control program at ARPA. Under Jack Ruina, who had left the agency in September 1963, and Ruina's successor, the physicist Robert Sproull, Lick had had the freedom to turn that program into a visionary quest for human-computer symbiosis. In fact, he had taken it so far in that direction that ARPA had felt obliged to give it a new name: the Information Processing Techniques Office, or IPTO. Lick was well aware, though, that his own successor was going to have exactly as much freedom as he himself had had—which meant that he had to get the right person in place, or his whole movement could collapse.

Fortunately, one of ARPA's traditions was that outgoing office directors were allowed to recruit their own successors; all Lick had to do was to come up with an experienced, senior person with good credentials, and everybody would be happy. So Lick, characteristically, nominated a totally inexperienced and very junior candidate whose credentials were about the most awkward and inconvenient conceivable in the Pentagon. He was a skinny, intense fellow named Ivan Sutherland. He was twenty-six years old. And at the time he was a first lieutenant in the army.

It wasn't that he hadn't tried to find someone more senior, Lick protested in his 1988 interview; it was just that the answers he'd gotten ranged from "No!" to "Hell, No!" "Most of my colleagues would much rather spend the government money doing research back in the lab than coast another year or two or three in Washington," he said.

Still, you have to wonder how hard Lick really *did* try; the truth was, he'd had his eye on Lieutenant Sutherland for a long while. When they'd first met, the younger man was still an MIT graduate student finishing up his Ph.D. work under Claude Shannon. The subject of his dissertation was a computer graphics program he had created on the TX-2 computer out at Lincoln Lab. Sutherland called it Sketchpad.

Now, Lick had always loved graphics. Indeed, he considered high-resolution graphics to be as critical to human-computer symbiosis as communications or even real-time interactivity. Humans are visual animals, he would muse to anyone who would listen. Our eyes are "a high-bandwidth data channel" capable of absorbing information at the equivalent of millions of bits per second. Our brains are organized to recognize patterns and sense complex relationships at a glance. In fact, as Lick told Oliver Selfridge and Marvin Minsky over dinner one evening, he felt that once they got time-sharing working well, ARPA's next two major initiatives should be graphics and networking. That was why he had funded the RAND Tablet, and why he had been so adamant that the Kludge and other graphics applications be included in Project MAC.

But Sketchpad had been a revelation. "In 1962," Lick would later write, "at the Spring Joint Computer Conference in San Francisco, during the discussion period of a session on man-computer communication chaired by Douglas Engelbart, Ivan Sutherland mentioned his Sketchpad program and, at the end of the session, showed to a few lingering enthusiasts the most dramatic on-line graphical compositions that any of them had ever seen."[42]

To modern eyes Sketchpad would look quite familiar—like a computer-aided design system, perhaps, or a high-end drawing package. But that's only because

Sutherland himself single-handedly pioneered most of the techniques that are now used in such programs. Touch two points on the TX-2's little display screen with a light pen, and the computer would draw a straight line between them. Draw a rough curve with the pen, and the computer would "correct" the curve and smooth it out. Indicate that two components were attached at a given point, and the computer would ensure that they *stayed* that way, no matter how you moved or rotated them.

Lick was enthralled. It was as if his whole vision of symbiosis were coming to life before his eyes: dynamic modeling as it was meant to be. From then on he would point to Sketchpad as the program that had done more than any other single thing to convince the world that interactive computing was worthwhile. And from that day on, of course, Lick had kept close tabs on Sketchpad's creator.

Sutherland had never participated in Project MAC, unfortunately; after receiving his Ph.D., in January 1963, he'd been obliged to go straight into the army due to an ROTC commitment left over from his undergraduate days at Carnegie Tech. Still, when the time came for Lick to nominate his own successor, he didn't hesitate. So what if Sutherland was young? "Ivan was a true believer in the things I was a believer in," Lick explained, "and, in my view, he was better at [them]."

Actually, Sutherland himself hadn't been so sure about the ARPA job at first, but after some genial arm-twisting from Lick, he'd agreed to try it. So had ARPA. "I had lots of talks with Sproull, Charlie Herzfeld, and several of the others, about how it was going to be possible for such a young guy to make it in the Pentagon," Lick says. Until Sutherland finished his term of service in early 1965, the protocol of having a first lieutenant run a program in the Pentagon was going to be delicate, to say the least (among other things, he would need a new deputy; Buck Cleven was a colonel, and you couldn't have a colonel reporting to a lieutenant). But the effort was well worth it, Lick would insist in 1988: "I think maybe the best thing I did was to pick my successor. Ivan was surely more brilliant than I, and very effective. He carried it on."

In the late summer of 1964, just a few days before Lick was due to depart for IBM, his office at the Pentagon finally took delivery of the first tangible evidence of his efforts: a Teletype console that would let him hook into the time-sharing systems at Berkeley, SDC, and Project MAC. Alas, he later recalled, he didn't have much time to use it himself—"but when Ivan Sutherland succeeded me, he had a steady stream of military people coming to play with that console, including generals and admirals. He found that he had to put it in a little room that was big enough for just one person, because if an admiral was sitting at the console, and there were junior officers looking, the admiral was afraid to move his fingers for fear he'd reveal he didn't know what to do!"

As tangible output, one Teletype terminal wasn't much, perhaps. But then,

Lick never claimed to have accomplished much at ARPA—certainly not much that he could actually take credit for. "I went for one year and stayed for two," he said in his 1988 interview, "and the time scale for doing anything significant is longer than that. So I had to buy into things and finish them and make them demonstrable, so that there would be something for people to look at without realizing that I didn't do them. ONR [the Office of Naval Research] had had Corby [Corbató] started on the time-sharing system and the Computation Center, and then we came along and greatly increased the speed with which that project was going. But we could never have started it and gotten it going. Berkeley hadn't started time sharing, but they had a laboratory and a computer and everything else, so the time-sharing part was relatively simple. They had a time-sharing system [Project Genie] running before I left. The SDC time-sharing system ran before I left Washington, but they had a hundred or two hundred programmers working before I got there. We were just responsible for channeling that a bit. So I think that there's nothing I can point to with pride and say, 'I did that.' "

Lick did have a point: little was happening at the end of his ARPA term that hadn't already been under way when he started. And yet in a larger sense, he was also selling himself woefully short in this account—not unusual for him. "When you look at Lick's legacy, two very distinct things stand out," says Bob Fano. "One is that he was a very imaginative, creative psychoacoustics man. That's the first part of his history, and you shouldn't overlook that." Because Lick had come to computing from psychology, Fano explains—instead of through mathematics and engineering, like almost everyone else—he instinctively saw computers in relation to the workings of the human brain, rather than as an exercise in pure technology. And that, in turn, was why he was so quick to embrace computers as a way of enhancing human creativity and enriching human life. "It was a vision of man-machine interaction that was often unhampered by practical realities," says Fano, who had many an argument with Lick on that very point. "But he really had an understanding of the role the computer could play."

Second, says Fano, when Lick was presented with a miraculous, never-to-be-repeated opportunity to turn his vision into reality, he had the guts to go for it, and the skills to make it work. Lick had the power to spin his dreams so persuasively that Jack Ruina and company were willing to go along with him—and to trust him with the Pentagon's money. Once he had that money in hand, moreover, Lick had the taste to recognize and cultivate good ideas wherever he found them. Indeed, the ideas he fostered in 1962 would ultimately lay the foundations for computing as we know it today. The time-sharing technology he pushed so relentlessly would turn out to be the evolutionary ancestor of both personal computing and local-area networking, as well as a test bed for all the issues of on-line social behavior that would reemerge a generation later. The computer-graphics experiments he funded so lovingly would likewise turn out to be important steps along the road to our current generation of high-resolution computer displays with their windows, icons, menus, and so on. And of course, Lick's vision of an Intergalactic Network would be the direct inspiration for the Arpanet of the late 1960s, which would in turn evolve into the Internet of the 1990s.

Perhaps most important of all, however, Lick had the patience to take the long view. He couldn't get it all done in one year, or two years, or a lifetime. But by creating a community of fellow believers, he guaranteed that his vision would live on after him. When he arrived at ARPA, in October 1962, there was nothing more to "symbiotic" computing than a handful of uncoordinated development efforts scattered all across the country; by the time he left, in September 1964, he had forged those efforts into a nationwide movement that had direction, coherence, and purpose. Moreover, by putting so much of the agency's money into research at universities, where most of it actually went to support students, he neatly co-opted a substantial portion of the rising generation. "It seems to me that Licklider and ARPA were mainly about winning the hearts and minds of a generation of young scientists and convincing them that computer science was an exciting thing to do," says James Morris, chair of Carnegie Mellon's computer science department. "Remember, in the aftermath of *Sputnik,* the glamour field was physics, not computing. Lots of very smart people made a career decision to go into a field that didn't exist yet, simply because ARPA was pouring money into it."

Indeed, in his 1988 interview, that was the one accomplishment for which Lick was willing to take credit: "I think that I found a lot of bright people and got them working in this area," he said. "I got it moving. [And it was] a fantastic community. I guess that's the word. It was more than just a collection of bright people. It was a thing that organized itself into a community, so that there was some competition and some cooperation, and it resulted in the emergence of a field."

CHAPTER 7

THE INTERGALACTIC NETWORK

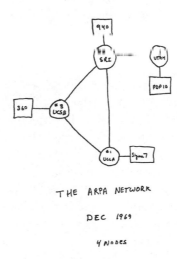

THE ARPA NETWORK

DEC 1969

4 NODES

It was around March of 1960, says Bob Taylor, thinking back on how he found his direction in life. He was at the Martin factory in Orlando, Florida, working as a systems engineer on the Pershing missile project, when he looked in his mailbox one day and found the premiere edition of *Human Factors in Electronics*.

The lead article leapt out at him immediately. J. C. R. Licklider was one of *the* famous names in psychoacoustics, and psychoacoustics just happened to be the field in which Taylor had pursued his master's degree back at the University of Texas. But what really grabbed him was the article itself. He knew a little something about computers. He'd even read Vannevar Bush's article about the Memex, and Norbert Wiener's *The Human Use of Human Beings*. But interactive computers? Networks of computers? Humans in *symbiosis* with computers?

The idea resonated deeply for him, says Taylor. He'd always had a feel for responsive machines. Cars, for example: he'd learned to drive in 1943, at age eleven, when the country roads of southern Texas had seemed smooth, wide, and empty. "The first car I drove was a friend's Model T," he says. "I discovered that I loved moving fast, not so much in a straight line but around curves, where the car was on the edge of control." That sensation may have appealed to a cer-

---

Sketching in the future: the first four nodes of the Arpanet

tain restlessness in his blood. As the adopted only son of Raymond and Audrey Taylor, he had spent those first eleven years of his life moving from one small Texas town to another: Uvalde, Victoria, Ozona, Mercedes—anywhere the Reverend Taylor's Methodist ministry took them. But driving definitely gave him a feeling of joyous power, a sense that this immense machine was responding to his will as if it were an extension of himself.

Likewise guitars: "I like to sing," says Taylor, "so I taught myself how to play acoustic guitar by ear. I know four chords and four hundred songs! It's just like a car, or a lawn mower, or a bicycle, or an airplane, or a boat, or any other tool one can think of: if the guitar is set up correctly, it is a delight to play. Otherwise, it's a struggle."

And yet, he says, before he read Lick's "Symbiosis" article, it had never occurred to him that a *computer* could be responsive. True, this was a machine to enhance your intellectual abilities, not your physical ones. But still—just imagine being able to feel all that power at your fingertips. Just imagine being able to feel that such a machine was an extension of yourself.

Just imagine being able to *create* such a machine.

This notion hit him at a critical moment in his life, says Taylor. Beneath that fresh-faced, minister's-son exterior of his, he had always harbored a Texas-sized streak of pigheadedness. And because of it, he and the barons of the University of Texas psychology department had recently reached a parting of the ways. They had insisted that any Ph.D. candidate of theirs would have to take numerous courses in clinical psychology and behaviorist-style animal experimentation—two branches of the field in which Taylor had no interest. He wanted to understand the *human* brain, thank you, all the way down to the neurons. So after the barons rejected his counteroffer—a customized Ph.D. curriculum built around mathematics and sensory psychology—he took his leave. Thus his stop-gap job at Martin's Orlando plant. And thus his response to "Man-Computer Symbiosis": "Lick's paper opened the door for me," he says. "Over time, I became less and less interested in brain research, and more and more heartily subscribed to the Licklider vision of interactive computing."

By 1961, in fact, his fascination with Lick's vision led the twenty-nine-year-old Taylor to NASA headquarters in Washington, where he headed up a new program that funded research in manned-flight control systems, displays, simulation technology, and—of course—computers. Shortly after that, moreover, his fascination led him to the man himself. In the fall of 1962, Taylor was among the first people invited to join Lick's impromptu caucus of computer funding officials.

The two men hit it off immediately. They were both southern boys, after all. They were both the only children of ministers. They had the psychoacoustics background in common. They even shared a passion for automobiles. "Lick owned a Messerschmitt three-wheeled vehicle for a while," Taylor recalls, laughing. "He enjoyed unusual cars." But most important, they shared a passion for human-computer symbiosis. Robert W. Taylor was easily the most enthusiastic disciple J. C. R. Licklider ever had in Washington, if not in the entire country. It was Taylor who pulled out all the stops to get NASA funding for Doug Engelbart. It was Taylor who put NASA money into a number of specific efforts within Project MAC—and even funded a study of interactive, computer-assisted air-traffic control at BBN. And it was Taylor who was brought over to ARPA in early 1965—on Lick's enthusiastic recommendation—to serve as a deputy to Ivan Sutherland.

Taylor had a grand time at ARPA. His young boss was smart, honest, forthright, and exceedingly well informed—a pleasure to work with. Moreover, Lick had bequeathed them a marvelous program. "I had always thought it was very well conceived," says Taylor. "But as I got to meet the people Lick had been supporting, I thought even more highly of the program, because these were very good people."

Taylor was equally impressed with the way Sutherland enlarged that circle. By 1966, the Information Processing Techniques Office was funding sixteen different "contractors," as the research groups were called. And with only one major exception—the Illiac IV project, an effort to create a parallel-processing supercomputer at the University of Illinois—virtually all of them were still focused on Lick's vision of interactive computing. Just as Lick had hoped, for example, the creator of Sketchpad had very quickly inaugurated a formal program of computer-graphics research, through which new research contracts were awarded to Lincoln Laboratory and several other institutions. Among these contracts, significantly, was a small one that Sutherland awarded as a personal favor to Dave Evans, who had just returned to his home state to chair a brand-new computer-science department at the University of Utah. Evans had done a superb job as head of Berkeley's Project Genie, Sutherland felt, and he had no intention of letting a guy of his caliber drop out of the ARPA community if he could help it.

Utah would go on to do great things, as Taylor would later have cause to know firsthand. But in the meantime—and this was Taylor's personal favorite—Sutherland was also expanding the ARPA orbit to encompass his mentor from the TX-2 days, Wesley Clark, who had at last led his team to a permanent home at Washington University in St. Louis. (The fact that it was J. C. R. Licklider's alma mater was pure coincidence; Clark and his coworkers on the LINC project had found their Promised Land only after two years of wandering in the wilderness, with a brief, tumultuous stop at MIT.) Happily, they had thrived there. Soon after arriving, in 1964, they had completed development on their LINC desktop laboratory computer and started to get the machines out into the hands of working scientists (eventually there would be some twelve hundred LINCs controlling experiments in laboratories around the world, including a commercial version manufactured by DEC). And with funding from Sutherland, they had embarked upon an effort to develop "macromodules," a series of electronic building blocks that would allow users to assemble and reconfigure specialized interactive computers as needed. This was what took Sutherland and Taylor out to St. Louis in 1965 for a site visit. And it was there, Taylor remembers, that he got his first good look at the LINC.

It was a revelation. Clark and his followers were as adamant as ever that the *real* future of interactive computing lay not in time-sharing, which they regarded as a lamentable mistake, but in individualized computing, wherein everyone would have a computer of his or her own. The LINC was their prototype. And in using it, says Taylor, he finally began to understand what they were driving at. "When I sat down at the LINC it had a little six-inch green and white display that looked like an oscilloscope, and a knob that let you adjust the speed of the machine. So you could turn up the speed, type an *A* on the keyboard, and the *A* would appear on the screen very quickly—the *A* was made of dots, like a dot-matrix printout— or you could turn the speed down, and the *A* would build up very slowly. Also, you had a loudspeaker: the machine would click slow or fast according to what it was doing. So you had a sense of scale. You could see that this machine was building up a complicated reality through millions and millions of *yes-no* decisions." For the first time, he says, he could really *feel* the interactivity. "A Teletype was one kind of interaction," he says. "Spacewar was another. E-mail was still another. But the LINC was somehow more dynamic than any of them. So I'm saying to myself, 'If something like this could be this entertaining and exciting all by itself, just imagine what it would be like if numbers of folks could have it. Then they could be playing together in this medium.' "

Ah, but that was the big *if*. The LINC was a stand-alone computer, with no connection to anything else; Clark and his team had left communication as a problem to be dealt with later. Yet Taylor had increasingly come to believe that communication was the heart of human-computer symbiosis—as important as, or perhaps even more important than, interactive computing per se. The on-line communities that had sprung up around the time-sharing systems at SDC, Berkeley, and Project MAC absolutely fascinated him. "As soon as each time-sharing system became usable," says Taylor, "the individual people who were interested in computing began to know one another. They began to share a lot of information, and to ask of one another, 'How do I use this? Where do I find that?' It was really phenomenal to see this computer become a medium that stimulated the formation of a human community."

This notion of an on-line fellowship was clearly what Lick had been getting at in his "Intergalactic Network" memo. With it, ARPA's researchers could take the dream of human-computer symbiosis to the next level. But without it, they would never be anything more than a patchwork of fiefdoms using separate languages, separate programs, incompatible machines, and incompatible software. And anybody who doubted that, says Taylor, simply had to look in his office at the Pentagon, where he'd had to install three different Teletype terminals to connect him with the three different ARPA-funded time-sharing systems. It was ridiculous, he says: "Anyone in that context would have quickly thought, Hey, wait a minute—why can't I get to any of these places from *one* terminal?"

Indeed, that question was much on Taylor's mind in June 1966, when Sutherland ended his tour at ARPA—he had accepted an offer from Harvard—and Taylor was named to succeed him, as the third director of the Information Processing Techniques Office.

Now, Taylor knew full well that he didn't command the kind of automatic respect from the ARPA community that Sutherland had commanded. In their eyes, Ivan Sutherland had been a technological genius—just the kind of man the agency needed in that position. But Taylor was, well, a psychologist. With no Ph.D. "Can you imagine going from J. C. R. Licklider to Ivan Sutherland to . . . Bob *Taylor?*" asks one PI, remembering his reaction at the time.

No one said anything overt, of course, since Taylor *did* control the money—for the time being. No, they just conveyed an ineffable air of condescension, an unmistakable message that the community considered him to be a mere administrator, a caretaker who was keeping the chair warm until a good technical person could be brought in. There was enough truth in that assessment to sting, if Taylor had been of a mind to listen to it. But he insists that he never let it bother him, if only because of that Texas-sized streak of pigheadedness. He had no intention of being a caretaker, and every intention of leaving his mark. Wasn't that the whole point of ARPA, after all—to invest in research that could change the status quo by an order of magnitude? Lick had made that kind of investment with time-sharing. Sutherland had done it with the Illiac IV and his initiatives in computer graphics. So what was Bob Taylor going to do?

Networking, obviously. But the question was, how? Should he just go the standard route and put money into small-scale networking experiments scattered around the country, with maybe a pilot project or two? Not a chance. For one thing, Lick had already gone the small-scale route with the networking experiment he funded at UCLA, where three computer centers were supposed to link up into a campus-wide community. Sutherland had enthusiastically continued that experiment. And they had all watched in disgust as the project fell victim to academic bickering: none of the three UCLA centers really wanted to work with the others, and certainly none wanted to take direction from ARPA.

Then, too, IPTO had just finished up a second experiment demonstrating that long-distance networking was technologically feasible, if damnably difficult. In the fall of 1965, Taylor and Sutherland had commissioned an experimental hookup linking the TX-2 computer at Lincoln Lab to the Q-32 computer at SDC in Santa Monica. With nothing more than an ordinary commercial line rented from Western Union, the bits had flowed—barely—and the computers had indeed talked to each other.

So forget the experiments and pilot plants, Taylor decided; he was the kind of guy who liked to drive fast, on the outer edge of being in control. If you're going to do it, he reasoned, *do* it. Link *all* the IPTO contractors—all sixteen of them. Take interactive computing to the next level. Make computers into a medium for worldwide communication. Create an integrated ARPA community.

Build the Intergalactic Network.

BLACKMAIL

Of course, this idea did have one or two little problems—starting with the fact that almost no one in the ARPA computer community especially wanted a nationwide network. It was one thing to hear J. C. R. Licklider fantasize about an Intergalactic Whatsit. It was even kind of fun to join in, especially when you were sitting around at a meeting somewhere and kicking back a few beers. But in the cold light of day, the Big Three time-sharing sites—MIT, Berkeley, and SDC—saw a nationwide network as a standing invitation for outsiders to tap in and poach their computer cycles. Even if it actually worked, which was a pretty big "if," why throw open the doors when you could barely accommodate your own people? Likewise, the smaller ARPA-funded sites saw the network as locking them into a lifetime of second-class citizenship. Why agree to be a permanent have-not, always begging computer time from the big boys, when you could aspire to be a have, with a shiny new ARPA-funded computer of your own? And all of them saw it as a distraction from their *real* work in AI, graphics, or whatever—just the sort of naive nonsense you'd expect from a nontechnical manager type like Taylor.

The upshot was that Taylor's pitch for a nationwide network was met with little more than cool politeness. Of course, being Bob Taylor, he insists that none of this fazed him. But even so, he was reassured to find that two of the most respected figures in the community were solidly behind him. One, surprisingly enough, was Wes Clark. The creator of the LINC stand-alone computer wasn't particularly enamored of the network idea, not if it meant building some kind of vast, centralized time-sharing machine and forcing everyone to use it. But if it meant thousands or millions of machines linked to a nationwide communications system that would allow them to function as equals, then great! Give all those machines some autonomy!

The other supporter—of course—was J. C. R. Licklider. After all, says Taylor, the only reason Lick hadn't tried to build the Intergalactic Network himself was that the technology wasn't ready when he was at ARPA. "So when I told Lick what I wanted to do, he said, 'Go for it.' " Indeed, says Taylor, he and Lick went through any number of bull sessions as they tried to imagine what people could do with such a network. On-line collaboration? Digital libraries? Electronic commerce? The two of them had such a marvelous time that they wrote up their speculations as a joint article entitled "The Computer as a Communication Device."[1]

If the support from Lick and Wes Clark was reassuring, however, Taylor did have a more immediate concern. Every project that IPTO had undertaken to date, from Project MAC to Engelbart's Augmentation Research Center, had been conceived and managed by the researchers themselves, according to a bottom-up approach that was perfect for small-scale, proof-of-principle efforts. *This* project, however, was going to be a nationwide effort requiring nationwide coordination. For the first time, Taylor realized, his office would have to initiate and manage a project from the top—which meant that he would not only have to build up a brand-new organizational structure within IPTO to do it, but find a brand-new pot of money to pay for it. (Although Taylor's budget was a bit bigger than Licklider's had been—at some $15 million per year—most of it was already committed to big-ticket projects such as SDC, Project MAC, and the Illiac IV.)

So there was no way around it. To get the extra money he needed, Taylor was going to have to go to the head of ARPA, the formidable Charles Herzfeld.

Actually, Charlie Herzfeld wasn't all that ferocious in person. Quite the opposite: he was a gregarious, affable bear of a fellow, a University of Chicago–trained physicist who spoke in a rich, deep-voiced Viennese accent and loved nothing better than to kick around ideas. No, Herzfeld was formidable because he was an old Washington hand who knew high-level B.S. when he heard it and wasn't afraid to call it by name. In 1965, in fact, after two years as head of ARPA's Defender antiballistic-missile program, and another two as ARPA's deputy director, he'd taken Bob Sproull's chair as director of the agency—and promptly appointed himself the Pentagon's gadfly. The services were intellectually hidebound, he believed. So he was going to use ARPA's scientific and technological expertise to question conventional wisdom at every level, no matter who was left sputtering. He held his ground arguing nuclear targeting strategy with air force chief of staff Curtis LeMay. And on more than one occasion, he even tried to tell Secretary of Defense Robert McNamara that the U.S. government didn't know what it was doing in Vietnam—not that his warnings had much impact.*

On a day-to-day basis, however, Herzfeld more often found himself dealing

---

* No one could call Herzfeld a dove on Vietnam. After escaping from Nazi-occupied Vienna in 1938 and then witnessing the Soviet takeover of Eastern Europe only a few years later, he had no love of nazism, communism, or any other brand of totalitarianism. Moreover, he was convinced that at some deep level, decision makers from Lyndon Johnson on down really believed they were trying to help the South Vietnamese people resist aggression, a goal he himself could support.

Nonetheless, Herzfeld had a clearer picture than most that the Vietnam involvement was headed toward disaster. One of his initiatives as ARPA director was greatly to expand Project AGILE, a research program on counterinsurgency that had been bubbling along at a low level since the early days of the Kennedy administration. Herzfeld was determined to mount a comprehensive, system-wide effort to understand the counterinsurgency problem as a whole, an effort that was to include an honest attempt to understand the Vietnamese people on all sides. "We studied the Vietnamese culture," he says. "Even the character traits that result from being brought up in a very strict family structure. And we studied the different mind-set that comes from the enormous importance placed on the ancestors, the racial group, and the family. That was much more important than ideology. Among other things, it meant that when you tried to move villagers out of Vietcong-controlled areas, that went over very badly; they wanted to stay near their ancestors."

In addition, says Herzfeld, every time he went over to Southeast Asia himself—ARPA now had field offices in Saigon and Bangkok—he found lower-ranking officials eager to feed him the kind of information that never seemed to make it back to headquarters. "I developed a theory that I was high enough in the pecking order that people wanted to show me stuff," he says, "but not high enough for them to want to lie. That's how I learned about the Vietcong's network of tunnels north of Saigon, which nobody in Washington would believe [existed]."

with an official deception promulgated by President Lyndon Johnson himself, who'd decided in 1965 to escalate the American involvement in Vietnam without raising taxes to pay for it, lest it threaten his new Great Society initiatives. The first waves of U.S. combat troops had landed on March 8, 1965, and more had soon followed; the predictable result was that Pentagon officials were now scouring the Department of Defense to find money. Every agency was expected to do its part, and no program could justify its existence unless it had an immediate, demonstrable payoff for Southeast Asia.

In that kind of environment, under a different kind of director, ARPA's little program in computer research might have vanished without a trace ("American boys are dying, and you're spending money on *mice?!*"). Fortunately, however, Herzfeld was not that kind of director. He did have to ask the various ARPA offices to "help out" wherever they could, which is why one office found itself developing laser-guided smart bombs, another went to work on acoustic sensors for monitoring troop movements on the Ho Chi Minh Trail, and so on. Nonetheless, through sheer bureaucratic bullheadedness, Herzfeld was able to keep such depredations to a minimum. He deeply believed in what ARPA stood for. And most especially, he believed in what the Information Processing Techniques Office was doing in computer technology. After all, he was something of a Licklider disciple himself: the two men had been the closest of friends and allies during Lick's time there.

So when Taylor came to Herzfeld asking for cash to start his network project, the ARPA director didn't need much persuading. "OK," he said after about fifteen minutes of Taylor's pitch. "You've got it. How much money do you need to get off the ground?"

Um—a million?

"No problem," said Herzfeld, with the decisiveness of a man who commanded a $250 million budget, of which maybe 10 percent was still fluid enough to dispose of as he pleased. "Go to it."

As Taylor left Herzfeld's office, he looked at his watch. The ARPA director took pride in the notion that someone with a good idea could get a million dollars within a day. Taylor had just gotten *his* million in twenty minutes.

He had no trouble finding uses for the money. If nothing else, it provided instant credibility. Now, when the various PIs asked the obvious question—"Who's going to pay for this thing?," meaning, "How much are you going to carve out of *my* funding?"—Taylor could point to that brand-new $1 million and say, "I will." The network wasn't going to cost them a cent.

More substantively, meanwhile, the new money allowed Taylor to start building an organization to run the project. First, to handle the contracting side of things, he hired the estimable Al Blue from the ARPA director's Program Management Office. "Al was wonderful," says Taylor. "He was interested in computing, and he knew the contracting stuff cold. So he kept this whole process running relatively smoothly for us." Simultaneously, Taylor was looking for someone to take the lead on the technical end: a program manager with the expertise, the credibility, and the clout to design the network and push it through. Indeed, he already knew exactly who he wanted: Ivan Sutherland's former Lincoln Lab officemate, Lawrence G. Roberts.

Sutherland himself rated Larry Roberts as one of the smartest people he'd ever met. And indeed, to most observers it was a toss-up as to which of the two young men was the greater genius at computer graphics. As an MIT undergraduate in the late 1950s, for example, Roberts had discovered the TX-0 shortly after Wes Clark and his group sent the machine down to campus in 1956, and had promptly set to work devising a program to recognize handwritten characters using what would now be called a neural-network algorithm. It had worked quite well, considering the limitations of the hardware, and had served as the subject of Roberts's first published research paper.

After moving on to Lincoln Lab, where he was a staff associate during his graduate-student years—and where he shared that office with Sutherland—Roberts used Wes Clark's TX-2 computer to experiment with compression algorithms for digital images. Those algorithms would later be used by NASA for space probes designed to send back images from the Moon, Mars, and other planets. For his Ph.D. dissertation on the perception of three-dimensional objects, Roberts likewise used the TX-2 to formulate pathbreaking algorithms for the representation and display of 3-D solids. Along the way, moreover, because it seemed to need doing, Roberts endowed the TX-2 with a time-sharing operating system that would remain in use for many years thereafter. And then in the summer of 1962, when the departure of Clark and his coworkers on the LINC project left the Lincoln TX-2 group in limbo, Roberts filled the void. "I never officially got appointed," he recalls. "On paper the group was virtually leaderless. But even though I was still a graduate student, I took over. I wound up going to ARPA and getting funds from Ivan for a project in graphics."

Indeed, Roberts's group at Lincoln Lab quickly emerged as the leader of Sutherland's new computer-graphics initiative, much as Project MAC had become the flagship for ARPA time-sharing. Among other things, Roberts and his colleagues worked on what is now known as data visualization—the displaying of large masses of statistical data in visually meaningful ways—as well as graphical programming, or controlling a computer through drawings. (After Sutherland left ARPA for Harvard, moreover, Roberts would start a collaboration with him on what would now be called virtual reality, complete with the world's first 3-D virtual headset.)

Ironically enough, it was the graphics research that finally triggered Roberts's shift into networking—graphics, that is, plus a timely push from J. C. R. Licklider.

The occasion, Roberts explains, was another air force–sponsored meeting at the Homestead resort in Virginia: the Second Congress on Information System Sciences, in November 1964. One evening, after the formal sessions were over for the day, he found himself sitting around in a late-night bull session with

Lick, Fernando Corbató, Alan Perlis, and maybe a few others. "The conversation was, what was the future?" he says. "And Lick, of course, was talking about his concept of an Intergalactic Network." It was an idea Roberts was ready to hear: "You see, at that time, Ivan and I had gone farther than anyone else on graphics. But I had begun to realize that everything I did was useless to the rest of the world, because it was on the TX-2, and that was a unique machine. The TX-2, Corby's CTSS, and so forth—they were all incompatible, which made it almost impossible to move data. So everything we did was almost in isolation. The only thing we could do to get the stuff out into the world was to produce written technical papers, which was a very slow process."

But now, says Roberts, as he listened to Lick continue his rhapsody about digital communications, he became convinced that *this* was the future. "It seemed to me that civilization would change if we could move all this stuff [over a network]. It would be a whole new way of sharing knowledge. And that, to me, seemed the new challenge."

Roberts came away from the Homestead meeting determined to shift the focus of his research from graphics to networking. And in short order, his former officemate, then still at ARPA, handed him an opportunity to do so. "Now Ivan became excited about networking, too," says Roberts, "because he had worked with Lick. So he commissioned me and Tom Marill to do the study about link-ing the TX-2 at Lincoln to the Q-32 at SDC. We did the experiments in nine-teen sixty-five, and they showed that we could get the computers to work with each other. But at the same time, they showed that the [dial-up] telephone sys-tem was very slow and unreliable, and not an effective way for the computers to work together. So that was the problem which I then started thinking about."

Down in Washington, meanwhile, Sutherland's successor was thinking about Roberts. "Larry," Bob Taylor said when he made the call in the autumn of 1966, "I want you to come down to ARPA and be a program manager for this net-working project."

Hmm. Roberts was conflicted. On the one hand, Taylor's offer would mean he'd have the chance to build a network for real—no more of these small-scale, pilot-project efforts. But on the other hand, going to work for Taylor would mean he'd have to put aside all his research at Lincoln Lab, research he truly en-joyed. And worse, it would mean, well, going to work for Bob Taylor.

Roberts had more reason than most to consider Taylor an "interim" director of IPTO. Earlier that year, when Ivan Sutherland had been planning his depar-ture from that post, he had first asked Roberts to take over the office. "Ivan had concluded that the only person who could replace him technically was me," he says. He'd said no because he wanted to stay at Lincoln, which had left Suther-land with Taylor as his distant second choice (a fact unknown to Taylor himself, apparently). Having turned down the top job in the spring, Roberts decided that he wasn't about to reverse himself in the fall just to go be an underling to a chair-warmer. So he gave Taylor the same answer he'd given Sutherland: no.

Taylor, however, was as Texas-stubborn as ever. "So Larry turned me down," he says, "and I would think of other people to get instead. But nobody really satisfied me, so I'd go back to Larry. He turned me down again. This happened frequently. And then one day in about September or October of nineteen sixty-six it dawned on me. So I went to see Herzfeld, and I said, 'Doesn't ARPA still support fifty-one percent of Lincoln Lab?' "

ARPA did indeed. Herzfeld made the phone call on the spot. And in short order, the director of Lincoln was pointing out to Roberts that a stint at ARPA might be a very good career move for him.

That appeared to resolve Roberts's inner conflict. Within two weeks of Herzfeld's phone call, he said yes. And by the end of the year, he would report for work at the Pentagon.

Of course, there was a price. Roberts would build Taylor's network for him, all right. But he had also decided to consider Taylor's offer an extension of Sutherland's. Whatever his official title—Roberts was now the "chief scientist" of IPTO and was *not* reporting directly to Taylor; the agreement was to list him instead as a special assistant to the deputy director of ARPA itself—he felt that he was being groomed to replace the caretaker. "Bob was never the technical lead the office should have had," he says. Taylor was welcome to keep on doing the paperwork—but only until Larry Roberts had learned the ropes.

That view was widely shared by the research community, where news of Roberts's appointment was greeted with sighs of relief. "Larry was the obvious successor to Ivan Sutherland," says one ARPA principal investigator. "So the fact that Larry was called chief scientist made things OK." A similar view was taking hold in Washington, where Roberts was already becoming known as the fastest man in the Pentagon. Intense, crew-cut, dressed in a white shirt and tie, he could most often be seen striding through the labyrinthine corridors along the quickest possible route from point to point. (Rumor had it that he had scouted out the building with precisely that in mind.) He had the air of a man who knew precisely where he was going. And because of that, not surprisingly, a lot of people just assumed that he already *was* the director.

But of course, he wasn't. Seeing the deference being given to Dr. Roberts must have been a particular kind of purgatory for Mr. Robert W. Taylor. Nonetheless, Taylor was determined to make this work. It was a testament to his management skills that he endured the situation without ever saying a word in public. Neither he, nor Roberts, nor anyone else describes their working relationship as being anything but cordial. Taylor was willing to let Roberts and the rest of the world believe anything they wanted—so long as the network got built.

STORE AND FORWARD

It was definitely a tough audience, says Larry Roberts, thinking back to March 1967 and the last day of the IPTO principal investigators' meeting at the University of Michigan, when he got up to introduce the agency's new networking plan. Out of the twenty or thirty people in the room, no more than about half a dozen were actually interested in the project; the rest were just as dubious as ever. "Although they knew in the back of their mind that it was a good idea and were supportive on a philosophical front," says Roberts, "from a practical point of view, they each wanted their own machine."

Still, this was what IPTO was going to do.

In working out the basic architecture of the network, Roberts told them, he'd been heavily influenced by the experiment in 1965, when he and Tom Marill had linked the TX-2 at Lincoln with the Q-32 in Santa Monica. First, he said, the experiment had convinced them that AT&T's telephone system, while beautifully optimized for voice communications, was all wrong for digital communications. Unfortunately, ARPA was pretty well stuck with using that system, since nobody was going to give the agency a few billion dollars to string its own lines across the country. So in order to make it work, they would somehow have to circumvent the system's basic architecture, starting with that blasted dial-up process.

Basically, the 1965 experiment had tested a refinement of Lick's original conception of the Intergalactic Network, which was really just time-sharing writ large. "In those days people were doing time-sharing by logging in on dial-up telephone lines at a hundred and fifty to three hundred bits per second," Roberts says. "What we wanted was to set up an automatic dialer—which was very rare then—so that when the TX-2 needed a matrix multiply [a mathematical operation], for example, it could dial up the Q-32 and get the answer, just like calling a subroutine from its own hard disk."

In practice, however, the automatic dial-up was a train wreck. On a human time scale, the one or two seconds required to make a long-distance connection were negligible. But on a computer time scale, those seconds passed like geologic ages. Even with the TX-2 and the Q-32 making the connections automatically, said Roberts, "the system was so slow, with so many restrictions, that we just couldn't do all that we would have liked."

Fortunately, he added, this was one problem that could be solved through the simple application of money: in effect, ARPA would make a series of massive long-distance calls and just never hang up. More precisely, the agency would go to AT&T and lease a series of high-capacity phone lines linking one ARPA site to the next. A diagram of the resulting network would thus look something like a road map of the interstate highway system, with the leased lines corresponding to the highways, and the various ARPA computers that sat at the intersections or "nodes" of the network corresponding to the cities. Dial-up delays would be nonexistent in such a scheme because the computers would always be connected.

A second conclusion was a bit more esoteric, said Roberts, but less of a surprise: digital messages could not be sent through the network as a continuous stream of bits. Instead, they would have to be broken into segments, with some fixed number of bits in each one (think of a long letter written on a series of postcards).

The problem was noise, he explained. Continuous signals were fine for voice communications, and even for logging into a nearby time-sharing computer through a modem. But continuous signals were not so effective for delivering bits across the country. The farther a message traveled, the greater the chances that one or more bits would be garbled by static and distortion on the line. And in the digital world, one erroneous bit might easily spell disaster.

Thus the digital postcards, or "packets," in modern parlance. While they wouldn't eliminate the noise, they would isolate the errors and give the sender a chance to correct them. Specifically, the sender could have his or her computer label each outgoing packet with a special "error-checking" code, which would raise the digital equivalent of a red flag if any bits got corrupted in transit. That would allow the computer at the other end to sort through the incoming packets, throw out the ones that had been flagged as having garbled bits, and then signal the first computer to send new copies. This process would be repeated until all the incoming packets were deemed acceptable, at which point the receiving computer would string them together and reconstruct the original message.

Complicated? Perhaps, said Roberts. But the basic techniques were well understood; people had been segmenting the data on tape drives and other error-prone media for more than a decade. And more to the point, the use of packets in networking had been thoroughly analyzed in the 1962 Ph.D. thesis of Roberts's MIT classmate Leonard Kleinrock,* who was now at UCLA. He and Kleinrock had discussed the issues extensively when he was planning the 1965 experiment, Roberts said. And the experiment itself had proved that the packet idea would work: the packets arriving on the other side had been reconstructed quite well.

Now, those first two conclusions were comparatively straightforward, said Roberts. Taken together, however, they led to a third issue that was not straightforward at all: routing, or getting the data packets where they were supposed to go.

In one sense, of course, this problem was self-inflicted. If they'd stuck with the dial-up connections, the telephone system would have given the packets only one path to follow. That was what it meant to dial a number: in effect, those pulses from the spinning dial were asking the telephone system to create a complete circuit between your telephone and the one you were calling. (By the mid-1960s this was being done by automatic switches, but in the old days you

---

* Indeed, since Kleinrock was using the TX-2 to simulate networks at the same time Roberts was using it for visual recognition and Sutherland was working on Sketchpad, the three of them constituted a kind of TX-2 triumvirate. Among other things, they all presented the oral defense of their Ph.D. theses on the same day, before the same faculty committee, made up of Claude Shannon, Wes Clark, and Marvin Minsky.

272 THE DREAM MACHINE

literally asked; the connections would be made by a telephone operator sitting at a plug board.) The circuit was a temporary one, to be sure; it would disappear as soon as you hung up and the switches were reset. But until that happened, the circuit would lead voice or packet straight to its destination. The packets might be garbled, but they would not get lost.

However, once you abandoned the telephone system's "circuit switching" architecture for the network's open-highway model, the packets most assuredly *could* get lost, and be left to wander through the network forever.

One way to keep that from happening was a kind of Grand Central Terminal strategy: connect every machine on the new ARPA network to one gigantic computer in, say, Nebraska. That way the packets could flow straight in, get sorted, and then flow straight back out to their respective destinations. The "highway map" of the resulting network would just be a continent-sized star, with all the rays converging on that one central point.

This Grand Central Computer scheme had a glaring drawback, however: one blown transistor and the machine could go down, taking the whole network with it. For that reason, said Roberts, his preferred alternative was a digital variation on the old "store and forward" technique used in the early days of telegraphy. (Telegraph lines were so noisy that transmission was limited to short hops; the clerks in each telegraph office had to write down messages as they arrived, then key them in again to send them farther down the line.) The idea was to keep the complex, highway-map structure of the network, with lines running every which way and an ARPA site sitting at each intersection. But the individual sites would share routing responsibilities equally. That is, the computer at each site would first read the digital address on each packet as it came in. (The packets would actually carry the digital equivalent of an entire bill of lading, with destination address, return address, error-checking codes, message identifiers, and so forth.) Then the site computer would either accept the packet, if the address was local, or send the packet off again on the next stage of its journey. The result would be a network that operated collectively, with little or no central control.

So in sum, said Roberts, that was the plan: full-time access, messages divided into packets, and distributed control. Now, who wanted to help?

It was a delicate moment, Roberts remembers. ARPA *was* going to build this network, and these guys *were* going to use it. To focus their attention, in fact, he and Taylor had made it very clear that the PIs weren't going to get one more dime out of ARPA for new computers until the computers they already had were hooked into the network and fully utilized by the community as a whole. Nevertheless, he didn't dare come across like a dictator, not with this crowd. They'd have told young Dr. Roberts exactly where to stuff it, funding or no funding. So instead, he says, he tried to hold out a carrot. "We invited them to work on both sides of the problem. Not only were their computers going to be involved, but those PIs who had ideas for contributions they could make to the network were encouraged to participate."

He had no idea how well his carrot was going to work. As he started taking questions, most of the PIs in the room seemed as dubious about the ARPA network as ever, if not more so. And a few of them, starting with Marvin Minsky and John McCarthy, were happy to tell him why. Their computers were straining to keep up with local time-sharing demands as it was, the two AI researchers pointed out at length, supported by many others. And now Larry Roberts wanted the machines to handle packets, too? How were they supposed to get any work done? And speaking of work, who was going to do all that programming? Running packets straight through everybody's computer meant that they were going to have to do major surgery on their operating systems. How many years would *that* set them back? Furthermore—

Good questions all. Still, the audience wasn't actively hostile, which was something. And in a few cases—Doug Engelbart, for example, marching as always to drumbeats that only he could hear—the reaction was even quite the opposite. "I was thrilled," Engelbart remembers. "We had just been given the OK from ARPA to bring up a real time-sharing system at SRI, which meant that we could finally get multiple people working together. So at that point we were really oriented toward this 'groupware' idea." The new system would consist of a brand-new computer running a copy of the Project Genie time-sharing software developed at Berkeley. With it, Engelbart says, he wanted to explore on-line collaboration at every level. He envisioned on-line reference libraries, on-line address books, on-line technical support services, on-line discussion centers, on-line transcripts of decisions and debates—a massively cross-linked record of all the collective knowledge amassed by the collaborators over time. And as usual, his imagination didn't stop with the different groups in an individual workplace. "My whole thought was, How do we get a *big* community connected to experiment with? Well, here we were at this meeting in Michigan, and oh gosh, they're going to make one! The other guys were negatively excited, but [it] was very exciting to me." So after listening to the grousing for a while, Engelbart got up and volunteered to make SRI the home of groupware—or, as it was later known, the Network Information Center (NIC)—for the net as a whole.

Meanwhile, the network's many critics around the room were somewhat mollified by a promise from Roberts and Taylor that ARPA would give them enough additional computer power to keep anyone from being handicapped because of the network. ("Of course," notes Roberts, "I didn't promise *where* that computer power would be"—meaning that people might have to access it through the network itself.) Then, too, Roberts's invitation to participate in the creation of the network was having its intended effect: the PIs were becoming intrigued with the technical challenge, almost in spite of themselves. By the time the session was drawing to a close, the debate was not about *whether* they should do the network, but about *how*.

Out in the audience, in fact, Wes Clark was beginning to wish they would all just shut up. "I was kind of bored," he remembers. "The details they were going into were a bit abstruse; in fact, I just didn't follow it all. But toward the end, just before we broke up, I do remember suddenly realizing what the meta-problem was. Well, I didn't want to keep the discussion going. I wanted to go home. So I passed Larry Roberts a note saying that I thought I saw how to solve their problem. He collared me as we left the meeting and wanted to hear about it." After all, Wes Clark had been Roberts's mentor on the TX-2, and the younger man had always found it wise to listen very carefully to what the older one had to say. So since they all had planes to catch, Roberts hustled Clark into a taxicab to the airport, along with Bob Taylor, Al Blue, and Dave Evans; they could talk on the way.

Clark's point was simple, once he'd caught his breath: all the PIs back there had been upset about the problem of programming their machines for the network because it really *was* a problem. Roberts was proposing to run the packets right through their main computers, which was like trying to run an interstate highway right through the main street of every little town in its path. All you'd get that way would be havoc in the towns and nonstop traffic jams on the highway. "That was the wrong way to go about it," Clark remembers saying. "So my idea was simply to define the network to be something self-contained." That is, make the ARPA network into the digital equivalent of a limited-access highway, with an "interchange" located just outside each town. Each of these digital interchanges would actually be a small computer, of course, separate from the main computer. But like its asphalt counterpart, the digital interchange would handle all the routing chores. It would provide an on-ramp to the network for new packets coming out of the main computer; an off-ramp for incoming packets addressed to the main computer; and traffic directions for packets passing through on their way to other computers.

Now, said Clark, the beauty of this scheme was that it would simplify life for everybody. ARPA could take responsibility for designing and implementing the network proper—meaning the information highways and the digital interchanges—without having to worry that some contractor somewhere would mess up his site's programming and thereby bollix up the whole system. And the contractors, for their part, could focus on one comparatively simple task—establishing a link from their central computer to the routing computer—without having to worry about all the ins and outs of all the other computers on the network.

So, said Clark, that was the idea: small, independent routing computers.

Roberts liked it. And not for nothing was Larry Roberts known as the fastest man in the Pentagon. By the time they got to the airport, his decision had been made: the ARPA network would be based on small routing computers à la Wes Clark. Interface Message Processors, they came to be called, or IMPs.

Over the next six months, with input from Len Kleinrock, Dave Evans, and a number of other enthusiasts, Roberts crafted a preliminary network plan containing specifications for the response time, the reliability criteria, the number of bits in each packet and their layout, the transmission speed in bits per second—all the engineering nitty-gritty. He then gave the initial plan its debut presentation in

October 1967, at the ACM Symposium on Operating System Principles in Gatlinburg, Tennessee. And as he sat down afterward, he had reason to feel satisfied: this was something genuinely new in the world, a way to organize communications that was radically different from anything that had gone before.

Except that shortly thereafter, an Englishman named Roger Scantlebury got up to give a paper on a system being developed by Donald Davies's telecommunications research group at the National Physical Laboratory (NPL) in Teddington, outside London—and proceeded to describe essentially the same idea: packets, IMPs, distributed control, the works.

What the . . . ?

The story, as Scantlebury would explain it to the disconcerted Roberts later that day, was both ironic and sad. The irony was that Donald Davies had gotten his original inspiration when he was hosting a conference on time-sharing back in late 1965 and fell into an impromptu discussion about networking with J. C. R. Licklider and Roberts himself. Almost immediately after that, Davies had been struck by the notion that a store-and-forward system with very short message segments would be perfect. By June 1966 he had expanded this idea into a formal proposal, calling for a United Kingdom–wide digital network running at speeds of up to 1.5 megabits per second. This paper was where he introduced the concept of an "interface computer," the equivalent of Wes Clark's IMP, and coined the term "packet" to describe the message segments (at that point Roberts and his ARPA colleagues were still calling the segments blocks). Davies's name for the scheme as a whole was "packet switching."

From there, said Scantlebury, Davies and his group at Teddington had continued to develop the packet-switching idea with computer simulations. They had even scraped together enough money to build a "one-node" network, consisting of a single Honeywell computer connected to a lot of terminals through a special interface. It wasn't much, admittedly. But it did demonstrate the switching principle: you could type in text on one terminal and have it print out on any other terminal you specified.

And that, explained Scantlebury, was the sad part of the story: the powers-that-be at the British Postal Service, which had absolute control over the U.K. telecommunications system, had flatly refused to fund Davies's vision of nation-wide packet switching. They couldn't even see the point of a demonstration. So, said Scantlebury, having gotten there first, the NPL group would now have to sit back and watch as the Americans did a packet-switched network for real.

That hurt—and Roberts could certainly sympathize. Still, the frustrations weren't personal. Scantlebury and his companions from the NPL group were happy to sit up with Roberts all that night, sharing technical details and arguing over the finer points. Take this business of the data rate, Scantlebury said at one point: why on earth are you bothering with a crummy 9,600 bits per second? Why not slam them through at ten times that rate, or a hundred times? The phone lines could handle it.

Because 9,600 was what ARPA could afford, replied Roberts. Because—

Hmm. "I had been thinking of doing the network with 9.6-kilobit lines because that was what worked in terms of the traffic levels we expected," says Roberts. "But the British made me realize that I had not looked at the question seriously. So [after getting back home] I started thinking about the possibility of what the phone companies called a 50-kilobit line. The idea was that you bought a very expensive modem that tied twelve lines together. Stepping up to that would be a big expense. But I did the numbers over and over for myself, and it looked like it would trade off economically: you could get more through the higher-speed lines, so you would not have to buy as many, and you would get better response time. So that was a good idea."

Once again, the fastest man in the Pentagon made his decision without hesitation: ARPA's network would run at 50 kilobits per second (later upped to 56 kilobits).

Meanwhile, Scantlebury had told Roberts one other thing in Gatlinburg. While the NPLers were doing their homework, they had come across the work of an American who had—again—invented the packet network independently. Baran was his name (pronounced to rhyme with "Sharon")—Paul Baran, out at RAND.

Oops. Roberts knew Baran slightly and had in fact had lunch with him during a visit to RAND the previous February. But he certainly didn't remember any discussion of networks.

How could he have missed something like that? As soon as he got back to the Pentagon, Roberts dug out the office copy of Baran's final network design, which had been circulated as a RAND report in August 1964, just before Roberts's own shift to networking. The "report" proved to be a small library: eleven volumes that covered everything from routing algorithms to estimated cost. (Two other volumes, dealing with cryptography and vulnerabilities to the network, were classified.) Reading through it, Roberts found that Scantlebury was right. True, his plan and Baran's did have many differences in technical detail, stemming largely from the fact that Baran had put more emphasis on digital voice communications than on computer communications. Moreover, their motivations were very different. Whereas Roberts was designing a civilian network that would connect the ARPA research centers, Baran had been after a nationwide command-and-control system that could survive a thermonuclear war.* But Roberts

---

* The motivation was not to improve the Pentagon's ability to fight such a war, but rather to lower the likelihood of its having to. Vulnerable communications systems were highly destabilizing, went the argument: a commander in chief who feared the sudden loss of communication would feel obligated to fire off every weapon he had at the first sign of attack, lest they quickly be rendered useless. Conversely, a commander in chief blessed with survivable command and control could afford to wait it out, see what developed, and make some effort at a measured response. Indeed, Baran and his colleagues even advocated sharing the packet-switching technology with the Soviets, on the grounds that having survivable communications on both sides would be the most stable configuration of all.

could see that the essence of Baran's network—packets, a decentralized architecture, computer routing—was the same as his.

So why hadn't Baran's plan been adopted already? Because it was too far ahead of its time, apparently. AT&T engineers, most of whom had spent a lifetime perfecting their circuit-switching network, found Baran's packet-switching concept ludicrous ("Son," Baran remembers one telling him with exaggerated patience, "this is how a telephone works . . ."). Worse, Pentagon politics dictated that the network would have had to be implemented by the newly organized Defense Communications Agency, which was also staffed by old-line telephone engineers and which simply did not have the technical competence to pull it off.

Baran's network had accordingly fallen into limbo, on the grounds that *no* implementation was better than a *bad* implementation. Baran himself had gone on to do other things. And there the matter had rested—until the ARPA project came along.

Once again, Roberts could sympathize. He'd been encountering a lot of the same kind of flack himself, and often from the same people. "The Defense Communications Agency people around me at the Pentagon basically said I was crazy," Roberts remembers. "They even stood up in meetings when I made speeches, and booed and hissed and made nasty comments. You know, 'The buffers are going to run out,' or 'There's just no way this can work.' People just were caustic, because they could not get their minds into a new focus."

However, Roberts did have one enormous advantage: he could ignore what he called the channelized thinkers, referring to their fixation on the circuit-switching paradigm. Unlike Davies, he didn't have to work through the British Postal Service. And unlike Baran, he didn't have to work through the Defense Communications Agency. Roberts was backed by ARPA, whose whole reason for existing was to cut through the bureaucracy. His bosses were giving him a free hand. And he meant to exercise that freedom.

He meant to get this network ready to *go*.

WAITING FOR THE GROWN-UPS

Of course, Roberts was making a big assumption here: namely, that ARPA's computer research would continue as presently planned—or that ARPA would continue to exist at all.

Neither of those things was a given, not with more and more American troops headed for Vietnam every day. And especially not after March 1967, when IPTO's most ardent protector, Charlie Herzfeld, abruptly stepped down as ARPA's director.

Herzfeld wasn't forced out, exactly. But he had recognized that it was time for him to go. For one thing, his brand of gadfly activism had left too many gen-

erals and admirals fuming. For another, ARPA's support of university-based research, at a time when half the campuses in the country were erupting in protest
over Vietnam, looked suspiciously like giving aid and comfort to the enemy.
And perhaps most important, quite a few people out on the Pentagon's E Ring—
not to mention on Capitol Hill—were wondering exactly what this agency was
for. There was a general feeling that the independent, freewheeling ARPA that
had flourished under Ruina, Sproull, and Herzfeld hadn't actually achieved very
much for the Department of Defense. It had sponsored great technology, certainly, but too much of it was, well, academic. The ideas just weren't making it
out to the services for deployment, at least not at a rate that justified the expense. And it didn't help that some of ARPA's projects, such as behavioral science, had become extremely controversial. (Lick's successor in that office had
launched an ill-advised "Camelot" program to study foreign cultures, and in so
doing elicited screams of protest not only from university social scientists, who
saw it as an imperialist plot, but from the barons of the State Department, who
saw it as an intrusion on their turf.)

The upshot was that by the beginning of 1967, Secretary of Defense Robert
McNamara had visibly lost confidence in ARPA. His deputy secretary, Cyrus
Vance, was in favor of shutting it down altogether, and even its strongest defender within the secretary's office—John Foster, DDR&E since 1965—had concluded that the agency needed a complete overhaul. Add to that a fierce
argument over Foster's plan to transfer the Defender antiballistic-missile program to the army—Herzfeld wanted to keep it; Defender accounted for half his
budget—and yes, it was definitely time for Herzfeld to go.

So he went, a departure that marked the beginning of years of turmoil for the
agency. First it was plunged into an immediate state of bureaucratic limbo: since
Herzfeld had left no obvious heir, ARPA would have to function under an acting director for the next seven months. And when Herzfeld's successor finally
did take office, in November 1967—the post went to the industrial engineer
Eberhardt Rechtin of NASA's Jet Propulsion Laboratory—he quickly set about
killing programs deemed expendable, and putting pressure on the remainder to
be more "relevant." Rechtin's basic strategy was for the agency to keep a low
profile, to build allies among the services, and to give them technology they
could use. Operational technology.

That wasn't a category that included much on IPTO's agenda—not Project
MAC, not Engelbart's center, and certainly not the network. With the technical
design for the network still in draft form at that point, the entire effort could easily have been stopped right there, leaving Larry Roberts just as thoroughly stymied
by the bureaucracy as Donald Davies and Paul Baran had been before him.

And yet somehow it didn't happen. ARPA's computing program continued
to lead its charmed life, rather like a person sleepwalking through a battlefield
without getting a scratch. One reason was DDR&E Johnny Foster, who resisted
any and all attempts to focus ARPA's efforts purely on Vietnam. The DoD had

other concerns during this period, after all, starting with the Soviets' rapidly expanding nuclear capability. Underlying the strategic balance, Foster knew, was the technological balance. And since ARPA had been founded precisely to help maintain that balance, it couldn't be allowed to degenerate into a counterinsurgency office.

If that provided some cover for computing, however, what provided even more was that Rechtin didn't particularly care about the program one way or the other. He was much more interested in ballistic-missile defense, nuclear-test detection, and counterinsurgency—the big-ticket, applied-research ARPA programs that were ready to be transferred to the services. The smaller, longer-term, basic-research programs—a group that included behavioral science, materials science, and computing—were left to Rechtin's deputy director, the former MIT physicist Stephen J. Lukasik. And it just so happened that Steve Lukasik was fascinated by computers. "I was a grad student at MIT when Whirlwind was built," he explains, "and I knew fellow students who were using it. Then I went to Westinghouse and I used computers to do reactor calculations. I learned Fortran at the Stevens Institute of Technology, and I used computers in my research. So I've always had a natural affinity and inclination for that sort of thing."

It also just so happened that in Lukasik's earlier incarnation at ARPA, as head of the VELA nuclear-test-detection program, his office had been right across the hall from IPTO. "So those were the people that I ran into and talked to," he says. "And when you mixed in that circle, you were dealing with some of the smartest people in the agency."

Certainly Lukasik was intrigued by the network project, albeit for reasons that weren't quite the same as Taylor and Roberts's. "Why did ARPA build the network?" Lukasik asks. "There were actually two reasons. One was that the network would be good for computer science. We were going to share files, wire ourselves together, and so forth. Now, this rationale leads in a more or less linear way to the Internet as we understand it today. And it was true: this was by far the dominant reason among the researchers. But there was also another side to the story, which was that ARPA was a Defense Department agency. And after Eb [Rechtin] came in, defense relevance became the dominant notion. Everybody was writing relevance statements. Besides that, we did not have a lot of money. Johnson was funding the war without doing anything different on the surface, which meant that every time you turned around, they were raiding you for another ten to fifteen million dollars. So in that environment, I would have been hard pressed to plow a lot of money into the network just to improve the productivity of the researchers. The rationale just wouldn't have been strong enough. What *was* strong enough was this idea that packet switching would be more survivable, more robust under damage to the network. If a plane got shot down, or an artillery barrage hit the command center, the messages would still get through. And in a strategic situation—meaning a nuclear attack—the president could still communicate to the missile fields. So I can assure you, to the extent that I was signing the checks, which I was from nineteen sixty-seven on, I was signing them because *that* was the need I was convinced of."

Now again, Lukasik emphasizes, both explanations were true. In fact, they were complementary: "The best way to develop the network for the Defense Department was to develop it for the IPTO guys," he argues. It was a notion he pushed hard whenever he was making the case to Rechtin, Foster, and the other higher-ups. "So, how do you build a technology like this?" he would ask them, echoing Taylor's original thinking. "Just build some phony terminals and send dummy traffic around? That's pretty stupid. You don't really test a network until you have real users. But who? You aren't going to have real military users fiddling around with this stuff, because they're too busy doing real missions to worry about testing things for ARPA. Oh! Brilliant management move! We will use our principal investigators. They will generate a lot of traffic. They will do all the screwball things real users do. And they have a lot of motivation to solve all the problems, because *their* operation is going to be enhanced. So it's really a very interesting bootstrap management scheme."

Rechtin and the higher-ups bought it. At a time when most other ARPA offices were being cut, restructured, or spun off—ballistic-missile defense was finally transferred to the army in 1968, instantly cutting ARPA's overall budget by half—IPTO's budget steadily, miraculously kept rising. By the end of the decade it would stand at something like $30 million a year, double what it had been when Taylor took over. Indeed, says Taylor, his budget reviews with Rechtin and Lukasik were a walk in the park. "Steve would take the lead," he says, "and Eb would just sit in the background, occasionally asking a managerial question. But it was almost pro forma. I can recall defending a twenty-five-million-dollar budget in a thirty-minute meeting. We were doing important stuff, and when we needed money, we got it."

Out on the university campuses, meanwhile, the success of the ARPA network was resting on another big assumption: that the ARPA computer community would last long enough for anybody to use it.

This was also not a given, not in the sixties. In a decade that was justly famous for turmoil, the years that brought the ARPA network into being—1966 through 1969—were enough to make one wonder what shreds of the social fabric could possibly survive. The summer of 1967, when Larry Roberts was drafting his preliminary plan for the network, was the Summer of Love for the hippies in San Francisco's drug-soaked Haight-Ashbury district. But it was also a Long, Hot Summer of rioting in the black ghettos of Newark, Detroit, and more than a hundred other cities; the National Guard was called out for the first time since World War II. October 1967, the month when Roberts announced his network plan in Gatlinburg, was also a month that saw some fifty thousand antiwar demonstrators rally at the Pentagon, where protesters in psychedelic face paint stuck flowers in the gun barrels of guardsmen sent to control them. By that December, some 486,000 U.S. troops were in Vietnam; nine thousand had been killed in that year alone, and many others were mired in the siege of Khe Sanh. And 1968, famously, was even worse, from the Tet Offensive on January 30 to the assassination of Dr. Martin Luther King, Jr., on April 4 to the assassination of Senator Robert F. Kennedy on June 5.

Taking these events together, the era was like a giant primal scream, the sound of a nation tearing itself apart. On the face of it, there seemed every reason to believe that the ARPA community, too, might tear itself apart. ARPA was an arm of the Department of Defense, after all, in a period when academics had begun to use the word *Pentagon* as an expletive. And while it was true that computer researchers in general were notoriously apolitical, they were hardly unconcerned about the war. Many of the students were eligible for the draft, if nothing else, and at least some of them were vehement activists. The question was heard again and again: Is it right to take this money?

And yet the answer was almost always yes. To senior researchers such as Fano, Newell, and Feigenbaum, IPTO was not part of some abstract military machine, but an extension of the research community itself. Licklider, Sutherland, Taylor, Roberts, even Lukasik—these guys were peers, colleagues, people they had known and worked with for years.

The students, meanwhile, felt much the same way. Given the nature of interactive computing, in fact, even the fieriest activists among them could argue that *they* were co-opting the Pentagon, and not vice versa. They were striking a blow for intellectual freedom. They were liberating human potential. And not incidentally, they were liberating a few dollars that might otherwise be spent on B-52s. "That was the lie we told ourselves," recalls a wry Bob Metcalfe, then an undergraduate computer-science major at MIT: "Our money was bloody on only one side."

Then, too, many of these kids were nothing if not escapist. The obsessive techno-weenies who haunted the terminal rooms of Project MAC and the other ARPA sites were children of the sixties, all right. But the worlds they were creating for themselves had less in common with the angry antiwar movement out on the streets than with the psychedelic, peace-and-love communes of Haight-Ashbury. Their passion for a newly popular epic fantasy called *The Lord of the Rings* was equaled only by their obsession with science fiction—and with the ubiquitous computer game Spacewar, which sprang into existence wherever there was a PDP machine to run it on. For most of them, "ARPA" didn't mean "Pentagon." It was more like a magic word that opened a door into the Land of Faerie itself, a magical realm full of machines that could do anything their imaginations could contrive, while providing a refuge from the all-too-real tumult around them.

In any case, the upshot was that the ARPA community stayed together through the Vietnam era—with the rising generation in particular beginning to coalesce as never before. Not only were they not alienated by ARPA's Pentagon connection, but they were becoming exactly the kind of community that Lick had envisioned from the beginning.

Of course, it didn't hurt that Bob Taylor was doing his behind-the-scenes best to foster that development.

The idea had come up in late 1967, Taylor remembers, when the principal investigators were meeting in Alta, Utah. Now, the meeting wasn't held there because it was ski season, he hastens to add—or not entirely, anyway. Mostly it was because of Alta's proximity to Salt Lake City and the University of Utah, where Dave Evans was rapidly building his new computer-science department into a powerhouse. Indeed, Taylor had already given Utah formal designation as a Center of Excellence for graphics. "So anyway," says Taylor, "in the Alta meeting I invited some of the Utah grad students to observe. Then at the end, I asked the grad students for suggestions. It was John Warnock who asked, 'Why not have a meeting for the grad students the way you do for PIs?' Well, I thought, That's a great idea!"

In short order Dan Slotnick, head of the Illiac IV project at Illinois, had volunteered to host the meeting at that university's Allerton conference center, outside Champaign-Urbana. Each of the contractors had picked out two of its best students to send to the meeting, at ARPA expense. And the master-of-ceremonies role had fallen to the twenty-four-year-old Barry Wessler, an electrical-engineering student who had come down from MIT in the fall of 1967 to serve as Taylor and Roberts's assistant in IPTO. ("I said I wouldn't go [to Allerton] because I was over thirty!" jokes Taylor, who was then an old man of thirty-six.)

The grad-student conference started in July 1968, with no one knowing quite what to expect. "It turned out to be a really amazing group of people," recalls Steve Crocker, who was just in the process of returning to UCLA after eighteen months in Minsky's AI Lab. "It was probably the smartest, most intense group of young computer scientists who had ever gotten together up until that time. We just cooked for about three days. We sized each other up. We got to know who was working on what. And at the end of those three days, we all knew each other very well."

Vinton Cerf, Patrick Winston, John Warnock, Danny Cohen, Bob Balzer—the roster of that meeting would eventually read like a who's who of modern computing, and many of the friendships they forged in the cornfields would endure down to the present day. But for those three days in the Illinois summer, they were young, wild, and crazy. And by all accounts the wildest of the bunch was Utah's Alan Kay, a guy who was so far out in the future that not even *this* crowd could take him seriously—yet so funny, so glib, and so irrepressible that they listened anyway. When it was his turn to give a presentation, Kay told them about his idea for a "Dynabook," a little computer that you could carry around in one hand like a notebook. One face would be a screen that would display formatted text and graphics and that you could draw or write on as if it were a pad of paper. The Dynabook would communicate with other computers via radio and infrared. It would be so simple to program that even a kid could do it. And it would have lots and lots of storage inside—maybe even a hard disk!

"Come on, Alan!" Barry Wessler remembers saying with a laugh, holding out his arms as if to embrace the three-foot platter of a 1968-vintage drive—which then cost twenty-five thousand dollars. "A *hard disk?!*"

"No, no, no!" Kay replied, curving his fingers around to indicate something about an inch in diameter. "A *little* hard disk!"

Well, it was absurd. But who cared? "It was great," says Kay as he recalls the conference. "The spirit at that meeting was about as good as it gets—the most civilized, the most ecumenical, the most incredibly supportive atmosphere you could want. People would cheer you on even if they didn't agree with you, just because they loved the fact that you were good. It was like in tennis, when somebody beats you with a great shot: you congratulate them. That was the spirit that made the ARPA community work. And that was the spirit that was fostered by the grad students."

Indeed, by all accounts, it was there in the cornfields of Illinois that the young hotshots of this rising generation first began to feel that they were part of something larger than merely this or that research group. They were part of a nationwide community—the ARPA community. Certainly the twenty-three-year-old Steve Crocker thought so. Before that meeting, he says, "I had been keeping track of where I thought the hot research was: BBN, MIT, CMU, Stanford—I had my tastes and biases, and I thought I knew it all. But then I walked into this room, and I was struck very forcefully that there was a nearly hundred-percent match between where I thought the good work was being done and the people and the places represented—which meant that somebody else had figured this out, too. There was a lot more structure here than I thought. The adults at ARPA knew what they were doing. And I thought that was absolutely wonderful. So at the end I went up to Barry and said, 'When are you leaving?'—meaning, 'I want your job.' He just looked at me as if to say, 'Where did *you* come from?' "

Actually, Wessler was just as thrilled as any of them. "For me it was a life-shaping event," he says. "The senior ARPA PIs had just as much intellectual horsepower—maybe more. But they were relatively set in their thinking, versus these young people, whose ideas weren't necessarily polished but who had the raw enthusiasm. They'd say, 'Wow!' and then move in some totally new direction. The energy at that meeting was palpable."

Still, says Wessler, there was one sense in which the meeting fell short. Bob Taylor had had two goals in sponsoring it. One—to make the young researchers feel like they were part of an ARPA community—had been fulfilled beautifully. But the other—to drum up enthusiasm for the ARPA network—had gone precisely nowhere. Wessler had tried at every opportunity to get the participants talking about it, but he had found no takers.

True, agrees Crocker: the young hotshots in Wessler's audience were bored stiff by networking. "The thing I always have trouble explaining to people nowadays is how intense all the *other* stuff was," he says. "AI. Graphics. Time-sharing.

Architecture and distributed processing. Programming languages, with Lisp being the core of a lot of things. Experiments with high-level languages that eventually transformed the way we were programming. ARPA was sponsoring all these vibrant activities that were deep and transformational in their own right. Networking came across to us as only one of many activities—and a rather pale and dry activity, at that. The network itself was only a piece of connective tissue."

The irony, Crocker adds, is that so many of the participants at that first grad-student meeting, himself included, would eventually discover that this little piece of connective tissue had defined their professional lives. They would likewise find that the very act of building the network had done more to forge them into a real community than any number of bull sessions in the cornfields. But then, he says, that wasn't clear until later. And in the meantime, being the arrogant young know-it-alls they were, they needed a good swat with a two-by-four to get their attention.

It wasn't long in coming.

In June 1968, about the same time that Wessler, Crocker, and the rest were assembling in Illinois, Larry Roberts got the formal go-ahead for his network plan. He immediately started preparing a formal request for quotations, which would invite companies to bid on construction of the ARPA network. The RFQ, as it was known, listed all the specifications that Roberts and the community had been hammering out for the past eighteen months: one IMP computer at each site, a data rate of fifty-six kilobits per second, a fully distributed architecture, a response time of 0.5 seconds, and so on. However, following the walk-before-you-run principle, Roberts's RFQ called for starting the operation with a four-node network, with an option to expand it to all the ARPA sites only if those first four were successful.

The first of the four nodes, almost inevitably, would be UCLA. Two years earlier, the remnants of the ill-fated networking experiment there had been redirected into a related project to measure how bits actually moved from place to place inside computer systems, including networks. From there it had been an obvious step for Roberts to designate UCLA as the ARPA network's Network Measurement Center—especially since the place was also home to one Len Kleinrock, who had done his best to make sure that Roberts would do so. "Larry brought a bunch of us back to Washington to help him specify what this network would look like and what performance characteristics it would have," says Kleinrock, remembering the one and only meeting of Roberts's networking advisory committee. "I banged my fist on the table and said, 'We've *got* to put measurement software in there! If this network is going to be an experiment, we must make measurements!' " Moreover, they needed data from day one: How many packets actually flowed through the network? When and where did they pile up into digital traffic jams? How often did they get lost? What kind of data rates did you actually achieve? What traffic conditions broke the network?

So UCLA had to be the first node. And from that decision, the rest of the initial configuration had followed almost automatically. Engelbart's group had to be node number two, since he had volunteered SRI to be the Network Information Center. And then from there, well, since UCLA and SRI were both on the West Coast, and since Roberts didn't want to hassle with transcontinental lines right at the start-up, the other sites would have to be there, too. But Stanford was out; John McCarthy was as unenthusiastic about the network as ever. Likewise Berkeley, for much the same reason. And that left Utah and the University of California at Santa Barbara, which fortunately were gung-ho.

So there it was: one four-node packet-switching network going into UCLA, SRI, Santa Barbara, and Utah. The RFQ came out at the end of July 1968. At least a dozen companies large and small at once started writing their proposals. Submissions were due that fall, and the contract promised to be a lucrative one.

At the four pioneering sites, however, that RFQ was the two-by-four. The people who took care of the operating systems and the basic hardware at each location, which in most cases meant the graduate students, greeted the document with a bafflement that bordered on horror. There wasn't a word in there about how they were supposed to *use* this network after they got it. To take a modern analogy, it was as if ARPA were proposing to go through a corporate office sticking network cards into each computer and hooking them up to an Ethernet cable without doing anything about communications software, user training, or even an "installation" process to tell the computer's operating system how to recognize the card. Just, Here it is, guys, use it!

Within a few weeks, SRI's Elmer Shapiro had organized an impromptu meeting in Santa Barbara so that representatives from each site could begin to figure out just how they were going to handle the network. "We were mostly second- and third-level people—meaning grad students, not PIs," says Steve Crocker, who had been drafted to be a delegate for UCLA. Indeed, he found a lot of familiar faces there from Illinois. "But we were all system programmers. We had access to, and in some cases direct responsibility for, the operating systems."

The meeting itself was almost surreal, he remembers. What the participants knew was that the winner of the contract, whoever it might be, would install an IMP at each of the first four sites and make the bits flow between them—period. But what they were now having to face was that bits alone weren't communication, any more than the letters of the alphabet alone were communication. Before anything meaningful could be communicated with those letters, everyone first had to agree on a vocabulary: how letters were formed into words. Then everyone had to agree on a grammar: how words could go together to form sentences. Then everyone had to agree on rules of discourse—appropriate ways for sentences to be used in making a request, say, or framing a reply. And it continued from there, through level after level of convention and protocol.

Clearly, says Crocker, an analogous set of protocols had to be agreed upon at each level of digital communication. Or rather, that was clear in retrospect. At the time, about the only thing he and the others really knew was that they were facing chaos. How would the IMPs and hosts be connected? What sort of applications would run on the network? How would those applications interact with what was already going on in the various operating systems?

"We just started asking some very basic questions," says Crocker. "But the answers were not forthcoming in any way. In fact, my recollection is that not only weren't there any answers, but there wasn't even a strong acknowledgment that this was the set of questions that we needed." Worse, he says, they couldn't even be sure that there was any real point to it anyway: "We had gotten no external directive [from ARPA], and that was an absolutely central concern. We had no idea whose toes we were stepping on, if anybody's. In fact, I envisioned that a group of professionals from 'the East' would come along by and by to tell us how it was supposed to be done. We had this sort of magical view of 'the East.' "

Still, two things *were* apparent to the group. One was that the prospects were fantastically interesting, now that they had actually taken the time to think about the network. "We found ourselves imagining all kinds of possibilities," Crocker says. "Interactive graphics, cooperating processes, automatic database query, electronic mail." The other thing was that they were having a great time. "There was a kind of cocktail-party phenomenon," he says, "where you find you have a lot of rapport with each other."

So in the end, the Santa Barbara participants reached no conclusions whatsoever, other than that they wanted to keep on meeting. "Over the next several months," says Crocker, "we managed to parlay that idea into a series of exchange meetings at each of our sites. We met at SRI, at Utah, and at UCLA. We said that the first effect of networks, which were designed to make it possible to collaborate at a distance without all this travel, was to increase travel enormously!"

It was a blast, he remembers, and they let their imaginations run riot. What did it matter, after all? They were just grad students; professionals from the East would come around to tell them about the real protocols soon enough. So they started fantasizing about things like mobile interfaces, little chunks of software that would migrate through the net and allow you to interact with an application on a remote machine as if it were right in front of you (nowadays such mobile programs would be called applets or agents). Crocker and his companions even started playing around with programming languages that would make it easy to write such applications. And as is the way of such things, without ever quite deciding to do so, they found themselves beginning to take it seriously. In the spring of 1969, in fact, after a particularly delightful meeting in Utah, they decided that they'd better start transcribing their discussions.

Crocker volunteered to draft the first set of notes, and promptly discovered that it was going to be a lot harder than he'd thought. "I stayed up all night trying to get the wording right," he recalls. "I was most concerned that we not tread on someone's toes—someone who'd been officially assigned to oversee these things. So I used humble wording such as, 'These notes are unofficial, and they are just to stimulate conversation, and anybody is able to write down anything, and they have no status.' There was also an issue as to whether these were formal publications, and so we said they were not."

And then, of course, there was the question of what to *call* the notes. Crocker struggled for a while, trying to think of something that would emphasize the informality and not seem too presumptuous to the real authorities. Finally, somewhere deep in the night, he typed a title he hoped would work: *Request for Comments*.

ARPA'S WOODSTOCK

By the early afternoon of Monday, December 9, 1968, the Fall Joint Computer Conference in San Francisco was already well under way. Everybody was there: mainstream engineers from IBM and the other manufacturers, funding officers from the various research agencies, far-out ARPA contractors, enthusiastic graduate students—*everybody*. In those days, the fall and spring joint conferences were the only real computer meetings going. And the word was already out: the auditorium was standing-room-only for the afternoon session. People were craning to get a glimpse through the doorways.

This, everyone could tell, was *not* going to be your standard presentation.

Down on the right-hand side of the stage, all alone, sat Douglas Engelbart. He was wearing a fresh white shirt and tie for the occasion, plus a set of headphones that made him look for all the world like a NASA launch-control officer. Across his lap—pivoting from the arm of his chair, actually—he had a kind of console that featured a built-in keyboard in the middle, a tray on the right for holding an odd little box with some buttons on top and a cord coming out the end, and an identical tray on the left for holding an equally odd gadget with five metal keys. Looming over Engelbart's right shoulder, dominating the stage, was a twenty-two-by-eighteen-foot display screen that magnified his every expression to the proportions of the Jolly Green Giant. And behind *that*, invisible to the audience but very much a part of the show, was a jury-rigged chain of cameras and video links and telephone lines stretching thirty miles down the peninsula to Menlo Park.

With a setup like this, no one knew quite what to expect. But Engelbart definitely had their attention. "The research program that I'm going to describe to you," he began in that soft, strangely compelling baritone, "is quickly characterizable by saying, 'If, in your office, you as an intellectual worker were supplied with a computer display backed up by a computer that was alive for you all day, and that was instantly responsive to every action you had, how much value could you derive from that?' "

Well, that *was* the question, wasn't it? And it had been ever since Lick first arrived at the Pentagon. Engelbart's answer, of course, was the oN-Line System, or NLS. In the eighteen months since the Michigan PI meeting, he and his team had gotten it working very well indeed, especially once they'd taken delivery of their SDS 940 time-sharing system. NLS now offered all the attributes of what would later be called a word-processing program: full-screen text editing with automatic word wrap; cutting, pasting, corrections, and insertions; automatic formatting and printing—the works. It likewise offered the functionality of what would later be called an outline processor, enabling the user to organize the text as a series of headings and subheadings; collapse or expand headings to any level of detail; move a heading and all its subheadings as a unit; automatically number the headings; and so on. It also implemented what would later come to be known as hyperlinks, Vannevar Bush's old notion of user-defined associations between one written concept and another. And for those situations where users wanted something that wasn't already built in, it offered a programming language that allowed them to construct the functionality themselves. NLS, like Fernando Corbató's CTSS, was an open system.

Meanwhile, Engelbart and his team had also been working on the collaborative features of NLS: on-line documentation, on-line discussions, and all the rest. By the fall of 1967, in fact, they'd been far enough along to do a little show-and-tell for their three funding agencies, ARPA, NASA, and the air force's Rome Air Development Center. The occasion was a project review held on October 12 and 13. Bob Taylor, who was there representing ARPA, described the scene in an article published the following spring: "Tables were arranged to form a square work area with five on a side. The center of the area contained six television monitors which displayed the alphanumeric output of a computer located elsewhere in the building but remotely controlled from a keyboard and a set of electronic pointer controllers called 'mice.' Any participant in the meeting could move a near-by mouse, and thus control the movements of a tracking pointer on the TV screen for all other participants to see. . . . The computer system was a significant aid in exploring the depth and breadth of the material. More detailed information could be displayed when facts had to be pinpointed; more global information could be displayed to answer questions of relevance and interrelationship. A future version of this system will make it possible for each participant, on his own TV screen, to thumb through the speaker's files as the speaker talks—and thus check out incidental questions without interrupting the presentation for substantiation."[2]

This kind of "groupware" is fairly advanced stuff even now; in October 1967 it bordered on science fiction. Certainly Bob Taylor thought so. The SRI group's working implementation was by far the best example Taylor had ever seen of what it meant for a computer to be a communication device, not to mention its being a superb demonstration of human-computer symbiosis. You couldn't imagine a better advertisement for the ARPA vision.

And yet that was just the problem. Engelbart and his crew hadn't even published a research paper on NLS until earlier that year. And they were buried out in the boondocks of California, where precious few people had ever seen what they were up to. "Even in the ARPA community, there were a lot of people who thought that Engelbart's work was silly," says Taylor. "To them, Engelbart was not doing serious science. They just didn't understand the importance of writing and communicating, or the human implications of a computer's acting as a communications device."

The only way they could turn that perception around, Taylor told the SRI team, was to show people what NLS could do—and really *show* them, not just talk about it. They would have to shock the brains of a not-very-receptive computer profession, and with any luck set a few minds running off in wholly unfamiliar directions. In short, said Taylor, they would have to *sell* NLS, together with all the ideas that went into it.

Actually, says Engelbart, he and his colleagues had already reached the same conclusion. "We knew that if we didn't get people's attention, ARPA would have a hard time maintaining its support for us. So we talked among ourselves about how to do this." The 1968 Fall Joint Computer Conference was an obvious opportunity: San Francisco was only thirty miles north of Menlo Park. All they had to do was figure out how to demonstrate NLS for several thousand people at once—when they were thirty miles away from the only time-sharing computer that could run the program.

That job fell to Engelbart's chief engineer, Bill English, who once again exercised his genius for turning his boss's brainstorms into working hardware. "First," English remembers, "we discovered that we could borrow an Eidophor display from Ames Research Center in Mountain View. Eidophor was the brand name. It was a Swedish thing—an arc-light projection for video—and *very* expensive, so it was used by the military almost exclusively. Inside it had a parabolic drum coated with oil, and an electron beam that deflected the surface of the oil, so that when an arc light reflected off the surface and onto the screen, it formed an image. Next, we arranged with the phone company to get two video links from SRI to San Francisco. These were straightforward video relay links, which were already in common use for TV." The idea, he explains, was to have these links carry NLS's screen output—the image would be captured by a standard TV camera aimed at one of the computer's CRT displays—as well as head shots of various people at SRI. Onstage in San Francisco, meanwhile, there would be another camera pointing directly at Engelbart to produce a head shot, and still another staring down at the console, to show the audience how he used the keyboard, mouse, and keyset (input from these devices would be carried down to the SRI computer over leased telephone lines, independent of the video links). Finally, all of these TV feeds would pass through a video controller set up backstage inside a makeshift control room, which would be English's command post on the day of the demonstration. Armed with the script, he would be giving people their cues, mixing the images, fading from one scene to the next, splitting the screen, and generally serving as the broadcast engineer.

"So it was all fairly complicated and took a lot of work to set up," says English. "The whole process from the concept to the completion must have taken several months at least. And obviously we had to pray that it would all hold together. We did have redundancy built in. But not for the Eidophor; it was the only one on the West Coast. If that failed—well, that was the real weak link in everything."

Weak link: that wasn't exactly a reassuring notion for Engelbart, who was going to be out there performing without a net. It wasn't too reassuring for the conference's program committee, either; before giving Engelbart's proposal a final OK, they had sent a delegation down to SRI to assess the presentation's feasibility, a very unusual step. And it most definitely wasn't reassuring for Engelbart's sponsors at ARPA, NASA, and the air force. The SRI crew was paying for this extravaganza with money that it technically had no right to spend: federal research funding was earmarked for research, not proselytizing. So before the demonstration, as Engelbart remembers it, Taylor and the other sponsors insisted on plausible deniability: "They told me, 'If something goes wrong, I've never heard about this. I couldn't legitimately authorize it.' So if any of the computer hardware failed, or the mouse, or the modem, or the telephone lines, it could have been the demise of the whole program right there."

Not surprisingly, Engelbart was a bit nervous as the demonstration got under way; he couldn't stop himself from looking over his shoulder about every five seconds to make sure his image was still up there on the screen. It didn't help that his headphones carried the background chatter between English and the others as they struggled to keep the show on track. Unbeknownst to the audience, Engelbart would spend the entire presentation listening to the murmur of impending disaster: "Bill, we've lost XYZ!" or "Bill, we're not going to be ready in time!"

Nonetheless, Engelbart managed to keep his voice reasonably steady as he forged onward, putting NLS through its paces. To begin with, he typed out some text—STATEMENT ONE: WORD WORD WORD WORD—and English lost no time in putting his video mixer to dramatic effect: the letters appeared on the screen as if magically suspended in the air beside his boss's head. Through the overhead camera Engelbert then showed the audience the mouse and demonstrated how it could make a "bug," or cursor, move around on the display to do full-screen text editing. He likewise showed them the keyset, with which he (and almost no one else) could do one-handed typing nearly as fast as on the standard keyboard. He went on to demonstrate word-wrap, cut-and-paste, outlining, and hypertext, whipping out commands with the mouse and the keyset with near-incomprehensible dexterity. (The system confirmed every action with a soft beep or buzz, like something straight out of the TV show *Star Trek*.) He demonstrated on-screen windows, with his face on one side of the display and the text he was editing on the other. He demonstrated on-line collaboration, with two or more people working on the same document at the same time (they each had control of a "bug"). He showed how NLS could be programmed to serve as an on-line user manual, an on-line project-planning system, even a kind of prototype E-mail system—on and on.

Engelbart had no idea how anyone was reacting to all this.

No, that wasn't quite true. Right there in the front row Engelbart could see his wife, Ballard, along with his three daughters: Gerda, age fourteen, and the twins, Diana and Christina, age thirteen. He was pretty sure that the girls weren't terribly impressed; this was just what their dad did. But as for the rest of the audience well, at least they were laughing at his little jokes. Also at his flubbed lines, of which there were many. They were staying put, with no obvious rush for the exits. But other than that, Engelbart heard only his own voice booming out into a vast silence.

Sooner than he would have thought possible, Engelbart saw the meeting organizers signaling frantically that he'd run out of time: he'd used up his allotted question-and-answer period and was threatening to delay the next scheduled talk. He wasn't nearly finished, but there was nothing for it but to throw out the remaining items on his list and wrap it up. Engelbart invited everyone to visit Menlo Park for the group's three-day open house—a chance to test-drive NLS and kick its tires, so to speak. Looking down into the front row, he announced that he was dedicating this presentation to his wife and daughters, as an inadequate thank-you for their putting up with an in-house lunatic for so many years. He thanked everyone for coming. He gave one last glance over his right shoulder at the screen—the Jolly Green Giant was still up there, looking over *his* right shoulder—and then he was done.

Engelbart waited for he knew not what.

Actually, Engelbart's audience had sat transfixed during the entire presentation, united in the conviction that that Eidophor display was one very cool piece of technology. Few of the attendees had ever seen anything like it. Certainly none of them had ever seen a demonstration carried out with more showmanship and flair.

As for the content, well, the response basically came down to whether or not a given audience member was ready to hear what Engelbart had to say. Many of them had a reaction similar to that of Robert E. Kahn, then a young communications theorist at BBN: "What was *that*?!" It would take time for people to grasp the significance of something so new.

Still, Engelbart's audience in December 1968 was considerably more receptive to his message than it might have been even a few years earlier. Not only had interactive computing become far more than an academic curiosity in that time, but in certain segments of the marketplace, it was downright hot. Indeed, mainstream computer manufacturers had begun to take time-sharing seriously as far back as 1964, thanks in large measure to the publicity surrounding Project MAC and its vision of information utilities—not to mention its dramatic announcement of a partnership with General Electric. By 1965, DEC had debuted its PDP-6 machine, the first general-purpose time-sharing computer on the market; IBM had declared its intention to build a time-sharing computer compatible with its System/360 line; and similar announcements had come from Control Data Corporation, Sperry Rand, Burroughs, and Scientific Data Systems.

From there it was just a short step to computer service bureaus, or "computer utilities" that sold computer processing by the minute to small users that couldn't afford their own machines. Before 1965 was out, GE was offering just such a commercial time-sharing service based on the Dartmouth College system, which included the new interactive programming language BASIC. IBM was likewise offering a commercial time-sharing service based on Quiktran, an on-line version of Fortran. And *Datamation* magazine had concluded that "the broad acceptance of time-sharing" was perhaps the most important trend of the year.[3] In practice, it's true, computer utilities hadn't always lived up to glowing expectations. The bulk of their market was actually in batch processing, for the simple reason that their business customers needed it for payroll and accounting jobs. Moreover, the time-sharing services that the utilities did offer were almost always limited, special-purpose systems along the lines of Dartmouth BASIC: many users could tap in and work interactively, but they had access to one and only one program. "Real" computer utilities, in which many independent users could run whatever programs they liked, remained a dream for the future.

Nonetheless, the momentum had kept on building. In September 1967, for example, DEC had introduced a much-improved successor to its PDP-6. Known as the PDP-10,* it would become a popular choice for many of the later computer-utility installations and an equally popular fixture at universities, where researchers found its thirty-six-bit architecture to be uniquely powerful and elegant. By the time of Engelbart's presentation in December 1968, the computer utility market had attracted at least twenty separate firms, old and new. GE alone was operating in twenty cities. And one of the largest commercial concerns, University Computing Company of Dallas, had terminals in thirty U.S. states and a dozen countries—as well as a stock price that had soared from $1.50 to $155 per share.[4]

In short, the success of the computer utilities had long since convinced most of the computer professionals in Engelbart's audience that time-shared, interactive computing was a legitimate activity. The utilities boom had likewise popularized the notion that computers—or at least terminals—might one day have a place in the home. As Lick himself would write in 1970, he'd already been hearing for

---

* There is no point in trying to make sense out of the PDP numbering scheme, because there wasn't any; the numbers were simply assigned in sequence as the various machines were designed. So the PDP-6 and -10 constituted one evolutionary sequence, the PDP-5 and -8 formed another, and so on.

years about "home information systems" that would be linked to data centers via telephone lines or antennas. "The picture that has been built up in my mind is a montage of television, hi-fi, and microfilm. . . . Through the home information system, the family receives its news and entertainment, does its shopping, browses the library, makes reservations for travel and theatre, and so on."[5]

And perhaps most important of all in the long run, the utilities had offered thousands of *non*professionals a glimpse of what it meant to interact with a computer. Just a few months before Engelbart's presentation, for example, in the fashionable Seattle suburb of Lakeside, Washington, the local mothers' club had used some of the proceeds from its annual rummage sale to buy computer access for the science and math students at the Lakeside School. The system was rudimentary: just an ASR-33 Teletype terminal that could dial in to a nearby GE Mark II time-sharing system, which offered little more than GE's version of BASIC. But the kids loved it. And one of the most ardent was an undersized, freckle-faced eighth-grader who taught himself BASIC as fast as he could. Bill, his friends called him—William G. Gates III.

Of course, the significance of that bit of democratization wouldn't become apparent until later. Nonetheless, by December 1968 it didn't seem quite so crazy to believe that computers could enhance human creativity, or that computers could help build human communities, or that computers—this was the sixties, after all—could embody human freedom. So perhaps that was one reason so many people at the Fall Joint Computer Conference in San Francisco sat transfixed until the very end, watching Doug Engelbart shoot a last, over-the-shoulder smile at the projector that had not failed him. And perhaps that was why they started clapping when Engelbart was done, and then rose to their feet to keep on clapping: on and on, in a spontaneous standing ovation.

"I hadn't expected that," says Engelbart, still sounding touched and faintly astonished by the reaction some thirty years later. "It surprised me. And . . . I didn't know what to do."

Actually, what he had to do right then was get off that stage; the conference organizers had a schedule to keep. But his point had been made. "It was *great*," remembers Chuck Thacker, who was then a graduate student at Berkeley. "One of the most impressive things I ever saw: Engelbart on stage, dealing lightning with both hands!"

"The audience was blown away, and I was on cloud nine," agrees Bob Taylor. "I had been on pins and needles the whole time, not because I thought Doug's program was in any danger, but to see if the computer community would be awakened by this way of using computers. So when I saw them, finally, realize what Doug was talking about, I was happy."

So was Bill English. Not only had the equipment worked, but for the first time, he and his colleagues felt accepted. "We had been on the fringes, even at SRI," he says. "But now we got the attention of the whole community. And as time went by, we got more visitors. We got a lot of good press. We even got the attention of SRI management!"

Of course, much of the initial, when-can-*I*-get-one enthusiasm abruptly faded once people found out what an NLS installation would cost them: the SDS 940 was a million-dollar machine, not including the cost of all those video terminals. Nonetheless, with the clarity of more than thirty years' hindsight, computer professionals of a certain age have come to look back upon Engelbart's presentation as one of the defining moments of their generation. By December 1968, the future as envisioned by the ARPA community was beginning to take shape, however sketchily. And now here was Doug Engelbart, expanding by an order of magnitude their sense of what that future might really mean. The name Woodstock wouldn't acquire its modern connotation for another nine months, not until the weekend of August 15–17, 1969, and a certain rock concert in the hills of upstate New York. But if ever there was an event that qualified as ARPA's Woodstock, this was it.

THE IMP GUYS

In January 1969, barely a month after Doug Engelbart's extravaganza in San Francisco, Larry Roberts formally announced that Engelbart's group at SRI would serve as the nascent ARPA network's Network Information Center. No surprise there.

A short time earlier, however, just a few days before Christmas, Roberts also had ended the suspense over who would *build* the network. The winning proposal, he announced, had been submitted by Bolt Beranek and Newman of Cambridge, Massachusetts.

This was considerably more of a surprise, since BBN had been one of the smaller firms in the competition. Right up until the day of the announcement, in fact, Roberts had seemed to be leaning toward one of the big firms, Raytheon, which had lots of engineers on staff, lots of DoD contracting experience, and lots of clout in Washington. But then, size wasn't everything. Roberts had finally decided that the Raytheon team was top-heavy in the extreme; he could envision having to spend hours on the phone working his way up and down the hierarchy for every little problem. He much preferred the BBN group, which was maybe a dozen guys, with no hierarchy at all. More to the point, they were absolutely first-rate, as was the technical content of their proposal. And—not the least of the attractions—they were backed by a company that had clearly made a serious commitment to computer research: what had once been little more than a sideline to BBN's core acoustics business had now grown to encompass roughly half the firm.

"That was a tribute to Lick," says John Swets, who had taken a leave of absence from MIT to help fill the gap when Lick went off to ARPA in 1962, and ended up having so much fun that he has stayed at BBN to this day. "There wasn't any decline in activity and interest when he left; he'd set up something that could maintain itself on its own intellectual energy." Indeed, BBN had become a mecca for Harvard and MIT types who wanted to pursue whatever interested them without having to worry about students and teaching. By the time of the network award, the firm had long since taken on the look of an academic commune. There were nerdy MIT dropouts wandering the halls, occasionally in shoes and socks. There were dogs and bicycles in the offices, along with employees playing guitars and recorders. But the important thing was that there also were people hacking code and hardware at all hours of the day and night. BBN had a substantial ARPA contract to explore the applications of time-sharing, which covered activities ranging from artificial-intelligence research to the development of advanced programming tools. The company likewise had contracts for work in fields such as "medical informatics," an exploration of what physicians could do with on-line access to medical records, pharmaceutical information, and the like.

BBN even had that rarest of animals, an in-house expert on packet-switched networking. Robert E. Kahn was compact, dark, and intense, with a notable fondness for Wagnerian opera, amateur photography, and young ladies (he seemed to have a different date for every party). In the early 1960s, when he was working toward his electrical-engineering Ph.D. at Princeton and simultaneously serving as a staff member at Bell Labs, Kahn had focused his considerable energies on the theory of communications. After joining the faculty of MIT's Research Laboratory for Electronics in 1964, moreover, he had continued in much the same vein. After a time, however, Kahn's boss at RLE had suggested that the young theorist might benefit from a little practical experience. "It was some of the best advice I ever got," says Kahn. "I took a leave of absence in nineteen sixty-six and went to BBN."

His background made him quite an anomaly in the computer group there: computer people and communications people still moved in very different orbits in those days and rarely had occasion to talk to each another. So in the interests of getting his hands dirty, Kahn zeroed in on the one obvious area of overlap: networking.

Now, this choice had nothing to do with ARPA, he says, since the agency's networking project hadn't even started in 1966—and since he'd barely heard of the place in any case. "But at one point after I had started to work on networking," he remembers, "Jerry Elkind came up to me and said, 'You know, this is really interesting; you really ought to let Larry Roberts at ARPA be aware of it.' " Well, says Kahn, he had no idea who Roberts was. But he knew that Jerry Elkind was his boss. So he contacted Roberts and went down to talk to him, he says, "and I found out at that point that he was interested in creating a real network! Having been a mathematician and a theoretician, it really had not occurred to me that I might ever get involved in something that could become real. It was a revelation of sorts for me."

This was sometime in 1967, Kahn remembers, not long after Roberts's arrival at ARPA. In the months that followed, Kahn sent all his papers to Roberts as soon as he finished writing them (they were among the many inputs that the latter used in designing the ARPA network) and plunged even deeper into his own research at BBN. Networking was virgin territory back then, he says: "Nobody had any experience in knowing how networks would function in practice." Indeed, he discovered, the behavior of real networks was hopelessly beyond the reach of pencil-and-paper math. So he and BBN's Warren Teitelman developed an on-line graphical simulation system that would help them visualize what was going on.

That simulation was up and running by early 1968, says Kahn. So it couldn't have been too many months later that a guy he didn't know dropped by his office, wanting to talk about networks. He introduced himself as Frank Heart and said he was the head of BBN's medical-information group. And they proceeded to have what Kahn recalls as a very pleasant chat.

Actually, Kahn was just one of the many BBNers whom Heart was talking to that spring. Heart hadn't followed networking development very closely, as it happened. Back at Lincoln Lab, where he had started in 1951 as a graduate student on the Whirlwind project, his focus had been on computerized radar and related technologies. Since coming to BBN in December 1966, moreover, he'd been preoccupied by his medical-information group. But his interest in the subject had been piqued the previous fall, after Larry Roberts came by BBN to give a talk on the ARPA network. (Roberts was on the road incessantly during this period, trying to get potential bidders interested.) "This was just an obviously interesting area of R and D," Heart remembers thinking. He began to look into the matter more seriously—thus his chats with Kahn and the others that spring. And when Roberts's formal request for proposals finally appeared in July 1968, he immediately got hold of a copy.

Hmm. Packets. The IMP routing computers. Four sites to begin with. Roberts had sketched out the basic architecture of the network pretty thoroughly. Of course, that still left an infinite number of details to be worked out by the contractor; implementing this thing would not be a piece of cake. Still—

You know, Heart realized, BBN could do this thing. In fact, the group that he had brought with him from Lincoln Lab was almost uniquely qualified for the job. "In that period of time," Heart explains, "the world was not full of people who knew how to make computers run in real time and connect to real-time systems. But by the time I left Lincoln I had been in charge of sizable efforts to build computer-based control systems and computer systems for antennas and radars. So the group that ended up at BBN probably knew more about connecting computers to communication lines and to real-time systems than any other group in the country."

True, the clock was already ticking louder by the hour. ARPA, having waited until the end of July to release the request, now wanted all the proposals in by September. But that just added to the adrenaline rush. Severo Ornstein, a Lincoln Lab veteran who had worked on the TX-2 and LINC under Wes Clark, plunged into the hardware details of the proposal, while David Walden, another former Lincoln Lab man, started figuring out how the programming would work. Then, since Walden felt a little uncomfortable about having sole responsibility for the software—he'd been a programmer for only four years at that point—Heart took the opportunity to do what they'd been wanting to do anyhow: bring Will Crowther over from Lincoln, where he had been Heart's subordinate and Walden's direct boss. Dreamy, otherworldly, and whimsical, Crowther was a classic hacker, the kind of programmer who wore sneakers to meetings at the Pentagon and was just as interested in rock climbing as he was in coding. But he was undeniably brilliant. "Both Frank and I knew that the program, if we ever came to write it, would have to be a really tight, real-time program," says Ornstein. "And we knew that Will Crowther thoroughly understood how to write such programs in machine language code."

Bob Kahn joined the team as well. "Writing something like this proposal seemed to me to be the epitome of practical experience," he says. And he wasted no time in getting that experience. Of course, as Ornstein points out, Kahn the theoretician could sometimes get on people's nerves: "Bob wasn't politically in our group," he recalls, "and he got along relatively badly with some of our people at first because he really wasn't a computer person. Some of the things that he was suggesting were, we felt, off the wall—just wrong. And of course, people were busy and impatient and didn't want to take time to explain." Nonetheless, Ornstein admits, you had to give Kahn credit for perseverance. "Bob wanted to know *everything*," he says. "So he would take me aside and say, 'OK, explain exactly how it's going to work.' He wanted to know all about the hardware design in great detail. He's a very thorough person, so he learned a lot."

Before it was all over, in fact, any number of BBN's people got involved. "I think that more man-hours were charged to that proposal than had ever been done for any proposal at BBN before," says Ornstein, "because there were so many people who were really, seriously interested. We knew that we were competing with large companies that would pour enormous resources into their proposals and that obviously had much larger reputations to offer. If we were going to stand a chance of winning, ours was going to have to be the glistening, clear, technical best."

Take this business of the IMP computer, for example. Since bidders were free to choose whatever hardware they wanted, the BBN team had quickly settled on a new machine known as the Honeywell 516. Not only was the 516 "the best machine for the job that year," as Dave Walden remembers it, but Honeywell was willing to work with them to build the special circuit boards that would connect the IMP to the rest of the network. However, instead of just specifying the Honeywell 516 and leaving it at that, the BBN proposal defined that special-purpose interface hardware very precisely. "I would say I did ninety percent of the design of those interfaces while we were writing the proposal," says Ornstein. "It was just a matter of laying in the gates after that."

And so it went. While the rest of the world was focused on the Soviet invasion of Czechoslovakia and the bloody street demonstrations at the Democratic National Convention in Chicago, Heart and his crew kept their attention focused squarely on networks. By the time they sent their proposal off to ARPA on September 6, they were fairly confident that their work really was the glistening, clear, technical best. Nonetheless, says Heart, "we knew it was extremely competitive, and that most of the other organizations bidding were larger."

Yet BBN kept making the cut. Of the twelve proposals submitted, ARPA deemed four to be worthy of serious consideration. In November 1968, it started final negotiations with two. And then the ultimate decision came like a wonderful Christmas present: the ARPA network was BBN's to build.

Of course, the award was also slightly terrifying. Now they had to deliver—and the clock was still ticking. ARPA wanted this network up and running by September 1969. "Everybody started working like mad on the code and on getting Honeywell actually to start building the machines," says Ornstein.

Heart did expand the team somewhat, bringing in several new recruits from a computer-hardware course that Ornstein had been teaching at Harvard. Still, the ARPA contract was going to support only so many people—fewer than a dozen, all told. And in truth, that was fine with Heart. "We felt strongly that only a very small group could do the project on that time scale," he says. "If we'd had a much larger group that had to start interacting with memos and formal documents, it would not have been feasible."

Certainly the "IMP guys," as they took to calling themselves, were anything but formal. "We all sat in offices that were right next to each other," remembers Dave Walden, "and we worked together constantly. I don't remember such a thing as a weekly progress meeting. We probably were more in tune with progress than that; we probably did it hourly." Frank Heart himself mostly stayed above the fray, since he still had responsibility for BBN's medical-information effort. Nevertheless, says Crowther, Heart set the tone and established a standard of excellence. "Frank had technical control," he says. "When he ran a project, he wouldn't let go of anything until he completely understood every little piece of it. This was, I thought, a very good thing because it meant everybody had to explain everything to Frank, and by the time he understood it, everybody else understood it, too."

Bob Kahn played a somewhat similar role, notes Crowther. As the group's resident theorist, he focused on the global, architectural issues of networking. "But he wasn't really interested in the implementation," says Crowther. "So he was like a consultant: we talked to him a lot and had grand little fights about how things should be done, but then we actually implemented it."

Happily for Crowther and Walden, the two main programmers, that implementation was comparatively straightforward. "It had become an engineering problem as opposed to a theory problem," says Walden. "The computers had now gotten fast enough and had good enough memories and were cheap enough that you could actually implement packet switching in software algorithms that ran fast enough to get the job done." At one point, adds Crowther, "Dave and I sat down, worked out the algorithms, and figured out that it was only going to take a hundred and fifty lines of code to process a packet through one of these switches! Of course, there [was] a whole lot of other stuff that you [had] to have, too. But this was the kernel, the part that really mattered, the thing that took in a packet, figured out what to do with it, and pushed it back out the line."

That "other stuff," it goes without saying, still gave Crowther and Walden plenty to worry about. How could they incorporate measurement software, for example, so that network operators could monitor the flow of packets and keep track of how well the system was working? How should they write the code so that users could hook up more than one host computer to a single IMP? And speaking of hookups, exactly how much processing should they allot to the IMP, and how much should be left up to the users and the host computer? "There wasn't much theory for how you build a packet-switching network," says Walden. "So we just got out there and did it. All the stuff that is now taught in courses about networks and protocols and all of that, I would say we were mainly inventing it."

Still, the software guys definitely had a smoother time of it than Ornstein and Ben Barker on the hardware side. In principle, their task should have been equally straightforward; as Ornstein says, "things had been so completely designed that by the time we got the proposal and were awarded the contract, it was mostly just a matter of doing what we [had] said we were going to do. There was relatively little invention left at that point to get the IMPs connected together." In reality, however, dealing with Honeywell was not always easy.

"One of the reasons we chose the Honeywell 516 was that we thought it was a mature machine that was not going to give us grief," Ornstein says, sighing. "Well, we were wrong. We pushed the 516 very hard; Honeywell had never had so much real-time traffic coming into the machine before, and we uncovered a design flaw, a synchronizer problem. Now, synchronizer problems are very, very subtle. The program would run fine for days on end, but at the end of three or four days suddenly the machine would just die, inexplicably. Well, Honeywell could hardly believe it. So we had to dig and dig and dig at them, and finally from their back room they produced a *really* smart enough guy—they did have a few—and I finally had someone who could understand what I was talking about and believe me. Fortunately, it was a fixable problem. We showed the Honeywell people the trivial fix they could make to the machine. But they had to make it to all their 516 machines, which was a major undertaking."

Alas, Ornstein continues, that was hardly their only problem with Honeywell. "I had to take a fairly heavy hand at times," he says. "Honeywell was behind schedule and mostly they sent us cabbages instead of computers. For example, they hadn't understood the interfaces that we had designed. And when the machine finally came, [the interfaces] didn't work. So I would send back drawings with corrections, and they would incorporate three quarters of the corrections and the other quarter were overlooked. At one point, with Frank standing by with his jaw dropped down, I actually turned a truck around at the loading dock and told them to take the thing back. In fact, I had to be really quite nasty at times, and beat on the table, and shout and scream to get them to fix the troubles they were having. Eventually they shaped it up, but it took a long time, like the fourth or fifth machine."

This was getting serious: the BBN team had promised to deliver the first IMP to UCLA in early September 1969. As the summer wore on with no working machine in sight, that deadline was looking less and less feasible every day. Even the software guys were feeling the pinch: Crowther and Walden had finished their routing code, but without a working machine, they had no way to test it. As a stopgap, they did create a program to simulate the 516 inside another computer, which meant they could at least *try* to run their software in the simulated state and get it debugged. But of course that assumed that there were no bugs in the simulator program.

So there was nothing else for it: all they could do was keep their fingers crossed and watch the calendar count down toward Labor Day.

Out at UCLA, Steve Crocker was watching the calendar, too, and if anything, he was even more apprehensive than the BBNers. To begin with, he explains, there was this business of the user protocols. He and his fellow grad students from the first four sites were still groping in the dark, with only the vaguest idea of what they were doing—or even if they were supposed to be doing it. They still had not heard from the Wise Men of the East about the official plan. And they were beginning to have a horrible suspicion that there really *was* no official plan—that it was entirely up to them to formulate it.

It was a realization that dawned in stages, says Crocker. The first clue came on Valentine's Day of 1969, when representatives from the first four host sites were invited to BBN to meet the people who had just won the bid to build the network. "I don't think any of us were prepared for that meeting," Crocker later wrote in his retrospective of those days, circulated as *Request for Comments (RFC) 1000*. "The BBN folks, led by Frank Heart, Bob Kahn, Severo Ornstein and Will Crowther, found themselves talking to a crew of graduate students they hadn't anticipated. And we found ourselves talking to people whose first concern was how to get bits to flow quickly and reliably but hadn't—of course—spent any time considering the thirty or forty layers of protocol above the link level."

Clearly, says Crocker, the Wise Men of the East had to be somewhere else, and would announce themselves soon enough. But then came the second clue: in April, Crocker and his ad hoc group released that first, ever-so-carefully-worded *Request for Comments* circular about the brainstorming they'd done to date—and no Wise Men rose up to complain. Nor was there any protest in response to *RFC 2*, or *RFC 3*, or any of the numbers that followed. "We just got more people wanting to play on our team," says Crocker. Soon, in fact, as the *RFCs* spread the word, meetings of their group were drawing upward of fifty people.

Finally, says Crocker, at about that same time, BBN released its internal report number 1822, the document that formally defined what the IMP software would and would not do. Specifically, it stated that the software would do as little as possible. The network as installed by BBN would deliver the packets, period. Everything else would be the responsibility of the host computer.

Now, admittedly, says Crocker, this did make for a clean interface, with minimal complication for BBN. Crowther and Walden could program the IMP side of the interface without having to worry about the vagaries of the various host computers. (In that sense, the IMP would function a bit like a standard electric wall socket, delivering AC power at a standard 110 volts and sixty cycles per second, regardless of whether the user plugs in a lamp or a toaster.) "But that left a total vacuum on the protocols," explains Crocker.

In short, Crocker and his contemporaries *were* the grown-ups. Sobering thought. Suddenly, what they did *mattered*. "Over the spring and summer of 1969," he would write for *RFC 1000*, "we grappled with the detailed problems of protocol design. . . . It was clear we needed to support remote login for interactive use—later known as Telnet—and we needed to move files from machine to machine [later known as File Transfer Protocol, or FTP]. We also knew that we needed a more fundamental point of view for building a larger array of protocols. Unfortunately, operating systems of that era tended to view themselves as the center of the universe; symmetric cooperation did not fit into the concepts currently available within these operating systems."

By now, time was pressing very hard indeed. Not only did they have to get the host-to-host protocols working, Crocker knew, but on or about September 1, 1969, ARPA was going to wheel in a Honeywell 516 IMP, plunk it down next to the UCLA group's computer, and say, "Talk to it." And when that happened, the UCLA computer—a big Sigma 7 machine from Scientific Data Systems—had better have something to say.

Fortunately, the home team didn't lack for manpower. The "UCLA Mafia" of graduate students included Crocker himself, who seemed to have a knack for becoming the ringleader of any group he was in; his old high school buddy from Van Nuys, Vint Cerf; Jon Postel, who would eventually take over the editorship of the *RFC* (and would continue in that role until his death in 1998); and a number of others—not to mention a flock of associated secretaries, programmers, managers, undergraduates, and faculty members. Len Kleinrock's Network Measurement Center at UCLA now comprised some forty people, a small army by academic standards.

In any case, one of the UCLA Mafia's most urgent worries was how to connect the Sigma 7 to the IMP when it arrived. Remember, wrote Crocker, they couldn't just go out and buy a standard cable. "SDS wanted a large sum—

$19,000, as I recall—and a lot of time to build a custom interface. But we didn't have the time; we were under pressure. So Mike Wingfield, another graduate student, took it on. Wingfield said he could do it in six weeks, and it would cost four or five thousand dollars. Now, Wingfield is a classic, clean, organized, careful guy, and in five and a half weeks he had built a box with colored buttons and lights on it. He showed me how it worked. You could sequence a bit at a time through this thing. And when I plugged it in—it worked! It was fine!"

One problem down. On the software front, meanwhile, in an effort to make their Sigma 7 computer do what ARPA was paying them to do, which was gather real-time data on network performance, Vint Cerf and several others were grafting a series of hurried modifications into their homegrown time-sharing system, designated SEX, the Sigma Experimental system (official handbook: "The SEX User's Manual"). And Crocker was in charge of the software that would actually talk to the IMP. "We were naturally running a bit late," he would recall in *RFC 1000*. The software still had bugs. But not to worry: "September 1 was Labor Day, so I knew I had a couple of extra days to debug the software. Moreover, I had heard BBN was having some timing troubles with their software, so I had some hope they'd miss the ship date. And I figured that first some Honeywell people would install the hardware . . . and then BBN people would come in a few days later to shake down the software. An easy couple of weeks of grace."

Not quite. On Saturday, August 30, Crocker learned to his horror that the IMP was already sitting on the loading dock. Against all odds, it developed, the BBN crew had gotten their problems fixed ahead of time and had immediately trundled the first IMP onto an airplane for shipment out to California via air freight. And they had sent along Ben Barker, the junior hardware guy, to ride shotgun. Together with Truett Thach, a technician at BBN's Los Angeles office, he was waiting now to get the IMP set up. "Panic time at UCLA," Crocker noted.

Kleinrock was in a sweat, too. Everybody gathered in the computer room to watch the hookup, he remembers: "My staff, the computer-science chairman, the School of Engineering administration, somebody from the chancellor's office, somebody from AT&T long lines, local telephone-company people, Honeywell, the ARPA guys, the BBN guys—and everybody was ready to point the accusing finger!"

They wheeled in the IMP—a steel-colored metal box about the size of a refrigerator—and set it up next to the Sigma 7. Someone plugged it in; lights started flashing. The Sigma 7 was connected to the IMP through Mike Wingfield's magic box. More flashing lights. Barker typed in a few characters at the Teletype console. The bits flowed from there to the IMP to the Sigma 7, where they were instantaneously sent back out through the IMP to be retyped on the same console. Essentially, Barker was sending a message to himself. But the packets were moving.

"It didn't look like anything," says Crocker, who remembers feeling more relief than wonder at the time. "But all the basic software had to be up and running even to get an echo. So it was dramatic, just to plug in the IMP and get the Teletype to respond."

The ARPA network—or the Arpanet, as people were now beginning to call it— was finally on line. It had gone from contract award to equipment on site and running in less than nine months.

"It was a good time to go," says Bob Taylor, thinking back to his departure from ARPA in September 1969.

After all, he explains, the success of that first Arpanet node at UCLA made for a wonderfully symbolic moment: the Intergalactic Network, his one overarching goal from the day he'd taken over IPTO, was finally becoming a reality. And with that, he says, he felt he'd been there long enough. "I don't think it's good for someone to stay in that office for very long, really," he says. "There's sort of a czar mentality that creeps in: absolute power corrupts absolutely."

Then, too, adds Taylor, he badly needed a period of decompression. If nothing else, he had a second office to worry about. In 1968, the head of the behavioral-sciences program had resigned, and ARPA director Eb Rechtin, rather than hire someone new, had simply asked Taylor—like J. C. R. Licklider before him—to do double duty.

But mainly, Taylor says, he had to decompress because of Vietnam. For three years now, ever since Charlie Herzfeld was first forced to ask the various ARPA offices to "help out" in Southeast Asia, Taylor had been making periodic trips there to help the U.S. forces set up an integrated, computer-based reporting system. ("There had been discrepancies in reports from the army, navy, and Marine Corps about enemy killed, supplies captured, bullets on hand, and logistics of various kinds," he recalls. "Now the White House got a single report rather than several, which pleased them. It was still a pack of lies, but at least they were consistent lies.") Like many others, however, Taylor had become so profoundly disillusioned about the war that he was reluctant to continue working for the Defense Department in any way—no matter how benign IPTO's research might be. The last straw, he says, was the election of Richard Nixon to the presidency in November 1968. He was fed up and started looking for a way to leave.

Happily, the perfect opportunity quickly arrived in the form of a job offer from Dave Evans and Ivan Sutherland at the University of Utah. "In nineteen sixty-eight Dave had recruited Ivan from Harvard to come help him with the graphics research at Utah," Taylor explains. "They decided to form a company called Evans and Sutherland, to make graphics terminals. But they both stayed at the university for a while. Well, Dave Evans told me that he needed somebody at the university to collect a bunch of federally sponsored computing projects under one umbrella. So I went out and organized a laboratory for him."

Back at the Pentagon, of course, Larry Roberts immediately became the official head of IPTO. That was fine by Roberts, who barely noticed the transition; he had considered himself the de facto head all along. And besides, he didn't have time to worry about it. One node was up and running, but there were a lot more to come. Larry Roberts still had a network to build, and a schedule that was tight.

THE TROUBLE WITH MAC

Sometime in the spring of 1970, says David Burmaster—in March, he thinks, or maybe April—he was sitting in his office on the ninth floor of MIT's Technology Square when suddenly, out in the hall, he heard a *whoosh* of people rushing past.

Curious, Burmaster got up and followed. The ninth floor of Tech Square was pretty much wall-to-wall machine room, he remembers, with raised flooring and cables running every which way underneath it. He could see a lot of commotion and installation activity, as well as a whole bunch of telephone lines.

"What's going on?" he asked someone.

"A new box just came in!" came the reply, meaning a new computer. "It's the new IMP! This is our first node on the Arpanet!"

David Burmaster was Project MAC's very young business manager, and not any kind of a computer person. "What's an Arpanet?" he asked.

Here, someone pointed. Maps were being passed around. Burmaster could see how the planned network would extend across the country, connecting some military sites and the major computer-science departments. Just a few of the very best departments, he was assured. "And MIT would be one of the first!"

Actually, the first four sites were already up and running: UCLA, the Stanford Research Institute, Santa Barbara, and Utah had come on line as scheduled, one per month. And more to the point, the packets were flowing. Larry Roberts was pleased and had lost no time in OKing further expansion. IMP number 5 had gone into BBN, of course. And the IMP that Burmaster saw coming into Tech Square was number 6.

Not that the leading lights of Project MAC were particularly interested. Downstairs in the AI Lab, for example, Marvin Minsky and his crew were as suspicious and as xenophobic as ever. And Fernando Corbató's Multics crowd, who hadn't gotten their ambitious new operating system up and running until the previous fall—more than three years late—were still preoccupied with issues of performance and reliability.

Fortunately, however, IMP number 6 did have a home. However dubious the rest of Project MAC might be, the Arpanet was getting a warm embrace from the brand-new "Dynamic Modeling" research group, which had started up in Tech Square just a year earlier. Significantly, this new group was being headed up by David Burmaster's boss: Professor J. C. R. Licklider, the director of Project MAC.

There is no way to sugarcoat this: Lick's tenure at IBM had been a disaster.

"He knew it was not a good match almost as soon as he arrived," says Louise Licklider, remembering their move to Westchester County and the Thomas J. Watson Research Center. He couldn't even get the executives at the lab to call him Lick, for example; they insisted on Joseph, or Joe, a name he hadn't used even as a boy. ("That one," he would whisper to Louise at parties, pointing out one or another of the glad-handers who were forever trying to hustle their way to the top without knowing a thing. "That one's a 'Joe.' ") Then there were the pictures, she says. "They asked him, 'Which do you want hanging in your office: the "Think!" sign, or the portrait of the chairman?' There were only two choices." Well, she says, Lick thought to himself, You've got to be kidding! But he mulled it over, perhaps remembering the Impressionist prints he had hung in his Pentagon office, and finally he said, "I guess I'll take the portrait." But when the portrait was hung, he couldn't help noticing that there in the picture behind chairman Thomas J. Watson, Jr., as big as if he'd hung the thing separately, was a one-word sign: "Think!"

Still, he did his best. As soon as he arrived in the fall of 1964, he set up his lab, hired some bright kids, and went to work on his chosen research topic: finding better ways to create computer programs. Ultimately, of course, he still wanted a graphical, intuitive, responsive programming system that anyone could use with little or no training. And he also knew that that dream was as far away as ever. But he was bursting with ideas about how to get started. He and his young research associates tried to implement a real-time debugger for Fortran: type in a statement, and the computer would tell you immediately if there was a syntactical or logical problem with it. They tried to create a system that would take a schematic flow chart indicating the overall logic of a program and turn it into detailed computer code. They even tried (with some difficulty) to get a Lisp interpreter running on a System/360 computer. They tried everything they could think of. And in the process, says Penny Nii, who joined the group as a programmer after working on IBM's exhibition for the 1964 New York World's Fair, they had a blast.

"Lick certainly didn't fit the IBM mold that I knew," she recalls. "IBM was a very stuffy, very hierarchically organized place. There was not much mixing between managers and workers. But Lick was very accessible. By that point he was quite an old man—at least to me. [In fact, he was then forty-nine.] Yet he liked to talk to young people. He liked exciting ideas; his eyes always twinkled when he talked about ideas. And he always sounded enthusiastic, like a young kid in a toy store. He was a relatively soft-spoken person, but he got people excited about his ideas because *he* was excited and seemed to be having fun. That's probably why I remember him so well; I don't remember any of the other people I came across there."

Unfortunately, however, Lick's aura of excitement didn't seem to inspire very many people beyond his own group. Quite the opposite: dedicated IBMers seem to have grown just as impatient with Lick's visionary, academic manner as had their counterparts at the Pentagon. At IBM, "academic" was considered, well, not irrelevant, perhaps, but certainly not real-world; most IBM engineers still hated the very idea of time-sharing. And worse still, Lick had actually tried to get his group a computer from *DEC!* Never mind that he wanted to base his research on interactive computing, and IBM was making nothing remotely suitable for that purpose. In the land of the crisp white shirts and ties, you never, *ever* bought a computer made by someone else.

In any case, it didn't take Lick long to realize that he and IBM were a hopeless mismatch. "He was so miserable there," remembers Louise Licklider. "Intellectually, it wasn't a very stimulating environment for him. And there was just more mental inertia to push against than he had energy for. The powers at IBM were simply too rigid. They didn't have the imagination to see how any computers other than mainframes could be useful. So he was going crazy, butting his head against the wall."

The final straw came one day in early 1966, when Louise confessed how miserable *she* was. "Our stay in Westchester County was very traumatic for me," she remembers. "I had had a very active life in Cambridge and in Washington. [And now,] all of a sudden, I was living in this big, very isolated house in the woods, not knowing a soul. There were no close neighbors, and I had never lived in the country before. Meanwhile, IBM put Lick on the road constantly, talking about computers. And the IBM hierarchy was such that you did not make contact with anybody above you or equal to you in the hierarchy until you were entertained by somebody above you or equal to you. So we sat there for months, totally isolated.

"Then," she continues, "I felt a great responsibility for Lindy. Tracy had gone off to Harvard, but Lindy was just fifteen or sixteen—not an easy time for anybody. So I was constantly anxious about her when she was off with the other kids her age. Around there the kids were used to a lot more freedom and leniency than I thought she was old enough for. Also, the mothers of two friends of hers were dead drunk when I met them—in the middle of the afternoon."

The result was that Lick immediately went to the powers that be and told them that one way or another, he and his family were going home. He was polite about it, as always. But he was adamant: either they could transfer him to the IBM office in Cambridge, or he would walk.

So they transferred him. By mid-1966, Lick was ensconced at the IBM Cambridge Scientific Center, which was located on the fourth floor of 545 Technology Square, conveniently enough. And to no one's particular surprise, that setup was even more short-lived. "As soon as he was back on the MIT campus," says Louise Licklider, "it became clear there was no point in his staying with IBM at all. He couldn't move fast enough there. And he always wanted to move fast."

Very fast. In short order Lick had not only moved upstairs to become a member of Project MAC but arranged to serve as a visiting professor in MIT's electrical-engineering department—the same department that had been his official home a decade earlier, before he left for BBN. On October 11, 1967, moreover, that temporary arrangement was made permanent: Lick formally severed his last ties with IBM and rejoined the MIT faculty as a tenured professor of electrical engineering. The prodigal son had returned, remembers Bob Fano, "and they were very happy to have him back; Lick was highly regarded at MIT."

Fano himself was even happier. The previous spring, he explains, he had informed the university that he intended to step down as head of Project MAC at the end of the 1967 1968 academic year. (Five years was enough, he felt; he wanted to get back to real research while he was still young enough to do it.) By the time Lick rejoined the faculty that fall, Fano had come to an understanding with the dean of engineering and with Lick himself that there was only one logical successor to him. On January 31, 1968, Fano accordingly named Lick to be his assistant director, as a way of preparing him for the job. And on August 13, 1968, he sent around the official announcement: J. C. R. Licklider would be the new director of Project MAC, effective September 1.

At the time, of course, David Burmaster knew nothing of this; he was a biology and biophysics major just starting his senior year at MIT. But shortly after the semester began, he recalls, "a friend of mine approached me to help him set up a completely separate course, a so-called student-run course. This was the heyday of self-asserting students. So we checked around, and we found out that the dean would let us teach our own course if we got a faculty member to sit in and say that it was bona fide. We did that. Then, as we were getting organized, the six or seven of us who were involved in the course decided that we wanted to do a project. And someone—not myself—suggested that we undertake a project to 'expand computer access to all undergraduate students at MIT.'

"Now, in the fall of 'sixty-eight, this was a pretty radical idea. MIT was famous for computation, but as an undergraduate, you just couldn't get your hands on the machines. So we started going around, meeting all of the senior faculty, and trying to persuade them to do something. And somehow, by the luck of the draw, I was one of two students who were supposed to see the head of Project MAC."

They certainly weren't planning to present nonnegotiable demands or anything like that, says Burmaster; radical confrontation had never been his style (by 1968 standards, in fact, his shoulder-length hair and earnest, dark-rimmed glasses made him seem downright clean-cut). Still, in the fall of 1968 you couldn't be twenty-one years old without harboring at least a few stereotypes about the over-thirty crowd, most of whom seemed exceedingly confused and angry about the sixties, not to mention reactionary. And at first glance, this Professor Licklider appeared to fulfill every one of those stereotypes. "He was an old guy—maybe fifty-five or sixty—and he was sitting behind his desk wearing a coat and tie," Burmaster remembers. (Lick was then fifty-three.) "He was a big man—

I'd say six feet one or two, maybe. And he was sort of portly, or heavy—certainly not trim. Basically he had that rumpled, professorial look."

As he and his friend launched into their appeal, however, Burmaster began to realize that Professor Licklider deserved the benefit of the doubt, and more. "He just had the most fascinating, thick glasses," says Burmaster. "And behind them was the most fascinating part of Lick, which was his bright-blue eyes. He could just sit there and look at a person and understand them, and empathize with them, and listen to them—and then ask some of the best questions possible, and make some of the best suggestions. It was amazing." Indeed, he says, once Lick saw what they were getting at, he seemed just as enthusiastic about the idea as they were, and was soon waxing eloquent about the possibilities; he appeared to love the idea of expanding the on-line community.

Then, says Burmaster, Lick asked them to wait there in his office a moment. "He went out of the office and dictated a one-paragraph letter to his secretary, who typed it. And he came walking back in and said, 'Here, how does this letter strike you?' It was addressed to the MIT provost, Jerome Wiesner. 'Dear Jerry,' it began, and went on to say that Professor Licklider was earmarking a hundred thousand dollars of funds from the next fiscal year's budget of Project MAC to get this project off the ground for students who wanted access to computers.

"I was stunned," says Burmaster. "In nineteen sixty-eight, a hundred thousand dollars was worth a great deal more than it is today. And a few days later, there we were standing in the provost's office, talking to Jerry Wiesner himself. And he was blown away, too. He immediately grasped that universal access for undergraduates was a powerful idea and should go forward. 'If Lick thinks this is a good idea,' he said, 'let's do it.' "

And they did, says Burmaster. He and his fellow students continued to take the lead, "but Professor Licklider really put a lot of muscle behind it. There was an all-undergraduate group formed that evaluated student proposals for computer-related projects and granted access to the computers during off-hours. This would work for the next fifteen years, until PCs came along, and it was an enormous success—a perfect expression of Professor Licklider's commitment to students, to computers, to access."

In short, David Burmaster's admiration of Professor Licklider now knew no bounds. In turn, Lick seems to have developed a high regard for the younger man as well. And then, as the fall of 1968 turned into the spring of 1969, Burmaster found himself about to graduate—a fact that had not gone unnoticed by his draft board (the draft lottery would not begin until the following autumn). "I began to realize with great horror that I was on track to graduate and go to Vietnam," he says. "That was something I did not want to do, for both political reasons and personal reasons. Well, I mentioned that to Professor Licklider at the end of a meeting. And he sort of stored it away."

In fact, Lick was deeply sympathetic; his own son was in precisely the same situation. Tracy Licklider was due to graduate from Harvard that May with a degree in mathematics, which meant the end of his student deferment. The family was worried sick about it; at least once during this period, Lick was heard to muse that they might all have to move to Canada, and he wasn't joking.

In his own quiet way, in fact, Lick was already in the antiwar camp. "He was very idealistic," says Tracy Licklider, who had many, many conversations with his father on this subject (and eventually resolved the draft issue by going into the Peace Corps). "His image of the U.S. military was as the savior of the world from totalitarianism. So to have his government be so misguided in Vietnam was very upsetting." Like many other academics, moreover, Lick had developed a visceral loathing for his government's new chief executive. Having won the presidency in 1968 with the promise of a "secret plan" to end the war, Richard M. Nixon now seemed to have no such intention; in the name of "national prestige" and "credibility," he was allowing the conflict to drag on and on and on, with little more than a vague scheme of "Vietnamization" as a way out. Lick, like millions of others, felt betrayed. One indication of his sentiments was his decision to join Jerry Wiesner and many other top-level scientists in publicly opposing the Nixon administration's plan to deploy an antiballistic-missile system. This didn't win Lick any applause at the Pentagon, but he was adamant. ("I'm sorry," he was overheard telling old colleagues who called to ask what the hell he was doing, "but I wrote what I believe to be the truth. I believe in a strong defense, but in this case, I think that the ABM treaty will be destabilizing, and I think it is my duty to speak out.")

Another indication was Lick's willingness to testify at the draft-board hearings of a number of Project MAC students, where he made the argument that they were doing much more good for their country where they were. And still another sign was a call he made to MIT senior David Burmaster just a few weeks after that young man mentioned his own draft problems.

"He said he was looking for someone to work as his assistant at Project MAC, in the director's office," remembers Burmaster. "So I went and had the interview and was offered the job. Now, one of the reasons that this was such an attractive offer was that under the rules of the U.S. government at the time, Project MAC, being funded by ARPA, automatically carried with it a military deferment. So Professor Licklider essentially gave me that as a personal gift."

Burmaster took it, with a sense of gratitude and wonder that lingers in his voice to this day. Lick had not only freed him from having to make an agonizing choice—among exile, jail, and serving in a war he despised—but given him a way to serve his country in what he regarded as an honorable and worthwhile manner.

Unfortunately, Burmaster's hero worship was all too soon tempered by the reality of watching Lick deal with paperwork, meetings, budgets, and the day-to-day business of running an organization. Lick despised it all and would do anything to get out of it. "The good news was that his secretary, Dottie Scanlon, could run the personnel side," says Burmaster. "She did all of his scheduling and handled all of the paperwork for graduate-student and faculty appointments. And I put a lot of effort into learning the business side. So between us, Dottie and I kept that part of Project MAC functioning quite smoothly. But the administrative part that was more substantive in nature, that couldn't be delegated to a secretary—it just languished. For example, Professor Licklider would carry around one of these big thick briefcases crammed full of papers, and letters, and correspondence, and all kinds of stuff that he had to make decisions on. And he would occasionally 'lose' the briefcase. Now, if that had been me, I would have moved heaven and earth to try to find the thing. But for him, it's fair to say that this was a liberating experience. He felt relieved of certain duties. He once confided in me with a sort of twinkle in his eye, When something like this happens, maybe it's an act of Providence. If a piece of paper was really important, then that person would contact him and ask him to do whatever it was again."

Not surprisingly, this sort of behavior infuriated Lick's colleagues, who found that they had to badger their new director for days, weeks, or months to get anything done. By now Bob Fano was quietly kicking himself and doing his best to keep his mouth shut. The near-universal enthusiasm that had greeted Lick's accession the previous fall had become near-universal exasperation. At ARPA, his colleagues now realized, Lick had succeeded because he never *had* to be a manager; the job had left him free to do what he did best, which was be a visionary. But at Project MAC, where he did have to be a manager, the top administrators had long since quit in frustration—the accountant, the assistant director, the publications director, all of them. That was why Project MAC was being run largely by a secretary and a biology graduate. Lick's former colleagues generally hate to talk about this side of his personality, much preferring to remember him as a warm friend; a superb, meticulous experimenter; and a visionary of extraordinary imagination and insight. But there's no sugarcoating this fact, either: J. C. R. Licklider again and again took jobs that required him to be an administrator—and then was almost contemptuous of the skills it took to do those jobs right.

It was perhaps his greatest failing. And sadly, it was a failing that was not confined to paperwork. For someone who was so warm and empathetic face-to-face, Lick could be remarkably maladroit in dealing with Tech Square's delicate political sensitivities. Take his steadily deteriorating relations with the AI Lab, for example. It's true that Minsky and his crew were infuriating at the best of times. They were more or less *in* Project MAC—sharing its office space and administration, though separately funded by ARPA—yet they had never really been *of* Project MAC. "It was like the Hatfields and the McCoys," says Stanford University's Terry Winograd, who was an AI student during this period. In the ninth-floor machine room of 545 Technology Square, he recalls, the Multics computers were on one side and the AI computers on the other, with a door in the middle—"and you were either a winner or a loser, depending upon which side of that door you were on."

What characterized the "winners," of course, was their fervent, almost religious conviction that they were on the verge of creating truly intelligent machines—not just understanding intelligence, or modeling intelligence, but *creating* it. It was a belief that flowed from Minsky himself, says another graduate student of that era, Patrick Winston, who would later succeed Minsky as the director of the lab. "I think he wanted to make a contribution as fundamental as Darwin's and Freud's. So everyone had a mission, everybody worked hard, everybody stayed up all night."

They never doubted for a minute that they could do it, agrees Winograd, who has since come to entertain a great many doubts. Like their counterculture contemporaries out on the streets, they felt that the wisdom of their elders was old, dead, out of date; the future was what they were inventing right there in Tech Square. "It was very much of an attitude—and I think Minsky still has this to a large extent—that nothing that anybody had ever done before would be relevant to your research," says Winograd. "Everything would have to be invented from scratch, and if you were smart you could do it. Background wasn't relevant; it was pure cleverness that counted."

But that, of course, was precisely why everyone who wasn't a winner was a loser: to Minsky and his followers, the "computer utility" vision that animated the rest of Project MAC was *boring*. At best it was a means to an end, a way to get the computer power they needed for their real work. Likewise this stuff about on-line communities: if the outside world was full of losers, why bother talking to it? And as for Lick's vision of human-computer symbiosis, what was the point? Why waste your time augmenting human intelligence, when humans were virtually obsolete?

In short, Minsky and his crew were very much children of the sixties: brilliant, touchy, self-congratulatory, and deeply suspicious of the establishment—meaning the likes of Fano, Corbató, and now Lick. Handling them required exquisite sensitivity and tact on the part of the director. Bob Fano had managed the task fairly successfully. Lick, sadly, did not. The irritants were little things, mostly, like his having the offices painted without consulting the lab directors beforehand, an act they found enraging. But the little things soon began to add up into bigger things, such as the AI Lab's conviction that Lick was cutting back on their funding in favor of Multics. And relations steadily went from bad to worse, until a final blowup came later in 1970. The last straw was so utterly trivial that even Minsky doesn't remember it clearly anymore. (It had something to do with overcrowding in Tech Square and Lick's proposed reallocation of space; in effect, he was asking Minksy et al. to tear down the wall and mingle with the "losers.") But the upshot was that Minsky marched himself into the office of Walter Rosenblith, who was then the MIT provost, and demanded that the AI Lab be granted a divorce. Rosenblith, who was sick of the bickering, gave it to him: henceforth, Project MAC and the AI Lab would go their separate ways, as completely separate organizations.

By that point, the separation was probably as much a relief to Lick as it was to Minsky, but it was deeply embarrassing nonetheless. And in the meantime, ironically, Lick hadn't been doing any better with the Multics crowd.

In truth, Multics had been Lick's chief concern from the day he became director. Originally planned for completion in 1966, the second-generation time-sharing system was already more than two years late. Indeed, as Fernando Corbató himself had to admit, it was beginning to take on a distressing resemblance to the Vietnam involvement. There always seemed to be light at the end of the tunnel, yet somehow the end never came.

Fundamentally, says Corbató, Multics was just too complicated. In their naïveté, he says, he and his colleagues had drawn up specifications that called for everything an information utility might conceivably need, from elaborate security schemes to an ability to coordinate multiple processors and multiple memory modules. "We were top-heavy with too many ideas," says Corbató, "and we failed to recognize that they would not always go together. So a lot of the rewrite process consisted of pruning things down to the bare essentials and trying to find a more elegant way of doing things."

And then on top of all that, Corbató adds, there was a whole series of complications and delays that were self-inflicted. First there was the too-many-cooks problem: in the three-way agreement among MIT, General Electric, and Bell Labs, no one was really in charge, which made for confusion and slow decisions. Second, the hardware needed a massive upgrade. Corbató and his colleagues had chosen the GE 635 in the first place because it seemed already very close to what they needed, but in practice, the "minor" modifications required were so extensive that the end result was a new machine, dubbed the GE 645. And that, in turn, meant that none of the software from the original machine was usable. "It was like building a house without even having a foundation to start from," says Corbató. "We had to start over and redesign everything."

All of which led to the final problem: PL/1, the new and supposedly all-purpose programming language that IBM had introduced along with the System/360. "In retrospect, PL/1 was a monstrosity to be trying to use to design a system," says Corbató. But at the time, when all they had to go on was IBM's written definition of the language, it seemed like the state of the art in power and flexibility. The Multics team accordingly made plans to buy a PL/1 compiler being written by a commercial software company that had had some previous success with Fortran. But alas, says Corbató, "what we all failed to foresee was that the language had gotten terribly complicated in its specification. And when the company failed to deliver, having strung us along for twelve or eighteen months, we were suddenly in deep, deep trouble, because we were totally dependent on this thing. So we hastily rammed some things together and built a very crude PL/1 compiler, which we began to use to get off the ground."

In short, says Corbató as he looks back on the experience, "I think the naïveté was shared by all of us. Everyone thought it was easier to build systems than it turned out to be." Of course, they were hardly alone in that: in a stroke of truly poetic justice, IBM was running into equally severe problems in its crash program to develop a time-sharing computer compatible with the System/360 line. The first 360/67 machine, as it was known, was delivered by mid-1966, but the first version of its software, designated TSS, for Time-Sharing System, wasn't ready for another six months—and then it was labeled for "experimental, developmental, and educational" uses only. IBM eventually went through eight revisions of TSS, most involving the complete redesign of one or more major components. And yet, as Alan Perlis later groused, "it never succeeded in doing what it was supposed to do." Perlis, who had been gulled into obtaining a 360/67 for Carnegie Tech, swore he'd never use an IBM product again.

Shared naïveté or no, however, Lick had begun to see Multics as a horrendous opportunity cost for Project MAC, a diversion of talent and effort that had left the actual users in limbo. He also knew that ARPA's patience wasn't going to last forever, given the new emphasis on "relevance." Indeed, he could foresee Multics' dragging down the whole ARPA dream with it. "The present time is . . . a critical one in the history of digital computing," he had declared in an unpublished speech he gave at the Technical University of Berlin in August 1968, just a few weeks before he took over at MAC. "Many are waiting to see how the second generation of general-purpose multiaccess systems [that is, Multics and TSS] turns out. [If they] turn out to be too complex to implement, or to operate efficiently, there may be a general retrenchment to noninteractive applications. . . . That would be a great loss, I think, to the universities and, indeed, to all who want to use computers as aids or partners in learning, experimenting, modeling, solving problems, making decisions, and other intellectual activities."

Given that latter scenario, the new director of Project MAC had to wonder if Multics was still worth it. After all, it wasn't the only game in town anymore. The time-sharing system developed at Project Genie might have been somewhat less ambitious, but it was very capable nonetheless: just look at what Engelbart and his crew had done with it. Then there was DEC's big PDP-10, which came with a time-sharing operating system that was likewise workable, if not great. And if those systems were available now, one had to wonder, was there any real point in waiting for Multics? Was this just another case of "perfect" being the enemy of "good"?

Lick thought it was high time for Project MAC to get some answers. But that, alas, was where his maladroitness came into play. His way of dealing with the problem was to have ARPA call for a full-scale, outside review of Multics to determine if the time had come to terminate it.

Opinions differ as to whether Lick was genuinely trying to kill Multics or just attempting to save it with a swift kick in the backside to focus people's minds. But either way, he definitely got their attention. The committee gathered at Tech Square starting December 18 and ultimately ended up granting Multics a stay of execution. "I felt that there were some excellent ideas in Multics, trapped in this overgrown, baroque monster," says committee member Ed Fredkin, who had joined the AI Lab the summer before. "I thought that those ideas would be discredited if Multics was killed, and that would cause general harm to the progress of computer science." Even that favorable decision, however, came at a tremendous cost to Lick's relationship with his colleagues, to whom the whole process had seemed ham-handed, insensitive, and autocratic. Worse, they felt betrayed. Lick, as the director of Project MAC, was supposed to be on *their* side, protecting them from the bureaucrats, the way Fano had always done. Lick's colleagues today insist that they never lost their affection for him as a person. But nonetheless, a coolness had set in, an estrangement that would last for years.

It didn't help that Multics itself in the end proved to be something of a disappointment. Although Corbató and his team did get a usable version up and running by October 1969, some ten months after the review, they would have to put in several more years of effort before the system was really fast enough and stable enough to be usable. And by that point GE had long since lost interest. In 1970, in fact, the company sold its entire computer division to Honeywell, where perplexed managers found that they now owned marketing rights to a product they'd barely even heard of. Only after considerable prodding from Larry Roberts at ARPA did Honeywell finally offer Multics for sale. And while some customers did indeed buy it—an eventual total of seventy-seven copies were sold worldwide, which was fairly respectable for such a major installation—most people in the ARPA community ended up turning to the PDP-10 and Tenex, a new time-sharing system created for that machine by BBN. And then, of course, came the microcomputer revolution of the 1980s, which made Multics look like a complete dinosaur.

If Multics became a dinosaur, however, it was a remarkably influential one. Witness the fact that the Association for Computing Machinery would award Corbató its 1991 Turing Award, computerdom's most prestigious honor. For all its complexity and delay, Multics gave living proof that a grown-up operating system was possible—that sophisticated memory management, a hierarchical file system, careful attention to security, and all the rest could be integrated into a single, coherent whole. In that sense, Multics was a prototype for virtually all the operating systems to follow. And along the way, not incidentally, the project became a model of planning and documentation. At the 1965 Fall Joint Computer Conference in Las Vegas, Corbató and his team spelled out their preliminary designs for Multics in unprecedented detail, giving six separate papers that would occupy some sixty pages of the proceedings volume. Once the project was history, moreover, they would be equally open about the painful lessons they'd learned. In that sense, Multics was a significant contribution to the emerging discipline of software engineering. "I think we probably got people to think at a more sophisticated level about systems design and the need to articulate the design process," says Corbató. "And to some extent I think it's been a consciousness-raising experience for them to realize the kind of problems and hazards you can get into in building systems." If nothing else, he adds, the Multics project was the training ground for a cadre of top-notch students who would soon be taking the lessons they'd learned to DEC, Data General, Prime,

Apollo—all the leading-edge companies of the 1970s. And then, of course, there were two young Multicians, named Dan Bricklin and Bob Frankston, who would go on to galvanize the emerging microcomputer industry with a little program called VisiCalc, the first electronic spreadsheet.

So Multics definitely had an impact. And that's not even counting its biggest, if least direct, legacy, which came in the form of a backhanded compliment.

In January 1969, just a few weeks after the harrowing ARPA review at Tech Square, Bell Labs announced that it was dropping out of the Multics partnership, effective immediately. The timing may or may not have been coincidental, but no one was surprised. Bell Labs' managers had grown increasingly impatient with Multics—they needed a computer system *now*, not in some far-distant future—and work on it there was already close to a standstill. The January decision simply shut down that last little bit of activity and put the laboratory on track to buy a conventional GE 635 computer that would operate in a conventional batch-processing mode.

But then a funny thing happened: two of Bell's former Multicians, Ken Thompson and Dennis Ritchie, began suffering from withdrawal pangs. As baroque as Multics was, it was responsive, interactive, *alive*. And once they no longer had it, Thompson and Ritchie found themselves craving that interactivity. So, good hackers that they were, they decided to write a little time-sharing system of their own. They had to sneak it in on the side, since their superiors weren't about to OK another time-sharing project, and they had to make do with a discarded, obsolete PDP-7 computer that Thompson had scrounged from somewhere. But they persevered, taking some ideas from Multics, others from Project Genie—Thompson had graduated from Berkeley in 1965—and adding a good many of their own. Their prime criterion was that the result had to be lean, clean, and simple, with none of Multics's gargoyles and gingerbread. By mid-1970, they had a preliminary version up and running. Somewhere along the way, moreover, their homebrew operating system had acquired a name. According to one version of the story, the name signified "one of whatever Multics was many of." According to another, it stood for "Multics without balls." But either way it came out the same: Unix.

So there it was: having inherited an admittedly difficult situation in the fall of 1968, Lick had managed to offend almost everybody by the middle of 1969. The crew at the AI Lab thought he was selling them out to Multics, and the Multics group thought he was selling *them* out to ARPA.

But that, alas, was not the end of Lick's troubles. In the fall of 1969 he found himself embroiled in a controversy over the Cambridge Project, a separate ARPA-funded effort that he had helped to organize in collaboration with Bob Taylor, Jerry Wiesner, and a number of psychologists, sociologists, and political scientists from MIT and Harvard.

The inspiration for the project had come from the fact that behavioral scientists had a problem—masses of statistical data that were becoming impossible to analyze by hand—for which ARPA had a solution: the time-shared computers at Project MAC, which they could use to analyze their data interactively. By providing the behavioral scientists with money to buy time on the MAC system, Lick and his colleagues had reasoned, the Cambridge Project would presumably build up a large and vocal new constituency for interactive computing. At the same time, they hoped, such a manifestly benign use of the Pentagon's money would be reassuring and would help to reestablish a dialogue with a segment of academia that had become particularly hostile toward all things military.

Well, it was a nice try. Bob Taylor approved the funding in his capacity as the interim head of ARPA's behavioral-science office, and the Cambridge Project got under way in 1969. But Harvard withdrew almost immediately, after bitter faculty meetings denounced the notion of having any contact with the Pentagon whatsoever. And by the fall of 1969, as MIT was being roiled by its first and only serious bout of antiwar protests, the project was under fire on that campus as well. Rumor had it that the Cambridge Project was a front for the CIA, that its true intent was to develop software that would keep track of people's liberation movements around the world, that giant computers to run that software were already waiting in the basement of the Pentagon, and so on.

To Lick, this was the worst part of what Vietnam was doing to the country: a younger generation was being driven into cynicism, paranoia, and a kind of furious despair. Nonetheless, as a sponsor of the Cambridge Project, he felt bound to stand up and defend it. His tactic was to reach out to the protesters, to try to bridge the gap—and to argue his case on the basis of fact and rationality. "Professor Licklider felt that if he just told the story and made copies of the original proposal available," says David Burmaster, "then eventually the truth would win out. I saw what must have been two hundred and fifty copies of the proposal piled around the Project MAC office. It was free to anyone who wanted it. It was literally handed out on the street corners."

Burmaster also remembers accompanying Lick to perhaps half a dozen faculty and student meetings, where he tried to make the case that the Cambridge Project's research was worthwhile and that the Pentagon's money was as good as anybody else's. "He'd hold up the proposal and say, 'Here it is: show me the page number that you find offensive. We'll just remove that from the project and let the other parts go forward.' And since nobody could ever point to anything, they would just say, 'This is really a front—a sham.' So there was no way to satisfy the critics."

Yet Lick kept at it. One day in October 1969, remembers Douwe Yntema, an old friend from Lincoln Lab who had become the director of the Cambridge Project, Lick attempted to defend it to a raucous group of protesters in front of the student center. "The group was hostile," says Yntema. "But he was pretty cool about it. At one point, in fact, they had a copy of the proposal and tried to set fire to it—not very successfully. Well, after a few minutes Lick said, 'Look, if you want to burn a stack of paper, don't just try to light it. Spread the pages out first.' So he showed them how, and it really did burn much better!"

It's impossible to say just how much difference Lick made. To this day, what MIT old-timers mostly remember about the Cambridge Project (if anything) is that it was "supposed to be some kind of front for the CIA." Still, it's a testament to Lick's manifest reasonableness that he was never vilified personally, despite his long-standing Pentagon connections. It's also telling that the protests against the project soon died down; perhaps Lick's efforts at full disclosure and open debate had found more of an audience than he realized. In any case, the Cambridge Project did go to completion; during its five-year lifetime, in fact, it would become one of Project MAC's largest users.*

Back in Tech Square, meanwhile, reactions to the director's extracurricular activities had been somewhat muted. No one at Project MAC had actively opposed the Cambridge Project, but then, no one had helped Lick defend it, either. And in any case, their director had already gotten himself into fresh hot water. Lick had been so eager to further his visions of graphical programming and on-line community that he had secured ARPA backing to do it himself. The funding for his new Dynamic Modeling research group had come through at the same time as the Cambridge Project; by 1969 the program was well under way.

His colleagues were not amused. "Lick didn't understand what a director should *not* do," Bob Fano says with a sigh. "A director has to allocate resources within the larger project. So when Lick started his own research group, he immediately created a suspicion of conflict of interest."

Nonetheless, he persevered. After all, he reasoned, ARPA was backing Dynamic Modeling not just because *he* thought it was worthwhile, but because professional programmers needed all the help they could get. As Lick himself pointed out in his 1969 article about software for the antiballistic-missile system, the agonies that afflicted Multics, OS/360, and IBM's 360-compatible Time-

---

* Project MAC itself became a target only once, and then very briefly, as the MIT protests reached their height in November 1969. "The politics of it," explains Burmaster, "was that the local chapter of the Students for a Democratic Society wanted to stop all DoD funding at MIT, for any purpose. So at one point they fastened on MAC and the AI Lab. Lick and all the rest of the professors were distraught. They had heard that a group was supposed to rally at such-and-such a time, on such-and-such a street corner, and march on Tech Square. So there was a widespread fear that they might actually have battering rams and smash into the computers. In preparation for that, Lick had guards posted in the building and extra locks put on the outside doors. And he had wood paneling put over the doors leading to the computers."

Given what Project MAC was all about, the symbolism of blocking out the world like that was too painful for words. But fortunately, says Burmaster, the appointed time came and went with no protesters in evidence. The paneling came down, and Project MAC was soon back to normal.

Sharing System had long since become the norm in *every* large software under-taking. No matter what the project was about, it seemed, and no matter how many programmers were assigned, deadlines continued to slip, costs continued to skyrocket, and bugs continued to proliferate. "The presence of such errors in a program is not evidence of poor workmanship on the part of the programmers," he wrote. "[The fact is] that no complex program can ever be run through all its possible states or conditions in order to permit its designers to check that what they think ought to happen actually does happen."[6]

To Lick, of course, the obvious solution to the software crisis was to apply more and better interactive computing, à la Dynamic Modeling. But to pro-grammers who worked in the commercial sector—which was to say, most pro-grammers—the answer was very different. In their world, software was a product to be gotten out the door, on time and within budget. So their instinctive reac-tion was to adopt an industrial approach, with an ever-increasing emphasis on planning, discipline, documentation, coordination, and control. Perhaps not surprisingly, this instinct also fitted in perfectly with the batch process–oriented, think-it-through-ahead-of-time school of programming. By the early 1970s, moreover, it would be enshrined in the doctrine of "top-down" programming, which was pretty much what it sounds like: Start from the top, getting the over-all design of your program right before you do anything else. In particular, strive to divide your programs into modules, or subroutines, that can operate more or less independently and that can be written by separate teams. Then, once that's done, look at each subroutine and get *its* structure right, and so on down. The actual programming code should be the last thing you write. Meanwhile, by no coincidence, the top-down philosophy was being enshrined even further in "structured" computer languages such as Niklaus Wirth's Pascal, which were de-signed to make it hard to write programs in any other way.

Lick himself didn't really seem to have any objection to this approach; to his mind, top-down programming was simply a commonsense method of delegat-ing responsibility. He just didn't think it should be the *only* approach. Before you can delegate responsibilities, he argued, you have to figure out precisely what those responsibilities are and how you should carve them up. And when you're really out there on the edge, as he knew after a professional lifetime of re-search, that process is anything but clear-cut. "On the frontier," he had written in 1965, "man must often chart his course by stars he has never seen. Rarely does one recognize or discover a complex problem, formulate it, and lay out a procedure that will solve it—all in one great flash of insight."[7]

When the systems are truly complex, in short, programming has to be a process of exploration and discovery. That had been the whole point of interac-tive languages such as Lisp, as well as interactive-design tools such as Sketchpad: they made it easy to explore new solutions by making it easy to formulate and then reformulate ideas on the fly. And that was the whole point of Lick's Dy-namic Modeling project: he wanted to push exploratory programming as far as he could in every direction. As Bob Fano explained it in a 1998 biographical memoir[8] of his friend, "Lick's goal was a self-contained, coherent, interactive computer system that could be used by researchers with moderate computer skills to investigate the behavior of a variety of models with little programming required on their part. [So] the software scheme conceived by Lick was very different from any existing problem-oriented language. Briefly, it was akin to a software Tinkertoy, based on a vast library of software modules that readily could be assembled into specific models and programs for investigating them. New modules could be constructed and added to the library by users to meet their special requirements, so that the library would eventually become the repository of the work of the community. . . . The implementation effort started in 1969, and Lick personally developed a significant fraction of the library, which grew over time to some 2,000 modules."

Granted, Lick's software Tinkertoy scheme was wildly impractical for the hardware of the day; he was basically gambling that computer speeds and memory capacity would increase rapidly in the coming decade—as they did. Nonetheless, his approach anticipated the "run-time libraries" of subroutines that are now a standard part of programming languages such as C++. Moreover, because users could add modules at will, his scheme had some of the same open-system quality that had made Corbató's CTSS such a success. And, of course, he hoped to make the assembly of those modules as intuitive and as transparent as possible: "Users were to interact with the computer system mostly through visual displays with light pens or similar input devices," Fano explained.

Lick was never quite able to achieve that goal. (Nor has anyone else done so.) But in the process, he did use his ARPA money to buy a number of commercial graphics terminals known as IMLACs, which immediately persuaded the students that Dynamic Modeling was one of the coolest projects in Tech Square.

"The IMLACs were these really interesting little machines that looked like personal computers on steroids," according to David Lebling, one of the early students. "Basically, each one was a minicomputer with a display processor built in so you could do pictures. They didn't have much memory, and they weren't very fast at computing by themselves, so we mainly used them as graphics terminals for our PDP-10 up on the ninth floor. And they were fabulously expensive; each one cost about forty thousand dollars—in nineteen sixty-nine dollars! But it's difficult to convey today what an incredible advance they were. You could edit your text on the screen! You could make a change, and it would happen before your eyes: it was actually a kind of WYSIWYG [What You See Is What You Get, a feature of more recent word-processing programs]. In fact, you could do fonts on the IMLAC; I did a Russian font just for fun." Another student named Stuart Galley used the IMLACs to produce a graphical simulation of what was happening inside the group's own PDP-10. The simulation allowed a user to monitor the instructions being executed, the contents of the PDP-10's registers, and the contents of its memory. Indeed, it was one of the first "graphical debuggers," a category of tools that has since become standard for professional programmers. And from the very beginning, of course, Lick's students programmed the IMLACs to play games such as Pong, and Spacewar, and Maze—the last being a time-shared, multiperson game in which you wandered around through a three-dimensional labyrinth and tried to shoot all the other players. Students would come from all over Tech Square to play, and would occasionally find themselves in a shootout with the director of Project MAC himself.

But then, just as their counterparts at IBM had discovered some half a decade earlier, Lick's students were learning that *this* boss was not your ordinary boss. He could definitely expand your mind, says yet another former student, Christopher Reeve. "For example, he thought that computers should have unlimited amounts of memory. Now, this was way back when memory was very expensive, and a computer with a megabyte was very uncommon. But Lick always had a certain level of enthusiasm that was quite contagious."

"Lick was always ten steps ahead of everybody else," agrees Al Vezza, a young Project MAC staffer who was now Lick's deputy on the Dynamic Modeling project. And that was especially true when it came to the Arpanet, which Lick regarded as central to his vision of an on-line community. When the Arpanet started expanding, in early 1970, Lick immediately took on the responsibility of getting MIT connected. Never mind that it gave the rest of Project MAC one more thing to be irritated about. The AI Lab, the Multics team, and all the others were welcome to drag their feet if they wanted. Lick was quietly determined to get his PDP-10 on the network *now*.

SHOWING THE WORLD

Up in the director's office, Project MAC's business manager didn't quite know what to make of all this networking stuff. But as it happened, David Burmaster had a roommate, named Bob Metcalfe, who was very interested indeed.

"Harvard wouldn't let me do what I wanted to do," says Metcalfe, explaining that even though he was an MIT man—he had graduated from there in the spring of 1969, with Marvin Minsky himself serving as his senior-thesis adviser—he had gone over to Harvard that fall to begin work on a Ph.D. "The Arpanet was just starting up, and I wanted to get Harvard connected. But Harvard thought that was too much responsibility for a grad student, so they gave the job to BBN—which immediately turned around and gave it to another graduate student. Anyway, I needed a job. So when Dave told me there were openings at MAC, and said I should talk to Al Vezza, I went down the river to MIT. Vezza said, 'Sure, I'll pay you to connect us.' Other people were working on that, too. But my job was Lick's computer: get IMP number six hooked into the PDP-10."

He did—and very soon, since he was now one of the few people on the planet who had successfully connected anything at all to the network, Metcalfe found himself becoming an "Arpanet facilitator." "That was a group that included me, Jon Postel, Steve Crocker, Vint Cerf, and a bunch of others," he says. "Our job was wandering around helping people get on the Arpanet. Now, this was the Vietnam era, so we all had these huge beards, and here we were on air force bases! In fact, Postel and I were once asked to leave an Officers Club because of our beards—although Steve remembers it was because of our blue jeans!"

Actually, the demand for hand-holding was pretty high in 1970. The network was growing fast, with roughly a dozen contractor sites on line by the end of the year, not to mention the odd air force base. But the startup was not going smoothly. Up at BBN, for example, junior software guy Dave Walden found it prudent to keep a printout of the IMP routing code right next to his home telephone, because users would call him day and night, desperate to know why their IMP had just stopped. "My telephone number was quite literally on the front of the original packet switches [IMPs]," he says.

In truth, the first release of the IMP routing software had been a very long way from perfect; as the field reports came rolling in, Will Crowther, Walden, and their fellow software mavens found themselves chasing down bug after bug. And in the unlikely event that anyone at BBN ever forgot those imperfections, Len Kleinrock and his students at UCLA were happy to remind them. As the official Network Measurement Center, they had been probing the network, calibrating the network, stressing the network, and trying to crash the network from the day they hooked up IMP number 1. They had done everything they could to identify weak points and problems in the network. And they'd found plenty.

So had BBN's Bob Kahn, who would camp out at UCLA for weeks at a time. "He wanted to force the network into certain modes in which he predicted it would fail," says Vint Cerf, who first met Kahn during these visits. "Other people at BBN didn't believe it would fail this way. So he came out to UCLA, and we struck up a very productive collaboration: Bob would ask for software to do something, I would program it overnight, and we would just blast the thing and the network would break. Then Bob would be happy." One of the first and most dramatic such demonstrations came in October 1969, when they showed that under certain circumstances, as few as twelve errant packets could cause the network to lock up solid: every packet would just sit there in some computer's memory banks, waiting for another packet to move first.

Even as the IMP guys at BBN were struggling to get those software problems fixed, unfortunately, they also found themselves dealing with the site representatives, who were more frantic than ever: not only were more sites being added to the Arpanet all the time, but Larry Roberts was putting considerable pressure on them to start *using* the thing. BBN did its best. After all, as Frank Heart points out, "it was in our interest to have them have a happy connection." Nonetheless, the role of Arpanet facilitator increasingly fell to Crocker, Metcalfe, and the other students in what was now being called the Network Working Group. Certainly it had become something like a full-time job for Crocker, who was the group's de facto ringleader. "As I started to get heavily involved with this, I started to visit a lot more places," he says, "and my travel costs started to go up. Well, one day Kleinrock asked me what my plans were for the next couple years. So I made up what seemed to me an outlandishly large number—thirty thousand dollars a year—and he just wrote it into the budget!"

The Network Working Group would continue to be a big part of Crocker's workload even after he moved to ARPA in July 1971 (Barry Wessler had left to finish his Ph.D. work at the University of Utah, and Crocker got his wish: Wessler's job). He didn't get to work in the Pentagon building itself, as it happened; a year earlier, having lost out in the DoD's never-ending office-space wars, ARPA had been relocated to an anonymous glass tower in Rosslyn, Virginia, a high-rise office district just across the Potomac from Georgetown. But Crocker rarely got to sit still anyhow. "I had a set of things that I wanted to do at ARPA," he says. "The AI program, the speech-understanding programming, and so forth. I was trying to extricate myself from the day-to-day business of the network. But the Arpanet was growing during this period of time, and it was a lot of work to add each node to the network. So because of the time I had spent in the field, and the expertise I'd developed, I continued to work with a lot of the new host sites to give them advice, help get them connected, and so forth."

And in the meantime, of course, there was still that little matter of getting the higher-level protocols straightened out. Larry Roberts was very supportive of the working group's efforts in this area—as well he might be, since its members were helping him solve a problem he'd barely known he had. But having given them that support, he now expected them to deliver. "Steve was sort of the putative head of the Network Working Group," says Vint Cerf. "So he would report progress to Larry, and if Larry was unhappy with something, he would holler at Steve. Steve would take the heat, then he would come back, and we would all caucus and figure out how we were going to fix it."

Progress was painfully slow. The protocols were originally supposed to be ready by the summer of 1970, but that deadline had come and gone with no sign of completion. Meanwhile, the Arpanet was a network in name only. "It was like picking up the phone and calling France," says Frank Heart. "Even if you get the connection to work, if you don't speak French you've got a little problem."

"The reality was that the machines that were connected to the net couldn't use it," agrees Bob Kahn. "I mean, you could move packets from one end to the other. You could run all the test programs you wanted. The network nodes could even send test traffic. I could sit at the Teletype at one place, connect it to an IMP, and talk to somebody at the other end. But none of the host machines that were plugged in had yet been configured to actually use the net."

No, they hadn't, concedes Vint Cerf. But it wasn't for lack of trying. "Initially," he says, "progress was sluggish because there was an incredible array of machines that needed interface hardware and network software. By nineteen seventy-one there were about thirty different university sites that ARPA was funding. We had Tenex systems at BBN running on PDP-10s, but there were also PDP-8s, PDP-11s, IBM 360s, Multics on Honeywell machines—you name it. So we had to implement the protocols on each of these different architectures." And once they'd done that, of course, they had to make sure that everyone's implementation worked with everyone else's.

It was Al Vezza who suggested the solution: a software "fly-off." On a given date, he said, everyone should come together in one place—at MIT—and try to log into everyone else's site. Then they would each pound on the implementation software until it worked. (Lick was happy to go along with his deputy's idea, of course—but then, by that point he was a much happier man in general. Having recognized that his effectiveness as director of Project MAC was approaching zero, and that his colleagues were simply going around him, he had resigned earlier that same year. Much to everyone's relief—including his own—he had handed over the reins to his onetime protégé Ed Fredkin and gone back to running his research group full-time, which was all he'd really wanted to do in the first place.)

The date for the fly-off was accordingly set for October 1971: one solid week of hacking in the warrens of Tech Square. "We invited them all down to the bullpen on the second floor, where they could use the IMLACs," says Vezza. "And that workshop was really a critical turning point in ironing out the difficulties, because everyone was right there together. People could shout down the hall—'Why doesn't your program do this or that?'—and get an answer right away."

The fly-off was seminal, agrees Metcalfe. "I remember we had a big blackboard where we drew a matrix with all the participants listed down one side, then listed again across the top. And we'd ex out the boxes as each one got their implementations of the protocols to talk to another." The matrix started out mostly blank, depressingly enough. But slowly, as the week went on and as people kept pounding on the software, the Xs began to accumulate. And finally, as Crocker would write in *RFC 1000*, "with the exception of one site that was completely down, the matrix was almost completely filled in, and we had reached a major milestone in connectivity." The Arpanet was up and running for real—the perfect moment for what Metcalfe remembers as "a great big party."

And yet as exhilarating as the moment was, they knew even then that they weren't through. They had convinced *themselves* that the Arpanet worked and was worthwhile. But out in the mainstream, the old-line communications engineers were as skeptical as ever about packet switching. The basic reaction was "Go away, little boy, there's no revenue there," notes Len Kleinrock. The little boys of the ARPA community could win that argument only if they *showed* people what the Arpanet could do, publicly and convincingly.

Amen to that, said Bob Kahn, who had been making much the same case all along. The fly-off had finally gotten the protocols working, he pointed out. But the protocols by themselves were just the network's operating system, so to speak. What was needed now was a fly-off writ large: a public demonstration of the Arpanet that would give researchers an incentive to get their applications up and running on top of the protocols—and that would be so compelling that no one, not even the phone company, could ignore it.

Larry Roberts thought that sounded like a great idea and quickly came up with the perfect venue: ICCC, the first International Conference on Computers and Communications, scheduled to be held in Washington in October 1972. Everybody would be there: techno-weenies from the ARPA community, computer engineers from the mainstream companies, communications mavens from the telephone companies—everybody. All the ARPA crowd would have to do for the demonstration was organize it.

And that, Roberts told Kahn, will be *your* job.

Meanwhile, another little demonstration was getting under way on the banks of the Potomac. It was sometime around April 1972, remembers Steve Lukasik, who had become director of ARPA in 1970 after Eb Rechtin stepped down. "I always kept an open door to my office so that people could come and go any time they wanted," he says. "So one day around four o'clock Larry came in and said, 'Hey, Steve! I'd like to get you on the Arpanet.' "

"But what am I going to do on the Arpanet?" Lukasik replied. He was a manager, for heaven's sake. What use did he have for file transfers and remote log-ins?

"Well," said Roberts, "you could do E-mail."

E-mail? Lukasik knew what it was, of course. Every time-sharing system in the country had E-mail, starting with Project MAC in 1965; it was one of the main things people *did* with time-sharing. From the beginning, moreover, Bob Taylor and Larry Roberts had always assumed that E-mail would spread to the Arpanet; every time they'd had to explain what the network could do to their bosses in the Defense Department or in Congress, they had put it on the list. It was just that they'd always put it way *down* on the list. And then when Crocker, Cerf, and the others had gotten down to business with the Network Working Group, other things had always seemed more urgent. At the fly-off in October 1971, E-mail hadn't even been an issue; no one had found the time to implement it yet.

Until now, apparently. Earlier that year, Roberts explained, a BBN engineer named Ray Tomlinson had sent a message between the company's two PDP-10s—the same ones that BBN was using to run the Arpanet itself, as it happened. It wasn't a difficult experiment; Tomlinson had already written an E-mail utility for Tenex, BBN's new time-shared operating system for the PDP-10, and had also begun to experiment with a new version of the Arpanet's file-transfer protocol. So putting the two together seemed a natural step.

It was: Tomlinson found that he could transfer messages between the two

PDP-10s flawlessly. And of course, since the messages were actually flowing through the Arpanet itself, getting them across the room was tantamount to getting them across the country. Tomlinson had accordingly packed up the results of his experiment into an elegant little utility program he called sndmsg, for "send message." Invoke it, and the thing would ask you for a *To:* field, a *From:* field, and a *Copy:* field, plus a few other options. Then you would type your message and hit Control-Z, and off it would go. At the other end, the recipient could see what you'd sent by invoking a companion utility that Tomlinson called readmail. In writing these utilities, moreover, Tomlinson had come up with an elegant way to define E-mail addresses: take the "user name" that the person typed when logging in to his or her host computer, and simply link it to that computer's "host name" on the network with an "at" sign: username @ hostname. But the bottom line, said Roberts, was that Arpanet E-mail was *here.* So if Lukasik was willing to try it . . .

Lukasik knew exactly what Roberts was up to with his invitation. "Larry wanted me to experience the Arpanet," he says. "He knew that you could explain it, wave your arms, show charts, and so forth. But until you experience it, you don't *know.*" And more than that, he says, Roberts wanted him to bear witness. "Larry knew that his researchers had done a wonderful thing in creating the Arpanet," he says, "and that it was something that could be very important to the DoD. But he also knew that the researchers could go only so far in penetrating the defense community. So he wanted me to be the central interface with the military. That's what was really going on—not just getting the boss playing with your toy."

In any case, continues Lukasik, "within a day or so, a model 33 Teletype terminal appeared behind my desk. It had its own stand. There was a roll of paper coming out the back. And I was on E-mail."

Of course, as Larry Roberts would soon have cause to know, it wasn't a completely great idea to get Lukasik up and running on an Arpanet that was still so experimental. If the software utilities were bug-ridden hacks, the boss knew about it. If the documentation was miserable, the boss knew about it. And if the network crashed in flames, the boss *really* knew about it. "I'd immediately be on the phone complaining to people," says Lukasik, who admits that he's never been noted for his patience or his sweet temper in such situations. "Larry subsequently said that he'd probably made a mistake—that he'd put me on the net six months or so too early—because I saw all of the flakiness!"

Flakiness or no, however, the ARPA director was soon firing off messages to Roberts and anyone else with a terminal. He was the first person outside of IPTO and the research centers even to be on the Arpanet, and he had fallen in love with E-mail. He loved it so much, in fact, that he quickly decided to bring ARPA itself on-line: from now on, he decreed, *everybody* at the agency would have a terminal.

This move did not exactly meet with universal acclaim at ARPA. Even in so technocratic an agency as this one, there were a fair number of technophobes who loathed E-mail, not to mention the Teletype terminals. But no matter: "The way to communicate with me was through electronic mail," says Lukasik. "So suddenly the Strategic Office understood its utility, and the Tactical Office understood its utility, and my old Nuclear Monitoring Office understood its utility. Almost all the offices got on the net."

Indeed, Lukasik soon began to realize that he'd succeeded a little *too* well: he was getting flooded by E-mail messages, to the point where he couldn't make sense of them all. "What you got in your mailbox with readmail was just a list of messages in chronological order, oldest first," he explains. "So I went to Larry: 'Larry, what do I do with all these messages? They just accumulate.' Well, he literally went home one night and wrote a little system called rd, for 'read mail.' It had folders for saving mail in this category or that category. It allowed you to selectively retrieve messages. It allowed you to throw messages away. It was the first beginnings of a mail *system,* as opposed to just sending telegrams."

Lukasik loved this program, too, of course. But then, so did everybody: rd and its derivatives were soon proliferating over the Arpanet just as rapidly as Tomlinson's utilities had ever done. Indeed, E-mail had already begun to trigger a profound shift in the ARPA community's attitude toward the network. What had once been a nuisance, accepted only grudgingly, was now wildly popular. "E-mail was the biggest surprise about the ARPA network," says Len Kleinrock. "It was an ad hoc add-on by BBN, and it just blossomed. And that sucked a lot of people in. It still is the biggest use of networks today."

"What the Arpanet did was knit this community together more tightly than it had ever been knitted together before," agrees Bob Kahn. "Remember, before the network was built, people in the computer-science community didn't necessarily know one another. It seems hard to believe, but at that time, if you would ask the MIT faculty who's on the computer-science faculty at Stanford, I bet you some of the names would have been a blank. It was not a tightly knit community because there was not much basis for interaction." That's why the idea that you might want to send E-mail across the country hadn't really entered into the discussion of the Arpanet, he says. But once E-mail was there, the process that had started in the 1960s with the PI meetings and the graduate-student meetings suddenly accelerated enormously. Because of Arpanet and E-mail, the ARPA community was rapidly transformed into a community in fact as well as in name.

But all that would happen later. For now, Kahn and his colleagues still had the ICCC demonstration to take care of—especially if they wanted the Arpanet ever to be accepted by the *rest* of the world.

Kahn had needed no further encouragement from Larry Roberts: if the ICCC demonstration was his job, then it was going to get done. For the next twelve months he would devote himself almost exclusively to this one task.

The basic idea was straightforward: just fill up the cavernous interior of the Washington Hilton ballroom with dozens of Arpanet terminals, then invite the conference attendees to come in and tickle the keys. The devil, as always, was in the details. Kahn spent much of that year cajoling researchers from all over the country to get their applications up and running so that the key-ticklers would actually have something to look at. When he wasn't doing that, he was arranging for a professional film crew to create a thirty-minute-long 16mm movie explaining the Arpanet, complete with an electronic-synthesizer soundtrack to set an appropriately futuristic mood. (The film covered everything from the basics of packet switching—as diagrammed on a blackboard by Kahn himself, looking very young and earnest in his long black sideburns—to the potential social and economic implications of the network, as expounded by J. C. R. Licklider at his most avuncular.) In between, Kahn was commuting to Washington to give periodic progress reports to Larry Roberts. And of course he was also arranging to have a TIP installed in the Hilton's basement—TIP being short for Terminal Interface Processor, a kind of second-generation IMP that could connect many terminals at once to the Arpanet without needing a stand-alone host computer.

The job was overwhelming, but fortunately for his sanity, Kahn also found himself joined by a small army of helpers: in one way or another, most of the Network Working Group pitched in. "Bob managed this whole project," says Vint Cerf, who was among the most active. "It was his baby. But we took advantage of our work together in the working group to build a team. I remember that Bob Metcalfe led the group that wrote the manuals for how to use the system. Literally, it was, 'Sit down, type the following things, and this will happen.' Ken Pogran, who was instrumental in developing an early ring-based local-area network and gateway, which became Proteon products, narrated the slide show. Jack Haverty, who was an MIT undergraduate and who later became chief network architect of Oracle, was there with a holster full of tools. Frank Heart was there, along with David Walden, Alex McKenzie, Severo Ornstein, and others from BBN. A whole set of people who are now legendary in networking history were involved in getting that demonstration set up."

His own job was mainly coordination, Cerf continues: "I remember doing things like maintaining the mailing lists, making sure everybody knew who was responsible for what, trying to organize what the demos were, who was going to do what demos." In the end, that task devoured virtually all his time from March 1972, when he finished his Ph.D. dissertation at UCLA, until a few days before the demonstration in October, when he flew into Washington. And then, says Cerf, things *really* got hectic: "I can remember being up early, pacing the floor, walking around trying to make sure there weren't any loose ends dangling in the application models, trying to make sure that everybody knew what they were going to demonstrate and when. A lot of other people were doing that sort of thing, too; I don't mean to mistakenly argue that I was at some central point of it all. But I certainly was busy, along with fifty other people, it seems like, trying to get that whole thing functioning well."

They succeeded.

"It was a major event," declares Kahn, who still remembers the demonstration as one of the high points of his career. "We had many different vendors contributing terminals. We had just about everybody involved in networking at the time there. Over a thousand attendees at the conference came through. It was a hands-on, live demonstration. . . . It was a happening!"

It was indeed, confirms Len Kleinrock: "People were pulled out of the hallway, handed a handbook, and told, 'Sit down, we'll help you use the Arpanet'— and they *could*." Moreover, he says, the stuff those first-timers saw was really amazing. "One of the things that was demonstrated there was a distributed air-traffic-control system [created by BBN]. The idea was there would be some air traffic on the system. And as a plane moved out of its region, it would be picked up by another computer in the next region, data would be exchanged, and collectively the computers would be managing this whole air space over a large region. This package also had the ability to freeze the simulation at any given point, take the program on machine A, squeeze it over to machine B, and then continue the simulation with the user being none the wiser. So there were really some sophisticated things going on there."

Another demonstration he'll never forget was put on by Jon Postel, says Kleinrock. "In this demo," he says, "you would sit down in Washington at a Teletype, log on to a machine at BBN, pull up some source code, ship it across the country to a machine at UCLA, compile and execute it, and then bring back the results to be printed on the Teletype right next to you in Washington. So Jon sat down to demonstrate this thing: log on to BBN, and so forth—and nothing happened! He couldn't figure out what was wrong. He kept looking around until suddenly he saw a turtle on the floor." Now, this "turtle" was actually a little wheeled gadget that was a part of an artificial-intelligence demonstration developed by MIT, explains Kleinrock. It was supposed to roll around the room, bumping into obstacles and thereby learn to avoid them. But in this case, Postel could see that the turtle was twitching spasmodically. "It turned out that Jon's output was going to the turtle," Kleinrock recalls, laughing. "Someone had messed up a connection. So there was Jon's output, jumping around on the floor!"

Meanwhile, Vint Cerf was watching the attendees who were seeing all this for the first time. "I thought there were three kinds of reactions," he says. "The first came from the die-hard circuit-switching people from the telephone industry, who didn't believe packet switching could possibly work. And they were stunned because it *did* work. The second group were the people who didn't know anything about computer communications at all to speak of, and who were sort of overwhelmed by the whole thing. But then there was the third group: the people who were just as excited as little kids, because of all these neat things that were going on."

Cerf was as excited as any of them. "My God," he says, "one of the demonstrations involved Chinese-character displays! As I recall, it was Susan Poh from MITRE who did that work. We also demonstrated our own tools for network measurement: people would be doing things on-line, and we would show them displays of where all the traffic was going, and which IMPs were talking to each other, and things like that. We had simple demonstrations of E-mail, text editing, file transfers, and remote log-ins. But we also had access to the oN Line System at SRI International: you could get into the databases there and get documents, *RFCs*, and things like that. So it was just a remarkable panoply of on-line services, all in that one room with about fifty different terminals."

Of course, Cerf adds, the unsung heroes were the BBN guys, who kept the network up and running through the entire ICCC meeting with only one brief crash–though that one failure did manage to leave Bob Metcalfe twisting in the wind.

Metcalfe remembers it well. As editor of the demonstration handbook, he explains, he became the designated "demogod," the guy who gave the tour to visiting VIPs during the meeting itself. "So picture this grad student with his huge red beard and Birkenstocks," he says. "And now picture these ten pinstripe-suited AT&T executives who arrive in a group. They were in the land of packet switching, which was viewed as the opposite of the phone companies' circuit switching, and there was definitely a technological animosity. So I started walking them through the ballroom doing demos. At one point they were clustered around looking at the screen as I typed, and the damn thing crashed. The only time the Arpanet crashed in the whole three days of that meeting was right then.

"Well, I sat there for a moment like you always do, hoping that something would actually work. But it never does. So I looked up at the guys in the pinstripes, and I found them all *smiling*. They were *happy* that it crashed. And it came alive for me in that moment: they were hostile. In fact, from that moment forward I carried a grudge against guys in pinstripes. I did eventually learn to wear pinstripes myself–but I still carry that grudge."

Happily, however, the glitches didn't really matter in the end. The ICCC demonstration did what it was intended to do, which was make the world sit up and take notice of packet switching. It was what Metcalfe calls the Arpanet's debut–its coming-out party, its coming of age. "Up until that point you couldn't *see* it anywhere," says Kahn. "All you could do was read an arbitrary abstract paper somewhere that said, 'Here is this new way to do computer communications.' But ICCC was the watershed event that made people suddenly realize that packet switching was a real technology."

DIASPORA

Looking back on it, there were any number of ways that the Arpanet project could have failed. It could have been snuffed out by the Vietnam-era budget crunch before it even got started, as Pentagon officials scrounged for money high and low. It could have been crushed by the mainstream telecommunications community, which saw packet switching as utterly wrongheaded at best and a competitor at worst. Or it could have been ignored and left to wither away by researchers who couldn't really see any use for it.

And yet the Arpanet succeeded beyond all expectations, thereby pioneering the technology (and becoming the first piece) of that all-pervasive entity known today as the Internet. "It was one of the great experiments in science," declares Len Kleinrock. "It completely changed the way things are going now—commerce, government, industry, science, everything."

A good part of the credit for that triumph goes to the leadership of ARPA, notably Charles Herzfeld and Steve Lukasik at the agency director level and John Foster at the DDR&E level. These men not only understood the vision that animated their computer office but protected and encouraged what that office was trying to do. And perhaps most important, they continued to foster ARPA's extraordinarily un-federal-government-like management style—one that might be summarized as allowing "the freedom to make mistakes."

An even larger share of the credit goes to the successive directors of IPTO itself: J. C. R. Licklider, Ivan Sutherland, Bob Taylor, and Larry Roberts. Although the ARPA management style granted them enormous, almost unfettered authority to dictate the course of research, they all almost invariably exercised that authority with taste, tact, and restraint. To the researchers, they were not dictators so much as protectors, intermediaries who would keep inquisitive congressmen, senators, and generals out of their hair—not to mention paperwork off their desks. "They wanted progress," notes Kleinrock, "not progress reports."

If there was an exception that proved the rule, it was Larry Roberts. On the one hand, he definitely did the Arpanet *his* way, whether the intended beneficiaries wanted the thing or not. On the other hand, he was also quick to seize an unlooked-for opportunity—the Network Working Group—to get users involved with the network (the younger ones, anyway). "For a lot of reasons both technical and political," says BBN's Alex McKenzie, "Larry knew that if the host organizations were ever going to take the network seriously, and really connect to it, and really begin using it, and really begin finding applications for it, then they had to be involved, too. So I think the working group was Larry Roberts's way of saying, 'We don't want to put all our eggs in one basket, with one contractor [BBN] telling everybody how to do everything. Everybody has to be involved because there are good ideas everywhere, and because ARPA wants people to feel like they have a stake in the outcome.' "

Through his support for the working group, in short, Roberts was coaxing the people who used the Arpanet to take charge of their own destiny, to learn how to govern themselves. And almost without realizing it, the young researchers who heeded the call ended up creating a parliament that was about as democratic as anyone could imagine: if you wanted to be a member of the Network Working Group, you were. Almost by accident, moreover, they created a transnational forum for comment and debate, in the form of the *RFC* series, which was soon being promulgated around the world over the Arpanet itself.

Their parliament was in session everywhere, all the time. And yet by technological necessity, they almost inevitably found themselves arriving (eventually) at a consensus; they knew full well that a protocol had to be universal to be a protocol at all. They had to agree, or the bits wouldn't flow. "There was a mixture of competitive ideas," says Steve Crocker as he thinks back on those meetings. "But people were also talking to each other. There weren't armed camps saying, 'You've got to do it my way or not at all.' And all this was happening without anybody having to summon people to a meeting or let formal contracts." Indeed, this brand of hyperdemocratic, bottom-up decision making proved to be so effective that it would later become the model for the governance of the Internet. Contemporary standard-setting bodies such as the Internet Engineering Task Force are essentially the Network Working Group writ large.

So Roberts was definitely that exception. And in proving the rule, not incidentally, he opened the way for the group that deserves perhaps the largest share of the credit for the Arpanet's success: the rising generation of researchers who participated in the Network Working Group, who built the hardware, who debugged the software, who got their hands dirty, who made it happen. Like their parents in the World War II generation, they had a sense of participating in epochal events. "One thing that Roberts, Kahn, and those people who are real thinkers realized early on," says Dave Walden, "and what the rest of us certainly realized quite quickly, was that putting in this infrastructure would change the way the world worked—not just the communications world, but the way *people* worked. From the first time we sent a message across the network or wrote a paper across the network, none of us had any doubt that what you are seeing today with the thousands of distribution lists and virtual networks and worldwide queries was going to happen." Like their parents before them, moreover, the members of this Arpanet generation often found themselves exercising responsibility at a remarkably early age. And perhaps that is why so many of them became major figures in the later development of the network.

Certainly they were already moving out into the world. By the time of the ICCC demonstration in October 1972, Steve Crocker had long since gone to ARPA, of course. And Bob Kahn was about to follow him; Roberts had already made him an offer. After half a decade of obsessing about networks night and day, says Kahn, he was getting a little sick of the subject. Going to work for Roberts seemed like a chance to try something *new*. "I went there intending to make a clean break," he says wryly, knowing how ironic this sounds in retrospect. "In fact, the agreement I had with Larry was that I would set up a program in flexible automation in manufacturing."

That sounded good to Kahn; he left BBN for ARPA almost immediately after wrapping up the ICCC demonstration. His buddy Vint Cerf, meanwhile, was on much the same schedule: as soon as the demonstration was over, he left UCLA to take a faculty position at Stanford. And Bob Metcalfe . . .

Well, Metcalfe had left Harvard quite a while back, as it happened. He had gone out to the West Coast to work for—of all people—Bob Taylor. In fact, a truly amazing number of ARPA alumni had been doing the same thing of late; Taylor was snapping them up for a brand-new laboratory funded by—of all places—Xerox, the copier company.

PARC, they called it: the Xerox Palo Alto Research Center.

CHAPTER 8

LIVING IN THE FUTURE

It all began one day in May 1969, explains Jacob E. "Jack" Goldman, a large, self-confident New Yorker who still can't go more than fifteen minutes without lighting up a cigar ("My trademark," he says through a cloud of smoke). The Xerox Corporation had just signed on the dotted line to buy a computer manufacturer called SDS: Scientific Data Systems, of El Segundo, California. And Jack Goldman was feeling a trifle agitated about it—not an unusual condition for him.

Now, on paper, Goldman concedes, the purchase did make a certain kind of sense. Only a decade earlier, the Haloid Company of Rochester, New York, had been a modest-sized photographic paper and supply firm struggling to perfect xerography, a radical new copying technology it had acquired from the inventor Chester Carlson all the way back in 1945. But now, as Xerox, the company was racking up worldwide sales of more than $1 billion per year. It was a superstar of the Fortune 200, the darling of Wall Street, the creator of a whole new industry. It had revolutionized the way people communicated; the tens of billions of copies now being created each year included everything from bridal-gift lists to brokerage firms' hot tips. The company had succeeded to the point where its

Getting personal: one of PARC's new Altos

brand name had followed "Kleenex" and "Aspirin" into the everyday language: to photocopy a document was to *Xerox* it.

And yet, for all its meteoric rise, Xerox was still a one-product company, a fact that had long troubled its former president Joe Wilson, now chairman of the board, as well as its current president, C. Peter McColough, who had master-minded the equally explosive expansion of Xerox's sales and service forces. Both men feared that the company's days of glory might be very short-lived indeed un-less it moved now to diversify, to transform itself into a much more broadly based communication business. And the key to that, they believed, was computing. Xerox would use its prodigious cash flow to create the electronic office of the fu-ture—or, as McColough would put it in March 1970 in a speech to the New York Society of Security Analysts, it would pioneer the "architecture of information." Having decided to get into computers, moreover, Wilson and McColough had been determined to do so as quickly as possible, which meant buying an existing computer company instead of developing the machines on their own. They had begun to put out feelers as early as 1965. And after establishing that several likely firms were not for sale at any price—DEC, for one—they had finally reached an agreement in 1969 to buy SDS for $900 million in Xerox stock.

So indeed, says Goldman, the deal made perfect sense on paper. What trou-bled him about it was the reality. For one thing, this was the first he had heard of it. The Xerox top brass had just spent nearly $1 billion for a computer com-pany without even mentioning their intentions to him, their own vice president for research. "They used management consultants," he says, making the term sound like a swearword. But let that go. He was the new kid on the block; he'd started at Xerox only the previous December. His real concern was that SDS was a turkey, the Edsel of computer firms. The world was shifting toward the small, inexpensive "minicomputers" being pioneered by companies such as DEC and Data General, yet those guys out in El Segundo were still obsessed with making mainframes, mainframes, and more mainframes.

Worse, Goldman knew for a fact that Xerox was manned by a pack of com-puter illiterates. Back at Ford Motor Company, he says, where he had been founding director of the automaker's in-house research laboratory, it had been all he could do to hold his scientists back: "Every one of them wanted a PDP-8, or a Nova—a minicomputer that he could have for himself, and not have to tap into a time-sharing machine," he says. But when he'd first arrived at Xerox and his predecessor had given him a tour of the company's lead research lab, in Rochester, New York, the only computer he saw there was a UNIVAC in payroll.

In short, Goldman could see, Xerox's move into computers was shaping up to be the blind leading the blind. Not good. So he fired off a memo to Mc-Colough that very day. Yes, he wrote, Xerox did indeed need to go digital, and maybe something could even be made of SDS. But to accomplish either of those objectives, he insisted, Xerox needed a research center that could attract the new breed of computer guys—the kind of people who *understood* computing, who could really make progress in computing, who could carry Xerox to the forefront of computing.

It was chutzpah, admits Goldman, but he sent the memo anyway. Whatever mistakes they'd made in buying SDS, he reasoned, McColough and Wilson had gotten the essentials exactly right: computers *were* the future. Moreover, they seemed like just the kind of people who could tackle that future head-on. "Joe Wilson was one of the great men of the world," he says, "just a very forward-looking individual. In those days, Xerox was a rare company."

And indeed, says Goldman, even though his proposal did catch a bit of flak from the new Xerox board members from SDS—"They wanted to use the money to build another Sigma mainframe, like the ones that weren't selling," he says—Wilson and McColough liked the idea; they had been a bit troubled by SDS's mainframe mentality themselves. Go for it, they told Goldman.

And that was pretty much that, he says—except for a few zillion details, such as figuring out how this new center would be organized, where it would be located, and who would run it.

Ah yes, who would run it? This was a critical question, Goldman knew. The corporate world was full of executives who believed that everything came down to numbers on a spreadsheet, and were convinced that a tough-minded guy who knew how to work those numbers could equally well run an automobile manufacturing plant, a copier company, a research lab, or anything else. That had been the attitude Robert McNamara promulgated at Ford when he headed the company in the 1950s. That was the attitude his protégés had retained throughout the 1960s, after McNamara went off to Washington to apply the numbers mentality to the Vietnam war, with such brilliant success. And that was the B.S. they were forever trying to impose on their corporate research labs. In company after company, you saw the numbers guys trying to justify the cost of research by demanding *results* in the form of new products this quarter, next quarter, and every quarter after that. And that was fine, says Goldman, if all you cared about was calculating cost/benefit ratios and building a slightly better widget. But if the medical community had tried to conquer polio that way, we wouldn't have gotten a vaccine that stopped the epidemic in its tracks, because that vaccine was discovered only after years of failure, frustration, and blind alleys, none of which could have been justified by cost/benefit analysis. Instead we would have gotten the best iron lungs you ever saw.

No, Goldman knew—if Xerox really wanted to change the equation, then he had to find a lab director who really understood how to make that happen. In his own experience, the prescription was simple enough: hire the smartest people you can find and give them their head—but let them know who's paying the bills. Putting that formulation into practice, however, was hard, expensive, and risky. It wasn't enough just to hire a bunch of supersmart individuals. You had to build a community, a culture, an environment of innovation. You had to give your people the kind of challenge that would light a fire in their eyes, that would generate an atmosphere of nonstop intellectual excitement, that would let them feel in their gut that *this* is where the action is. You had to provide them with lavish resources—everything they needed to do the job, without stinting. And through it all, most important, you had to keep the bottom-line guys at bay so *your* guys could have the freedom to explore and make mistakes. Somehow you had to make the higher-ups accept that none of this would necessarily result in products the following year, or maybe even in five years. But in ten years you might just change the world.

In any case, says Goldman, "A lot of names came up, including some very famous ones." None of them seemed quite right. But then one day that fall, he heard that Ford, which was still trying to fill his old job as director of the research lab there, had offered the post to a physicist named George Pake.

Hmm. Goldman knew Pake fairly well, as it happened. They'd first met in the closing days of World War II, when they were both greenhorn physicists doing war research at the Westinghouse plant in Pittsburgh. Pake was a slender, unassuming fellow, courteous and well-mannered to the point of courtliness. But he was very sharp indeed, and as open and direct as they come. Goldman had followed Pake's career ever since: a physics Ph.D. from Harvard; a six-year stint at Stanford; hired away in 1962 to become provost of Washington University in St. Louis; appointed to the president's science advisory panel in 1964. Moreover, Goldman knew from their conversations over the years that he and Pake were on the same wavelength: "He understood what research was about," he says.

Perfect. And best of all, to Goldman's way of thinking, was that Pake had turned down the research directorship at Ford without a second thought. Thank you, he'd told the astonished Ford people as diplomatically as possible, but I think I'll stick to academia.

Now *that*, says Goldman, was the sign of a man who had his priorities right. Ford's bonuses alone were worth far more than Pake's academic salary. But the numbers guys hadn't fooled Pake for a minute: they had everything at Ford so organized and quantified and controlled that you could barely move. Goldman could remember fighting his way up through five layers of bureaucracy before he could so much as *talk* to the CEO. In fact, for him that had been one of the big attractions of Xerox, where he would report directly to McColough.

But in any case, Goldman's search was clearly over. He made the call over Thanksgiving weekend. And on the following Sunday—December 7, 1969—on his way to give a speech in Chicago, Goldman diverted the Xerox corporate jet for a layover at the St. Louis airport so he could make the pitch in person.

Pake came out to the tarmac to meet him; Goldman was an old friend, after all. But as they sat there talking it over in the cabin of the little jet, Pake made it clear that he wasn't really interested. For starters, he said, he didn't want to move his family out of St. Louis, a city they all loved. "And besides," he added, "you don't want me—I don't know anything about computers."

Actually, that wasn't quite true. Pake's own research into the physics of "condensed" matter—meaning the behavior of matter in its liquid or solid form—gave him an excellent background for understanding the ongoing revolution in solid-state electronics. As provost at Washington U, moreover, Pake had taken the lead in bringing Wes Clark and his LINC team to St. Louis, which meant that he had been fully exposed to Clark's fervent belief in interactive, individualized computing—not to mention time-sharing, networks, and this visionary whatever-it-was that they called the ARPA community.

In any event, George Pake was Goldman's man, and Goldman meant to get him. He sensed that the real problem, other than the fact that Pake really did love St. Louis, was that other recent offer from Ford: the bureaucratic miasma in Detroit had left Pake feeling leery of the whole idea of industrial research. "So my purpose was to assure him that the attitude at Xerox was different," says Goldman. "I told him that we had a much more progressive group of leaders."

Well, maybe. As they parted—Goldman had to get on to Chicago—Pake did promise to think about it. A few weeks later, in fact, he even agreed to let Goldman fly him back to Rochester to meet McColough in person. "McColough and I interviewed each other," Pake remembers with a laugh, "and I guess we each passed!" Indeed, the CEO used all the right words: Xerox wouldn't expect a product tomorrow, he promised. Nor would the lab be treated like the corporate fire department, with researchers' constantly being asked to drop everything and cope with crises elsewhere in the firm. Pake's job would be to invent the company's future.

Once he was back home, says Pake, he "did a lot of soul-searching over the Christmas holidays." He certainly didn't want to leave St. Louis—but then, he says, "how many chances do you get in a lifetime to found a new institution? Especially when it seemed at the time to have very enlightened corporate support?"

Not many, he realized. Then, too, he was forty-five years old already. "I had just finished six or seven years as provost and executive vice chancellor of Washington University, and I was starting to receive feelers for university presidencies. That was flattering, but I didn't want to go that direction—it would mean living in an academic-political goldfish bowl." But if not that, Pake wondered, then what? Go back to hands-on research? Presumably, though that's not so easy when you're in your midforties. Or he could say yes to Xerox.

"So I said yes."

Pake and Goldman hit the road almost immediately, looking for a place to locate the new center. That was priority number one, says Goldman, even ahead of hiring the researchers: the kind of topflight people they hoped to recruit weren't about to cut themselves off from their profession for the sake of Xerox. Unless the area around the new center featured a heavy concentration of cutting-edge computer firms, plus easy access to at least one major university that was strong in computer science, they simply wouldn't come.

Unfortunately, as Goldman had pointed out in his original memo, that criterion eliminated Rochester, New York, which wasn't a computer center, as well as Goldman's own favorite university choice, Yale, where the lab would have been right next door to Xerox's brand-new headquarters in Stamford, Connecticut. In 1970, the Yale computer-science department didn't have anything like the strength it would later attain. Of course, that still left MIT, Princeton, and many other institutions nearby. But Pake felt strongly that the lab should be on the West Coast, if only to be within easy commuting distance of the company's new computer division down in the Los Angeles Basin. Besides, Pake argued, the more he thought about it, the more he liked the idea of Palo Alto, California, and the area around Stanford University. For one thing, he had been a professor of physics at Stanford from 1956 to 1962, and he still had good contacts on the faculty. But much more important was the industrial ferment just south of Palo Alto: down in "the Valley of Heart's Delight," as the Santa Clara had once been known, the prune and apricot orchards were sprouting semiconductor companies on every side. True, the area wasn't yet a major computing center in a class with, say, Cambridge. But in the coming age of integrated circuitry, Pake maintained, it would become the epicenter of computing.

"Now, here's where you have to give George credit," says an admiring Goldman. "He really saw the Silicon Valley phenomenon emerging."

In truth, it would have been pretty hard to miss. No one was calling the area Silicon Valley yet, except maybe as a kind of in-joke; the name wouldn't appear in print until January 1971, when the periodical *Electronic News* ran a three-part series on the region entitled "Silicon Valley USA." But it was already high-tech heaven by the time Goldman and Pake went out to visit it—and indeed, it had been developing that status for more than three decades. In 1938, for example, two young Stanford graduates named Bill Hewlett and Dave Packard had founded one of the earliest of the high-tech start-ups in a Palo Alto garage. (Their first big customer was Walt Disney, who bought eight of their "audio oscillators" for use in his company's new animated feature film, *Fantasia*.) Another milestone had come in the early 1950s, when Stanford began to lease out some of its open land to high-tech companies that could form beneficial alliances with its engineering school. By the time George Pake arrived at the physics department in 1956, Stanford Industrial Park was booming, with a rapidly expanding list of tenants that included Varian Associates, Eastman Kodak, General Electric, Lockheed, and an ever-more-vigorous Hewlett-Packard.

Meanwhile, silicon was coming to the valley, albeit by a decidedly indirect route. In that same year of 1956, Bell Labs' William B. Shockley, coinventor of the transistor, returned to his hometown of Palo Alto, where he promptly organized the Shockley Transistor Laboratory. Alas, Shockley's management style proved to be an inimitable mix of arrogance and paranoia (when a secretary gashed her hand on a broken pushpin embedded in the door to her office, he suspected sabotage and forced two men to take a lie-detector test). Worse, he insisted on placing his bets on germanium, a semiconductor that had more than its share of technical problems. So in 1957, having had their fill of Shockley, and convinced that they could do far better working with a more tractable material, silicon, eight of his brightest young associates left to form a company of their own: Fairchild Semiconductor.

They were right. Fairchild didn't invent the integrated circuit itself; that honor goes to Jack Kilby of Texas Instruments, who demonstrated his first device in Dallas on September 19, 1958. But less than a year later, in mid-1959, Fairchild cofounder Robert Noyce did devise a way to mass-produce integrated circuits by etching thousands of transistors simultaneously onto the surface of a single silicon wafer. This was the invention of the "chip" as we know it today. Fairchild prospered accordingly, especially after NASA selected its chips for the on-board computers in the Gemini spacecraft. By 1967 the company had twelve thousand employees and revenues of $130 million per year.

Along the way, moreover, Fairchild achieved an even greater significance as the seedbed of Silicon Valley. Thanks to some overly meddlesome management by the company's parent firm, Fairchild Camera and Instrument Corporation of New York, one group of employees after another started striking out on their own, eventually creating some fifty spin-off companies. Among these émigrés was Noyce himself, who teamed up in 1968 with another Fairchild cofounder, Gordon Moore of Moore's law, to found a company they called Intel, short for "integrated electronics" (fortunately for posterity, they rejected other candidate names such as Elcal, for Electronics of California, and Ectek, for Electronic Computer Technology). Intel's first product, a sixty-four-bit solid-state memory chip, was not quite cheap enough or compact enough to compete seriously with magnetic-core memory—yet. But by the end of 1970, the firm would come out with a 1,024-bit memory chip that would be a very serious competitor indeed. And that was clearly just the beginning; the way was open for a revolution in the economics of computing.

Indeed, by that point the company had achieved an even more significant milestone, almost by accident. In 1969, according to the now-famous story, a Japanese calculator company called Busicom asked Intel to design twelve custom chips for its newest model. Because that seemed a daunting task given Busicom's deadline, Intel engineer Ted Hoff suggested a quicker way: design *one* chip that could be programmed to perform twelve different functions. The ultimate result was the legendary Intel 4004, the first "microprocessor" to contain an entire central processing unit on a single chip—and indeed, the direct ancestor of all the 80x86s, Pentiums, and such that fill the world today. The 4004 went on sale to the public in November 1971. It was a feeble thing by today's standards, containing only 2,250 transistors as opposed to the ten million or so in a Pentium III, and equipped with a processing power no better than ENIAC's, but it dramatically demonstrated the possibilities of microprocessor technology. In the very near future, there would be complete computers on a single chip.

And in the meantime, Jack Goldman and George Pake had long since decided that their search was over. In June 1970, Xerox's new Palo Alto Research

Center–PARC–opened for business in a small rented office building at 3180 Porter Drive, in Stanford Industrial Park.

NEVER HIRE GOOD PEOPLE

Now–who was actually going to do the research at PARC?

The answer depended on which part of PARC you were talking about. Pake and Goldman had already agreed to organize the center into three separate laboratories, with the expectation that each would eventually grow to include about fifty people. And in the first lab, at least, Pake felt he knew what he was doing. The General Sciences Laboratory, or GSL, as they called it, would focus on basic research into magnetism, materials science, semiconductor physics, and the like–the experimental foundations of computer technology, which was Pake's home turf. He knew exactly what kind of staffers he wanted there, and better still, he had a ready supply of them from within Xerox itself. When word had gotten back to the Rochester lab that the company was planning a new research center in a place that did not know the meaning of the word *snow*, there had been a sudden wave of enthusiasm for transfers.

However, that still left the other two divisions of PARC, both of which would be devoted to computing itself. The idea was to have one of them–the Computer Science Laboratory, or CSL–concentrate on the deep principles and theoretical foundations of the field, and the other–the Systems Science Laboratory, or SSL–focus on the creation of large-scale applications.

Now right away, stresses Pake, you can see how naive he actually was about computer science. When he used the word *system*, for example, he was still thinking in terms of something like a space shuttle or a jetliner: a complex, interdependent assemblage of hardware. It would be a long time before he understood that computer systems were critically dependent on software as well. Worse, says Pake, in trying to divide the two labs between basic and applied computer research, he was being led badly astray by his background as a physicist. At the time, he remembers, "I was a bit baffled. I kept looking for these underlying principles of computer science"–the analogs of Newton's laws of motion, say–"and I couldn't find them. There were certainly some deep ideas–for example, information theory, or the Turing machine. But those ideas had not led to a large body of theory as there was in physics." The upshot was that his plan for the separation of the two labs just didn't work, because "both ended up doing applications." And in the long run, he admits, that confusion of responsibility would lead to some of his worst administrative headaches.

But again, that was later. For now, in the summer of 1970, Pake's challenge was to staff these two computer labs. And he felt more strongly than ever that he was completely out of his depth. He had tried; as soon as he had signed up with Xerox, even as he and Goldman were looking at sites, he had made a point of visiting the Big Three computer schools: MIT, Carnegie Mellon, and Stanford.

He had come away from those visits with a reasonably good grasp of the issues, he thought. But mainly he had come away with the conviction that he needed a guide, "somebody to help me recruit people and then determine the structure of a computer-science lab." As he talked about his concerns with Wes Clark and a number of other computer-savvy colleagues, moreover, he kept hearing one name come up as someone he really ought to consider for that position. This guy certainly wasn't a technical person. But what he did have was a wonderful sense of what needed to be done, who the good people were, and what opinions to believe. Taylor was his name—Robert W. Taylor

Hmm. Pake knew Bob Taylor a bit, as it happened. They had occasionally had lunch together during Taylor's ARPA days, when he'd come out to Washington University to visit Wes Clark's lab. Pake had found Taylor downright monomaniacal on the subject of interactive computing, obsessive to the point where it wore people's patience thin. Nonetheless, Clark's good opinion of him was no small matter. In addition, from what Pake himself could remember, Taylor did seem to be a canny judge of people and projects. And best of all was that after being the director of ARPA's computer office for three years, he knew *everybody*.

In short, Taylor was precisely the kind of guy Pake was looking for—except for one little detail. "As I talked to Wes and several other people," says Pake, "they all agreed that Taylor was a good man, but probably not the man to head up the lab. Bob had had a stellar record as a young administrator at ARPA, but he had no track record in research of his own. And to get the respect of researchers, you need to have a track record. So I remember discussing this with Wes and the others at length: how could we weave Taylor into the group?"

Pake and his friends mulled it over for quite a while. And by the end of the summer, they believed they had found a solution.

Bob Taylor got the call from Pake in early September 1970, just as he was finishing up his first year at the University of Utah in Salt Lake City. Would he like to fly out to Palo Alto for an interview? Pake asked.

Taylor accepted, albeit with mixed feelings. On the one hand, he was definitely ready for a change. "The Utah period was pleasant for me in some ways," he says. "It was a period of decompressing from Vietnam, and from Washington. I learned how to ski. I enjoyed the western ambience, as opposed to the eastern. But intellectually, no. There was too much Mormon influence for my taste."

On the other hand, Taylor wasn't at all sure that this was the change he wanted. His reaction to Xerox's move into computers was much the same as Goldman's: if the company really wanted to go digital, it had gotten off to the wrong start by buying a lemon like SDS. He had seen SDS up close during his ARPA years, when it provided the computer for Berkeley's Project Genie, and the view hadn't been pretty. SDS founder Max Palevsky and his board had shown zero interest in time-sharing, on-line communication, or computer utilities. Where was the market? Palevsky had wondered. The only reason they'd gotten involved in the first place was to sell that one machine to Berkeley; all the rest of it struck them as a government boondoggle, a chance for university kooks to play in the sandbox. And even after Taylor finally did manage to talk SDS into marketing the Project Genie system, in 1966, the company had continued to treat it as a one-shot product, with no serious investment's being made to engineer a second generation.

Taylor was only slightly more inspired by the new laboratory's director-designate. Judging from their earlier conversations at Washington University, Pake did seem to have a reasonably good understanding of the hardware aspects of computing. But Taylor wasn't at all convinced that Pake understood the other half of the story, which was, well, not software exactly, but *systems*—how people actually used computers. "I always thought that the physical side of computing was so damn boring because it was so predictable," says Taylor. "Even in nineteen sixty-nine, you could know with great accuracy what the cost of an integrated circuit would be in ten years. But you couldn't predict with any accuracy what you could *do* with it."

That was what people like Vannevar Bush, J. C. R. Licklider, Wes Clark, and Doug Engelbart had always perceived so well, he thought. The real significance of computing was to be found not in this gadget or that gadget, but in how the technology was woven into the fabric of human life—how computers could change the way people thought, the way they created, the way they communicated, the way they worked together, the way they organized themselves, even the way they apportioned power and responsibility. That was what had resonated so deeply in Taylor's mind. And that, in the end, was why he now found himself hoping against hope that this interview with Pake would work out. If this new laboratory could be kept from a fatal infection of SDS-itis, and if Pake could be made to understand what computing was *really* about, then there could be a tremendous opportunity here to make the dreams into something real.

Indeed, Taylor even had some notion of what the results might look like. Although he has no clear recollection of where the idea came from—at the time it was more like an intuition, an instinct, one of those gauzy visions that enter your brain so slowly that you don't even know they're there—he can look back on it now and see that one inspiration must have been Doug Engelbart's augmenting-the-intellect ideas. Another must have been the on-line communities that he had watched springing up around the various time-sharing projects. And still another must have been that first encounter with Wes Clark's LINC, back in 1964. But if he had to name any one thing that did it, says Taylor, it would have to be the sight of those three typewriter terminals in his office at ARPA. "You already had your very own terminal behind your desk," he explains, "so why not your own small computer on *top* of your desk? I started thinking about this around nineteen sixty-five or nineteen sixty-six. From the vantage point of the midsixties, you could already see computers getting faster, bigger, and, for a given size, less expensive. So the notion of sitting down and having your own machine was not a big leap."

Indeed, Wes Clark had made that leap a long time before—as had Lick, Engelbart, and Vannevar Bush in their own ways, and no doubt many others as well. Even so, this notion of individual small computers wasn't one that Taylor wanted to push very hard just yet. For one thing, in those ARPA days he was having enough trouble pitching his network idea. For another, he could remember twenty years back, when pundits were predicting that every garage would soon house a personal airplane; twenty years from now, he figured, personal computers might sound just as silly. After all, falling hardware prices didn't necessarily mean small computers; they might just as easily mean million-dollar computers that were very large indeed. In twenty years we might even all be tapping into the giant computer utilities that Project MAC was dreaming of. And in all honesty, Taylor had to admit that that might make more sense. A world of desktop computers would sacrifice what he considered to be the very heart of interactive computing: connectivity, and the kind of on-line community that time-sharing provided for free.

And yet the idea wouldn't quit niggling at him, especially after Taylor had launched the Arpanet project. Once you got the network up and running, he wondered, why couldn't you tie in desktop machines just as easily as you tied in big time-sharing mainframes? That would give you back the connectivity you'd lost. And in the meantime, since you would no longer have to have all your computing done on a central mainframe, you could be giving every one of those desktop computers a powerful graphics capability, like the Kludge terminal up at MIT, or the IMLACs, or the displays that Evans and Sutherland were building in Utah. You could begin to get the kind of "high-bandwidth," graphical interface that Lick was always talking about, along with quick response times, private storage space, and all the other advantages of stand-alone machines.

So that was how the idea had grown, says Taylor—still nebulous, still a fantasy, but somehow getting more and more real all the time. While he was at Utah, in fact, he did a few back-of-the-envelope calculations: if you really took Moore's law seriously, then computers small enough to sit on a desktop would be affordable in, what—ten years? Soon, at any rate. Then how long after that before you had a computer small enough to sit on your lap? Another ten years? How long before there was one small enough to fit in your *hand*?

And how long before time-sharing was totally obsolete?

The exercise was an eye-opener, says Taylor. Individual computers with high-resolution graphics, connected by a network—this could really be the future. You just needed someone to take the lead and make the huge commitment of time, money, and manpower it would require. But it could be done. And this new research center at Xerox, if only the company could be kept from squandering the opportunity, might be just the place to do it.

Although Taylor and Pake had never hit it off very well—and indeed, never would—they do seem to have kept their interview quite cordial. To Taylor, Pake represented probably the best chance he would ever have of fulfilling his vision of computing. And to Pake, Taylor represented exactly the kind of management talent he needed to make this lab work. He was willing to put up with almost any amount of prima donna behavior to get that talent.

Sitting in on the conversation was Bill Gunning, who was an old computer hand himself; he had helped to build the JOHNNIAC at RAND in the early 1950s and had later cofounded a commercial data-processing firm known as Astrodata.

Gunning would be heading up the applications-oriented Systems Science Lab, said Pake, after he explained to Taylor about PARC's three-part organization. Pake himself, for the time being, would be in charge of the physicists at the General Sciences Lab. And they wanted Taylor for the Computer Science Lab— but *not*, Pake emphasized, as its director. Without a research record of his own, Pake told him, he just wouldn't command the necessary respect and authority. So instead, Pake proposed, why didn't Taylor come to PARC as the *associate* director of CSL, and then help recruit his own boss? While he was at it, in fact, why didn't he help recruit the computer research teams for both his lab and Gunning's? Pake stressed that the position would give Taylor a big influence on the research at PARC, especially since he would be there long before his boss was. And then, he added, once the lab was up and running, he could start building up that elusive track record by starting a little research group of his own in, say, computer graphics.

Now once again, says Pake, this proposal looks incredibly naive in retrospect. Quite aside from the implied insult to Taylor—that he was good enough to recruit the team members but not good enough to lead them—it promised to be extremely awkward for his eventual boss. People tend to be loyal to whoever hired them, not to "outsiders." Still, it seemed like a good solution at the time. And in any case, Pake adds, "my recollection is that Bob had no problem with it."

No, he didn't—though, as is so often the case with Bob Taylor, the reasons were more complex than they seemed on the surface.

To begin with, while he very much liked the idea of having a big influence on PARC's research, he considered Pake's notion of a "graphics research group" a complete nonstarter. Sure, graphics technology was a critical part of this whatever-it-was he wanted to create. But so were text display, mass-storage technology, networking technology, information retrieval, and all the rest. Taylor wanted to go after the whole, integrated vision, just as he'd gone after the whole Intergalactic Network. To focus entirely on graphics would be like trying to build the Arpanet by focusing entirely on the technology of telephone lines.

And yet Pake did have a point, damn it. At age thirty-eight Taylor had spent his entire adult career funding computer research, but he had never actually *done* computer research. And that mattered. Just look at what had happened back at ARPA: he had hired an "assistant"–Larry Roberts, Ph.D.–and then watched the whole community defer to Dr. Roberts as the Boss.

So Taylor knew that he would have to bow to the inevitable on this one. He would come to PARC as associate director of CSL or not at all. On the positive side, he realized, there was a lot to be said for being the power behind the throne. After all, Taylor points out, "if I were perceived as the lab manager, I would then have to spend a lot of my time with visitors and Xerox management–people from outside the lab. And in the early years I didn't want that; I knew I was going to have a lot to do, and many pieces to put in place. So if I could hire a manager for the Computer Science Lab, then I would be left free to work on issues that I cared about which were internal to the lab."

Besides, Taylor reasoned, if he chose the right kind of manager, meaning one with ambitions to rise within Xerox, then the problem would solve itself in a few years. The guy would step up, and the mantle would fall to–well, by that point it wouldn't matter whether or not Bob Taylor had a "research track record." His team would have done such fabulous things that even Pake would see who the real leader was. And in the meantime, all he had to do was quietly recruit the right kind of people, set the right kind of agenda, and make PARC into the right kind of laboratory. *His* kind of laboratory.

So Pake and Taylor shook on it. They had a deal–albeit not quite the *same* deal in each man's mind.

OK, first things first. "I had been at the University of Utah, which was a center of graphics research," says Taylor. "So within a few weeks after I got to PARC, in late September nineteen-seventy, I brought two U of U researchers with me–Jim Curry and Bob Flegal–both of whom were interested in graphics."

That took care of his "graphics research group." Now for his "boss."

That turned out to be almost as easy. Pake had already gotten a recommendation from Wes Clark and several others that he hire Jerry Elkind, who had recently left BBN and gone to work at Project MAC for his former professor, J. C. R. Licklider. The suggestion sounded fine to Taylor; he and Jerry had always gotten along pretty well, he thought, and he had every reason to expect that that would continue to be the case at PARC. So in short order this deal, too, was done. Elkind would start at PARC in July 1971–which of course left Taylor a substantial amount of time in which to shape the lab to his liking.

No mystery about where to start: Taylor's motto was "Never hire 'good' people, because ten good people together can't do what a single great one can do." And Butler Lampson was among the greatest–"my choice as the finest computer scientist of the century," says Taylor.

Others in the field might not have put it quite that strongly. But still, when George Pake had been soliciting names the previous spring, the twenty-seven-year-old Lampson had been on *everyone*'s list. He was invariably described as a man who could think and talk faster than anyone you'd ever met—and in the computer world, that was saying something. "You couldn't even parse the words," recalls Stuart Card, who would work with Lampson a few years later. "You couldn't respond to his arguments, because the whole processing capacity of your brain was spent on trying to understand him!"

Moreover, behind the mouth lay a brain. In 1965, for example, as a graduate student at Berkeley, Lampson had designed the operating system that had elegantly turned a batch-processing SDS 930 computer into the Project Genie time-sharing machine. Then, in 1968, after he and the rest of Genie's best and brightest got fed up with SDS's foot-dragging and left the university to commercialize their ideas in a start-up company known as the Berkeley Computer Corporation, Lampson had been the lead architect of an even more elegant computer. Known as the BCC 500—because it was intended to handle five hundred users simultaneously—it featured such innovations as multiple central processors and highly sophisticated terminals that could reduce the load on the main computer by doing a good bit of the computation locally.

Fortunately for Taylor, however, the BCC 500 was also exceedingly complex, with technology that went a little too far beyond the cutting edge to be commercially viable. By the end of 1970, with the Berkeley Computer Corporation rapidly going broke and the outlook for its rescue bleak (the country was just then sliding into a recession), Taylor saw his opportunity to hire not just Lampson but the entire crew: a research lab imported wholesale. He made the call to Lampson that December. And once Lampson had agreed to come—he was as dubious as anyone about Xerox's connections with SDS, but he was intrigued by the possibilities of a computer lab backed by Xerox's money, and delighted by the prospect of working for Bob Taylor—the others followed almost as a matter of course. Lampson, Chuck Thacker, Peter Deutsch, Jim Mitchell, Dick Shoup, Willie Sue Haugeland, Ed Fiala—Taylor and Pake spent the holiday season of 1970 interviewing like mad. And by January 1971, the first of the Berkeley crowd were headed south toward Palo Alto.

Number 3180 Porter Drive wasn't much to look at. It was just a small, rented office building "with rented chairs, rented desks, a telephone with four buttons on it, and no receptionist," remembers David Thornburg, who had just been recruited to the General Sciences Lab. The place was pleasant enough, in a California-modern kind of way; the offices and conference rooms formed a rectangle that surrounded an interior courtyard and an attempt at a garden. But with the new recruits from Rochester and BCC only just now beginning to straggle in, PARC was *empty*.

Too empty for Gary Starkweather. He'll never forget that first day, he says, when he found himself standing alone in a vacant room, staring at the bare cinder-block walls, and wondering if he'd just made the biggest mistake of his life. Back in Rochester, after all, he'd had friends, family, and roots, not to mention a beautifully equipped optics lab. Here in Palo Alto he was three thousand miles from where he'd grown up. His wife and their two little children were strangers in the community. They were still in shock from the housing prices. And now here he was in this barren "laboratory" with the reality hitting him square in the face: he was going to have to rebuild *everything*. From scratch.

Still, as Starkweather kept reminding himself, his prospects back in Rochester hadn't exactly looked brilliant. When he'd first arrived there about a decade earlier, he'd been an eager twenty-something with a brand-new undergraduate degree in physics from Michigan State and a brand-new job with Bausch and Lomb, the optical company. "I found out I was in a world of sophisticated optics technology I hadn't imagined," he remembers. He found optics so fascinating, in fact, that he went back to school at the University of Rochester to get a master's degree in the field. "Then about a year into it," he says, "I went over to work at this little start-up copier company called Xerox."

He found plenty to fascinate him there as well. From the outside, the operation of a photocopying machine is simplicity itself: just place an original document facedown on the input glass, press the button, and out come the copies. On the inside, however, the machine is an elaborate system of lights, lenses, moving mirrors, and stationary mirrors, all working together to project an image of the original document onto the surface of a metal drum, the internal "printing press" that actually produces the copies.* So an optics maven like Starkweather could have a marvelous time in there, endlessly crafting new ways to make the optical system faster, better, cheaper, and more robust. And then when he wasn't busy with that, he could pitch in on the long-range projects, where he could *really* get creative.

One such project that had caught Starkweather's eye early on was an effort to use xerographic technology to make a high-speed facsimile machine. The basic idea was straightforward enough: scan *here*, make copies *there*. And indeed, it was almost feasible in practice. Starkweather and his colleagues at the Rochester lab soon showed that they could convert the optically scanned image into an electronic signal, transmit that signal over a telephone line, and then feed the signal into a television-like CRT display on the far side to reconstruct the original image. What they couldn't do very well, unfortunately, was take the final step—which was, in effect, to photocopy the CRT display for the final output. "The CRTs we were using to make the image didn't have a bright enough output," he explains. "As you turned up current for more brightness, the beam

---

* The surface of the drum is photosensitive, meaning that the image of the document creates a corresponding pattern of electrostatic charge. The charged areas then attract and hold a dry powder known as toner, which is analogous to the liquid ink used in conventional printing. Just as in conventional printing, moreover, this "ink" is then pressed onto the surface of a sheet of paper to make the copy. The difference is that the toner is dry at all times—whence the name xerography, from the Greek words for "dry writing."

'bloomed.' So you could only get about eighty dots per inch, which was a pretty low resolution."

As Starkweather continued to struggle with the problem, however, an answer suddenly hit him: "I realized that a laser was ten thousand times brighter than a CRT display. Now, this was nineteen sixty-four, when the laser was still a lab toy"—it had been invented in 1961—"but I realized that if I could devise an optical system to scan the beam back and forth across the photosensitive drum—you can't move a light beam electronically; you have to do it with mirrors and lenses—I could write an image directly." All he had to do was make the laser flicker on and off as it scanned, he says, and he could fill up the page with a pattern of tiny dots: an image.

Of course, this wasn't a new idea even then. A team of Xerox researchers had demonstrated the laser image-writing process two years earlier, in 1962. "The problem was that the scan system had to be extremely precise, which meant high-quality bearings on the scanning mirrors, and so forth," says Starkweather. "The cost of a complete system was about ten thousand dollars, which was prohibitive. What I figured out was a way to do this with some much cheaper components. So I did some test devices, and I was able to demonstrate the principle."

By this point, as it happened, Xerox had lost interest in the high-speed facsimile system. But Starkweather had already come up with a better idea. "About nineteen sixty-seven or 'sixty-eight it occurred to me," he says: "Why don't I let a computer generate the information to drive the laser, and use this as a printer?"

Alas, there turned out to be some very good reasons that he couldn't. Yes, a "laser printer" would be a tremendous advance over standard computer printers, which were essentially glorified typewriters. The tiny dots generated by the laser could form pictures, line drawings, mathematical symbols, text with letters of arbitrary size and typeface—anything you might want to put on paper. But lasers were still very, very expensive. Besides, says Starkweather, "if you printed at a hundred and fifty pixels per inch, the computer would have to generate and store a couple of megabits per page, which was an enormous amount of memory in the late nineteen sixties. So figuring in the cost of memory, plus the cost of the laser, a laser printer was completely impractical."

In short, the laser printer was going nowhere. And neither was Gary Starkweather—until July 1970, when he happened to pick up a company newsletter: "Lo and behold, Xerox was starting up a research center in Palo Alto!"

When he learned that several of the Rochester physicists he knew were already planning to transfer, Starkweather got on the phone. "One of them was John Urbach," he says. "He talked to his new boss, Bill Gunning, and I got invited out to Palo Alto for an interview. This would have been September or October nineteen-seventy. I told them about laser printing, and they fell in love with it."

Bob Taylor, in particular, was enchanted by the notion. A printer that could generate pictures and text with equal ease, that could work with paper in exactly the same way a computer graphics screen worked with pixels—it fitted into his ideas perfectly. "Get him here!" he had declared as soon as Urbach told him about Starkweather's idea. "Get him here and *we* will build a digital laser printer."

"Bob," Urbach had warned him, "Gary doesn't know much about digital technology."

"We'll take care of that! Just get him out here."

The various higher-ups were happy to go along, especially George White, Starkweather's boss in Rochester and the man who had led the team that did the first laser copying experiments back in 1962. White had always felt bad about holding his protégé back from any serious development of laser printing—"My job was to develop the next generation of copiers," he says with a sigh, "and there was no way that generation would have lasers"—and the PARC offer seemed like a good chance to atone for it.

Which was how Gary Starkweather had come to be there now, standing in an empty room, staring at the silent cinder block, and feeling very, very far from home. What if I can't cut it out here? he remembers thinking. These computer jocks operated in a world of bits and bytes that was utterly alien to him. He was just a physicist who knew a little about optics. A physicist with no Ph.D.

"Fortunately," Starkweather says, "that didn't last long. PARC turned out to be very friendly!"

Indeed, he soon learned that Taylor had already told the Berkeley contingent that laser printing was a great idea and that they ought to give this guy Starkweather a hand. So within a few days of his arrival, one ferocious computer jock after another started showing up in his doorway to ask, "Can I help?"

Sure! Starkweather knew full well that lasers, mirrors, and xerography were just the beginning. A full-fledged laser printer would also have to know how to take commands from the computer—"Print the letter A," for example—and translate them into a precise sequence of dots on the page. And that meant, in turn, that the printer would have to carry quite a lot of computational power on board. Thus Taylor's encouragement of the computer group. And thus Gary Starkweather's welcome into the fold: the collaborations he began with Lampson and the others during those first, lonely days at PARC would continue for years thereafter.

Meanwhile, the Berkeley contingent . . .

Well, the Berkeley contingent was finding "meanwhile" to be a bit of a problem, actually. Most of them were still commuting from their homes on the far side of the bay (Chuck Thacker put twenty thousand miles on his car that first year, at which point he decided it was time to move), and they had arrived to find an empty-laboratory problem of their own, in every sense. Lampson remembers a lot of arguing, brainstorming, and general milling around in those early days. Maybe Gary Starkweather knew what *he* was doing there, but what were the rest of them doing there? Creating the electronic office, presumably. But exactly what did that mean? Office-scale time-sharing? Corporation-wide computer utilities? What?

On one issue, though, there was virtually unanimous agreement: with no terminals, no time-sharing system, no Arpanet connection—nothing but those bare, rented desks—PARC needed to get a PDP-10 time-sharing system, and fast.

Now, that does sound a little strange from the modern perspective, admits Lampson, especially when you consider what PARC would later accomplish. "But we felt—I think correctly—that in order to do computer research in the nineteen-seventies, you really had to have time-sharing. There was already a sizable computer-research community supported by ARPA, with essentially everybody on PDP-10s running the Tenex operating system. We felt it was very important to be part of that community. There was also a sizable base of software written for Tenex: networking, standard mail software, AI. We needed that base. So there was no significant debate about this."

Not at PARC, perhaps. But when the folks at Xerox's new computer subsidiary down in El Segundo heard that Lampson and company were about to buy a big-time computer from their arch rival, DEC, the screams could be heard all the way back east at Xerox headquarters. And in short order, the inevitable edict was heard out west in Palo Alto: no PDP-10.

It didn't help when Pake, under what he described as strong political pressure, suggested that they buy a Sigma 7 from SDS—which was now, significantly, known as XDS: *Xerox* Data Systems. Lampson and the other Berkeleyites gagged at the thought. The Sigma 7 couldn't run Tenex or any of the Tenex software; they'd be isolated. Sure, they could probably kludge together some kind of Tenex-compatible operating system. But that would take years. You could *build* a PDP-10 faster than that! You could start from scratch and—

Well, why not? The more Lampson, Thacker, and the others thought about it, the more the idea made sense. Having recently built one computer, the BCC 500, they felt they could do a PDP-10 clone pretty quickly. And by doing it at PARC, they would simultaneously be creating the in-house infrastructure they would need in any case. Machine shops, test equipment, well-stocked laboratories, a network of suppliers—they couldn't do *anything* until they had all that in place.

So they decided to forge ahead. They would soon take to calling the new computer MAXC, which officially stood for Multiple Access Xerox Computer, but which they deliberately pronounced with a silent *C*, in a not-so-subtle dig at SDS head Max Palevsky.

In the meantime, however, Bob Taylor was biting his tongue, hard. *Timesharing?* This was not at all how he had expected things to go. "Listen," he would tell people at every opportunity, "if you think this is the right thing to do, I'll support it. But consider the alternative." And then he would try to explain his idea about small machines, graphics displays, and on-line communities.

Anyway, he would say in the ensuing embarrassed silence, it was just a suggestion.

Well, yes, thought Lampson, Thacker, and the others: it was definitely a suggestion. "Bob tried it out all over the place." Thacker laughs, recalling what would later be a favorite story for the Young Turks to tell on themselves: "Essentially it was what Licklider had been proposing for a decade. But from a technical point of view, the idea was still in a chaotic state. And Bob was simply not able to elucidate it in a way that we could understand it."

So they ignored him. After all, they told themselves, Taylor was just a manager, not a technical type. He would learn. And in the meantime, adds Thacker, "we had enough pressure building MAXC. We felt in a hurry—we were hiring a lot of people, and they needed computer facilities right then. So I was just trying to finish the damn machine."

Taylor got the message. Just shut up and let it go for now, he figured. His group did have a point about building up their capabilities in the short run. And in the long run? Well, he was willing to bide his time.

Bob Taylor has often been described as a complex man, which is an understatement. But for someone who was admittedly feeling his way as a manager, making it up as he went along, he was a remarkably quick study. Those who found themselves under his wing at PARC almost universally remember it as the warmest, most supportive, and most exciting research environment they've ever experienced.

"Bob did have an enormous desire for control," says Gary Starkweather. "Yet he was also one of the best managers I ever knew, because unlike some half-time researcher with a yen for power, he realized that his *job* was management. He took it seriously. And he knew how to handle people." Indeed, he says, even after Jerry Elkind arrived to take over formal control of the lab in July 1971, it was abundantly clear that the CSL crew still regarded Taylor as their leader. They liked Elkind well enough, but they were Bob's people. And it was equally clear that Bob Taylor took care of his own. If he liked you—and that could be a very big "if," says Starkweather; either you bought into his vision or you were an outsider—he would go to the wall for you. Resources, money, equipment, fierce advocacy before the higher-ups—they were yours.

As authors Douglas K. Smith and Robert C. Alexander described it in a later account of the PARC experience, "Taylor had a magical effect on the scientists—they cooperated and thrived in an environment they suspected could not exist without his leadership, yet they had difficulty articulating how and what Taylor did to keep the enterprise together."[1] Certainly that was Butler Lampson's impression: "Taylor is very good at getting . . . a collection of extremely intelligent and opinionated egomaniacs to work together reasonably well without fighting each other," he noted. "Damned if I know how! I can't do it, but he does."[2]

In fact, Taylor did it in part by knowing when to keep quiet. The kind of peo-ple he was hiring needed *lots* of room; witness their determination to go ahead with MAXC regardless of anything he could say. "Bob didn't want people who had to be managed," says Alan Kay, who had served as a consultant to PARC since the previous fall. "He liked people who were outspoken, who were very confident, who would argue back with him. And true to his word, he did not meddle with people's technical decisions. His idea was that he was there essen-tially to manage the personalities, to keep people from killing each other."

In practice, this meant a management style so carefully noninterventionist that visitors were often mystified by it—especially visitors used to the cool order of the executive suite. Under Taylor, the CSL wing of PARC sometimes seemed to have no organization whatsoever. There was no such thing as a standard workday, for example; many of his recruits had immediately reverted to hacker's hours, so that the building was sometimes as busy at 4:00 A.M. as it was at 4:00 P.M. Nor, for that matter, were there any ID badges in the Computer Science Lab (Pake's physicists had them). And there most certainly was no CSL organization chart, with formal team leaders and explicit lines of authority; everybody was equal and reported directly to Taylor. What PARC had instead was posters and cartoons on the doors, rock tunes in the air, and ten-speed bicycles in the hall-ways—dozens of bicycles. As Richard Stroup would later recall, "I would ride up to the lab and down the sidewalk and right in through the front door. Still on my bike, I would ride down the hall, park outside my office, work for the day, then get on the bicycle and ride down the hall and out the front door."[3]

But appearances could be deceiving. Taylor's hands-off approach was just the most visible element of an exquisitely delicate balancing act, modeled on the one practiced by Lick and all his successors at ARPA. Yes, went the argument, people needed the freedom to create. But their creations had to add up to some-thing—and not just another bunch of unconnected new gadgets, either. At ARPA that "something" had been human-computer symbiosis, broadly defined. Now, at PARC, it was the "electronic office," whatever that might turn out to be. Yet in either case the goal was a *system* of information technology, a whole new way for human beings to work together. All of the various gadgets had to be part of that system. And to achieve that goal, Taylor knew, he somehow had to get all these maverick geniuses moving in the same direction, without forcing everyone to move in lockstep. Somehow he had to give them a sense of purpose and group cohesion, without crushing spontaneity and individual initiative. Some-how, in short, he had to set things up so they would freely follow their own in-stincts—and end up organizing *themselves*.

This is arguably *the* fundamental dilemma of modern management, not to mention the fundamental political challenge in any democracy; leaders have been grappling with it for centuries. Fortunately for Taylor, however, he had just spent the better part of a decade at NASA and at ARPA, which had provided him with any number of models. "I'd traveled a lot," he says, "talking with people, studying the culture of different sites, and learning how they functioned. I'd especially spent a great deal of time talking to the youngest people in the ARPA groups—the ones who were doing the work and who had most of the ideas. I had learned about their values, about what worked and what didn't. So at Xerox, I put into practice what I liked and threw away the rest."

One thing he definitely liked and put into practice was a style of research that could be paraphrased as, "Don't just *invent* the future; go *live* in it." This had been the philosophy, too, behind Projects MAC and Genie, in which the time-sharing system was supposed to be simultaneously the main research tool, the primary object of experimentation, and the tangible product. For that matter, it had been Taylor's own philosophy in pushing for the full-fledged Arpanet instead of just a few demonstration projects. By all means, Taylor told his recruits, let's get way out in front of the curve—five years for sure, ten years if we can. And forget about the cost: Xerox is signing the checks for now, and Moore's law will solve the problem soon enough. But whatever you build, *use* it. In fact, get everybody in PARC to use it. Get them pounding on the technology every day, writing reports, writing programs, sending E-mail—anything and everything, so they can see for themselves what the problems and the possibilities are. And then use what they learn to build better technology.

That philosophy also reinforced another of Taylor's initiatives: ARPA-style communication. Back when he'd taken over IPTO, in 1966, Taylor explains, the challenge had been to maintain a sense of common purpose among research groups scattered across a continent. His solution then had been to revive Lick's idea of the principal investigators' meetings, and later to expand it to include the graduate-student conferences. And his solution now, at PARC, was to do the same thing, but more frequently: once a week the computer group would assemble, someone would talk about his work for an hour or so, and then the others would have at him. Taylor considered these meetings so important, in fact, that he made them mandatory, the one thing that CSL members actually *had* to do. Visitors from the other labs were welcome, but for CSL, Tuesdays at 11:00 A.M. were sacrosanct.

In practice, of course, Taylor also tried to keep the proceedings as easy and as informal as possible, to the point of having the conference room furnished with beanbag chairs. He even let the speakers set the rules for how each meeting would proceed, much as a card dealer could call the game in Las Vegas; thus their nickname, Dealer Meetings. And when the arguments got heated, which they often did, the minister's son would do his best to convert a "class-one" disagreement—one in which the combatants were simply yelling at each other—into a "class two" disagreement, in which each side could explain the other side's position to the other side's satisfaction. You don't have to *believe* the other guy, he would tell them. You just have to give a fair account of what he's saying. And it worked. As one CSL member later explained it, Taylor's class one/class two exercise was amazingly effective at clarifying unspoken assumptions and ferreting out facts that one person knew and another didn't. "So by the time you get done," he said, "you all know the same set of things, and you end up concluding the same thing."[4]

Yet another technique that Taylor put into practice was a tech-weenie version of fraternity rush: once CSL was established, everybody there had to vote "yes" to let the next person in. "Hiring was not a top-down thing at all," says Jerry Elkind, who was vetted this way himself, boss or no. "It was very much a community thing. People were proposed. And if they had enough initial support, they were invited out to do some talks. Then we had a lot of discussion, and we had to reach a general consensus that this was a good hire."

Granted, the hazing process could be nerve-racking: audiences at PARC were *not* polite (Butler Lampson, for one, would sometimes storm down front to rip into the speaker face to face). But the feeling was that if you couldn't take it, you didn't belong there. And besides, notes Alan Kay, if you made it through and were accepted, you knew that everybody wanted you there: "You were a colleague from the word 'Go.' So the arguments could go deeper without hurt feelings. For example, Butler and I argued just constantly. Yet we were great friends."

Certainly there was no shortage of people willing to run the gauntlet. Taylor continued to work the phones, bringing in a steady stream of ARPA's best and brightest for the group's consideration. Close on the heels of the BCC contingent, for example, came Ed McCreight from Carnegie Mellon University, a close friend and fellow graduate student of BCC alumnus Jim Mitchell. Then, in June 1971, in a coup that ranked right up there with his acquisition of the BCC team itself, Taylor arranged to have PARC hire some of Doug Engelbart's best people away from SRI—including chief engineer Bill English, the man who had always had such a genius for turning Engelbart's dreamy ideas into working hardware.

Meanwhile, Jerry Elkind was joining in the spirit of the thing. As soon as he arrived, in July 1971, he started recruiting a strong contingent from BBN, among them language maven Danny Bobrow; Interlisp developer Warren Teitelman; and eventually Bert Sutherland, the older brother of Ivan and successor to Bill Gunning as head of the Systems Science Lab. Elkind was also instrumental in luring Carnegie Mellon's Allen Newell to PARC as a part-time consultant to the Systems Science Lab, where Newell began laying the foundations for a real science of computer-user interface design: What principles, he asked, what design techniques, what facts about human beings should designers take into account that could make computers easier to use? In 1972, moreover, Newell's students Stuart Card and Tom Moran arrived at PARC with their freshly minted Ph.D.s and went to work in Bill English's group, doing human-factors studies of the mouse, the graphics displays, and much else—and in the process pioneering the discipline now known as human-computer interaction.

And so it went. By January 1972 the Computer Science Lab was up to fifteen people, with most of the new recruits' having come from MIT, Utah, and

Carnegie Mellon. The Systems Science Lab was staffed to a similar level, as was Pake's GSL. And more were on the way. The word was out: something special was happening at PARC.

A CLEAR, ROMANTIC VISION

In Bob Taylor's conception of PARC—with the emphasis here on "Bob Taylor's conception," since it wasn't necessarily George Pake's or anyone else's—the Computer Science Lab and Bill Gunning's Systems Science Lab would function as two parts of a whole. CSL would focus on developing new hardware and operating systems, broadly speaking, while SSL would focus on applications. And everyone would be working toward that final, integrated *system*—whatever that turned out to be. Of course, with no formal authority in SSL, Taylor couldn't actually enforce that vision. But since he had been told to help Gunning build his computer team, and since he had little faith in the other man's ability to provide the right kind of leadership—Gunning was "not a computerist at all," he says—Taylor could and did fill the lab with right-thinking people. These included Bill English and his crew, for example, and Alan Kay.

Make that *especially* Alan Kay. Brilliant, irrepressible, and blessed with a world-class gift for gab—not to mention a handlebar mustache and a grin that gave him the look of a mischievous hippie—the thirty-one-year-old Kay was Taylor's most gleefully subversive agent in SSL. Indeed, "Think different" had been etched into his genetic code at least a century before it ever became an ad slogan. "My maternal grandmother was a schoolteacher, suffragette, lecturer, and one of the founders of the University of Massachusetts, Amherst," Kay wrote in a later account. "My maternal grandfather, Clifton Johnson, was a fairly well known illustrator, photographer, and writer (100+ books). He was also a musician, and played piano and pipe organ. He died the year I was born [1940], and the family myth is that I am the descendant most like him, both in interests and in temperament."[5]

Kay's mother was herself an artist and musician who taught her son well: Kay was a boy soprano, a soloist in his school choir, and a fine guitarist. His Australian-born father, meanwhile, was a physiologist who designed prostheses for arms and legs. Thus, Kay wrote, "the atmosphere during my early years was full of many kinds of ideas and ways to express them. I did not distinguish between 'art' and 'science' and still don't."[6] Having learned to read at the age of three, he continued, he had already read several hundred books by the time he entered school. "[So] I knew in first grade that they were lying to me because I had already been exposed to other points of view. . . . They don't like having different points of view, so it was a battle. Of course, I would pipe up with my five-year-old voice. . . ."[7]

As a teenager in New York City, where his father had taken a hospital position in 1949, Kay attended the prestigious Brooklyn Technical High School—until he was suspended in his senior year for insubordination. Fortunately, he already had enough credits for graduation. So it was on to tiny Bethany College in West Virginia, where he majored in biology with a minor in mathematics—until he was thrown out for protesting the college's Jewish quota.

Since this was clearly getting to be a pattern, Kay took a year off to teach guitar in the high mountain air of Denver. Then, with his draft board beginning to express an interest in his nonstudent status, he hastily signed up for the air force to avoid the army. And in the air force, for a change, Kay did *not* get cited for insubordination. Instead he simply took care to remain an enlisted man (he'd trained to become an officer until he discovered that as an officer he would have to stay in for four years rather than two) and cast about for a specialty he could enjoy. He found it when he passed an aptitude test to become a programmer. "This was back in the days when programming was a low-status profession and most of the programmers were women," he wrote. "My boss was a woman. They also were taking linguists, so it was actually a pretty interesting bunch."[8]

At home among the outcasts, Kay had a fine time in the air force. Before his hitch was up, moreover, he came across at least one really neat idea: a trick for transporting data files from one air training command center to another. Nowadays, of course, that's just a matter of mailing a disk, or sending the files over the Internet. But in the early 1960s, says Kay, when there was no such thing as a standard operating system or a standard file format, the transfer involved a tedious process of programming and reformatting. So some unknown air force designer had found a way to finesse the problem entirely. Instead of sending just a tape full of data records, he or she had realized, you could send a tape full of data records *plus* a complete set of procedures from the old machine—all the routines that anyone on the far side would need to access the records, copy them, or update them. Furthermore—and this was the truly brilliant part, Kay thought—you could package everything together in such a way that no one on the far side would ever have to worry about how the data was formatted or how the procedures did what they did. The programmers would just have to provide the package with a standard command—"Copy record 37," say—and the package would automatically find the correct procedure, do all the work internally, and present the results. In effect, the data file would now behave like an intelligent assistant: it would "know" how to do its job without having to be guided step by step.

This was an idea that would later explode in Kay's own brain and become central to his work at PARC. At the time, however, he simply filed it away with all the other neat ideas he'd come across, then turned his attention to more urgent matters, such as, what was he going to do with his life after the air force?

His initial answer was to finish up his math-cum-biology degree at the University of Colorado in Boulder—though he admits that he spent most of his time there in the university theater, writing stage music (among other things, an original musical based on *The Hobbit*). In the fullness of time, however—1966—Kay did indeed graduate. So now what?

Well, he thought, computing was something he could do. In fact, he'd paid his way through the rest of college by doing database-retrieval systems for the National Center for Atmospheric Research in Boulder. True, Colorado didn't have a Ph.D. program in computer science. But the University of Utah did. And Salt Lake City, at more than four thousand feet above sea level, lay at the base of some of the most beautiful mountains in the country. So in the fall of 1966, says Kay, dead broke and totally unaware of ARPA, the ARPA community, or Utah's role as a center of excellence in graphics, he walked into Dave Evans's office looking for a job.

"On Dave's desk was a foot-high stack of brown-covered documents," Kay would recall, "one of which he handed to me: 'Take this and read it.' Every newcomer got one. The title was *Sketchpad: A man-machine graphical communication system*." The author was Ivan Sutherland; this was his 1963 Ph.D. dissertation. And the message, according to Kay, was very clear: Dave Evans took this Center for Excellence stuff seriously. His two-year-old department might be small—Evans was one of three professors, and Kay one of six grad students—but he was holding out for the best. And this was the acid test. To be accepted as a real person at Utah, Kay said, you had to *understand* Sketchpad.

"Head whirling," he continued, "I found my desk. On it was a pile of tapes and listings, and a note: 'This is the Algol for the [UNIVAC] 1108. It doesn't work. Please make it work.' " This was another Utah tradition, Kay learned. The newest graduate student got the latest dirty task to do.

"The documentation was incomprehensible," he wrote. "Supposedly, this was the Case-Western Reserve Algol, but it had been doctored to make a language called Simula; the documentation read like Norwegian transliterated into English, which in fact it was. . . . Finally, another graduate student and I unrolled the program listing 80 feet down the hall and crawled over it yelling discoveries to each other."

Eventually they figured out that Simula's authors—Kristen Nygaard and Ole-Johan Dahl, of the Norwegian Computer Center in Oslo—had designed it as a means to make it easier to create computer simulations, just as its name suggested. Their design incorporated two key ideas. First, like Sketchpad, Simula allowed the programmer to define data structures in terms of templates, or "masters," that could then be used to produce as many special "instances" as needed—without the code's having to be rewritten for each one. (Think of the master plans for a Boeing 747, say, which can be used on the assembly line to turn out specific aircraft outfitted for United, All-Nippon Airways, and so on.) Second, like that unknown genius of an air force designer, Simula put each of those data structures together with all its procedures in a tightly integrated package, so that each structure "knew" how to respond to commands. In practice, this meant that a Simula programmer could model an oil refinery, say, in much the same way that he or she thought about a real refinery: not as a list of abstract data structures and equally abstract procedures, but in terms of valves, pipes, tanks, and whatever—tangible objects that had well-defined properties and characteristic behaviors. The potential gain in conceptual clarity was enormous.

In a purely technical sense, of course, neither idea changed anything; down at the level of bits and bytes, the computer would still have to grind through binary commands one step a time, just as computers had always had to do. To Alan Kay, however, Simula was a revelation—a whole new way to think about computation. "For the first time," he recalled, "I thought of the whole as the entire computer and wondered why anyone would want to divide it up into weaker things called data structures and procedures. Why not divide it up into little computers, as time-sharing was starting to? But not in dozens. Why not thousands of them, each simulating a useful structure?"[9]

Simula was not just a better old thing, Kay realized. It was almost a new thing—an entirely new way to structure computations. True, several more years would pass before he could clearly define what that "new thing" was, and how to make it real. But "object-oriented programming," as Kay came to call it, was already there in embryo.

To help his students get out into the real world as quickly as possible, Dave Evans would often point them toward consulting jobs in local technology firms. Thus, wrote Kay, "in early 1967 he introduced me to Ed Cheadle, a friendly hardware genius at a local aerospace company who was working on a 'little machine.' It was not the first personal computer—that was the LINC of Wes Clark—but Ed wanted it for non-computer professionals." In retrospect, of course, Cheadle was about ten years ahead of his time. But why let a little thing like technical infeasibility stop you? Kay and Cheadle promptly entered into a very pleasant collaboration on what they called the FLEX machine, a prototype desktop computer that would feature a (tiny) graphics screen as well as Kay's first attempt at an easy-to-use, object-oriented programming language.

Shortly after that, Kay recounted, even as he and Cheadle were pondering how to achieve their goals, given the "little machine's" severe lack of horsepower, the Utah group received a visit from Doug Engelbart, "a prophet of Biblical dimensions." Engelbart gave them a progress report on NLS, and once again, Kay said, it was a revelation. Hypertext, graphics, multiple panes, efficient navigation and command input, interactive collaborative work, Engelbart's whole notion of augmenting human intellect—taken together, they made for an electrifying vision of what computing ought to be.

But that just led to an even more electrifying thought. Two years earlier, said Kay, he had read (and promptly forgotten) Gordon Moore's original article about the exponentially falling price of integrated circuits. But now Moore's insight came rushing back, and "for the first time," Kay wrote, "I made the leap of putting the room-sized interactive TX-2 . . . on a desk. I was almost frightened by the implications; computing as we knew it couldn't possibly survive—the actual meaning of the word changed." Instead of a world in which there existed at most a few thousand mainframe computers, all controlled by large institutions, Kay could now envision a world boasting computers by the millions. Computers under the control of no one but their owners. *Personal* computers. "It must have been the same kind of disorientation people had after reading Copernicus," Kay wrote, "[when they] first looked up from a different Earth to a different Heaven."[10]

Hardware constraints be damned; Kay immediately adopted as many of Engelbart's ideas as he could for the FLEX machine and, with Cheadle's enthusiastic cooperation, did his best to make it a prototype for computing's new world. But then came still more revelations. In the summer of 1968, Kay attended the first ARPA graduate-student conference at Allerton in Illinois. "My FLEX machine talk was a success," he recorded, "but the big whammy for me came during a tour to the U of Illinois where I saw a 1" lump of glass and neon gas in which individual spots would light up on command—it was the first flat-panel display. I spent the rest of the conference calculating just when the silicon of the FLEX machine could be put on the back of the display. According to Gordon Moore's 'Law,' the answer seemed to be sometime in the late seventies or early eighties. A long time off—it seemed too long to worry much about it then."

But later that year, Kay saw the Grail system at RAND. Based on the original RAND Tablet, with its stylus and digitizing pad, Grail could respond to a "language" of strokes and gestures in much the same way that modern-day pen computing does. "It was direct manipulation, it was analogical, it was modeless, it was beautiful," Kay wrote. "I realized that the FLEX interface was all wrong."[11]

So, more changes. And in the meantime, said Kay, his brain was taking still another direct hit. This one came about a month after he first saw Grail, when he paid a visit to Seymour Papert and his colleagues at the MIT AI Lab. The South African–born Papert, a disciple of the pioneering child psychologist Jean Piaget, was in open rebellion against the then-prevalent doctrine that learning had to be a matter of endless drill and rote repetition, as per the behaviorists. He likewise rejected the Industrial Age notion that education could be accomplished by systematically pouring knowledge into an empty vessel, as was implicit in the assembly-line organization of most American public schools. True learning, he insisted, required the active participation of the learner. True learning was a matter of curiosity and exploration—and the joy of discovering how each new experience fitted in with the web of memories, ideas, feelings, and sensations already in the mind.

True learning, he was convinced, was what computers could bring to everyone.

"The computer is the Proteus of machines," Papert later asserted in his 1980 book, *Mindstorms*. "Its essence is its universality, its power to simulate. Because it can take on a thousand forms and can serve a thousand functions, it can appeal to a thousand tastes." Indeed, he argued, with enough processing power and the right kind of interface, a computer could simulate marks on paper, paint on canvas, the "motion" encoded in film and television—any medium of expression that humans had yet devised. But much more than that, the computer could be dynamic and responsive in a way that no other medium had ever been. It could execute programs. It could respond to questions and experiments. It could engage the user in a two-way dialogue.

To prove that assertion, Papert and collaborators in the BBN education-research group had devised a Lisp-like computer language known as Logo, which included commands to guide a robot "turtle." The turtle's great talent was to draw lines as it rolled around on a big sheet of paper; the idea was to write a program telling it how to turn this way and that and thereby generate a picture of a house, say, or a flower. Logo also had commands for guiding a screen version of the turtle—less chance of a mechanical breakdown there—as well as for operating a simple music generator. By the time Kay came to visit, in fact, Papert and his colleagues were teaching Logo to elementary school children in the Lexington, Massachusetts, system.

Kay, the lifelong rebel against convention, authority, and schools run like factories, was enthralled, especially when he saw how the kids were actually responding to Logo. "First," he would write in a 1977 article coauthored with Adele Goldberg, even though computing is supposed to be a quintessentially "grown-up," intellectual activity, "the children really can write programs that do serious things. Their programs use symbols to stand for objects, contain loops and recursions, require a fair amount of visualization of alternative strategies before a tactic is chosen, and involve interactive discovery and removal of 'bugs' in their ideas. Second, the kids love it! The interactive nature of the dialogue, the fact that *they* are in control, the feeling that they are doing *real* things rather than playing with toys or working out 'assigned' problems, the pictorial and auditory nature of their results, all contribute to a tremendous sense of accomplishment to their experience. Their attention spans are measured in hours rather than minutes."[12]

By now, said Kay, his epiphany was upon him full force: "As with Simula leading to OOP [object-oriented programming]," he wrote, "this encounter finally hit me with what the destiny of personal computing *really* was going to be. Not a personal dynamic *vehicle*, as in Engelbart's metaphor . . . but something much more profound: a personal dynamic *medium*."* And that, in turn, led directly to a "clear romantic vision" of what a personal computer should be. The Dynabook, he called it. "I remembered Aldus Manutius who forty years after the printing press put the book into its modern dimensions by making it fit into saddlebags," he wrote. By the same logic, the Dynabook would have to be no larger than a notebook. "Now it was easy to know what to do next. I built a cardboard model of it to see what it would look and feel like, and poured in lead pel-

---

* To Engelbart, the difference between the batch-processing mainframes sold by IBM and the kind of interactive, "personal" computing that he was after (via time-sharing) was the difference between railroads and the private automobile. One represented schedules, standardization, and regimentation, and the other individuality, autonomy, and freedom.

lets to see how light it would have to be (less than two pounds). I put a keyboard on it as well as a stylus because . . . there still needed to be a balance between the low speed tactile degrees of freedom offered by the stylus and the more limited but faster keyboard. Since ARPA was starting to experiment with packet radio, I expected that the Dynabook, when it arrived a decade or so hence, would have a wireless networking system."[13]

Kay's direction was set. FLEX, alas, was only the first tiny step. (Or maybe not so tiny; Kay's "self-portrait" drawing of the FLEX, circa 1968, looks like a slightly overweight version of the Apple II microcomputer that would debut a decade later.) But no matter. Once Kay had completed his 1968 master's thesis and his 1969 Ph.D. dissertation at Utah—both devoted to FLEX—he went off to the Stanford AI Lab, where he spent far more time thinking about Dynabooks than he did about artificial intelligence. And then in September of 1970, when he already had a deal about where to go next—Allen Newell and Gordon Bell had invited him to come build Dynabooks at Carnegie Mellon—he received a visit from an old Utah buddy named Bob Taylor.

Kay remembers being dubious at first. Work for a *company?!* But he trusted Taylor and couldn't help being intrigued by PARC. So in the end he agreed to give it a try as a part-time consultant. But what should he do there? he asked.

"Follow your instincts," said Taylor.

Kay immediately set to work on designs for the KiddiKomp, a kind of interim Dynabook that would allow him to experiment with various approaches to the notebook's user interface. And then, he says, when Taylor recruited the BCC gang at the very end of 1970, things got very interesting. "In April of nineteen seventy-one, I called up Allen Newell and said I wasn't coming out to Carnegie Mellon, I was going full time at PARC. But Taylor was way ahead of me—he'd already hired Newell as a consultant to PARC!"

By this point, however, Kay had also had to admit to himself that he wasn't going to get anywhere with his Dynabook notion until he'd gathered a group of researchers who could help him; left to his own devices, he just didn't have the temperament to finish up one idea before going charging off to the next. Fortunately, Bill English took him under his wing that summer and helped him recruit a team that would eventually include such stalwarts as Dan Ingalls, Adele Goldberg, and many others. The goal, Kay told them, was to design a programming system in which simple things would be simple, and complex things would be possible. "I called it the Learning Research Group (LRG) to be as vague as possible about its charter," he explained. "I only hired people who got stars in their eyes when they heard about the notebook computer idea. . . . When anyone asked me what to do, and I did not have a strong idea, I would point to the notebook model and say, 'Advance that.' "[14]

During that same summer of 1971, meanwhile, Kay was refining his KiddiKomp idea into a new interim Dynabook design that he called the miniCOM.

Among other things, it would feature a brand-new object-oriented programming language that Kay now called Smalltalk (as in "programming should be a matter of . . ." and "children should program in . . ."). Thanks once again to Bill English, moreover, Kay and his brand-new group were also able to start experimenting with the miniCOM software without even having to build the corresponding hardware. It turned out that the émigrés from SRI were already working on their own vision of the electronic office. POLOS, as they called it–the PARC On-Line Office System–was partly a next-generation version of Engelbart's NLS and partly an experiment in the use of video as a display medium. But the really great thing about the system, from Kay's point of view, was that the POLOS terminals were actually high-resolution graphics consoles, modified by the addition of "character generators" that Butler Lampson and SSL's Ron Rider had designed to produce text in a form suitable for a laser printer. Kay and his team soon learned that with a little creative hacking, the terminals could be tricked into generating a "character" that was actually a complex graphic image. So they plunged in glee-fully and, during the fall, winter, and spring that followed, laid the foundations of the entire graphical user interface that would later be made famous by the Apple Macintosh and Microsoft Windows.

Kay himself, for example, designed a "paint" program so that the kids who would one day use his Dynabook could sketch their own computer graphics. His design was implemented by Steve Percell, a student intern from Stanford, and then integrated with a line-drawing system developed by PARC's John Shoch. Meanwhile, Bill Duvall of the POLOS team devised a "mini-NLS" that could be used for text processing on the terminals. Bob Shur built an animation system. Kay and POLOS's Jeff Rulifson began kicking around ideas for "iconic" programming languages, which would allow kids to write their programs in terms of graphical symbols instead of as text. And to keep the screen from get-ting too crowded, the team found a way to let documents appear in separate but overlapping "windows"–a brainstorm that had hit Kay one day while he was in the shower, his favorite place to think.

By May 1972, in fact, Kay was so enthused by their progress that he wrote a proposal arguing that the time had come for PARC to build some interim Dyn-abooks for real; he wanted to take them into classrooms and start working with actual children. Much to his dismay, however, the answer was no, on the grounds that a) the concept was still too nebulous, and b) MAXC still wasn't finished. The PDP-10 clone was costing a fortune, said the higher-ups, and PARC's budget was only generous, not infinite.

So Kay crawled away–"licking my wounds," as he put it–and instead made plans for his group to keep on doing as much as they could with the POLOS ma-chines. And that was still pretty much where things stood in September 1972, he says, when Butler Lampson and Chuck Thacker dropped by one day from CSL.

"Do you have any money?" they asked.

What are they up to? Kay wondered. Lampson was–well, Lampson. And Thacker, Kay knew, was one of the youngest members of the Berkeley contingent, having joined the others at BCC immediately after completing his undergraduate degree. At PARC he had taken the lead in crafting the MAXC hardware and was widely considered to be one of CSL's real wizards.

"Yes," Kay said. "I've got about $230K. Why?"

"How would you like us to build your little machine for you?"

GETTING PERSONAL

To hear Lampson and Thacker tell the story, it almost sounds as though the idea had crept up on them while their backs were turned. But then, in a sense their backs *had* been turned: for the previous eighteen months or so, they and their colleagues in CSL had been pretty much obsessed with MAXC, urgently trying to get the PDP-10 clone finished. Right in the middle of it all, moreover, they had been distracted by the chaos of moving. With PARC rapidly outgrowing the tiny Porter Drive site, and with the construction of PARC's permanent quarters only just beginning, George Pake had been obliged to relocate the center to a larger rented building around the corner on Coyote Hill Road, directly across from the construction site.

So it wasn't until the summer of 1972, when MAXC was pretty much done, that they had had a chance to look up, take a breath, and think about how they were actually going to achieve the electronic office. And in surprisingly short order, they found themselves in broad agreement: the electronic office should consist of small personal computers, running high-resolution graphics and linked together with high-bandwidth network connections—essentially the very same vision that Bob Taylor had been pushing from the start, the one they had once found so incomprehensible.

This had largely been Taylor's doing, says Lampson, though not through any obvious, do-it-my-way-or-else nonsense. Instead, Taylor had just subtly, but profoundly, shaped the whole context of their work. He had seeded PARC's computer science labs with products of the ARPA culture, people who felt the urge toward human-computer symbiosis in their bones. He had continued to expose them to provocative ideas ("Bob was always foisting Licklider's papers on us," says Chuck Thacker. "So by the time we finished MAXC, I had read the 'Symbiosis' article, which was where Lick had said it all, and which was probably one of the things that made Bob's idea less inscrutable"). And then, having done all that, Taylor was content to sit back in the Dealer Meetings and elsewhere and let his people function as a kind of self-exciting system. As Stu Card remembers it, "There was this thread of ideas that led from Vannevar Bush through J. C. R. Licklider, Doug Engelbart, Ted Nelson, and Alan Kay—a thread in the Ascent of Man. It was like the Holy Grail. We would rationalize our mission according to what Xerox needed, and so on. But whenever we could phrase an idea so that it fell on this path, then suddenly everybody's eyes would light up, and you'd hit this resonance frequency."

Take graphics, for example: everybody resonated with graphics. They had

Alan Kay right there, after all, constantly preaching his gospel of computers as the most richly expressive medium humans had ever known—and more to the point, showing them his group's prototype font editors, drawing programs, on-screen document windows, and iconic programming systems. It was living proof of what you could really *do* with computer graphics. They likewise had Gary Starkweather and his laser printer: living proof that you could build up any image you wanted just by arranging tiny dots on a piece of paper—or on a computer screen. And they had the living example of Doug Engelbart's NLS. "You got this feeling sitting in front of one of Doug's screens, and looking at his displays, that the computer image was as good as paper," says Lampson. "And that was a revolutionary idea at the time." In the electronic office, whatever that turned out to be, the computer screen would have to be able to display text, diagrams, formulas, annotations, doodles—anything paper could display.

If, of course, they could figure out how to make it do all that.

This was in fact the challenge that had eventually led them to the idea of small, personal computers—albeit with a heavy emphasis on "eventually." Although they'd all had a fine time listening to Alan Kay talk about Smalltalk and his Dynabook idea, says Lampson, "we saw it as interesting from an intellectual point of view, not for building systems." When it came to serious computing, he admits, most of the young computer jocks at PARC had grown up in the era of time-sharing, when the vision of democratized computing meant the creation of giant computer utilities. They couldn't help but have a tremendous emotional and intellectual commitment to time-sharing, he says—especially the Berkeley contingent, many of whom had now built three major time-sharing systems in less than eight years, counting Project Genie, the BCC 500, and MAXC. Moreover, MAXC itself had turned out to be a wonderful computer, everything they'd hoped it would be. It ran Tenex. It served as PARC's connection to the Arpanet. It had (by 1972 standards) a vast amount of memory and disk storage. Indeed, it was one of the first machines to use the Intel Corporation's brand-new semiconductor memory chips instead of standard magnetic-core memory.

Yet for all of that, says Lampson, it was MAXC as much as anything else that pushed them away from time-sharing. "Here we had a state-of-the-art central computer of the day," he says, "and it was perfectly obvious that it couldn't do what we wanted it to do." The computational muscle required for high-quality graphics just wasn't there, not when MAXC had to parcel out its memory and processing power to dozens of users at a time. In fact, says Lampson, he and his colleagues were increasingly coming to realize that the decision to focus on graphics displays undermined the most fundamental premise of time-sharing—namely, that computers are fast and humans are slow. As he would express it in a later account, "[This] relationship holds only when the people are required to play on the machine's terms, seeing information presented slowly and inconveniently, with only the clumsiest control over its form or content. When the machine is required to play the game on the human's terms, presenting a pageful of attractively (or even legibly) formatted text, graphs, or pictures in the fraction of a second in which the human can pick out a significant pattern, it is the other way around: People are fast, and machines are slow."[15] And once they had realized that, says Lampson, the conclusion was all but inescapable: the only way the machine could hope to keep up with its user was by devoting the bulk of its computational power to running the display. Thus one computer per person.

But was a stand-alone, graphics-intensive personal computer even possible? Or more precisely, was it possible at a price in the ten-to-twenty-thousand-dollar range, so PARC could afford to provide one for everybody in CSL?

Maybe. By 1972, notes Chuck Thacker, he and Lampson weren't nearly as skeptical on that score as they had been. Once they'd gotten a little hands-on experience with the Nova minicomputers being used in the POLOS* system, he says, "Butler and I became intrigued by minicomputers." For one thing, the current wave of minis was clearly a lot more capable than the first wave had been back in the late 1960s, when all you had was the twelve-bit PDP-8 or one of its competitors. In fact, the Nova, introduced by Data General in 1969, had been the first of the sixteen-bit minis, a group that now included DEC's own entry, the PDP-11. "And the potential of these machines was pretty attractive," says Thacker. Indeed, the market was booming: the second-generation minis were beginning to proliferate through laboratories, departments, companies—everywhere. DEC alone would eventually sell some half a million units of the PDP-11 in all its various models. But best of all, says Thacker, "by the early nineteen-seventies, minicomputers had dropped in price—which made them *really* interesting." Moore's law had begun to take hold in earnest, and the staffers at PARC had begun to feel it in their bones: almost any hardware they could imagine would be affordable in surprisingly short order. And if they could imagine a cheap, ultra-high-powered minicomputer serving a handful of users, then why not a cheap, ultra-high-powered minicomputer serving *one* user?

By the summer of 1972, says Thacker, "it was personal-computer time, just like it was railroad time in the eighteen-fifties."

So: low cost, good performance, high-quality graphics. How were they going to *do* all that?

---

* Bill English and his team were also after high-quality graphics, of course. But instead of going the one-computer-per-person route, they had attempted to replace that one central time-sharing machine with *many* central machines. Their idea was to divide up POLOS by tasks and then distribute the separate tasks among a web of interconnected Nova minicomputers. One Nova might run all the terminal displays, for example, while another did all the word processing, another did the overall time-sharing management, and so on. English and his colleagues also planned to have the Novas transfer functions to one another as needed, so that the system would be easy to expand: to get more capacity, you'd just plug in another computer.

Over in CSL, however, Lampson, Thacker, and their compatriots were dubious. They were convinced—correctly, as it ultimately turned out—that POLOS would be clumsy, overcomplicated, and a nightmare to program, with most of the Novas' computational resources' being eaten up by the effort to keep themselves coordinated.

Once he'd finished his work on the MAXC hardware, says Thacker—Lampson and the others were still wrestling with the MAXC software—he set out in earnest to find some answers. His first step had to be graphics, he knew, since they were clearly the key to everything else. The distressing thing was that PARC was already full of graphics terminals—"All of them expensive," he says, "and all of them, in one dimension or another, crummy."

There were the IMLACs, for example, which worked like an oscilloscope or, for that matter, like the TX-2 display that Ivan Sutherland had used for Sketchpad. That is, they would project a single bright spot on a dark screen and then electronically move that spot around to trace out a circle, say, or the letter *A*. By tracing and retracing the pattern very, very fast, the IMLACs could create the illusion of a solid outline. The result, as in Sketchpad, was beautifully sharp-edged line drawings that still represented the best that computer graphics had to offer. The problem was that the more complicated the drawing, the faster you had to wiggle that spot. So when you got to *very* complicated drawings—like a page full of text, with all the fine detail of all the individual letters—you were straining the physical limits of what the circuitry could do. The resulting output was sloppy, flickering, or both.

Then there were the "raster-scan" displays that Bill English had developed for POLOS, which worked very much like a standard television set. Instead of trying to wiggle the spot around to draw a picture directly, these displays would just sweep the spot back and forth across the screen at a uniform rate, first drawing one horizontal line, then drawing another line directly under it, and so on until they had to start all over again at the top. The screen would fill up this way about thirty times per second. And just as on a television set, the displays could vary the intensity of the spot as it traveled, thereby building up images line by line as a pattern of light and dark. The main difference was that TV sets used analog electronics, so that the brightness at each point on the screen could be white, black, or any shade of gray in between. The POLOS displays used digital electronics that were better suited to the binary world of computing: in effect, they would divide their screens into a fine grid of "pixels" and then make a picture by turning each pixel either on or off, as appropriate, with no shades in between.

Now, there was a lot to like about this approach, says Thacker. The displays could use much simpler circuitry than the IMLACs could, since they just had to move the spot back and forth at a uniform rate. Their images could have very high resolution, limited only by the size of the pixels. And the programmers would have a much easier time devising graphics software to generate those images, because all they had to do was define a chunk of computer memory to be a map of the screen, one bit per pixel, and then drop the appropriate bit into each memory location: 1 for white and 0 for black. The display would then read out the "bit-map" automatically.

Unfortunately, says Thacker, that use of the computer's memory was also the major difficulty with bit-mapped graphics: memory was very, very expensive in those days. At resolutions approaching 100 pixels to the inch, which was what you needed for comfortable reading, on a display that was roughly the size of an 8½-by-11-inch sheet of paper, which was what you needed in order to see what you were working on, a single screen full of graphics required something like 64 kilobytes of storage. Multiply that by the dozens or even hundreds of machines that would be needed at PARC, and, well, the cost just seemed out of the question in 1972.

Unless . . .

Thacker remembers that it was Kay's trick with the graphics on the POLOS terminals that suddenly brought it home to him. Yes, the cost of *magnetic core* memory was prohibitive. But Thacker had just finished building a whole new computer, MAXC, using Intel's new solid-state memory chips. And the cost of solid-state memory was going to fall fast, thanks to Moore's law. "That was the biggest 'Aha!' experience about this," says Thacker. "The realization that this semiconductor memory was inexpensive enough to devote a bit to every pixel and not go bankrupt."

But that wasn't the end of it. Having now produced an enormous simplification in the graphics with that first revelation, Thacker almost immediately experienced a second epiphany that would greatly streamline the basic computing hardware.

In conventional machines, he explains, most of the complication came not from the guts of the computer itself—the central processor and the memory banks—but from peripheral devices such as the keyboard, the tape reader, the hard disk, and the display, those gadgets required for getting data in and out. Furthermore, much of *their* complexity lay in the elaborate interface devices, or "controllers," that allowed each peripheral to communicate with the interior. But was all that specialization really necessary anymore? Thacker wondered. Given the way prices were falling and processor speeds were rising, why not just build a bunch of very simple controllers and then let the computer's central processor use software to do all the really hard work of input and output? The result would be a kind of internal time-sharing, Thacker realized. The processor would still cycle very, very quickly among all its users, but now only one of those users would be human; the rest would be input/output devices. "The payoff," he explains, "would be an enormous simplification of the machine."

You know, they could actually do this thing.

Certainly Butler Lampson thought so. "Once Chuck came up with the whole idea for the bit-mapped display, and how the machine could work as a stand-alone thing," he says, "we couldn't wait. It was just so obvious to us that this was the way to go, that this was the future. We didn't care about time-sharing anymore."

Alan Kay thought so, too. When Thacker and Lampson came by that September offering to collaborate on an "Interim Dynabook," he didn't hesitate for a minute. Here was the "vector sum" of his own dream of a graphical notebook computer, Lampson's vision of a five-hundred-dollar PDP-10, and Thacker's quest for a ten-times-faster Nova, all in a single, beautifully simple package. "It was perfect," he said of Thacker's elegant design.

And Bob Taylor *really* thought so. "As soon as I heard about the bit-mapped display," he recalls, "I said, 'Yes! That's *it!*' "

Of course, they were going to have to sell the idea to George Pake. And Jerry Elkind, for one, wasn't at all sure how warmly Pake would receive it. After all, Bill English and his group in SSL were going full blast with their POLOS project, which wasn't cheap. CSL itself had just dropped a bundle on MAXC. And now they wanted Pake to authorize twenty or thirty of these new personal computers at a cost of what, ten thousand dollars apiece? Twenty thousand? Add in network connections, laser printers, and all the rest, and you were looking at an outlay of around a million dollars. Personally, Elkind maintains, he was all for it.* But he insisted that the CSL crew had better be prepared to make a very strong case.

Fair enough. To do that, however, they would first have to get a real machine designed and prototyped. They would have to develop a clear idea of what kind of applications could run on the thing. And they would most definitely have to come up with a better name than Interim Dynabook.

That last point was one of Taylor's particular obsessions. "Good names for prototypes are very important, and very difficult to choose," Taylor explains. "They should be familiar, easy to pronounce, easy to spell, have a broad theme, and conjure up pleasant feelings." In fact, he says, he'd never found a better source for prototype names than the *Sunset Western Garden Book*. In this case, moreover, he thought it might be nice to give their new computer an association with their home in Palo Alto. So . . .

Well, that was easy. The Alto it was.

The Alto was certainly not the first personal computer; that honor has to go to Wes Clark's LINC, if not to Clark's TX-0, or even to Jay Forrester's Whirlwind. But it was the first machine that most of us would recognize as a personal computer. When Chuck Thacker outlined his preliminary design for the Alto in a memo dated November 22, 1972, he described a system that wouldn't look all that out of place in an office at the turn of the millennium. It would have:

- A bit-mapped display screen measuring 606 pixels horizontally by 808 pixels vertically, thus producing an upright rectangle about the size and shape of a standard 8½-by-11-inch sheet of paper. (Think

---

* Actually, Elkind concedes, some people at PARC may have heard such questions as opposition. Certainly Alan Kay did. To this day, Kay portrays the creation of this computer, the Alto, as a kind of guerrilla movement, with the workers pushing the idea forward in the face of fierce management opposition and Thacker rushing to get the prototype done before Elkind could return from an out-of-town assignment. The only problem is that neither Elkind nor Lampson nor Thacker remembers it that way. Lampson, for example, says he doesn't recall having had any particular problem getting the Alto approved. And Thacker says he was rushed only because he'd made a bet with a friend that he could have the machine up and running by a certain date.

of a modern 15-inch monitor turned on its side.) To enhance the similarity to paper even further, the display would produce black characters on a bright background.

- A standard typewriter keyboard, along with a three-button mouse. (The keyboard had no cursor control keys or numeric keypad, as modern office computers do. But it did have eight function keys whose use could be defined by the software.)
- A high-bandwidth communications interface, as yet unspecified.
- A system cabinet to house the hard disk, central processing unit, system memory, and other electronics. The cabinet would be a cube, about thirty inches on a side, and would sit on the floor under the user's desk.

Inside that system cabinet, the differences from today's computers become more obvious. The Alto would have a 10-megabyte hard disk, a 1.5-microsecond-per-instruction central processor, and 128 kilobytes of system memory—all of them shy of their year 2000 counterparts by roughly a factor of five hundred. (The Alto's CPU wasn't even a single microchip, since that was still beyond the state of the art in 1972; instead, it consisted of some two hundred separate integrated circuit chips, mounted on three printed circuit boards.) And in truth, Thacker admits, what he sketched out in that memo wasn't a high-performance computer even by 1972 standards—especially not with only 128K of memory, which would later impose some pretty severe constraints on software for the machine.* But then, it didn't have to be high-performance. The Alto simply had to respond to its user faster than a time-shared PDP-10—which, for most problems, it would. Because it was a stand-alone machine, moreover, the Alto would be predictable, in the sense that you would never have to worry about too many users' logging on and slowing the machine to a crawl. As Jim Morris put it, the Alto was wonderful because it wouldn't run faster at night. And best of all, the simple, straightforward Alto hardware would be "cheap." The initial estimate was that each new machine would cost about $10,500, which was about half

---

* In retrospect, says Thacker, he regards that 128K figure as "one of the two most embarrassing things about the Alto." And the other? Simply that his elegant, beautiful idea for simplifying the Alto hardware wasn't original. "Wes Clark stopped by one day after having read some of the Alto documentation," remembers Ed McCreight, who had eagerly signed on to help develop the Alto. "Wes said, 'Say, this Alto stuff is pretty interesting. I wonder if you could, in a few words, say what the relationship is to the TX-2 and, in particular, to the task structure of the TX-2?' Chuck and I looked at each other and said, 'Well, ah, well, ah—not really very well.' Wes said, 'Well, as it happens, the TX-2 papers were in the WJCC [Western Joint Computer Conference]. I have some copies here I can leave with you, and I could ask the question later.' So Chuck and I avidly read these things and compared notes and, as I said to Wes the next day when he came back for our answer, my only excuse was that I was in the eighth grade at the time." Thacker, McCreight, and their colleagues immediately acknowledged Clark's priority, of course, and withdrew their patent application on the idea.

what PARC had already spent on MAXC per CSL member. So everybody in CSL could, indeed, have one.

Once the memo started to circulate, the comments around PARC were favorable, if a bit lively. But Thacker was unfazed. "I was an old pro," he says with a laugh. "I had been in the field for five years!" He was also in a hurry: he had bet a bottle of wine with a vice president of XDS that he could get the new computer up and running within three months—and the clock had started ticking on November 22, the date of his memo. Happily, however, Lampson, Ed Mc-Creight, and several others volunteered to help, and by December they were well along with the detailed designs. By early 1973 they were hard at work with their soldering irons—all the laboratory facilities and technology they'd developed for MAXC paid off handsomely; the MAXC memory boards, in particular, went straight into the new machine without modification—and as winter turned into spring, the first two Alto prototypes were taking shape with astonishing speed.

True, admits Thacker, he didn't quite win his bet—there was some debate about when the machine actually booted up—and he never got his bottle of wine. But he came pretty close. On April 1, 1973, as GSL physicist David Thornburg would later tell the story, "I walked into the basement where the prototype Alto was sitting, with its umbilical chord attached to a rack full of Novas, and saw Ed McCreight sitting back in a chair."

In the upper-left-hand corner of the new machine's display, said Thornburg, in small type, were two words: *Alto lives.*

UP THE HILL AND DOWN THE HILL

Alto lived, all right—as an infant. But at least now it was an infant that had some hope of surviving long enough to leave the nursery. The concern about George Pake's approval had proved to be a nonissue: Pake liked the Alto as much as anyone—and indeed, would become a strong advocate for the machine in his annual budgetary tussles with the Xerox management. As a physicist, after all, he was used to spending lavishly on lab apparatus, well aware that the first investigators to invest in the next level of instrumentation are usually the first to make the next level of discoveries.

However, that approval still left the CSL team with the challenge of teaching their infant how to crawl, walk, and run—a task that promised to take a while. It wasn't until the summer of 1973, after several more months of testing and debugging, that Thacker and his coworkers were ready to commit the hardware design to printed circuit boards. And it wasn't until the end of that same year, after the construction of nine more prototypes, that they were ready to begin producing the thirty machines they would need for CSL as a whole.

As for software, there wasn't any. Or at least, not much. The original prototype Altos were able to run a few simple programs that exercised the disk and

the display, says Thacker, and he and his colleagues lost no time in running them. "The first recognizable image was the Cookie Monster from *Sesame Street*," he recalls. "One of Alan Kay's guys had digitized the thing. It was actually an animation with several frames—the Cookie Monster ate the cookie!" That summer, moreover, as a demonstration of how well the Alto could respond to its user in real time, they decided to grant Kay one of his many wishes by hooking up an organ keyboard, an amplifier, and some high-fidelity loudspeakers. With a little programming, the Alto functioned as a decent music synthesizer.

Still, when it came to more substantive applications such as text editors or graphics packages, progress was slow; no one wanted to start writing code until there was some real hardware to run it on. Lampson wouldn't even have the initial Alto operating system ready until the autumn of 1973, almost six months after the first two prototypes, and its first formal release wouldn't come until March 1974.

And as for communications—well, Thacker concedes, he, Lampson, and Mc-Creight hadn't really focused on that aspect of the system. They did have some vague idea of linking the Altos eventually, if only to give individual users a way to fire off files to the laser printer. And of course, connectivity had always been another of Bob Taylor's obsessions. The CSL deputy director kept insisting that their goal was not just *personal* computing but personal *distributed* computing: a web of interconnected machines that could support communication, collaboration, and the sharing of programs and resources—the kind of on-line community that had always been the best part of time-sharing. Somehow, Taylor told his people, you have to network these machines together; otherwise, you'll be taking a giant step backward, graphics or no graphics.

Fortunately, however, Thacker and his colleagues didn't really have to focus on communications. For something like six months now, PARC had had an experienced networking hotshot hard at work on the problem—a relatively new recruit by the name of Bob Metcalfe.

In an odd sort of way, admits Bob Metcalfe, a lot of the credit for the network he developed at PARC has to go to the applied-mathematics department at Harvard University: he hated it. In fact, he'd spent his entire graduate career there staying as far away as possible—which was why he'd actually done his Ph.D. research for money, working on the Arpanet as a full-time staffer for J. C. R. Licklider's group at Project MAC. And in the spring of 1972, with his Ph.D. dissertation ready for Harvard's formal approval, he thought he might finally be seeing the last of the place.

Since graduation was slated for June, Metcalfe remembers, "I went job-hunting. I got nine job offers, because I was in tight with ARPA and everybody wanted ARPA's money. But I didn't get the one I really wanted, which was to be an assistant professor at MIT. So instead I accepted an offer from Bob Taylor and Jerry Elkind to come out to PARC and put them on the Arpanet. My then-

wife began ramping down her job. We went house-hunting in Palo Alto. We started packing for the move. My parents started planning to come see their son get a degree from *Harvard*. And then I went in to defend my thesis before my Ph.D. committee."

To his utter horror, the committee rejected it. The professors wanted theory: mathematics, equations, lots of Greek symbols. What he had given them instead was practice: in effect, his subject was "How We Built the Arpanet."

Did he mention how much he hated Harvard? Why had they waited until *now* to tell him there was a problem?

But there was nothing for it. He would have to start over, and he would most definitely have to forget about graduating in June. "I had a lot of very embarrassing phone calls to make," says Metcalfe. "The worst was to my parents. But the easiest was actually the one to Bob Taylor. He just said to come on out to PARC and finish my thesis there."

At PARC, happily, the twenty-five-hour days left him no time to brood. "So here I am," he says. "As of July nineteen seventy-two, I'm the networking guy at PARC. I'm trying to write a new thesis. I'm putting MAXC on the Arpanet. I'm helping put together the ICCC Arpanet debut for October nineteen seventy-two. And I'm continuing to act as an Arpanet facilitator, helping other sites get on the Arpanet."

This last task, especially, kept him in a more or less constant state of jet lag, which was how he happened to find himself in Steve Crocker's guest room one night during a swing through Washington, D.C., tossing and turning on the sofa bed. Desperate for something to put him under, he happened to catch sight of a thick blue volume on the bookshelf next to the bed: *American Federation of Information Processing Societies Conference Proceedings*, volume 37, fall 1970. Perfect. Metcalfe started reading "The Aloha System," a paper by Norman Abramson of the University of Hawaii.

It did not put him under.

The Aloha system, he learned, was an experimental, ARPA-funded network that transmitted computer data via radio waves, instead of via the telephone lines used in the Arpanet. The University of Hawaii, he also learned, was a natural setting for such an experiment: its campuses on the various islands were separated by large stretches of open ocean, which made for telephone connections that were noisy, unreliable, and very expensive. Abramson's paper accordingly described a network in which the main IBM System/360 computer on Oahu sent packets of data back and forth to terminals out on the island campuses via radio. Serving as a front end to the 360 was Menehune, a small, packet-switching computer that handled the actual radio connections and that was similar in function to an Arpanet IMP. (In Hawaiian legend, the menehune were strong, skillful imps who would work all night without stopping.)

Now, on the surface, Metcalfe could see, this was a matter of straight substitution: anywhere Arpanet had a wire, Alohanet had a radio link. Beneath the surface, however, when you got down to the nitty-gritty of how the packet trans-

missions were regulated, the differences were much more interesting. On the Arpanet, where the bits flowed through telephone lines, an IMP with packets to send could wait for a break in the traffic, so to speak; that way, the packets never collided with one another. But on the Alohanet, where the bits were carried by staticky, interference-prone radio waves, a terminal with packets to send had no way of knowing what the traffic was like. It could transmit back and forth to Menehune (with luck), but it probably couldn't even *hear* what the other nodes in the network were sending. So, since waiting would be pointless, Alohanet allowed each terminal to fire off a packet to Menehune whenever it needed to, regardless of what the others were doing. If the terminal heard Menehune acknowledge receipt of that packet, then fine. But if it didn't—meaning that another terminal's packet had arrived simultaneously and turned the bits into gibberish—the first terminal would just back off, wait for a random interval of time, and then transmit its packet again. Since the second terminal would also be retransmitting, but with a different random interval, both packets now had a reasonable chance of arriving unscathed.

Beautiful! thought Metcalfe. It was control without control: the terminals were completely free agents, unregulated and unsynchronized by Menehune. And yet the packets got through anyhow.

Or did they? Alas, wrote Abramson, the system's beauty came at the price of instability. Using a branch of mathematics known as queuing theory, he argued that Alohanet couldn't use more than about 17 percent of the total capacity in its radio channel without causing a kind of chain reaction. Push it past that point, and each collision of packets would trigger the transmission of replacement packets, which would increase the probability of more collisions, which would generate more replacements—on and on until every packet was statistically guaranteed to hit another packet. The system would grind to a halt.

By now, recalls Metcalfe, he was wide awake: this couldn't be right. As he wrote about it later, "The Abramson paper . . . made two assumptions about the computer terminal user behavior that, on Steve Crocker's sofabed late at night, I found totally unacceptable. Abramson's model assumed that there were an infinite number of terminal users, and that each of them would go on typing whether or not they received answers to earlier inputs."

This was ridiculous, he thought. But it was also an opportunity. Those SOBs at Harvard wanted theory? He would give them theory. He would do this queuing analysis *right.* "That night," he remembers, "and in the weeks to follow, I worked hard on the less-tractable mathematics of Aloha channels with a few users, each of whom would insist on receiving a response to [his] input before typing a new one. I worked so hard, in fact, that Xerox sent me to work with Professor Abramson for a month—in Hawaii!"

By October 1972, says Metcalfe, just in time for the Arpanet "coming-out" party in Washington, he had produced a paper showing that an Aloha-type network could indeed be made stable under much heavier loads than Abramson had believed. That paper would later become a chapter in his new Harvard

"I remember Larry coming to me and saying 'Look, I know you didn't want to work on networking anymore. But you know more about it than anybody else around, and that's where our main efforts are going to be for the next several years. So why don't you just go do that?' "

Kahn shuddered at the thought. The compliment was kindly meant, he realized, if patently untrue ("*Larry Roberts* knew more about networking than anybody," he says). Still, he had to admit that Roberts did have a point. The Arpanet wasn't unique anymore; ARPA's networking research had already started to embrace new media and new technologies. Packet-switched radio networks were one example, as in Alohanet. Packet-switched satellite links were another, as in a contract that Roberts had recently signed with BBN to create an orbital connection with Arpanet nodes in Europe. Moreover, there seemed every reason to believe that these could grow into major ARPA programs in their own right. After all, if the Arpanet could be sold to the Pentagon as an experiment in survivable, land-based command-and-control systems—a subject of deep and perennial interest to the military—then packet radio and packet satellite could just as easily be sold as experiments in survivable command and control for mobile forces.

Hmm. Kahn began to get interested almost in spite of himself. "So in the end," he says, "I told Larry I would do it—but only the new stuff. Somebody else would have to handle the Arpanet."

Done, said Roberts. And Kahn plunged in. He mapped out new initiatives on security and voice transmission. He started expanding the packet satellite program, to a point where it would ultimately include an international partnership beaming packets back and forth through the Intelsat IV satellite. He likewise planned a major expansion of the packet radio program, which would later encompass portable terminals that may have been the very first application of microprocessors in a military system.

And in very short order, he found himself tripping over one of those trivial-seeming things that start as afterthoughts and end up being the most profound and far-reaching issues of all: How was he going to get all these satellite networks and radio networks talking to one another and to the Arpanet? How could he make the packets flow freely wherever they needed to go?

It certainly *sounded* easy enough. If you wanted to add a new packet satellite system, for example, all you had to do was connect a few ordinary Arpanet IMPs to satellite dishes, have the BBN team add a little bit of extra code telling the IMPs how to communicate with the satellites, and there you'd be, trading packets back and forth through the heavens. Indeed, this was essentially what Larry Roberts had had in mind when he commissioned BBN to do that first experiment with satellites. But then, Kahn wondered, what about packet radio links, as in Alohanet? What about local networks, fiberoptics cables, infrared transmissions, and the whole vast array of communications technologies that no one had thought of yet? Was he supposed to go back to BBN every time and ask for yet another software add-on?

Not if he could help it, thought Kahn. He could just imagine each new technology languishing in limbo until the BBN programmers got around to patching it in. The bottleneck would be horrendous. Worse, he knew, the various networks were optimized for very different environments. Arpanet, for example, lived in a world of comparative stability, with its packets flowing over fixed, reliable, land-based telephone lines. But packet *radio* lived in a world of chaos, with mobile transceivers forever moving in and out of range, getting cut off by hills and tunnels, and generally losing packets at every turn. The two systems had major incompatibilities in transmission speed, packet length, and almost any other parameter you could name. And even if the programmers could overcome such differences, the accumulating patches would only make the basic Arpanet software grow bigger, slower, more complicated, and more prone to bugs and failures—in short, everything it *shouldn't* be. Imagine what automobiles would be like if we'd tried to integrate our transportation system this way: they'd be vehicles with rubber tires to drive down the highway, wings to fly through the air, streamlined, waterproof hulls to cut through the water, steel wheels to move along railways, on and on. It would be a Rube Goldberg mess.

So no: total integration of the networks was bad management and bad engineering. Instead, Kahn realized, the right way to proceed was to *dis*integrate the networks. Start with the satellite portion, say, and carve it out completely. Make it into a separate network, in much the same way that air traffic is completely separate from the highway system. Give it its own IMPs, its own software, its own transmission protocols, everything. Then take that special satellite IMP that BBN was building and split it down the middle. On one side, put a standard Arpanet IMP, containing nothing but the standard Arpanet software. On the other side, put a new satellite IMP, containing nothing but the satellite software. And in between, put a "gateway," a third computer whose sole job would be to translate Arpanet packets into satellite packets and vice versa. Think of it as being something like an airport terminal, whose main purpose is to help passengers transfer back and forth from highway travel to airline travel.

More complicated? In some ways, yes, says Kahn. "There were lots of good reasons not to do it this way. With three boxes instead of one, the network would be less efficient, since each packet would have to go through three stages of processing instead of one. It would be more fragile, since now there would be three things to break instead of one. It would be more complex, more expensive—everything. But by doing this, you could constitute the satellite system as a separate network, which could be built, controlled, and operated by a separate entity." So long as both sides met the interface standards (which were still to be defined), neither would have to know anything at all about the internal details of the other. You could simply plug them together through a gateway—much as you'd plug an appliance into a standard electric socket—and the packets would flow as needed. Moreover, you could just as easily plug in a third network—packet radio, say—and a fourth, a fifth, ad infinitum. Instead of a *closed* system, Kahn realized—that is, a single Arpanet dependent on a single operator—you would now have an *open* system, a network of networks that could, in principle, accommodate anyone.

It would be several years yet before anyone actually used the word *Internet,* says Kahn; at the time his phrase was "interconnected network," or "internet-working." And even in retrospect, he says, he has no idea when or how the notion hit him. Maybe it was just something in the air. After all, he had recently helped build the Arpanet, which was already an open system with respect to individual computers. As long as you had an IMP and met the 1822 interface specification (written by one Robert E. Kahn), you could plug into the Arpanet with any computer you wanted, running any operating system you wanted, and the bits would still flow. A network of networks was the same principle, just one level up. Perhaps the idea also resonated with the open-software architecture of operating systems such as CTSS, Multics, and Unix, which gave users a standard interface for writing their own programs. Or perhaps it even resonated with Alan Kay's notion of software objects, which would present a standard interface to the world while doing their own thing inside.

But wherever the idea came from, it was an exceedingly potent one—as potent, in its own way, as the idea of an open marketplace guided by Adam Smith's Invisible Hand, or the idea of an open democracy in which all human beings are created equal. If you wanted to be a part of the Internet, you could be: it would be open to anyone willing to deal in the common currency and/or speak the common language, as defined by the interface standard. It was this architecture of openness that would enable the Internet to undergo its explosive growth in the 1990s, when it would expand from a handful of users to uncounted tens of millions in less than half a decade. And it was this same architecture of openness that would continue to allow the Internet to flourish with minimal central authority.

But all that would come later. In the early months of 1973, says Kahn, his chief concern was how to *define* this open architecture—how to determine precisely what the gateways should do and how the protocols should work on each side. And since there was no way he could do that himself in his spare time, he decided to seek the help of his old Arpanet buddy Vint Cerf, who had just joined the computer-science faculty at Stanford.

Kahn's pitch was very casual, Cerf remembers, with no mention of an ARPA contract. It was just, "Here's an interesting problem." But Cerf was intrigued even so. "I can remember sitting in a hotel in San Francisco," he says, "waiting for some conference to start. And literally on the back of an envelope I was scratching some ideas." The critical thing, he knew, was to keep the packets flowing through all the different networks automatically, without making the poor things carry along special instructions for each one. Otherwise, you'd be right back to that horrendous programming bottleneck every time you tried to add a *new* network.

Well, Cerf asked himself, why wasn't the flow automatic already? Because in the existing system, trying to send a packet from one network into another was like trying to send a postcard written in Japanese kanji characters through a post office in the United States: nobody there would know how to read it, and it would go straight to the dead-letter office. But then, Cerf realized, that sort of thing happened only because postcards traveled naked, as it were, never changing their outward appearance to accommodate changing conditions. What if they *could* change? Or more precisely, what if each postcard could be mailed inside an envelope addressed in the local language, and then moved into a *new* envelope addressed in the *new* local language when it crossed the border? End of problem: the local mail sorters would read kanji in Japan, English in the United States, Arabic in the Middle East, and so on.

So, wondered Cerf, how do we accomplish the same thing in the digital world? First, he decided, let's forget about the differences among networks for a moment, and imagine that every computer out there has agreed to use a specific, universal protocol for addressing its packets. (This would be the software equivalent of everybody's agreeing to address postcards on the right-hand side, using the Roman alphabet, with the name, street, city, state, country, and so on in a standard order.) Next, said Cerf, let's imagine that the computer doesn't just send its packets straight out into the local network, which might not understand the universal protocol, but instead places each one inside a wrapping of extra bits that the local network *will* understand. (This would be the software equivalent of taking a letter written in English, inserting it into an envelope addressed with Japanese kanji characters, and mailing it from Tokyo.) In that form, Cerf saw, each packet could sail through its home network until it reached the gateway to the next network in line, whereupon the gateway computer would simply strip off the extra bits, read the address written in the universal protocol, wrap the packet in a new set of bits appropriate to its new surroundings, and send it on its way. (Think of the letter inside the Japanese envelope arriving in Cairo and then being sent onward inside an envelope written in Arabic.)

Bingo. That was it, Cerf realized—"the simplest possible mechanism for making the system work." He and Kahn quickly made it a centerpiece of their internetworking design. The envelope approach promised to be quick, clean, and efficient, not to mention infinitely extensible: a packet could travel through two, three, or a dozen gateways just as easily as it could through one. Moreover, this approach would make plugging a new network into the system almost easy: you'd just have to provide a gateway computer that knew how to read the universal protocol and how to make envelopes for the networks on either side.

And the universal protocol itself? Kahn and Cerf had already started planning it out: a completely rewritten version of the existing Arpanet protocol that they called TCP, the Transmission Control Protocol. First and most important, they agreed, this new version would have to deal with real-world chaos. Under TCP, the network would *try* to get the packets through, but if some of them didn't make it, as they probably wouldn't, then TCP would make sure that the source sent out replacements, quickly and automatically. In addition, the two men agreed that the new protocol would have to have a much-improved addressing scheme. As things stood now, an Arpanet packet couldn't even specify an address on another network; if it did, it would essentially stop at the gateway, as if the gateway and everything beyond it had precisely the same address. "It was like saying, 'This letter is for the United States,'" recalls Kahn. TCP would fix that. (Ultimately, in fact, for technical reasons, the addressing component would be separately codified as IP, the Internet Protocol. Thus the modern name: TCP/IP.)

Kahn and Cerf continued to hammer out the details of TCP/IP all through the spring and summer of 1973, remembers Kahn. "Then at one point during the summer of nineteen seventy-three we sat ourselves down for an entire weekend, working in a little room at what was then the Cabana Hyatt in Palo Alto, and just wrote the paper."

"The paper," which would eventually be published in 1974 as "A Protocol for Packet Network Interconnection,"[16] gave the first architectural description of how the Internet would function as a network of networks, with TCP/IP as the glue holding it all together. Indeed, "the paper" is why Kahn and Cerf are so often hailed today as the inventors of the Internet, to the extent that any two people can be singled out for that honor: this was pretty much where the Internet began.

First, however, before the Internet could become real, their abstract definition would have to be turned into a working implementation. And to achieve that, Cerf had already begun the Stanford internetworking seminars that lured Bob Metcalfe down the hill. The seminars began in June 1973, Cerf explains, and were as much about consensus building as technology: "I was trying to get a large number of people to agree on a set of protocols, and every time we brought in a new player, we had to go through the argument again." The process was about as slow and contentious as might be imagined, he adds, since the guest list included students and visitors by the dozens, from all over the world. But it was effective: everybody got to have input, and many of the participants would go on to make major contributions to networking in their own right.

By December 1974, Cerf and his group had turned the long deliberations into a detailed design for TCP/IP, which they circulated over the Arpanet as *Request for Comments 675*. Shortly thereafter, Kahn awarded formal ARPA contracts for three independent implementations of TCP/IP: one to Cerf and his students at Stanford; a second to Peter Kirstein and his students at University College, London; and a third to Ray Tomlinson and his colleagues at BBN. And by the end of 1975, all three versions were sending packets to one another without a hitch. True, nearly a decade would pass before TCP/IP was stable enough for ARPA to shift the whole Arpanet over to it. Then even more time would pass before the Internet truly began to take on a life of its own. But it was definitely a beginning.

Unfortunately, Vint Cerf's internetworking seminars were not such a happy memory for one participant: Bob Metcalfe soon had to quit in utter frustration.

The discussions had started out well enough. Cerf would later single out Metcalfe and France's Gérard Lelann as the two attendees who had done as much as anyone to help him think through the ideas behind TCP/IP. If nothing else, says Metcalfe, he'd already had ample experience with internetworking at PARC, where he'd had to establish seamless communications among the Arpanet, the Ethernet, and an older Data General network. Moreover, he'd come up with a solution that seemed remarkably similar to what Kahn and Cerf were proposing. In the May 22 memo in which he first described the Ethernet, he had included a hand-drawn diagram showing how all three networks, and more besides, could be interconnected with "boosters," the analog of Cerf's gateways. And he had likewise designed a universal protocol known as PUP, the PARC Universal Packet, which would play much the same role as TCP/IP.

Nonetheless, says Metcalfe, the frustrations began to mount fairly quickly. For one thing, there was the Kahn-Cerf paper: after all he'd contributed to the design of TCP/IP, he says, "I do remember being miffed at not being included as an author. Kahn didn't attend these seminars. So including him but not me didn't seem fair." Then, too, he was constantly having to watch what he said. Because Ethernet was a proprietary development of the Xerox Corporation, his discussing it down at Stanford would have been a very good way to jeopardize the patent application. As it was, he had already had to tell Bob Kahn that he could no longer serve as a consultant on the packet radio project. Now, in the seminars, he and his fellow PARC delegate John Shoch felt obliged to phrase their own comments as oblique questions: "Have you thought about *this?*" or "Have you considered *that?*" (Cerf finally just looked at them and said, "You've done this before, haven't you?"). As a result, Metcalfe, Shoch, and their seminar-mates found themselves talking right past one another. The PARC view of the future called for hundreds, thousands, millions of networks—networks on the scale of individual companies, individual departments, and maybe even individual homes. But because they couldn't talk openly about what a local-area network *was,* says Metcalfe, he and Shoch found it almost impossible to challenge their colleagues' assumption that a network was something the size of the Arpanet. For example, he says, "in their original addressing scheme for TCP/IP, you could only have two hundred and fifty-six different networks; it would only work with one or two nets per country!"

But mostly, says Metcalfe, he was frustrated because he just wanted to get *on* with it. Down on the flats at Stanford, the seminars were all about talk and consensus building. But up on Coyote Hill Road, he and Dave Boggs were free to push the PUP scheme as far and as fast as the technology would allow. So after a while, he says, "there was a division in the road, and we went off to implement PUP rather than wait for Vint's committee process."

Now, in retrospect, Metcalfe admits, that was not such a good move. As part of their Ethernet development, he and Boggs would indeed get PUP up and running—well before TCP/IP, in fact. But because it was a proprietary system that they couldn't fully share with the outside world—unlike TCP/IP, which was completely public—they ended up largely cutting themselves off from any further role in the genesis of the Internet.

Nonetheless, that was later. And in the meantime, says Metcalfe, he and Boggs were much more concerned about the Ethernet implementation itself, because their beautiful concept kept running into ugly reality. First they found that bits were getting scrambled by distortion in the Ethernet cable itself; then they discovered that a different set of distortions got introduced every time they put a "tap" in the cable to attach another machine. At one point, Chuck Thacker became so exasperated by their lack of progress that he threatened to ditch Ethernet in favor of a new networking scheme of his own. "I think this is because he and I never quite got along," Metcalfe has suggested. "I was always giving him gas for the unreliability of his prototypes, first on MAXC and later on the Alto, which Dave Boggs and I found ourselves debugging long after Chuck declared it done. Or chalk it up to bad chemistry."

Still, says Metcalfe, in the fullness of time the packets were flowing. And not a moment too soon, either: the Altos were already proliferating. In late 1973, after building and debugging nine prototypes (and then being forced to listen to Metcalfe and Boggs gripe about having to debug them some more), Thacker and his crew had decided that it was time to start producing the machines for real. In May 1974, the first of the thirty Altos that would be needed for CSL were delivered to PARC from a local "garage shop," a small electronics firm on a backstreet in Palo Alto. And by the time the last of them were in place that September, it was clear that thirty was just the beginning: one look at the Alto and *everybody* wanted one.

Before the year was out, in fact, CSL director Jerry Elkind had hired John Ellenby, a brand-new Ph.D. from the University of Edinburgh, and asked him to come up with a more formal manufacturing plan. Ellenby, then one of the few people in the world who had come to computing from a business background, wasted no time in doing so. By June 1975 he had convinced Xerox headquarters to organize a Special Projects Group that would salvage the brightest people from the wreckage down in El Segundo.* Ellenby then supervised the group in the creation of the Alto II, which was essentially just Thacker's Alto with the packaging and components redesigned to make it more reliable, easier to maintain, and, above all, cheaper. The first Altos had cost eighteen thousand dollars apiece, nearly double the original estimate. The Alto II machines could now be turned out for a somewhat more reasonable twelve thousand per.

By decade's end the Special Projects Group would produce some fifteen hundred of the machines, a number that still scarcely met the demand. "People were

---

* The sad story of SDS rates a full chapter in *Fumbling the Future,* Douglas K. Smith and Robert C. Alexander's classic account of Xerox's missteps with PARC. Basically, Xerox headquarters lurched from one strategy to another without ever fully understanding the implications of any of its moves. By July 1975, "McColough's folly," as the former SDS was now known in the trade press, was hemorrhaging money so badly that Xerox decided to shut it down.

killing themselves to get the Altos," says Margaret Graham, a historian of technology who would work at PARC in the 1990s. "People were saving up their expense money so they could buy them. They were the most expensive items anybody had ever bought for personal use—but they were very prestigious. And they spread all over the company."

Indeed, computing on the Alto was computing as no one had ever experienced it before, except perhaps those lucky few who had used Sketchpad, or Grail, or one of their siblings. The combination of Alto's wonderful graphics screen and its strange little "mouse" took the arcane abstraction known as software and transformed it into something visible, tangible, almost tactile. Users had the weird and eerily seductive sensation of reaching *into* the computer, of being able to grab what they saw there and manipulate it by hand. Thanks to the programmers, moreover, they found plenty to manipulate: icons, pop-up menus, drop-down menus, scroll bars, and windows—even *overlapping* windows that seemed to be stacked on top of one another like real, 3-D objects ("I was in the seminar in which Dan Ingalls from the Smalltalk group demonstrated overlapped windows for the first time," says Stu Card. "We all said OOOH!").

Smalltalk, not surprisingly, was far and away the sexiest piece of Alto software anywhere. It was more than just a programming language, after all; it was the prototype interface for Alan Kay's Dynabook, the complete environment for exploiting this miraculous new medium to its fullest. It was also quite explicitly intended for children. "Early on," Kay noted in his history of Smalltalk, "this led to a 90-degree rotation of the purpose of the user interface from 'access to functionality' to 'environment in which users learn by doing.' " The resulting user interface grew in fits and starts, as the programmers juggled the various tasks of "feeding Smalltalk itself, designing children's experiments, trying to understand iconic construction, and just playing around." Practically everybody at PARC offered opinions at one time or another. So did outsiders, so many of whom were intrigued by the software that Kay and his crew in the Learning Research Group had to cut back on the Smalltalk demonstrations just so they could get their work done. Adele Goldberg counted more than two thousand visitors in 1975 alone. Nonetheless, said Kay, the end result—Smalltalk-76—was a marvel: "It was fast, lively, could handle 'big' problems, and was great fun."[17]

It also *looked* fabulous. Smalltalk's overlapping windows, its icons, its menus, and its mouse pointing device constituted a user interface that was more unified and more tightly integrated than that in any other Alto application—and that, moreover, would soon become very, very familiar to the rest of us. To see what it looked like, just glance at the screen of any Apple Macintosh, or any PC running Microsoft Windows: the descendants of that Smalltalk interface can now be found on tens of millions of computers worldwide.

Granted, Smalltalk wasn't the fastest programming language in the world. But then, that wasn't the point. Kay and his team tried to make the Smalltalk environment very much like an artist's studio, with all the tools so close at hand that users would never once have to break their concentration. With just a click of the mouse, for example, you could pop up a browser window to scan through a library of existing objects. You could likewise pop up an editor window to write new code defining new behaviors for those objects, then pop up another window to execute the code and watch how they actually responded, and then pop up yet another window full of debugging tools to help you ferret out any problems.

It was beautiful. Of course, the other coding wizards in CSL and SSL recognized a good idea when they saw one and were soon crafting similar programming environments for Lisp and other languages—even as they created a host of user applications for the Alto. Budding artists, for instance, soon had several choices. Markup was an illustration program that allowed users to "paint" on the Alto's screen with a set of virtual brushes, in much the same way that Macintosh users would later do with MacPaint. Draw, similar in spirit to Ivan Sutherland's Sketchpad, allowed users to build up drawings from mathematically defined lines and curves; it would be the direct ancestor of MacDraw. And Chuck Thacker's SIL gave computer engineers a fast and efficient way to create complex logic diagrams by calling up a library of predefined components, in a manner similar to modern computer-aided design programs.

For E-mail addicts, meanwhile, there was Laurel, a transparent and easy-to-use E-mail editor that allowed for reading, filing, and composing messages in much the same manner as modern programs such as Eudora. And for everyone, of course, there was the Bravo text editor, probably the most popular single application ever written for the Alto. Designed by Butler Lampson and Charles Simonyi, and then developed further by Simonyi and others, Bravo introduced What-You-See-Is-What-You-Get word processing: WYSIWYG. That is, the program would display text on the screen in virtually the same form that that text would have when printed. Italics, Greek letters, subscripts and superscripts, justified margins, various font sizes—all of it would be right there on the screen, changing and reforming itself as you typed. By no coincidence, moreover, Bravo was another program that would look very familiar to modern-day users: Simonyi would later take it to a young software house known as Microsoft, where it would become the foundation for Microsoft Word.

The popularity of the Alto and all its wonderfully graphical software resulted in a corresponding surge in the demand for laser printing. And fortunately, says Gary Starkweather, the technology to meet that demand was now well in hand. "By late nineteen seventy-one," he explains, "I had been able to prove on a lab setup that the laser printer would really work; there were no hidden showstoppers. Then I started working on a second version." This became SLOT, the Scanning Laser Output Terminal, which was basically a standard Xerox 7000 copier with its internal optics replaced by his laser engine. Making SLOT work was a bit tricky, in that the copier ran at the rate of one page per second and could not be slowed down; Starkweather had to make the laser run at that rate with a resolution of five hundred pixels to the inch. "But by nineteen seventy-three we'd done several versions," he says, "and we were convinced it would work."

His colleague Ron Rider then took the lead in creating a complete laser-printer system that would serve all of PARC. It would rely on Metcalfe's Ethernet to bring in print jobs from all the various Altos, and would use Rider's own personal Alto to handle the scheduling and other preprocessing tasks (he had named his machine Palo, which made it, of course, the Palo Alto). It would use his Research Character Generator, which he'd developed in collaboration with Lampson, to tell the laser how to render input as dots on a page. And it would use Starkweather's SLOT to do the actual printing. The resulting acronym wasn't particularly elegant—Ethernet-Alto-RCG-SLOT worked out to "EARS"—and the jury-rigged hardware was even less so. But when EARS was placed in service in the fall of 1974, just in time to print for the first contingent of Altos, it worked perfectly. In fact, it worked so well that it quickly ended any fantasy about the "paperless office." With everyone at PARC using the system daily, EARS's output was so prodigious that its successor, a next-generation laser printer developed in 1976, would be code-named Dover, for the White Cliffs of paper that would be required to feed it.

EARS likewise helped end any little fantasy that Ethernet would be merely an option on the Altos. It had been that at first, remembers Metcalfe: "Money was, as usual, tight, and so some people were ordering Altos without Ethernet and with 48K words instead of the full 64K of main memory." But that false economy lasted about a microsecond. That extra 16K made the Alto run *sooo* much better. And without Ethernet, how were you supposed to get your E-mail or print out your reports?

"I recall the day I accidentally removed a cable terminator from the Ether," says Metcalfe, "and I stood up among the cubicles only to find one after another of my colleagues popping up, wondering why the network was down. From that day on, no Altos were ordered without Ethernet."

CHANGING THE WORLD

Considered as an event, and not just as a place, the Xerox Palo Alto Research Center was unique, one of those rare, near-miraculous combinations of talent, luck, and timing that you couldn't reproduce if you tried. To begin with, PARC was the direct heir of the people, the ideas, the technology, and the culture that had been nurtured by ARPA since J. C. R. Licklider's time. Indeed, it's fair to say that PARC was ARPA continued by other means.

Second, PARC was established at precisely the right moment. Not only were microchips just beginning to make ARPA's visions feasible in the real world, but the first generation of ARPA students were just beginning to come of age. "In the history of art," says Alan Kay, "the most powerful work in any genre is done not by the adults who invent it, but by the first generation of kids who grow up in it. Think of perspective painting during the Renaissance. Well, we were that generation. We were the kids who'd had our Ph.D.s paid for by ARPA."

And third, PARC managed to bring a substantial portion of that generation together under a corporate sponsor, Xerox, that was rich enough to provide ample money and resources, yet patient enough to wait for long-term results. Moreover, PARC in its early years was blessed with a set of leaders—Jack Goldman, George Pake, and Bob Taylor—who deeply understood the dynamics of innovation. True, the chemistry between Pake and Taylor was far from ideal, but even so, they managed to cultivate an extraordinarily fertile environment for creativity. "The PARC of the nineteen-seventies was amazing," says Danny Bobrow. "Yes, it had a major fraction of the most brilliant computer scientists in the world at that time. But what really made it was the interaction of all these people. There was Alan Kay wanting a certain kind of on-lineness, Thacker and Lampson wanting a certain kind of machine, the whole notion of mice and hypertext coming out of the Engelbart tradition—even me arguing with Kay about what object-oriented programming meant. There was a sense that we could invent it all. We could do hardware, software, AI, printing, networking. And we had the freedom to do it. George Pake was providing the shield so that we could think hard and provide a new vision. And we did, in many dimensions."

They did indeed. In the space of about three years—an impressively short time even by computer standards—Kay, Lampson, Thacker, Metcalfe, Starkweather, and the other Young Turks who gathered at PARC pioneered all the basic technologies that have dominated the field ever since:

- a stand-alone personal computer with a bit-mapped graphics display;
- a recognizably modern graphical user interface (GUI) featuring windows, icons, menus, the mouse pointing device, and all the rest;
- object-oriented programming;
- WYSIWYG word processing and a host of other graphical applications;
- the laser printer; and
- a local-area network—Ethernet—to tie them all together.

In fact, given Bob Metcalfe's participation in the TCP/IP seminars down the hill, and his subsequent development of the PUP internetworking system in collaboration with Dave Boggs, it's even possible to argue that PARC helped pioneer the Internet. But either way, a good case can be made that nothing fundamental about computing really changed between PARC's 1970s golden age and the end of the 1990s, when the explosive growth of the Internet and mobile computing finally began a new paradigm shift. The intervening years brought drama and upheaval, yes, not to mention a very real revolution in our ability to deliver high-powered computing to a mass market. And yet for all the onrush of technology over that quarter century, and for all the many refinements and extensions of the basic ideas, there was very little fundamental innovation involved. The personal-computer revolution was mostly about mass-market technology's catching up with the Alto-Ethernet–laser printer–GUI system that PARC had in place by 1975.

For the PARC researchers themselves, meanwhile, their vow to live in the future had paid off better than they could ever have imagined. And so by the mid-1970s, with the Alto-Ethernet–laser printer–GUI system fundamentally developed, their main concern was, What's next?

For Alan Kay, not surprisingly, what came next was something a lot smaller than the Alto. As early as 1975, in fact, Kay had started sketching out designs for a "laptop" that he called the NoteTaker. And by 1978, through the use of a brand-new (and barely adequate) microprocessor from Intel, the 8086, he and his colleagues were able to build some prototypes. The NoteTaker that resulted was more like a suitcase than a notebook—indeed, it looked remarkably like the suitcase-style Osborne computer of the early 1980s—and would have crushed any lap. Nonetheless, it did come with a version of Smalltalk built in. It worked extremely well for its time, and it really was "portable," in the sense that it actually ran on batteries. "Several of us had the pleasure of taking NoteTaker on a plane and running an object-oriented system with a windowed interface at 35,000 feet," Kay would write.[18]

For the hardcore system designers in CSL, what came next was the Dorado, which was intended as a truly high-powered successor to the Alto. "It was difficult to think of the Dorado as a personal machine," wrote Chuck Thacker, who designed the computer in collaboration with Butler Lampson, "since it consumed 2500 watts of power, was the size of a refrigerator, and required 2000 cubic feet of cooling air per minute (while producing a noise level that has been compared to that of a 747 taking off). It was *used* as a personal machine, however."[19] The CSL crew began work on the Dorado in 1975; it would rapidly grow to become the largest engineering effort ever undertaken in CSL.

Increasingly, however, for almost everyone at PARC, what came next was the marketplace. They were very young, most of them, they were eager to change the world, and they just knew that the world was eager to be changed. True, the front-office types back at Xerox headquarters were a little dense about computing. But PARC had given them exactly what they had asked for: the electronic office of the future. Indeed, PARC had given them something magical: an Alto-Ethernet–laser printer–GUI system that was like nothing else on the planet. And not even a pack of total computer illiterates could fail to see how incredibly beautiful that was.

Could they?

THE END OF EDEN

On December 7, 1972—coincidentally, just a few weeks after Chuck Thacker had started work on the prototype Altos—*Rolling Stone* magazine ran a banner headline: "SPACEWAR—Fanatic Life and Symbolic Death among the Computer

Bums." In the article that followed, a thirty-three-year-old counterculture guru named Stewart Brand, the founder and publisher of *The Whole Earth Catalog,* recounted a wild-eyed night of Spacewar competition at the Stanford AI Lab, and then went on to proclaim the revolution. "Ready or not, computers are coming to the people," he declared. "That's good news, maybe the best since psychedelics."

Spacewar, Brand contended, was not just fun; it also symbolized a profound shift in the nature of technology. Instead of continuing to be oppressed by remote, cold mainframes grinding out data for the establishment, he said, we were now moving into an era of humane computing. Playful computing. Spontaneous computing. Even subversive computing—a new age of microelectronics that would democratize access to information, help individuals organize for action, challenge the established order, and generally serve the People and not the System. "Part of the grotesqueness of American life in these latter days is a subservience to Plan that amounts to panic," Brand wrote. "What we don't intend shouldn't happen. What happens anyway is either blamed on our enemies or baldly ignored. In our arrogance we close our ears to voices not our rational own, we routinely reject the princely gifts of spontaneous generation.

"Spacewar as a parable is almost too pat. It was the illegitimate child of the marrying of computers and graphic displays. It was part of no one's grand scheme. It served no grand theory. It was the enthusiasm of irresponsible youngsters. It was disreputably competitive ('You killed me, Tovar!'). It was an administrative headache. It was merely delightful."

Brand went on in this vein for quite a while, weaving a tale that his *Rolling Stone* readers undoubtedly found new and astonishing in 1972: The Pentagon funds the Revolution! Computers as liberation! Tech-weenies as counterculture radicals! (It probably didn't hurt that the photographer for the piece, Annie Leibovitz, captured some particularly hairy specimens with their locks in free flow.) Back at Xerox headquarters in Stamford, Connecticut, however, thanks to the passages in which Brand had prominently (and accurately) touted PARC as the intellectual epicenter of this revolution, the reaction was more akin to seventies-style Fear and Loathing. Those executives who had always taken offense at the overprivileged and underwashed hippies out in Palo Alto saw their chance to strike.

"OK," says a sighing Jack Goldman, who caught most of the flak on this one. "So along comes the *Rolling Stone* article. The people in Stamford and Rochester who are jealous of our budget, and who are eager to take potshots at those guys sitting out in the beanbags, they say, 'Hey, you've opened the door for Xerox to be criticized, to be made fun of. The shareholders will be upset: we're wasting their money.' Of course, they were using that as an excuse. It was really, 'Hey, you guys are wasting *our* money.' "

Now, this was never really a serious threat, insists Goldman, rumors at PARC to the contrary, and eventually it all blew over. "Nobody ever threatened to shut

PARC down, or to take the budget away," he says, "because our budget was protected by the CEO. It just meant I had to defend how we did research—when we didn't have any products yet."

But never mind that, says Goldman. What he found truly distressing about the episode was what it symbolized: Xerox had changed.

He'd been watching it happen almost from the day he arrived in December 1968, Goldman explains. "Remember, this was the period of Xerox's phenomenal growth. In nineteen sixty-nine, the company's sales crossed the billion-dollar mark. And because they were growing so fast, they had to create a whole new management staff to handle it." Indeed, the finance and administrative systems that Xerox had inherited from its slow-paced days as Haloid were nearing collapse. Copiers weren't getting delivered, salesmen weren't getting paid, logistics were breaking down everywhere. So, with their company on the verge of being ruined by its own success, Wilson and McColough had hired a string of top-rank executives from what were then considered the two best-run corporations in America. One was IBM, which contributed mainly marketing people; the other was Ford Motor Company. "Finance types," says Goldman with an almost audible shudder. "McNamara protégés." These were the guys he'd hoped to leave behind when he left Ford. And now these were the guys who had come pouring in right behind him at Xerox.

Now, in all fairness, Goldman concedes, the new recruits did give the Xerox management a much-needed overhaul, and they did keep the company from choking on its own growth. But that actually became a big part of the problem: success reinforced their own worst instincts. Talk about subservience to Plan— these were *numbers* guys, who thought that letting people pursue wild and crazy ideas was tantamount to letting them play in a sandbox; who thought that nothing was real until they could reduce it to entries on a spreadsheet; who thought that everything could be predicted, analyzed, controlled, and *managed*. At one point in the mid-1970s, Goldman recalls, "I tried to explain Moore's law to one of them. Twice the power for half the price, and so on. Well, the guy just couldn't understand it. He said, 'Nothing could follow that kind of law!' "

Still, he says, the numbers guys were little more than an irritation—until November 22, 1971, when the chairman of Xerox's board, Joe Wilson, suddenly died of a heart attack. The company was left in shock and mourning, for Wilson had been its true visionary, its guiding spirit since the Haloid days. Indeed, it was Wilson who had seen the promise of xerography in the first place, just as it was Wilson who had been willing to gamble his company on the technology, and who had then imbued the explosively growing Xerox with his own transcendent vision. "It is frightfully important for man to communicate with his fellow man," he declared in a film that was shown to newcomers all through the sixties, when that message seemed to have a special urgency. "And this is the very heart of our business." Indeed, it was Wilson who had first seen the need to move Xerox into computing, as a way of embracing a whole new form of communication.

daily operation. We could have had desktop publishing right *there*. But the schmucks in Rochester and Stamford and Dallas didn't understand that." In fact, he maintains, if he hadn't thrown a tantrum, they might just have rejected the laser printer itself—the one PARC-bred product that would eventually earn billions for Xerox and repay its investment in PARC many times over. By 1974, the Product Review Committee at Xerox headquarters finally decided that yes, the company would market a computer printer. But what kind? Well, says Goldman, "a bunch of horse's asses who didn't know anything about technology were making the decision, and it looked to me, sitting there a week before the selection, that it was going toward CRT technology"—that is, the photocopy-a-TV-screen approach that Gary Starkweather had abandoned a decade earlier. Apparently, some people felt that lasers were death rays that shouldn't be allowed in Xerox products. "It was Monday night," recalls Goldman. "I commandeered a plane. I took the planning vice president and the marketing vice president by the ear, and I said, 'Clear your Tuesday calendars. You are coming with me to PARC tonight. We'll be back for the eight-thirty meeting on Wednesday morning.' The guys at PARC, bless them, did a beautiful presentation showing what the laser printer could do."

It worked, sort of. Goldman went into the fight with a strong ally: Jack Lewis, the head of Xerox's printer division, who was already such an ardent champion of Starkweather's creation that he had twice refused direct orders to kill its development. And now, having kicked the tires, so to speak, the committee members did indeed endorse laser printing. Yet there were still delays, says Goldman. "They wouldn't let us get them out on 7000s. Instead they insisted on going with the new 9000 series, which were high-end machines that wouldn't come out until nineteen seventy-seven."[20]

Well, there it was. Goldman wasn't going to convert these people. All he could do was hang in there and hope for better days, while he shielded the research process as best he could.

Actually, says George Pake, Goldman did a remarkably good job on that last part. Not only did the research chief manage to keep the Fear and Loathing at a pretty safe distance—most of the time, anyway—but he also kept the money coming: PARC's research budgets remained as generous as ever during this period.

Nonetheless, Pake was deeply troubled by the growing gulf between PARC and the rest of the corporation. And while he certainly knew which side *he* was on—he had fought the Dallas decision as fiercely as he knew how—he was also convinced that the blame had to be shared on all sides. Take the infamous "culture clash," he says: all the Fear and Loathing back at headquarters was more than matched at PARC. "The people in the computer-science community who were 'hippielike' really wanted to believe that they were annoying the rest of the corporation," he says. "So I think there was a lot of 'oppression' that was imagined and hoped for." Take the identity-badge crisis, for example: "My recollection is that somebody at headquarters left a confidential document on an airplane, which led to a general effort to tighten up security throughout the corporation. And the result was that we had to have identity badges at PARC." Of course, he says, this restriction was no worse than what researchers had to live with at, say, Bell Labs. But before it was all over, the badges became a big symbolic issue at PARC, where they were widely reviled as an outrageous assault on free speech and the free flow of information. The badges were also seen as retaliation for the Stewart Brand article, which had appeared only a short time before. "I think a lot of the computer-science people *wanted* to believe the article caused the crackdown," says Pake. "But I never saw any evidence of it."

And then on top of the homegrown paranoia, says Pake, you had the arrogance factor. "The computer scientists had these incredible systems up and running," he says. "But they weren't very congenial to visitors trying to understand those systems. I spent a lot of time smoothing over ruffled feathers of people who felt they had been insulted or talked down to. I've always felt that if we could have kept that arrogant attitude under control and been more considerate of visitors, we could have avoided a lot of our troubles later on."

Of course, Bob Taylor and his crew had their own opinions about where the real arrogance lay. But there was no disagreement about the lack of communication. Witness the disastrous 1973 visit of Bob Potter, the thirty-eight-year-old IBM veteran whom McCardell and O'Neill had appointed to revitalize the Office Systems Division. "An idiot," snorts Taylor. "Potter came out and talked to us before setting up Dallas. He talked to us for an hour and a half about the technical program Dallas would pursue, and he didn't mention the word *software* once!" "We just sat there aghast," agrees Jim Mitchell. "We said, 'You don't have the faintest idea this is not going to work. This is useless!' And he just basically said, 'Screw you guys. You don't understand anything, and I'm going to go off and do this.' It was the strongest 'not invented here' I'd ever heard in the world. He knew nothing about computers, and he wanted to know nothing about them."[21]

Still, one of the things that the PARC crew didn't understand (or refused to understand) was that Potter was expected to come up with a product by the following year, not in five years. Nor did they seem to grasp that the electromechanical system favored by Potter was technically mature, familiar to customers, and much, much cheaper than PARC-style personal computing—which was still in the early prototype phase in any case. By subjecting Potter to one of their Dealer Meeting firestorms instead of listening to what he had to say, they quite effectively squandered any chance of influencing Xerox's second- and third-generation word processors. "I went out there and sat in their beanbags," Potter later recounted. "But I couldn't get anything out of them. I even told them I was their savviest, best customer in the corporation. But they were only interested in their own thing. They thought they were four feet above everybody else."[22] Visibly miffed, Potter went back to his own division and thereafter refused to have anything more to do with PARC and its self-satisfied princelings.

Somehow, Pake knew, this widening chasm had to be bridged, or all PARC's efforts would come to nothing. So when the opportunity came along to build such a bridge, he took it.

In late 1973, Pake explains, after several years of struggling with the problems of XDS—as well as with the looming threat of federal antitrust action—McColough and McCardell had decided that it was time to take a fresh look at Xerox's whole corporate strategy. Specifically, as McCardell would later put it, they wanted "some of our most imaginative people to look at Xerox with a wider perspective than we generally use."[23] Pake was accordingly asked to serve on a four-man committee that was to draft a strategy by which Xerox could continue its phenomenal growth rate while simultaneously diversifying into entirely new fields—computers most definitely included. And so it was that in October 1973, wearing his tweedy academic sports jacket, Pake arrived in Stamford for an extended stay in the buttoned-down, blue-suited world of Xerox headquarters.

Goldman, he discovered, had not been exaggerating.

It wasn't that the headquarters people were stupid or anti-intellectual, says Pake: "They were just *non*intellectual." The majority of them showed no interest in new ideas, or new technologies, or possibilities for the future—all the things that excited Pake. "The corporate culture was dominated by the sales force and the finance people," he says. "Long-term thinking didn't come easily for them. If you told them about a possible new product, yes, they were very interested. But not in the long-term investment that led to a product. Well, that was a shock to me. Joe Wilson had clearly understood technology." Indeed, if Wilson hadn't risked everything on a totally unproven new technology—xerography—Xerox wouldn't even exist. But Joe Wilson was dead. And his successors seemed unable or unwilling to change the underlying culture.

So there was nothing for it, says Pake. All he could do was hang in there, do what he could on this long-term strategy committee, and keep trying.

Back at ARPA headquarters, meanwhile, agency director Steve Lukasik had spent the middle part of 1973 trying to build a few bridges of his own, with equally frustrating results.

It had all started about a year earlier, when the top brass at BBN had finally decided to commercialize their Arpanet technology. Overtures were accordingly made to one Lawrence G. Roberts, architect of the Arpanet. And in May 1973, Roberts announced that he was leaving ARPA to become president of a new BBN networking subsidiary called Telenet.

Panic time. "It was as if Larry had been there forever, and nobody was prepared for his leaving," says Steve Crocker, recalling the crisis atmosphere that immediately descended over the ARPA computer community. "There wasn't any machinery in place for the transition." Unlike all the IPTO directors before him, Roberts had apparently felt no need to groom a deputy director to take over. And when it came to the obvious alternative, which was to promote one of the office's program managers—well, says Crocker, "in terms of people who were deeply involved in the community, that was just me and Kahn." But the reality was that he didn't even have his Ph.D. yet, and Bob Kahn was still very new at ARPA.

So it was clear that the new head of IPTO would have to come from the research community itself, a necessity that was soon putting a serious strain on Steve Lukasik's patience, never his strong suit. Having spent his entire time as ARPA director pushing the IPTO budgets skyward, even as the other offices were being cut back, and having made it abundantly clear that he considered ARPA's computer research to be the most exciting work in the agency—to the point where staffers in the other offices could be heard to grumble resentfully—he now couldn't find even one of those computer hotshots who was willing to take a little time off to come run the program. "All the bright Young Turks were too busy eating our money to contribute to the common good," says Lukasik. "I personally went after a number of the biggest stars in the IPTO community, and they just wouldn't come. I had a terrible time."

What made this indifference downright stupid, Lukasik adds, was that the agency's programs were under pressure as never before. In the aftermath of the Vietnam disaster—not to mention the sixties-era culture wars, and now, heaven help us, the unfolding scandal of Watergate—the miasma of suspicion, mistrust, and hostility lay thick over Washington. And for ARPA, at least, it had taken on a tangible form: the Mansfield Amendment. Passed by Congress in the fall of 1969 as a rider to the Fiscal Year 1970 Defense Department Authorization Bill, the amendment decreed that the funds in question could be used for research only if there was "a direct or apparent relationship" to a specific military function. "Of course, it was really a way of cutting the defense budget by alleging that a lot of it wasn't relevant," says Lukasik. "And of course, hardly anything was eliminated because, in fact, the department *had* been pursuing relevant research. But everyone who was in the research business stood on their head, and for a year we wrote relevance statements."

Technically, of course, the Mansfield Amendment was in effect only for that one year. Nonetheless, it sent a powerful message that Congress was watching, and that ARPA officials would no longer be trusted to make judgments on their own.* In retrospect, moreover, it came to be seen as the symbolic watershed for ARPA, the point at which it started its downhill slide from being a cutting-edge

---

* It also put many of ARPA's campus-based computer researchers in a very embarrassing position. "All that language about military rationale wasn't in the Stanford version of the proposals," explains Ed Feigenbaum: it was slapped on at the very end by the ARPA funding officers back in Washington. "The only people who ever saw it were the students who would later dig

agency that was blessedly free to take risks to being an ordinary agency that was cautious and risk-averse. "It was the end of Eden and the beginning of the real world," says Stanford's Ed Feigenbaum.

Another such moment came in 1972, when a bureaucratically inspired reshuffling of the Defense Department's organization chart gave ARPA the new name DARPA: the *Defense* Advanced Research Projects Agency. Of course, as Lukasik points out, this change was utterly meaningless in any practical sense. Certainly no one on the inside ever used the new name, except on the official stationery; he personally thought "DARPA" sounded like a dog food and told his people to ignore it as much as possible. Nonetheless, the import of that *D* was inescapable: "defense relevance" was the order of the day, and Big Brother was watching.

So in this environment, says Lukasik, IPTO clearly needed a capable leader. Yet no one seemed willing to brave the political minefields. "Finally," he says, "we were sitting around one day, racking our brains, and somehow the suggestion came up: 'How about Lick? He's such a good guy, he won't turn us down.' "

No one was eager to ask him, says Lukasik—"and I mean that as no reflection on Lick. He was our last resort because this was like asking somebody to take a second tour in Vietnam. But in the end we had to go back to him because nobody else would come."

Thus Larry Roberts's call to Lick in the summer of 1973: Would he? Could he?

Lick accepted without enthusiasm, undoubtedly thinking of all the administrative paperwork he'd wanted never to see again. In the two years since he resigned the directorship of Project MAC, he'd had a fine time working on dynamic modeling. Still, he told Roberts, if you really need me to do it for the good of the community, I will.

"So I told him that I'd keep trying to find somebody else," says Roberts, who was painfully aware of what he was asking of the older man—and equally aware of what a disaster Lick had been as head of Project MAC. "And I did try for months. But in September of nineteen seventy-three I had to take him up on his offer. Telenet was going to file with the FCC the next day for permission to be a carrier, and I had to be there." Roberts got on the phone again and at last tracked Lick down in Wales, where he was on a driving tour with Louise. "OK," Lick sighed. And it was set: Roberts's successor would officially start work in January 1974.

"I never saw Larry so overjoyed as when he finally got his replacement," says IPTO financial officer Al Blue, who would serve as acting director for the interim. Of course, Blue adds, he himself was pretty happy, too, albeit for a slightly different reason. "When you were working with Larry," he says, "you came into the office and you hit the ground running, and you did not quit running until the whistle blew, and then you ran some more. So as much as I loved

---

it up under the Freedom of Information Act. Then they'd bring it on campus and say, 'See, McCarthy is working on such and such.' McCarthy would say, 'What do you mean? I never heard of that!' "

working for Larry, I thought, Well, Lick is going to be different. Lick is relaxed. He is a gentleman."

But alas, says Blue, when Lick arrived in January 1974, the reality was more akin to a Christian's being fed to the lions: "We would go up to the director's office and it would be, 'Give me an analysis on this, and a complete budget rundown, and do this and do that and have it up here by four o'clock.' And we would go back to our office and Lick would say, 'This is not the ARPA that I knew.' "

It wasn't—and Lukasik, who could remember the old days, too, was sympathetic. "By Lick's second time around," he says, "we had budgets, line items, relevance statements, people crawling all over our back. I never asked him, but I'm sure he would have said, 'It's not fun like it used to be.' "

Then, too, Lick at age fifty-eight was not the man *ARPA* had known: physically, if not mentally, he was growing old and frail. His allergies had long since crossed the line into asthma—he never went anywhere without an inhaler—and his hands had a noticeable tremor. "Although he was extremely well motivated," says Steve Crocker, "it showed through: he just didn't seem to have the energy anymore."

Happily, though, that was one problem Lick's boss could solve. Reaching down into ARPA's nuclear-monitoring research office, Lukasik brought up an army colonel named David Russell to serve as Lick's deputy. The mild-mannered but very capable Colonel Russell soon had everything quietly under control. And Lick, now blessedly freed of doing any but the highest-level paperwork, was able to do what he did best, which was set the tone and direction of the office.

Of course, notes Steve Crocker, to people who were used to the hyperenergetic, do-it-all-myself style of Larry Roberts, their new office director looked suspiciously like a figurehead—honored, respected, and listened to, to be sure, but not really in charge. The real power, they believed, lay in the hands of Dave Russell. But whatever the truth of that perception, Lick himself probably didn't care: he'd made a promise to be here, and he felt duty bound to keep that promise. As Tracy Licklider points out, his father also had a very personal stake in the outcome: "His feeling was, 'I've invested a lot in creating the ARPA community, and I don't want to see it backslide.' "

Lukasik's push for practical applications didn't much worry him, so long as the really farsighted, exploratory research was protected in the budget—which it was. In fact, Lick was as eager as anyone to see ARPA-style computing make a difference in the world. That was why he had earlier agreed to serve as an adviser to ARPA's Speech Understanding Project, a five-year effort that began in 1971, after Lukasik started urging his AI researchers to come down out of the clouds ("Why don't you guys do something *interesting*, like fixing it so computers can understand speech?"). And that was why Lick had always been a big supporter of Ed Feigenbaum's pioneering work on expert systems, which was about as "practical" as you could get in the AI field. ("Call it nineteen sixty-eight, or early nineteen sixty-nine," remembers Feigenbaum. "I'm sitting in Lick's office at Project MAC, and I'm telling him about Dendral [the first expert system]. Lick is real enthusiastic, as usual: a big guy smiling behind his desk. And the gist of it is, 'You've just told me one of the greatest things I've ever heard. This is really wonderful stuff—what we all needed.' ")

What *did* worry Lick was what the Mansfield Amendment and all the rest had done to the spirit of the community. Take Ed Feigenbaum himself, for example. "I remember Lick taking a look at my nineteen seventy-four budget proposal to ARPA, which was about a hundred and fifty thousand dollars," says Feigenbaum. "Now, I liked to keep my head below the wall and not try to get shot at by asking for too much money, or in any way making funding waves; that way, whoever's shooting at us will hit McCarthy, and not me. Anyway, Lick said, 'Ed, this is ridiculous. You want to grow. Double your budget.' So we did. Eventually, in fact, Lick wanted to increase it to a million! It got carved back to eight hundred thousand. But still, that was our tremendous growth spurt." The result was Stanford's Heuristic Programming Project, and Mycin, the first of the modern expert systems and the forerunner of all the commercial expert systems that would follow in the 1980s. Along the way, moreover, Lick happily approved an Arpanet connection for the Stanford group's new PDP-10, even though that computer was being funded by the National Institutes of Health, not ARPA. "The idea was to support a community of investigators doing artificial-intelligence research in the areas of medicine and biology," says Feigenbaum. "It became a national resource called SUMEX-AIM. SUMEX stands for Stanford University Medical Experimental Facility. AIM stands for AI in Medicine, which was the name of the national community. And for Lick, it was like a dream come true: computers doing something real for the world. The idea that we would be sinking our hooks into medicine tickled him."

At the same time, however, Lick must have been deeply disturbed to hear an Ed Feigenbaum talk about "keeping his head down." He could certainly understand the impulse, since at various times almost everyone had drawn a bead on ARPA's artificial-intelligence program: Congress, the upper echelons of the Pentagon, even Lukasik, though the ARPA director had eventually become a cautious supporter. The payoffs just seemed to be so speculative, and far off. Nonetheless, the current climate was making researchers timid. And maybe worst of all was that it was making them view one another as rivals instead of colleagues. "In the early days," notes Vint Cerf, "ARPA could build a community because ARPA could afford to support everybody. But when there is scarcity, you don't have community; all you have is survival."

Still, J. C. R. Licklider was nothing if not an optimist. He was convinced that he could make a difference—as long as he hung in there and kept trying.

Actually, admits Steve Lukasik, when you got right down to it, he wasn't having much fun at ARPA, either. Not anymore. He was tired of answering dumb questions from Capitol Hill, tired of watching his paycheck get eaten away by inflation, and most especially tired of fighting with his boss, Malcolm Currie,

Defense Director of Research and Engineering since 1973. Currie did not like Lukasik's freewheeling, feet-on-the-desk style. Nor was he impressed with his efforts to emphasize applications. Currie wanted results out of ARPA *now*—and in any case, he was clearly anxious to have his own man in the ARPA spot.

To hell with it, then. In the summer of 1974, Lukasik announced that he would be stepping down, effective January 1, 1975. Within days, it seemed, Bob Taylor was on the phone: "Why don't you think about coming to Xerox?" he asked. "We need a guy with your track record."

Well, thought Lukasik, there *was* a certain inevitability about it: if you're hiring away ARPA's best young computer researchers, why not go after ARPA's director while you're at it?

PARC had exactly the same problem ARPA had, Taylor explained. His group was creating technology that was everything Lick and his heirs had dreamed of. But when you looked down the road and asked yourself how this technology was ever going to make it from the lab to the marketplace—well, the kindest way to say it was that Xerox wasn't set up for that. What the company needed was someone who understood PARC-style computing to the core, but who also knew how to talk to the corporate types. And frankly, Taylor didn't know anybody better at that than Steve Lukasik, who had spent the past decade getting ARPA technology transferred to the operational military.

Hmm. Lukasik was still a little miffed at Taylor for siphoning off so many of ARPA's best people. But he was also as susceptible to flattery as the next guy. And he was dying of curiosity: for more than three years now he'd been watching some of the country's best computer scientists vanish into PARC. What were they *doing* in there? This sounded like a good chance to find out.

They left it with Lukasik's saying yes, he might be willing to consider Xerox very seriously indeed. So Taylor was soon pitching the idea to Jerry Elkind, George Pake, and all the other higher-ups at PARC. What we need is a completely new division of Xerox to push *our* technology, he told them. And Steve Lukasik is the kind of guy we need to run it. All we have to do is persuade those people back at headquarters.

Exactly, agreed Pake. He was all in favor of the idea, too, now that his time at Xerox headquarters had come to nothing. His committee had worked hard, he explains, identifying major deficiencies in Xerox's manufacturing practices as well as in sales, engineering, development, finance—everywhere, actually; the much-vaunted management controls installed by McCardell and the Ford people were already obsolete. Looking to the future, the committee had unanimously recommended that the company pursue office automation as aggressively as possible, by building on the enormous head start achieved by PARC. The members had gone in to present their results to McColough and McCardell, expecting great things to come of it. And they had watched as the two leaders, distracted by the government's latest antitrust threats against the company, barely paid attention long enough to say, Thank you. We'll read over the report and consider it very carefully.

The ever-mild-mannered Pake was quietly furious, to the point where his soaring blood pressure—"brought on by exasperation with headquarters' lack of understanding of how to manage technology"—caused him to suffer a minor stroke. Blessedly, it left him with nothing more serious than a slight visual impairment in one eye. But the experience definitely made him receptive to this idea of a new development division.

Jack Goldman was receptive, too, albeit skeptical about how successful the effort could be. Politically, he knew, any such development division would have to come under the aegis of Xerox's Information Technology Group, itself under the thumb of one James O'Neill. But what the hell, said Goldman. Let's give it a try.

And so it was that in the fall of 1974 Steve Lukasik found himself with a fistful of invitations to visit Stamford, Rochester, and PARC, where Pake and his people were just moving into their brand-new building.

Getting there was nothing if not dramatic, says Lukasik. Starting from Palo Alto, you drove west out Page Mill Road, through the heart of Stanford Industrial Park, and up into the lion-colored foothills of the Santa Cruz Mountains. Then you made a left turn onto Coyote Hill Road and drove for a mile or so through more hills, with horse pastures and groves of live oaks on every side. You curved around a hillside and suddenly there it was: a lion-colored building cascading down a hidden slope like an outcropping of the hills themselves.

Of course, all that environmental sensitivity on the outside made for a labyrinthine layout on the inside. The hallways of the new Xerox PARC opened into a seemingly endless maze of offices, labs, common areas, and conference nooks, all on at least three different levels. But then, that was also an important source of the new building's power. You couldn't imagine a better environment for chance encounters, impromptu bull sessions, and the kind of serendipitous exchange of ideas that so often strikes sparks.

And they were definitely striking sparks, says Lukasik. Although most of the Alto software was still experimental when he first saw it, "my first reaction to the PARC technology was, 'WOW!' Remember, to someone who had come from the Defense Department, and who was used to fairly stodgy computing, the graphical user interface was nothing short of astonishing. Everything Jack Goldman, George Pake, and Bob Taylor said they would do—they had done it. It sounds like hype, but they had literally put together a new computing paradigm."

Unfortunately, says Lukasik, he could also see that all was not well in paradise. Taylor, for one, made little attempt to conceal his low opinion of Pake. The director was a big part of PARC's problem, Taylor insisted; he was an inoffensive little man who just didn't *get* it about computing. According to Taylor, Pake habitually surrounded himself with "noncomputerist" advisers such as Bill Gunning. He continued to pour money into his precious physics lab, which duplicated what Intel and the universities were already doing, when he ought to be supporting the Alto-Ethernet–laser printer system with all the resources he had. And maybe worst of all, he didn't even understand that the system was a *system*. Ask George Pake what PARC had accomplished, and the director would almost invariably talk about individual technologies like Ethernet or the laser printer. The transformational power of personal distributed computing escaped him completely; he couldn't see anything there but a collection of gadgets. And that was why Pake hadn't been able to sell the PARC vision at headquarters, Taylor maintained: he didn't understand it himself.

Pake himself felt that he was getting the message very clearly, thank you. And in the meantime, his own feelings about Taylor were mixed, to say the least. On the one hand, Taylor had done exactly what Pake had hired him for, which was to fill PARC with first rate computer scientists—who had come through brilliantly. But on the other hand, Taylor's "nondirectional" management style looked a little too much like manipulation for Pake's taste, just as his commitment to "group cohesion" and "a common goal" smelled a little too much like intellectual fascism. He feared that CSL's shared vision would ultimately harden into what would later be called political correctness, an ideology that would stifle all dissent and all further creativity. After all, says Pake, "creativity is a function of freedom."

In any case, Pake goes on, that was one of the reasons he was so eager for Steve Lukasik to join Xerox: he thought that as a longtime friend of Taylor's, Lukasik just might be able to ease the computer scientists' sense of isolation.

Lukasik, for his part, could see that a bridge was badly needed around PARC—several bridges, in fact. So the deal was soon set: Lukasik started with Xerox shortly before Christmas 1974. By May 1975 he was organizing what would soon become known as the Systems Development Division, with a charter to bring Xerox's new digital technologies to market.

And that, says Lukasik, was when he really began to understand what he was up against.

Lick obsessively followed it all, from the break-in at the Watergate to the resignation of the president. "My father had an almost morbid fixation on Watergate, and a fascination with the hearings," recalls Tracy Licklider. "He'd wound up with a violent hatred of Nixon, and I think he was desperate for good to win out so he could keep his positive ideas about what government could do."

Certainly Lick must have found a poignant symmetry in the moment. Almost eleven years earlier he and his family had stood on the balcony of their high-rise apartment in Rosslyn, Virginia, and watched John F. Kennedy's funeral procession wend its way toward Arlington National Cemetery. Now, on August 9, 1974, standing on the balcony of another apartment just a few blocks away from the first, he and Louise watched Air Force One lift off from Andrews Air Force Base in the distance, circle the city, and head westward, carrying the disgraced former president back to California. "I think he saw the process as a vindication of the system," says Tracy Licklider. "In some ways it was righting the ship that had been capsized by the Kennedy assassination."

Unfortunately, however, the ship didn't get righted at ARPA, at least not for Lick. That same month of August 1974 brought word of Steve Lukasik's impending departure from ARPA. And his successor, it soon developed, would be the thirty-eight-year-old physicist George H. Heilmeier. An impressive scientist in his own right—in 1968, at RCA laboratory, he had headed the team that developed the first practical liquid crystal display—Heilmeier was a bureaucratic pro at the Pentagon, having worked there in the office of the DDR&E since 1970. He was also most definitely not a man to play the ARPA game as it had been played all these years. Smart, energetic, tough, and decisive, Heilmeier brought an industrial-research sensibility to the agency, along with an impatience with open-ended academic research that bordered on contempt. He was frequently heard to say that if a technology couldn't be used in the first fifteen minutes of the next war, he wasn't interested in it. So after formally taking over the directorship in late January 1975, Heilmeier lost no time in pruning away the "crap," as he called it, and trying to get the agency back to its roots in the hard sciences.

In the process, moreover, he put ARPA's various program managers through a "wire-brushing," in his words. "DARPA was like a big cashier booth," he would say in a 1991 interview with the Charles Babbage Institute, "and my feeling was that the program managers did not have the kind of understanding of the programs that I felt comfortable with." So instead of just signing off on the agency's research grants, he actually started reading them before they went out. *All* of them. Then he would ask the office directors some pointed questions about what the research was trying to accomplish, why the researchers thought they could succeed, and what the military would get out of it if they did.

These questions came to be known as "Heilmeier's catechism." And a lot of the younger program managers in fact found it rather bracing, once they'd gotten over the initial shock. "As long as people were able to answer those questions—or if they'd say, 'Well, I need the money to go get the answer'—my reaction was, 'Fine,' " explained Heilmeier. "But the Heilmeier catechism was instituted, and we didn't have any problems after that."

No problems, that is, except with the Information Processing Techniques Office. Lick was appalled by Heilmeier's methods. To his way of thinking, the new director's obsession with near-term payoffs was a violation of everything ARPA stood for. Applications, sure—but applications that changed the world by an order of magnitude or more, not just another 10 percent. And besides—reading every contract? Grilling his own program managers behind his back?

"You have no *right* to do that!" Lick would tell his new boss with all the force his confrontation-averse soul could muster; this was micromanagement of the worst sort, he said.

"Lick," Heilmeier would reply, "I have a *responsibility* to do that."

To this day, Heilmeier makes no apologies for his management style. For all the wonderful technology IPTO had sponsored, he insists, it was the worst mess in the agency. And artificial intelligence was the worst mess in IPTO. "You see, there was this so-called DARPA community, and a large chunk of our money went to this community. But when I looked at the so-called proposals, I thought, Wait a second; there's nothing here. Well, Lick and I tangled professionally on this issue. He said, 'You don't understand. What you do is give good people the money and they go off and do good things and that's it.' I said, 'Lick, I understand that. And these people may be good people, but for the life of me I can't tell you what they're going to do. And I don't know whether they are going to reinvent the wheel, because there's no discussion of the current practice and there's no discussion of the implications, so I can't tell whether this is a wise investment for DoD or not.' "

Lick, in turn, found Heilmeier's skepticism almost insulting. "Lick didn't come across as a big ego guy," says Ed Feigenbaum. "But he had a very strong sense of self-confidence about knowing what research was worthwhile." It was painful to watch, agrees Bob Kahn, who had been a classmate of Heilmeier's at Princeton and who could sympathize with both points of view. Heilmeier, he says, was neither stupid nor unreasonable. "If anything, he was extremely incisive and pragmatic—perhaps more pragmatic than the community liked to deal with. Lick, on the other hand, had been trained as a psychologist. He was much more intuitive, much more driven by hunches." And the fact was that that both men were right. "There were some things that were more conducive to George's directed style," says Kahn. "Packet radio, for example, and maybe the Internet. But speech understanding was not amenable. In fact, almost none of AI fit in. The problem was that George kept looking for a kind of road map to the field of AI. He wanted to know what was going to happen, on schedule, into the future, to make the field a reality. And he thought it was quite reasonable to ask for that road map, because he had no idea how hard it would be to produce. Suppose you were Lewis and Clark exploring the West, where you had no idea what you were going to encounter, and people wanted to know exactly what routes you were going to take, where you would camp, and what you would do out there. Well, this was just not an engineering job, where you could work out the whole plan. So George was looking for something that Lick couldn't provide."

Unfortunately, when Lick tried to explain that to his boss, it was a bit like Seymour Papert's or Alan Kay's trying to explain exploratory education to a back-to-basics hard-liner. "IPTO really didn't have a program-management structure," declares Heilmeier. "They had a financial management structure, and they had a cheering section."

By Easter Sunday, March 30, 1975, when Lick began to type an E-mail message[24] assessing the situation for the principal investigators in artificial intelligence, he was obviously struggling with his conscience. First, he wrote, "Let me be clear that I am strongly in favor of ARPA's contributing maximally to the solution of pressing DoD problems." Indeed, he was even willing to concede that Heilmeier had a point: after ten years and some $50 million of ARPA support for artificial intelligence, he wrote, "it is natural for a new director, or even an old one, to ask, 'What have we gotten out of it in terms of improvements in na-tional defense?' . . . Unfortunately, most people who are asked that question an-swer 'I don't know' or even 'nothing.' "

However, Lick went on, he rejected the notion that the *only* criterion for re-search was whether some "buyer" in the military wanted the results right away. "The future," he wrote, "is not to be won by making a lot of minor technologi-cal advances and moving them immediately into the Services." Lord knows he'd tried to make the new director understand that: "The problem is that the frame of reference with which he enters the discussions is basically quite differ-ent from the frames of reference that are natural, familiar, and comfortable to most of us in IPTO—and I think, to most of you. In my frame . . . it is a funda-mental axiom that computers and communications are crucially important, that getting computers to understand natural language and to respond to speech will have profound consequences for the military, that the Arpanet and satellite packet communications and ground and air radio networks are major steps forward into a new era of command and control, that AI techniques will make it possible to interpret satellite photographs automatically, and that 10^{10}-bit nanosecond memories, 10^{12}-bit microsecond memories and 10^{15}-bit millisec-ond memories are more desirable than gold. In George's frame . . . none of those things is axiomatic—and the basic question is, who in DoD needs it and is willing to put up some money on it now? We are trying hard to decrease the dissonance between the frames, but we are not making good progress. As one of my colleagues put it Friday, 'I think we are slowly holding our own with George.' "

Indeed, Lick wasn't at all sure he could continue to be a party to it. He'd been planning to go back to MIT in September of 1975 in any case. Why not now?

Lick held off on sending his E-mail, since he wanted to see what would hap-pen in the budget meeting he'd scheduled with Heilmeier for the following af-ternoon. He had reason to fear the worst: when McCarthy's AI Lab at Stanford had recently come up for review in a separate action, Heilmeier had slashed its budget by 30 percent. And as it turned out, Lick's fears were confirmed: at the Monday meeting Heilmeier decreed that AI and everything else in IPTO would henceforth be "redirected" away from its open-ended, stream-of-research prac-tice. So on Tuesday morning, Lick added a postscript to his E-mail: "You can see clearly, I think, that we are at a watershed in the history of ARPA-IPTO. . . . I am deeply concerned about my own role in the redirection—whether to fight it, try to contain it, or join it wholeheartedly and try to steer it in such a way as to wind up with a larger, stronger, more productive enterprise. I have been about halfway between the first two alternatives, but neither the halfway-between point nor either of the first two is really a workable position within ARPA. It has to be either leave and fight or stay and join—and it is clear that to adopt the for-mer course precipitously would have a very bad effect on the program. And it is such an important—in many ways, absolutely crucial—program!"

No, he had now decided, there was only one responsible course open to him.

Satisfying though it would have been to resign in righteous fury, he would quietly stay on—while doing his best behind the scenes to keep Heilmeier's "redirection" from destroying what he had built.

On the morning of Wednesday, April 2, having mulled it over for yet another day, Lick added a second postscript explaining what he had in mind. His opening was characteristically polite: "Let me end this message with a proposal. . . ." But what followed was in effect a challenge to the whole ARPA community: Let's grow up, Lick told them. Let's admit that we've all been playing a little too much in the sandbox, and let's quit thinking that applications are somehow at odds with pure research. Instead, let's start making common cause with the engineers—and start taking responsibility for our own future. "[The] new modus operandi . . . will not, over a long period, be a zero-sum game," Lick wrote with determined optimism. "It takes advantage of every demonstrated success to increase both the basic research and application budget." Yes, he wrote, there will be a significant shift in the community's center of gravity, from pure research toward engineering. "[But] the shift will give the university research groups an engineering arm, a marketplace, customers, users. [That] interaction will strengthen the basic work because there will be more feedback from real tests of the new ideas and because every star needs an audience and performs best before a big and enthusiastic one. Moreover, ideas will really start to move into use."

In short, Lick said, a close alliance with the applications community could be healthy for all of them—a means of actually *living in* the future instead of just visiting there. And more to the point, he suggested, if they could give Heilmeier enough of what he wanted, then maybe they could get most of what *they* wanted, as well.

Lick finally sent his Easter message at 9:25 A.M. on Wednesday, April 2, 1975. And from then on out, his telephone and his Arpanet terminal were going incessantly. Making such a transition didn't come easily or naturally to any of them. But they did it. In fact, they even got some helpful hints from the ARPA director himself. "I went back to the community," Heilmeier remembers, "and I said, 'Look, I recognize the need to do basic research in AI, but I also recognize the need for you folks to sign up to some challenges.' So I gave them some challenges. One challenge was Morse code recognition, and another challenge was to do something about the interpretation of sonar signals and ASW [antisubmarine warfare] signals in general. The third challenge was command and control. I wanted systems that could adapt to the commander instead of forcing the commander to adapt. So I said, 'Look, if some of you guys would sign up for these challenges I can justify more fundamental work in AI.' And some did. Some of the folks at MIT stood up. Stanford did. SRI and RAND did."

They did indeed—with Lick doing his behind-the-scenes best to orchestrate it all. On the Morse code problem, for example, he put in a call to Al Vezza, who was interim head of Lick's own Dynamic Modeling group at MIT; within two years the group would be demonstrating a working system for a very pleased George Heilmeier. For the antisubmarine-warfare challenge, Lick got Ed Feigenbaum to revive a small effort in that area and turn it into a large one. Expert systems also seemed to show promise for adaptive command and control—"Mycin to the rescue," Feigenbaum says with a laugh—and by no coincidence, this was also the era when the budget of the Heuristic Programming Project started heading toward the million-dollar mark. Finally, Lick presided over a formal declaration that the Arpanet was fully operational, completing a previously planned transfer of day-to-day control to the Defense Communications Agency. ("George is not fully convinced about packet communications, yet," he had written in his Easter message, "but he thinks we may have something in there somewhere.")

Of course, Lick was also obliged to kill some projects, which hurt—especially since one of them was Doug Engelbart's. This didn't really have much to do with Heilmeier, except perhaps through the Heilmeier-era budget constraints. It was more that the whole enterprise seemed stuck in neutral. From Lick's point of view, Engelbart and the remaining team at SRI had lost something vital when Bill English and the others went to PARC—some indefinable gift for turning their visions into compelling reality. NLS just seemed to grow more bloated, more baroque, more difficult to use—and less interesting—every year. With ARPA headquarters now using NLS on a day-to-day basis, moreover, those deficiencies had become painfully apparent. So in 1975 Lick terminated the SRI contract.

A quarter of a century later, Engelbart still feels the pain of that moment and brings it up at almost every mention of Lick's name. "I realized that Lick had stuck his neck out to give me a chance in the beginning," he says. "I trusted him. Lick was like my big brother. And that's why it was so hard to handle."

Heilmeier, for his part, was quite pleased with the IPTO community's response to his ministrations. By the time he finished his tenure there in 1978, he said in the Babbage interview, "I thought we had essentially brought the IPTO program into DARPA again. It wasn't NSF West anymore."

Lick, however, apparently found this brand of "success" to be exhausting, depressing, and lonely. "What I remember most about that period was interacting with Lick on the network late at night," says Allen Newell, who often kept hacker's hours himself. "I would get a message from Lick at eleven P.M. saying, 'Tomorrow morning I've got to go before Congress, and I've got to have—whatever.' So now eleven to four A.M. was shot because you all of a sudden had to help Lick out. He was looking for all kinds of support: evaluating AI research, evaluating proposals, or often just getting some ideas for things to do next."

That relationship may have helped, and so may the IMLAC terminal that Lick kept in a corner of his office, where he could spend the odd moment or two hacking away at ideas he had about how to make programming more intuitive. But basically he just felt frustrated by the whole situation, and sick of it.

Lick departed for home in September 1975, as planned. (Many people in the community assumed that he had been forced out, but apparently that was not the case—though Heilmeier was certainly not sorry to see him go.) When he got back to MIT, he asked Al Vezza to go on being the director of the Dynamic

Modeling group. You're doing great, Lick told him. And the fact was, he'd pretty much had it with administration and politics. He was terrible at it, anyway. He couldn't wait to get back to good, clean research.

Back in the mid-1970s, says Steve Lukasik, remembering his sojourn at the Xerox offices in Rochester, New York, people used to keep a little plaque on their desktops:

Xerox Is a Xoo

"And it was true," he says with a sigh. In fact, it was an understatement. Closer to the mark would have been "Xerox Is a Bunch of Warring Camps"—and his brand-new Systems Development Division was caught in the crossfire.

"To begin with," Lukasik explains, "you had a divide between the analog-copier people and the digital-electronic people, who were still considered flakes. At that time Xerox was a five-billion-dollar corporation in sales, of which ninety to ninety-five percent came from copiers. These things were cash cows: the company built the machines and leased them, so that every time somebody hit the little button and a copy got made, Xerox got paid. And those copier people did not at all buy into the McColough-Goldman notion of an electronic office. They were *good* at copiers, and they were going to hand over a viable copier business to their grandchildren."

So that was the major division, says Lukasik: the xerographers versus the "electronickers," as he calls them. But he'd expected that. What he had *not* expected was to find the electronickers themselves so badly in conflict. Xerox actually had quite a few of them, he says, and they weren't all at PARC: "Every group larger than half a dozen had its own theories about the electronic office. And they were all competitors for resources. The main division was one I would call 'the systems people versus the stand-aloners.' PARC, for example, was almost all systems people. Now, the XDS people in El Segundo were also systems guys, but they saw themselves as good, solid businesspeople versus those flakes up in Palo Alto. And the people in Palo Alto saw the XDS crowd as stodgy, behind-the-times, corporate windbags. Then we had the fax people, who didn't want to play systems games at all; they just wanted to sell fax machines. We had the Dallas people, with their own stand-alone word-processing system. And we had the British people from Rank-Xerox, who had their own view of systems architecture. They kept writing things that were either incomprehensible or not lined up with the PARC views.

"So that was the group that I was brought in for," says Lukasik. "And I was told, 'Lead. Unify. Tell us what makes sense. Tell us what doesn't make sense. Tell us how much money to spend, using as a start the intellectual resources we've created at PARC.'"

Tackling the electronickers first, says Lukasik, "I organized the Systems Development Division in May nineteen seventy-five. I got a staff. I got a budget together. And I hired about two hundred researchers, including fifty to seventy-five people from PARC—all the people who were just itching to move the technology into the marketplace." Indeed, some of the sharpest engineers from PARC transferred to the new division, including Chuck Thacker (starting in 1977) and the laser-printing wizard Ron Rider. Xerox made the transfer easy for them: the PARC delegation just had to walk across Coyote Hill Road and set up shop in the same building they'd occupied when they first created the Alto. It was as if they'd never left PARC at all.

So given time, says Lukasik, he was pretty sure he could get the Xerox electronickers in line. The xerographers, however, were another story. He *hoped* he could bring them around eventually, when his Systems Development Division started coming up with some compelling products. The question was whether "eventually" would come before it was too late. "Remember," he says, "nineteen seventy-five had been the beginning of a downward slide for Xerox. The Japanese were doing a superb job in low-end copiers. IBM was getting into copiers. So was Kodak. Xerox was besieged in the copier market, its heartland business. And when your heartland business is threatened, your instinct is to pour your resources into that. So my problem was to unify and grow the electronic business fast enough to get a critical mass to go head to head with the xerographers."

This was a perennial topic of conversation on Coyote Hill Road. "We struggled," says Gary Starkweather, who was now the head of the newly organized Optical Sciences Laboratory at PARC. "We asked ourselves over and over: How can we make them realize what they have here? By nineteen seventy-five we *had* the office of the future! We could do our office memos on computers, we could transfer data electronically, we could print things. In nineteen seventy-six, we even built a scanner! But we could not get anybody in corporate to grok this model." (One top-level Xerox executive, after a day of being shown the wonders of PARC, had posed precisely one question to the researchers: "Where can I get some of those beanbag chairs?")

Maybe that was why they got so excited when they heard about Boca—the Xerox World Conference, scheduled for November 1977 in Boca Raton, Florida. As the first company-wide gathering of Xerox managers in six years, the four-day conference would showcase the company's newest and shiniest products, including its first commercial laser printer, the just-introduced Xerox 9700, which would soon become one of the company's hottest sellers. And it would build toward the grand finale of Futures Day, a celebration of all that had been achieved at the Xerox Palo Alto Research Center.

Back at PARC, says Starkweather, he and his colleagues saw this as their last, best chance. "The feeling was, 'If they don't get this, we don't know what we can do." So, he says, with John Ellenby coordinating an all-out effort, "we stripped everything out of PARC down to the power cords and set it all up again in Boca. Computers, networks, printers—the whole thing! I built a laser printer that did

color. Bill English had a word processor that did Japanese. We were going to show them space flight!"

They certainly tried. On Thursday, November 10, 1977, Xerox executives and their families swarmed through the Grenada Rooms of the Boca Raton Hotel for a hands-on demonstration of WYSIWYG editing in Bravo, graphical programming in Smalltalk, E-mailing in Laurel, artistry in Paint and Draw—the works. "The idea was a mental slam-dunk!" says Starkweather. "And some people did see it." The executives' wives, for example—many of them former secretaries who knew all about carbon paper, Wite-Out, and having to retype whole pages to correct a single mistake—took one look at Bravo and *got* it. "The wives were so ecstatic they came over and kissed me," remembers Jack Goldman. "They said, 'Wonderful things you're doing!' Years later, I'd see them and they'd still remember Boca."

Then there were the delegates from Fuji Xerox, the company's Japanese partner: they were beside themselves over Bill English's word processor. "Fuji clamored, 'Give us this! We'll manufacture it!' " recalls Goldman. In fact, he says, that was a near-universal reaction: "People from Europe, people from South America, marketing groups around the U.S.—everyone who went out of that conference was excited by what they had seen."

Everyone, that is, except the copier executives, the real power brokers in Xerox. You couldn't miss them; they were the ones standing in the background with the puzzled, So-What? look on their faces. "Oh," they would say as their wives waxed rhapsodic over on-screen cut-and-paste. "It does that?"

Chalk it up to bad timing, says Starkweather. "Savin had just come out with its desktop copier. So maybe that caused an attention deficit." Indeed, one of the corporation's purposes in calling this conference was to rally the troops for the coming era of ever-more-ferocious competition. In fairness, those executives in the background had to worry about defending the homeland *now*, not ten years from now. Or maybe they were simply too bound by the culture of the executive suite, vintage 1977. The xerographers lived in a world in which typing was women's work and keyboards were for secretaries. It was a rare executive who would even deign to touch one.

But whatever the reason, says Starkweather, it was disheartening. "Right afterward, it felt like a Moon launch: 'We did it!' But in the next few weeks and months that followed, when it became clear that nothing was going to happen, dejection set in. What did we have to *do* to get this across? But we all had some projects on the back burner we could go work on, and sulk."

George Pake was disappointed, too, in his characteristically measured way: "Boca Raton was not as much of a turning point as we'd hoped," he says.

Jack Goldman, meanwhile, was exasperated: Damn straight it wasn't a turning point, he says. "There was this feeling of exhilaration after Boca—then bingo, the meeting was over, and nothing happened!"

Bob Taylor, for his part, was disgusted: "I watched them just stand there," he says of the executives, still gritting his teeth all these years later.

And Steve Lukasik? He was long gone: Xerox had already stretched his slim reserves of patience past the breaking point. Yes, the company's core business was threatened, he says. But from his point of view, the reaction had been panic, and a rush to abandon the future. "Just about the time I came on board," he says, "money became tight. I'd tell them, such-and-such will cost ten million dollars. They'd say, We can't give you ten million, how about five million? Then in the next budget cycle they'd cut it to three million. After that happened five, six times, I realized they would never let me do what they'd hired me to do. In fact, all they wanted to talk about at the head office was the next-generation copier. Instead of developing the electronic office, I was supposed to tell them about how to feed the paper, do the stapling, and so on." So in June 1976, says Lukasik, with the Systems Development Division barely a year old, he walked.

"It had been fun," he insists. "I don't regret a single minute at Xerox." Nonetheless, he says, his eighteen months there have to be judged a failure—"a failure at the personal level by people like me, a failure of imagination on the part of the xerographers, and a failure at the highest level of management. Xerox never got organized for this. And perhaps Peter McColough didn't organize it."

But then again, he says, it's worth remembering that *everybody* in the computer business was stumbling its way through the 1970s, not just Xerox. Computation had once been a scarce resource; now, thanks to the microchip, it was fast becoming an abundant one. And not even the most imaginative visionaries had fully grasped the implications of that fact. Certainly nobody knew for sure what would fly in the marketplace—not PARC, not DEC, not IBM, and not even the hobbyists in the garages of Silicon Valley. In that environment, all anyone could do was try things out, see what worked and what didn't, and continue to explore the unknown.

CHAPTER 9

LICK'S KIDS

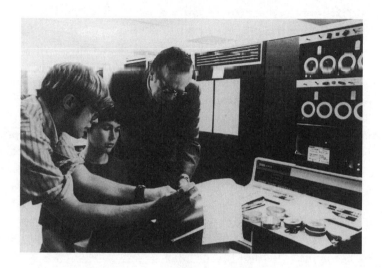

It took some getting used to, says Tim Anderson, thinking back to the mid-1970s and his time as an MIT computer-science student. But as often as not, you'd walk into the Tech Square terminal room and there he'd be: Professor J. C. R. Licklider, hacking away like a computer-struck freshman.

"He signed up for his two hours like everybody else," marvels Anderson, who had started there while Lick was away on his second ARPA tour and who had therefore never seen a sixtyish full professor typing code with his own ten fingers. "You'd come in and find this old guy sitting there with a bottle of Coke and a brownie. And it wasn't even a *good* brownie; he'd be eating one of those vending-machine things as if that was a perfectly satisfactory lunch. Then I also remember that he had these funny-colored glasses with yellow lenses; he had some theory that they helped him see better."

Anderson was never too clear about what Lick was working on—something to do with making computer code as intuitive as ordinary conversation, and as easy as drawing a sketch. Anyway, it never panned out: when it came to programming, Lick had great ideas but terrible execution. But that almost didn't matter.

The father of us all: Lick with MIT students during the late 1970s

To Lick, as Anderson soon realized, the important thing was to have fun with computers, to keep on pushing toward the future.

Being Lick, of course, he never said a word about what he'd done to bring computing into the future with him; a student had to pick up that bit of background through the grapevine. "When you learned what he had done, it was awesome," says Anderson. "He was clearly the father of us all. But you'd never know it from talking to him. Instead, there was always a sense that he was playing."

Always. "Lick was intellectually as alive as anybody could be," remembers Bob Kahn, who had seen those blue eyes twinkling often enough at ARPA, even through the worst of Lick's struggles with George Heilmeier. "There was not a topic you could raise that he hadn't sifted and sorted in his mind many times. And he was always conjuring up scenarios. 'Suppose we had one of these, and some of that, then we'd get something interesting over there, wouldn't we?' He'd come up with ideas that way that would knock your socks off, like his proof that the world couldn't possibly be a cosmic computer simulation: if it were, then surely God would have discovered a bug like Nixon and fixed it! And then he was always asking dialogue questions. This whole idea of interactivity was part of his demeanor. If you asked him something, he'd come back with, 'Well, have you heard of XYZ? Doesn't that bear on your question? What do *you* think about that?' You couldn't interact with him without participating."

Admittedly, Lick's scenario-spinning qualities had driven many people to distraction over the years, and not just Heilmeier. To anyone craving crisp analysis and decisive action, he could come across as an amiable doofus. But now that he was safely out of office, so to speak, those same qualities made him much in demand as a wise man. At the invitation of Bob Taylor, for example, Lick regularly went out to PARC and wandered through the computer labs as a visiting guru. He likewise served on the Committee on Government Relations of the Association of Computing Machinery; the Advisory Committee on the MIT Library System (as chairman); the Panel on the Social Security Administration's Data Management System (deputy chairman); and on and on. In 1978 he even spent another year in Washington, as a member of a task force examining the government's data-processing needs—part of the Carter administration's attempt at government reorganization.

Back in his natural habitat of Tech Square, meanwhile, he had reverted to being Lick at his best: not functioning as a manager and paper pusher, a role in which he was a disaster, but serving as visionary, teacher, mentor, and friend. "Lick became very important to me after I became director," says Michael Dertouzos, the Athens-born computer scientist who was named head of Project MAC after Ed Fredkin stepped down in 1974. The transition years were pretty stressful, says Dertouzos. If nothing else, there was the name change: Project MAC was now LCS, the Laboratory for Computer Science. (The argument was that with Multics as finished as it would ever be, and with the Tech Square crowd's moving on to a variety of new activities, neither "Multi-Access Computer" nor "Machine-Assisted Cognition" reflected what they were actually doing now.) Then, too, the faculty members were being their usual helpful selves. "In a place like MIT, nobody, but *nobody*, will tell you that you're doing a good job," says Dertouzos. "Lick was the only one who understood the loneliness, the only one who would come to me, regularly, after every meeting, and say, 'Mike, you're doing a great job.' So he gave me a kind of paternal reinforcement, a sense of someone's having faith in me."

Lick did much the same for the students in his Dynamic Modeling group. "I always felt that he liked and respected me," says Tim Anderson, "even though he had no reason to: I was no smarter than anybody else. I think everybody in the group felt that way, and that was a big part of what made the group the way it was." Lick was definitely the spiritual and intellectual leader, agrees Al Vezza, who was now the titular leader. Lick somehow made his people feel that by playing around and having fun with computers, they were actually building something much larger than themselves. He would happily sit for hours, spinning visions of graphical computing, digital libraries, on-line banking and E-commerce, software that would live on the network and move wherever it was needed, a mass migration of government, commerce, entertainment, and daily life into the on-line world—possibilities that were just mind-blowing in the 1970s.

Let's be optimists, he wrote in 1979, on one of those rare occasions when he committed such a scenario to paper.* Let's assume that Moore's law will continue to work its magic as it has in the past, and now let's imagine ourselves in the year 2000: "Waveguides, optical fibers, rooftop satellite antennae, and coaxial cables provide abundant bandwidth and inexpensive digital transmission both locally and over long distances. Computer consoles with good graphic display and speech input and output have become almost as common as television sets."[1]

Great. But what would all those gadgets add up to, Lick wondered, other than a bigger pile of gadgets? Well, he said, if we continued to be optimists and assumed that all this technology was connected so that the bits flowed freely, then it might actually add up to an electronic *commons* open to all, as "the main and essential medium of informational interaction for governments, institutions, corporations, and individuals." Indeed, he went on, looking back from the imagined viewpoint of the year 2000, "[the electronic commons] has supplanted the postal system for letters, the dial-tone phone system for conversations and teleconferences, stand-alone batch-processing and time-sharing systems for compu-

---

* The occasion was a series of essays on the future of computing, collected by Mike Dertouzos and his deputy Joel Moses and published as *The Computer Age: A Twenty-Year View*. Lick's forty-page chapter, entitled "Computers and Government," was one of the longest in the book. In addition to his fantasy about the Multinet/Internet, it included a very thorough overview of the policy issues raised by information technology—an analysis that stands up pretty well today. Among other things, Lick looked at the pros and cons of export controls on sensitive technology, the need to ensure privacy in a networked environment, the need to protect essential facilities from hacker attack, and the challenge of providing equitable access for the poor as well as the rich.

tation, and most filing cabinets, microfilm repositories, document rooms and libraries for information storage and retrieval."

This vision of the "Multinet," as Lick called it—the term "Internet" didn't really exist yet—was his synthesis of all the thinking and talking and writing that had gone on about the on-line world within the ARPA community since his "Symbiosis" paper in 1960. The Multinet would permeate society, Lick wrote, thus achieving the old MIT dream of an information utility, as updated for the decentralized network age: "Many people work at home, interacting with coworkers and clients through the Multinet, and many business offices (and some classrooms) are little more than organized interconnections of such home workers and their computers. People shop through the Multinet, using its cable television and electronic funds transfer functions, and a few receive delivery of small items through adjacent pneumatic tube networks. . . . Routine shopping and appointment scheduling are generally handled by private-secretary-like programs called OLIVERs which know their masters' needs. Indeed, the Multinet handles scheduling of almost everything schedulable. For example, it eliminates waiting to be seated at restaurants." Thanks to ironclad guarantees of privacy and security, Lick added, the Multinet would likewise offer on-line banking, on-line stock-market trading, on-line tax payment—the works.

In short, Lick wrote, the Multinet would encompass essentially everything having to do with information. It would function as a network of networks that embraced every method of digital communication imaginable, from packet radio to fiber optics—and then bound them all together through the magic of the Kahn-Cerf internetworking protocol, or something very much like it. It would be open in precisely the same way that Fernando Corbató's Compatible Time-Sharing System (CTSS) had been: anyone could join, and everyone could participate. Indeed, monolithic central control of the Multinet would be impossible, if only because its on-line data sources would be distributed so widely that no central authority could ever hope to keep up with them. Instead, Lick predicted, its mode of operation would be "one featuring cooperation, sharing, meetings of minds across space and time in a context of responsive programs and readily available information."[2] The Multinet would be the worldwide embodiment of equality, community, and freedom.

If, that is, the Multinet ever came to be.

That concern seems a bit mystifying nowadays, since Lick's vision of the Multinet was only a slight exaggeration of what the Internet has in fact become in the new millennium (except for a few items such as the pneumatic-tube network, which Lick may have meant as a joke anyway; his OLIVER, of course, would now be called an agent). It seems even more mystifying when you consider that from a purely technological point of view, all the essential components of the Internet were already in place by then: personal computers, local-area networking, packet switching, internetworking protocols—everything.

And yet the reality was that the Multinet/Internet was still a very tentative proposition in 1979. Lick found it all too easy to imagine an alternative scenario in which all his optimistic assumptions would fail to materialize—a tech-nological future in which "the year 2000 is new sheet metal on a souped-up 1970s chassis." For example:

- "IBM explores many new concepts and technologies but preserves functional compatibility with its established systems. . . . It stays out of the computer communication service field because it fears that initiatives there might attract government regulation to other parts of its business."
- "The Bell System offers point-to-point and dialed digital transmis-sion services but not packet switched services, which it seems to consider more a threat than an opportunity, as do the European national telecommunication authorities."
- "Electronic message systems have not replaced mail . . . because there were too many uncoordinated governmental and commer-cial networks, with no network at all reaching people's homes."
- "Controlled sharing of information in [government] computer systems and networks has been neglected because the military, diplomatic, and intelligence people (who control most of the gov-ernment's computer development funds) want to maintain their exclusive secrecy."
- "Social Security, Internal Revenue, law enforcement, and the cen-sus are the major non-defense users of computers [in the govern-ment]. They share the cost of maintaining the National Roster, the ultimate personnel data base with a file for every person, natural or corporate."[3]

Under this scenario, in sum, we would collectively stumble our way toward a fragmented, parochial, Big Brotherish kind of information system "characterized by supervision, regulation, constraint, and control." Moreover, given his view of the world in 1979, Lick had to rate this possibility as far more likely than his optimistic projection. An integrated, open, universally accessible Multinet wouldn't just happen on its own, he pointed out. It would require cooperation and effort on a time scale of decades, "a long, hard process of deliberate study, experiment, analysis, and development." That process, in turn, could be sus-tained only by the forging of a collective vision, some rough consensus on the part of thousands or maybe even millions of people that an open electronic commons was worth having. And *that*, wrote Lick, would require leadership.

Good luck. The pessimistic scenario, in contrast, would require nothing more than laissez faire: "[It] can merely evolve under the pressure of economic com-petition and the criterion of local gain." He did not underestimate the private sector's ability to innovate, Lick insisted; rather, he questioned its ability to *coop-erate*, especially for the sake of some ill-defined "electronic commons" whose payoff was nebulous, iffy, and still ten to twenty years away.

Take the situation in commercial networking as an example. The good news was that the market was flourishing. BBN's Telenet subsidiary (which would be sold to GTE in June 1979) had been offering an Arpanetlike networking service since 1974. IBM had simultaneously introduced its Systems Network Architecture, a proprietary packet-switching technology that allowed customers to interconnect their IBM mainframes. DEC had followed in 1975 with DECnet, its own proprietary packet-switching system, modeled in part on Arpanet. Xerox was planning to weigh in with Xerox Network Services, based on Metcalfe and Bogg's PARC Universal Protocol. And others were coming as well.

The bad news was that "open" wasn't even an option with these commercial networks. Not only were the various systems incompatible, but the vendors had no incentive to *make* them compatible—or to implement anything like the TCP/IP internetworking protocol, which would provide a kind of translation service among the networks. The vendors were primarily targeting Fortune 500–scale corporations that wanted proprietary networks for in-house use and that were endlessly paranoid about security leaks and industrial espionage; isolation was actually a selling point for such customers. And the fact that this isolation also short-circuited any possibility of E-commerce, E-banking, or anything else that required an electronic commons—well, it's hard to miss something you've never experienced.

Meanwhile, that long, hard process of deliberate study, et cetera, wasn't doing much better—not in Tech Square, anyway. You couldn't even get people interested in the problem, says Mike Dertouzos with a sigh. And God knows he tried often enough.

He and Lick had been independently thinking along the same lines, as it turned out. "I'd never liked time-sharing," Dertouzos explains. "Maybe it was the Greek in me, but it seemed like a socialist kind of sharing, with central control, like forcing everyone to ride a bus as opposed to driving a personal automobile. And I'd never liked the information-utility notion, which I first heard from Fano. Information isn't natural gas. It doesn't come from one place, or a few places. It comes from all over. It's much more of a commodity. But now, with the networks, we were moving away from a centralized-brain metaphor to a system without centralized control—a heterarchy. So in nineteen seventy-six or so I was looking for a metaphor for how the machines in such a system would interact with each other. Being Greek, I thought of the Athens flea market, where I used to spend every Sunday, and I envisioned an on-line version: a community with very large numbers of people coming together in a place where they could buy, sell, and exchange information. So every year I would push this information-marketplace idea to the lab members. And every year I would get lots of negative reaction. One, it was too far out conceptually. For most of them, it seemed completely far-fetched to go beyond the log-in/log-out interface for networking and try to establish some broader concepts. And two, the on-line marketplace just didn't seem like it was computer science; it was more a fruity sociopolitical thing. It wasn't so much that they fought the idea. It was more like—suppose we had just invented the car, and I was asking them to study the social implications when they were still focused on the engine and trying to make it go from twenty miles per hour to thirty miles per hour. In fact, my recollection is that until about nineteen-ninety, almost the whole world was oblivious to these ideas."

That was certainly Lick's impression in 1979. The Multinet/Internet/information-marketplace notion was encountering much the same kind of apathy that had greeted the Arpanet proposal a decade earlier. The difference was that this time around, ARPA wasn't in any condition to take the lead and make things happen. As Lick knew from painful experience, the agency that had been created to explore the furthest frontiers now had a "long-term" research horizon of something like two to three years.

Of course, as Lick undoubtedly was also aware, things had improved somewhat since he left. In the game of musical chairs that had followed his departure in September 1975, his deputy, Dave Russell, had moved up to become director of the IPT Office, and networking chief Bob Kahn had moved up in 1976 to become Russell's deputy. Kahn had then brought Vint Cerf in from Stanford to take *his* place, and gone to work doing what he could to restore IPTO's lost glory.

Such were the benefits of credibility, explains Kahn: because he'd learned to talk the language of "defense relevance" very early in the game, and because he was a pretty tough-minded program manager himself, ARPA director George Heilmeier had given him a relatively free hand. "If I really wanted something," he says, "George might grumble, but he'd usually give it to me."

What Kahn wanted for starters was to undo the damage. While he could agree with the director on one thing—that parts of ARPA's computer-research program had indeed gotten self-indulgent; Heilmeier's "wire brushing" had done the program good, Kahn thought—he had come to agree with Lick even more: ARPA's current obsession with "relevance" had come dangerously close to destroying what made the agency so special. Remember, says Kahn, when it came to basic computer research—the kind of high-risk, high-payoff work that might not mature for a decade or more—ARPA was almost the only game in town. The computer industry itself was oriented much more toward products and services, he says, which meant that "there was actually very little research going on that was as innovative as ARPA's." And while there were certainly some shining exceptions to that rule—notably PARC, IBM, and Bell Labs—"many of the leading scientists and researchers couldn't be supported that way. So if the universities didn't do basic research, where would industry get its trained people?"

Yet it was ARPA's basic research that was getting cut. "The budget for basic R&D was only about one third of what it was before Lick came in," says Kahn. "Morale in the whole computer-science community was very low. The challenge for me was, how do we rebuild what has been torn down, and put it on a more solid foundation?"

A big piece of Kahn's solution was a major initiative in cutting-edge microchip technology. Such a research program would be good for the universities, he knew, because they were literally being priced out of the market. Here was a technology that had become fundamental to nearly everything in computing, from major advances in AI to new kinds of massively parallel machine architectures. And yet with semiconductor fabrication plants now costing hundreds of millions of dollars apiece, and with the setup costs of producing a one-of-a-kind chip running to tens of thousands of dollars, only the IBMs and Intels of the world could afford to experiment with these new designs.

At the same time, Kahn knew, an ARPA initiative would be good for the computer companies, which weren't really looking at the long term. The technology's potential was obviously breathtaking. According to an ARPA-funded panel convened in 1976 by the RAND Corporation (and headed by none other than Ivan Sutherland), "the integrated circuit revolution [had] only run half its course" by then.[4] In the mid-1970s, the state of the art allowed for maybe a thousand transistors on a single chip, a level that would come to be known as medium-scale integration. Within a few years, moreover, the industry would be moving into "large-scale integration," with tens of thousands of transistors on each chip. But by the mid-1980s, said the panel—only ten years out—companies could reasonably expect to reach the level of "very-large-scale integration," wherein chips would hold millions of transistors each.

If, that is, the investment was made *now* to support such an advance. After all, integration on that scale meant getting down to microstructures less than a micron (i.e., a few millionths of an inch) wide, arrayed in patterns of bewildering complexity. And that meant that the chip manufacturers had to come up not only with new ways to carry out the fabrication process itself, but also with much better ways to design their chips in the first place. Unfortunately, that was precisely what the companies weren't doing. After an extensive tour of semiconductor firms, Sutherland and his fellow panel members concluded that "U.S. industry generally appears to persist in incremental development." Indeed, the microchip industry suffered from many of the same problems as private-sector networking: proprietary techniques, incompatible protocols (in this case, for variables such as the standard spacing between features), and a lack of incentive to share ideas or technology.

What was needed, in short, was leadership—which Kahn was determined to supply. By 1977 he had talked Heilmeier into approving a VLSI (for "very-large-scale integration") initiative that would encompass submicron chip design, and by 1978 he had begun to award specific grants—a process rendered somewhat easier by the fact that Heilmeier had left ARPA in December 1977 to become research chief at Texas Instruments. (The new director, Bob Fossum, took a much more relaxed approach, says Kahn. "Heilmeier dissected everything; he was a body puncher. Fossum would say, 'If an idea feels good to you, let's do it.'") And by 1979, when Lick's Multinet article appeared, Kahn was preparing to ramp up his VLSI program in a big way.

Well, actually, Kahn *had* been preparing to leave ARPA entirely, figuring that seven years there was long enough. But after Dave Russell retired as IPTO director that year, Fossum had insisted that Kahn was the man to take his place. So he'd finally agreed, says Kahn, because "I still had a very long list of things I wanted to see happen, and being director was probably an effective way to make them happen."

Besides, thanks to one of the first and most productive awards he ever made in the VLSI program—the award that went to Carver Mead of Caltech and Lynn Conway of PARC—another idea was beginning to crystallize in Kahn's brain.

Mead and Conway, he explains, had pioneered a way of teaching integrated-circuit design via an elegant and very general set of design principles. Draft chapters of their textbook, *Introduction to VLSI Systems,* had been circulating since 1977 and had already been used in courses at Caltech, Berkeley, Carnegie Mellon, and MIT. (The book itself, which was published in 1979, would go on to become a bible for VLSI professionals.) But just as important, says Kahn, Mead and Conway had conceived the notion of a "silicon foundry." The idea was that students in an IC design course would each prepare a chip layout, specified in a standard chip description language, and then send it over the Arpanet to a "silicon broker"—originally PARC and then later Hewlett-Packard. The broker, in turn, would compile dozens of individual designs and then arrange with a chip manufacturer to have them all etched onto a single silicon wafer, so that the cost could be shared. Finally, the chips would be cut apart, packaged individually, and sent back to the students for testing and experimentation. The system worked beautifully; in the fall of 1978, when Conway taught the design course at MIT for the first time, her students sent their designs off in early December and got their chips back just six weeks later.

By the time Kahn stepped up to the IPTO directorship in November 1979, this ad hoc silicon foundry network was being used by 124 designers from eleven universities. Indeed, it would soon become an institution: in January 1981 Kahn would make it a formal program known as the Metal Oxide Silicon Implementation System, or MOSIS. Run by the Information Sciences Institute at the University of Southern California and serving industrial and university researchers as well as students, MOSIS would flourish into the 1990s. And chip innovation would flourish along with it. MOSIS supported design experiments for advanced architectures such as Intel's Cosmic Cube and the Connection Machine from Thinking Machines, Inc. It supported experiments in reduced-instruction-set computing at Stanford and Berkeley, thereby providing a proof-of-concept for a number of cutting-edge commercial chips of the late 1980s and 1990s. It even supported the development of the "graphics engine" chip by Stanford's James Clark, who would soon be applying his expertise as a cofounder of Silicon Graphics, Inc. And most of all, says Kahn, "the MOSIS project produced an awful lot of people trained in VLSI design." Indeed, you could argue that MOSIS was as much responsible as any other single factor for the explosion in microchip technology during the 1980s and 1990s.

Even in 1979, moreover, Kahn could see that a whole community was coming into being around this MOSIS-in-embryo, just as a community had once coalesced around Project MAC—and indeed, around the Arpanet itself. And something about that fact deeply impressed him. Kahn was struck by the analogy to the way malls, stores, and office buildings will spring up along a new highway, or the way power-intensive industries like paper or steel will take root where power is cheap, as in Washington State or the Tennessee Valley. In his own way, in fact, Kahn was independently coming to much the same conclusion that Lick and Dertouzos had reached: networking had the potential to become not just a technology but an electronic commons, an information infrastructure that would be central to the economy of the twenty-first century.

Better still, Kahn realized, the technical wherewithal for such an information infrastructure was actually in fairly good shape, now that Vint Cerf was at ARPA and pursuing his "single-minded goal" of making the TCP/IP protocol suite as robust and as bulletproof as possible. Indeed, the Defense Department was getting ready to make TCP/IP the standard for all its digital communications, and Cerf was beginning to plan the mammoth task of converting the Arpanet itself to TCP/IP.

So there it was: even Lick had to admit that the situation at ARPA had improved significantly since his day. But for how long? Bob Kahn had done great things, granted. And after Heilmeier, a director like Bob Fossum was a joy. But in the current climate of budgetary and "relevance" constraints, who knew what the *next* director might be like?

Nonetheless, Lick wrote in his "Multinet" article, he was still an optimist. Even if you couldn't automatically look to the private sector or to the government for leadership, you still had—well, the People. At least when it came to computing, he wrote, "there is a feeling of renewed hope in the air that the public interest will find a way of dominating the decision processes that shape the future."[5]

Just look at E-mail, the Arpanet mailing lists, and all the rest, he said. Just look at the on-line communities that seemed to come into being wherever there was a network. Users of a modern computing system weren't just passive consumers; the medium itself drew them in. It gave them a forum, it made them active participants, it gave them a stake in deciding their own destiny. So if you could somehow expose ordinary people to this medium—if you could somehow get the technology out of the laboratory and into the mass market so they could experience it firsthand—then ordinary people might just create this embodiment of equality, community, and freedom on their own.

It was a vision that was downright Jeffersonian in its idealism, and perhaps in its naïveté as well. Nonetheless, Lick insisted, "the renewed hope I referred to is more than just a feeling in the air. . . . It is a feeling one experiences at the console. The information revolution is bringing with it a key that may open the door to a new era of involvement and participation. The key is the self-motivating exhilaration that accompanies truly effective interaction with information and knowledge through a good console connected through a good network to a good computer."[6]

THE SLAUGHTER OF THE WHITECOATS

He had a point.

Lick may have revealed a hopelessly unfashionable lack of cynicism in that last paragraph, but his "self-motivating exhilaration" was real enough—and not just in the Tech Square terminal room, either. The phenomenon had been gathering force for the better part of a generation, both in the marketplace and in society at large. Witness the public's eager embrace of computer utilities in the 1960s, when thousands of nonprofessionals had finally gotten the chance to tap in and experience the exhilaration firsthand. Or witness the rhetoric of counterculture gurus such as Stewart Brand (who'd called computing "the best news since psychedelics") and Ted Nelson, an independently wealthy computer activist who had declared that "hypertext"—a word he'd invented to describe the electronic links first imagined by Vannevar Bush—would at last allow us to break free from linear thought and hierarchical power structures. The ARPA vision of personal involvement with computers had resonated deeply with the head-tripping, antiestablishment spirit of the era.

Perhaps most of all, however, the power of Lick's self-motivating exhilaration had been reflected in the success of the Digital Equipment Corporation and its downright subversive efforts to democratize computing with cheap, affordable hardware.

Now, DEC was certainly no stranger to big machines. Its PDP-6 and PDP-10 time-sharing computers had cabinets that filled a room and price tags to match (staffers half jokingly called them personal mainframes). But for those very reasons, they had never gotten much more than a cool endorsement from company founder and CEO, Ken Olsen. He liked equipment that you could get your hands on, that you could tinker with yourself. Moreover, he had reason to believe that DEC's customers felt the same way. Back in the company's precomputer days, when it was still establishing itself as a manufacturer of transistorized circuit boards, Olsen and his colleagues had discovered that people were using their products in ways they'd never anticipated. "The laboratory customers bought all these little pieces from us," says Olsen—"modules, connectors, sockets, cables, piddly little mechanical things—and they'd put them together and they worked!"

In response, DEC had started printing up a product catalog, an inexpensive pulp paperback that included a tutorial on how to connect DEC's modules to one another and to other laboratory equipment. "When they first saw it," Ken Olsen's brother Stan later recalled, "our engineering people rolled over with laughter because they'd never seen anything so different from the real slick stuff. [But] we produced about sixty thousand of them. We would go to the IEEE show in New York City, the event of the year for us, and we'd hand out thousands."[7] Pages from the book could be seen blowing through the subways during the show, but dog-eared copies would also be found in laboratories and offices for years afterward.

DEC was clearly making a connection here from engineer to engineer. And when the company began its move into full-fledged computers with the PDP-1, it did its best to keep on making that connection. The PDP-1 was an "open" hardware system in much the same sense that more than a decade later, the Kahn-Cerf protocols would be open: all the details of the core architecture were publicly spelled out, so that users could add to the machine in any way they wanted. Nowadays, this kind of architecture is familiar to anyone who has ever opened up a personal computer and dropped in a new video card or added more memory. But in the 1960s it represented a relationship with the customer that was unique in the computer industry. IBM and the other mainstream manufacturers kept very tight control of their products in those days: computers were typically leased rather than sold, and the software was totally proprietary. And that was the way most of their customers liked it. Yes, they had to take whatever the vendor gave them, but in return, the vendor was always there to worry about maintenance, repairs, and upgrades—a form of technological handholding that most people in the business world found very reassuring.

As a start-up, however, DEC couldn't afford to provide that level of technical support; that was one reason it sold rather than leased its computers. Besides, the company was aiming not for businessmen but for the hands-on engineering crowd, the people who loved to tinker with their machines. As far as those customers were concerned, "interactive" computing had just acquired a whole new meaning.

By 1964, in fact, their response to the PDP-1 and the company's subsequent products led DEC to start developing a "tabletop" computer intended for small groups of users or even individuals. The PDP-8, as it came to be known, was a significant gamble for the firm: the machine that had inspired it—Wes Clark's LINC—was barely out of its development phase itself. No one knew how much of a market there really was for such a machine. As it turned out, however, the timing was auspicious. The watchwords for the PDP-8 project were price (low), size (small), and performance (high); semiconductor and storage technology was ripe for advances in all three areas. For many applications, in fact—the major exception being arithmetic on very large "floating point" numbers, where the machine's twelve-bit architecture proved to be a severe bottleneck—the PDP-8 would have no problem living up to DEC's trade literature, which proclaimed it "one of the fastest computers in the world."[8]

It would also be one of the cheapest to manufacture. "We set out to study all the technology used for home appliances and automobiles," says Olsen. "I went to several discount stores for hours, studying every single washing machine, drier, and stove to see how they built things and what we could use. The switch handles [on the circuit boards] came from a Maytag dryer. Using a glass panel without separate lights showing was the technology used in appliances. Inside, we used the slip-on connectors that were commonly used in automobiles. We tried every way we could to take the technology that made mass production possible in appliances."

The result was a computer so unbelievably small and light—at just 250 pounds—that it could be transported in the back seat of a Volkswagen Beetle convertible. (For those who *didn't* believe that claim, DEC ran an ad providing photographic proof: smiling model in front, PDP 8 in back.) More to the point, it had an unbelievably low price of eighteen thousand dollars, complete with Model 33 Teletype terminal. True, that wasn't exactly pocket change, especially not in the 1960s. But it was just under one sixth the price of the PDP-1, and far cheaper than anything in the IBM catalog. To customers, most important, it was irresistible. No sooner had the machines started shipping in April 1965 than they started flying out of the Mill, winding up in chemical plants, newspapers, laboratories, refineries, and schools. Much to the astonishment of DEC staffers and everyone else, even nontechnical users loved a computer they could get their hands on. One PDP-8 controlled the digital scoreboards at Boston's Fenway Park; another ran the streaming news display in New York City's Times Square. There was even a PDP-8–based time-sharing system: TSS/8, created at Carnegie Mellon University.

Indeed, the PDP-8 and its open architecture soon inspired an entirely new kind of entrepreneur in the computer industry. Known somewhat confusingly as Original Equipment Manufacturers, or OEMs, these firms would buy the raw PDP-8s, add peripherals and other equipment as needed, write a software package, and market the result as a complete typesetting system, say, or a complete scientific instrumentation system. DEC eventually sold at least half of its PDP-8s in this way, without ever having to make an investment in software, end-user training, service, or maintenance. In the process, moreover, the company proved just how powerful the open-systems approach really was. By giving up absolute control, by publicizing the kind of technical details that other computer manufacturers guarded with fanatical jealousy, by letting users take charge of their own machines, DEC could benefit from the creativity and entrepreneurial energy of a far larger community than it ever could have mustered in-house.

Before it was all over, in fact, DEC would sell some fifty thousand PDP-8s in all the model's versions—not counting the additional thousands it would later implement as one-chip microprocessors. By 1970, moreover, some seventy competitors were offering similar machines, though none could overcome DEC's head start. The PDP-8 *made* the company. DEC's revenues grew by more than 50 percent a year in this period, with its profits growing by a factor of six between fiscal 1965 and 1967 alone. And trade publications finally started taking notice, touting DEC as a rising young contender in an industry still dominated by IBM. True, IBM couldn't have cared less. Like the big American auto companies, which spent the 1960s being disdainful of the plebian Volkswagen Beetle and cheap Japanese imports such as Toyota and Honda, IBM was dismissive of the low-end niche markets being tapped by the PDP-8. The profit margins there were just too thin—and besides, there was no way to modify the System/360 architecture to accommodate such low-power machines.

But then, the DEC team couldn't have cared less, either. They saw the PDP-8 as the prototype for a whole new genre of computers. John Leng, head of DEC's operations in the United Kingdom, was particularly taken with that idea. Zigzagging through London in his Austin Mini Minor automobile, taking in the striking new fashion scene along Carnaby Street, Leng would dictate inimitable sales reports: "Here is the latest minicomputer activity in the land of miniskirts as I drive around in my Mini Minor."[9] The name stuck: the PDP-8 was the first minicomputer.

It would hardly be the last. By the early 1970s, thanks to Moore's law, that first generation of 12-bit machines had given way to a new generation of much more powerful 16-bit minis. The first, in 1969, was Data General's Nova, the machine that would so impress the hardware mavens at Xerox PARC. But DEC came back in the spring of 1970 with its own PDP-11, which quickly became *the* machine to beat in the 16-bit market. DEC would eventually sell nearly a million PDP-11s, running the gamut from small four-user systems to big sixty-four-user models; microprocessor versions would be marketed well into the 1990s. And then in 1977 DEC came out with the VAX, which was actually a whole new line of "Virtual Address eXtension" machines that extended the PDP-11 architecture to 32 bits. The first and largest of the line, the refrigerator-sized VAX-11/780, sold for $120,000 in its most basic configuration and was an instant hit. Every other minicomputer maker was left to play catch-up—especially Data General. (That company's struggle to produce an answer to the VAX would be beautifully chronicled in Tracy Kidder's 1981 book, *The Soul of a New Machine*.) By 1977 DEC's revenues had passed $1 billion per year.

Along the way, meanwhile, all of these increasingly high-powered minicomputers had rolled right over the faltering "computer utility" industry. After all, why should anybody pay upward of ten dollars an hour in connect charges per user—twenty thousand dollars for a year of eight-hour days—when that same money would buy a mini that could support a dozen users? (Most of the utility companies quickly folded; the idea wouldn't really revive until the 1980s, when modem-equipped microcomputers started tapping into newer, network-based utilities such as CompuServe and America Online.) And out in the labs, by no coincidence, those same new-generation minis were inciting users into open revolt against the tyranny of mainframes—a process that Steve Wolff calls the slaughter of the whitecoats.

Remember, explains Wolff, who was then a networking expert with the army's Ballistic Research Lab at Aberdeen Proving Ground, unless you were at one of the labs blessed by ARPA, computing was still a matter of typing up those same damn punch cards, trudging down to the computer center, and handing in your deck to those same arrogant SOB technicians in their white coats. Wolff had resented those guys since he was an undergraduate. "You really felt it when you'd been slaving away to write your programs in an un-air-conditioned dorm room, and you could feel the blast of cold air coming out of the window where you handed in your cards," he says.

Ah, but then the minicomputers came along, the high-tech embodiments of individual autonomy, hands on control, and *freedom*. Naturally, says Wolff, the computer centers tried to have them outlawed. "But there were many groups that stormed the bastions and slaughtered the whitecoats. They discovered that they could buy a computer cheaply—a PDP-8, a PDP-11, or, later, a VAX—and run it themselves. It wasn't as fast as a mainframe. But they could turn it on and let it run all weekend." Sure, he says, you still needed mainframes for big-time number-crunching; IBM wasn't going broke, and neither was Seymour Cray, with his "supercomputers." But for your small- to mid-size groups, such as university departments, research labs, and even many businesses, the in-house mini rapidly became the computer of choice in the 1970s.

It was glorious fun, Wolff remembers. Then, around 1974, Unix came along, and suddenly you could get your hands on the software, too. "Unix was the first really general-purpose operating system for minicomputers," he says. "It took off for two reasons. One, it was free. And two, Unix was the first operating system you could get source code [the full list of programming commands] for. You could hack it."

Actually, Unix was "free" only because AT&T wasn't allowed to sell it; until the breakup, in 1982, the company was a regulated telephone monopoly and as such was barred from doing much of anything in the computer business. But Unix was hackable because it was about as open as a system could get. Order the data tapes (nominal charge: $150 for materials), and you got the complete source code, which was yours to modify as you wished. Dennis Ritchie and Ken Thompson had created Unix for their own use back in 1969, after Bell Labs' withdrawal from the Multics partnership, and the project had retained that loose, hands-on feeling ever since.

Then, too, Ritchie and Thompson had created Unix with hacking in mind, deriving its basic structure from Fernando Corbató's CTSS via Multics—namely, a tool kit of utilities and software routines that the user could call upon to perform specific tasks, all built on top of an invisible "kernel" of software that functioned as the autonomic nervous system of the computer, seamlessly moving bits through the machine at a level below the user's consciousness. The difference was that the Unix kernel was a marvel of efficiency, simplicity, and conciseness. This was partly a reaction against the complexities of Multics (the designers kept the features that seemed truly useful, such as the file structure, and left out the more baroque elaborations) and partly a reflection of Ritchie and Thompson's ancient and underpowered PDP-7, the only machine they could scrounge for the project. But mainly Unix was tightly written because that was how the two men felt a kernel ought to be: sweet, quick, and clean. When Ritchie and Thompson were finally able to migrate their system to a PDP-11 in 1970, they weren't even tempted to add extra bells and whistles.

They followed the same aesthetic with the software tool kit. Each routine—"copy," for example (abbreviated *cp*), or "print" *(pr)*—was designed to do one specific thing as efficiently as possible. But the routines were also designed to be connected together like so much plumbing—literally. By February 1973, when they made their first public presentation of their system, Ritchie, Thompson, and the other members of their steadily growing Unix group had data flowing from one component to the next through software constructs known as pipes. The result was a Unix environment in which it was very easy to connect simple routines on the fly to form more complex utilities.

At the time of that initial presentation, there were just sixteen Unix installations in the world, all inside AT&T. Within six months, as word spread outside the company, that number had tripled.

Of course, sometimes you had to get down into the bits and bytes and compose a completely new software tool. But that was pretty easy, too, especially after Ritchie completely rewrote the Unix kernel in an elegant new computer language of his own devising. Since he'd based it on an earlier, experimental language by Thompson, code-named "B," Ritchie code-named his language "C." The name stuck. By July 1974, when the written version of their presentation came out in the journal *Communications of the ACM,* the team had rewritten key elements of the software tool kit in C, and Unix had taken a long stride toward the nirvana of "machine independence." In principle, all you had to do was take the free Unix source code, run it through a C compiler for your specific brand of machine, and you'd have a Unix operating system that worked just as well on your machine as it did on the PDP-11.

Once the article had been published, orders for the data tapes started pouring in. Hundreds and then thousands of minicomputer users junked their manufacturer-supplied operating systems, which tended to be pretty lame anyway, and started compiling Unix to run on their own machines. In practice, of course, this wasn't quite as easy as it sounded on paper. But for the techies of the world, of whom there were many, that just added to the appeal. They happily wrote their own "drivers"—that is, the routines that told Unix how to communicate with the specific brand of display terminal they were using, or the specific printer, or the specific kind of network connection—and then passed those drivers around to other users. They wrote new editors, new debuggers, new software tools of all kinds, and passed those around. And they devised little improvements in the kernel and passed *those* around, along with software patches for bugs they'd uncovered and fixed on their own initiative.

Indeed, the users quickly turned Unix into the People's Operating System, "open software" that would flourish through the 1970s in much the same way that its descendant, Linux, would do in the late 1990s. Moreover, at Ken Thompson's alma mater, Berkeley, where he spent a sabbatical year in 1975 spreading the gospel, a quietly brilliant graduate student named Bill Joy was soon leading a covey of like-minded Unix mavens on a quest to make Unix more usable by and more accessible to *non*techies. By decade's end, their Berkeley System Distribution 4.2 Unix was emerging as a semiofficial standard in the community, complete with an endorsement from ARPA itself. And Joy and his crew had been commissioned by that same agency to integrate TCP/IP into Unix.

To help with all this sharing of information, meanwhile, another group of Unix mavens were putting together a "poor man's Arpanet," a.k.a. Usenet.

That innovation began in the fall of 1979, when graduate students Tom Truscott and Jim Ellis, along with a number of their friends at Duke and at the nearby University of North Carolina in Chapel Hill, were looking for a way to communicate with their fellow Unix enthusiasts on-line (Truscott had just returned from a marvelous summer at Unix heaven, Bell Labs). Neither school was on the Arpanet, unfortunately, but Truscott, Ellis, and their friends realized that they could just cobble together a little bulletin-board utility based on the Unix version of E-mail. True, communications would have to run over 300-bit-per-second modems, but still, the system worked. After several revisions and a complete rewrite in C, moreover, not to mention some help from sympathizers at both Bell Labs and DEC, it worked quite well. In January 1980 the Duke-UNC cabal presented its "rapid-access newsletter" at a national meeting of Usenix, the Unix users' group. At first, Truscott noted in the written announcement, most messages would probably be concerned with "bug fixes, trouble reports, and general cries for help,"[10] but the system could also support other newsgroups if there was enough interest.

There was. Slowly at first, then more and more rapidly, Usenet became living proof of a prediction that Lick and Bob Taylor had made back in 1968, in their article on the computer as a communications device: "Life will be happier for the on-line individual because the people with whom one interacts most strongly will be selected more by commonality of interests and goals than by accidents of proximity."[11] Newsgroups proliferated. There was net.unix-wizards, for example, and net.chess, and fa.human-nets—the last devoted to the implications of worldwide networking. (The prefix *fa* stood for "from Arpanet." The naming scheme with the dots was invented by Duke graduate student Stephen Daniel; it has since been refined into now-familiar categories such as *comp* for computers, *rec* for hobbies, and *alt* for anything goes.) By 1981 the ad hoc Usenet encompassed something like 150 sites around the country. And among them, significantly, was Berkeley, where Usenet had gained its first portal to the Arpanet. This, in turn, opened it to the already flourishing world of Arpanet mailing lists, such as the legendary "SF-Lovers" list for science-fiction buffs. By the mid-1980s articles were flowing both ways, newsgroups were proliferating madly—today they number in the tens of thousands—and Usenet was well on its way to becoming Network News, an integral part of the Internet.

THE LOGICAL NEXT STEP

That would all come later, of course. Nonetheless, Lick's "self-motivating exhilaration" had clearly become a very potent force by the mid-1970s. And to modern eyes, at least, the next step in its evolution was blazingly obvious: just take a small minicomputer serving a handful of people and—in effect—shrink it down to a machine serving *one* person. Indeed, that was exactly what PARC had already done to create the Alto. But alas, Xerox was too busy being Xerox to know what it had. And elsewhere—well, just because a step seems "logical" in hindsight doesn't mean it automatically gets taken.

At DEC, most notably, there were at least half a dozen proposals for a personal-computer product line during the 1970s, including several that were actually built and marketed to specific customers. (Indeed, building them was almost trivial: it was just a matter of adding some chips to one of DEC's high-end display terminals, which were already the market leaders.) Yet CEO Ken Olsen seemed unable to shake the idea that "personal computers" and "home computers" were one and the same thing: toys. Gadgets that you hooked up to the TV and used for video games. Well, DEC didn't make toys. Besides, Olsen famously asked on a number of occasions in the 1970s, "Why would anyone want a computer on his desktop?"

In fairness, the context of those comments shows that what Olsen really meant was, "Why would anyone want a feeble little thing like that on his desktop when he could put a time-sharing terminal there instead, tap into a PDP-10 or whatever, and get some *real* processing power?" Moreover, at a time when the available microcomputers really were pretty toylike, he had a point. Nonetheless, Olsen's attitude explains why DEC didn't take the lead in personal computers, even though it had in place a corporate philosophy and a corporate culture that might have allowed it to do so easily.

Alternatively, there was another route to personal computing that to modern eyes seems just as obvious: simply combine one of the new one-chip microprocessors with the ubiquitous pocket calculator, thus creating a "microcomputer." And indeed, Hewlett-Packard had already established quite a bit of momentum in that direction, starting with the introduction of its $400 HP-35 in early 1972. Customers loved the device. Simple four-function, arithmetic calculators were one thing; the $250 Bowmar Brain had been a big seller just the previous Christmas. But the HP-35 offered logarithms, exponentials, trigonometric functions—anything a scientist or an engineer might need. Every techie in the world wanted one; countless trusty slide rules were already doomed to obsolescence. In January 1974, moreover, Hewlett-Packard had followed up that success with the first fully programmable pocket calculator, the $795 HP-65, which could store short sequences of keystrokes and play them back (indeed, the HP-65 was explicitly introduced as a "personal computer"). By the middle of the decade, competitors were pouring into the field, prices were dropping precipitously, and the pocket-calculator market was booming.

Certainly those devices set the stage for personal computers, the computer historian Paul Ceruzzi has noted: "Pocket calculators, especially those that were programmable, unleashed the force of personal creativity and energy of masses of individuals." Within a year, the HP-65 alone had sold more than twenty-five thousand units, "each one owned by an individual who could do whatever he or she wished to do with it. . . . Nearly all were adult professional men, including civil and electrical engineers, lawyers, financial people, pilots, and so on[, whose] passion for programming made them the intellectual cousins of the students in the Tech Model Railroad Club. And their numbers—only to increase as the price of calculators dropped—were the first indication that personal computing was truly a mass phenomenon."[12]

And yet the reality was that the major players out in Silicon Valley weren't even thinking about microcomputers. In April 1974, for example, Intel introduced its 8-bit 8080 chip, the first microprocessor to come within shouting distance of, say, a 12-bit mini such as the PDP-8. Intel could very easily have gone on to use it in a full-fledged microcomputer, if anyone there had made the connection. But the company was full of hardware guys who were used to selling to other hardware guys, not to end users. "Intel's mental model of its product," according to Ceruzzi, "was this: an *industrial* customer bought an 8080 and wrote specialized software for it, which was then burned into a read-only memory to give a system with the desired functions. The resulting inexpensive product (no longer programmable) was then put on the market as an embedded controller in an industrial system."[13] (Ceruzzi adds that Intel *did* build several microcomputers during this period; however, it marketed the ten-thousand-dollar Intellec "Development Systems" not as general-purpose computers but as tools to help customers write and debug software for embedded processors.)

And so it was that the field was left open for the hobbyists in their basements and garages, individuals who didn't have to worry about meeting payrolls and pleasing the stockholders but just liked to play around with electronics—and were used to making components do things their manufacturers had never imagined. These were the same kinds of tinkerers who'd gotten creative with DEC's first transistor modules in the 1950s, who'd fallen in love with the PDP-8 in the 1960s, and who were snapping up the programmable calculators now. They were the spiritual brothers of the MIT hackers and the freewheeling Unix mavens. They were the people who had been ham-radio operators and/or stereo buffs since they were teenagers, often using equipment they had built themselves from mail-order kits, or from scratch. They were the guys who had gotten intrigued by the minicomputers they'd encountered at work or at school. And for no "logical" reason whatsoever–certainly none that they could explain to their spouses–they were the people who wanted computers of their own at home, to play with, to experiment with, to experience.

Enter the Altair.

Now, the Altair was *not* the first commercial microcomputer; that honor goes to the Micral, an Intel 8008–based machine that was sold in France starting in May 1973. Created by Vietnamese immigrant Thi T. Truong, the two-thousand-dollar Micral sold fairly well for a time but was marketed only on the Intel model, as an industrial process controller.

The Altair wasn't even the first microcomputer kit to appear on the cover of a general-interest electronics magazine. That honor goes to the Mark-8, another 8008-based machine, designed in the fall of 1973 by Jonathan Titus, then a graduate student in chemistry at the Virginia Polytechnic Institute in Blacksburg, Virginia. Hard-core hobbyist that he was, Titus never even thought of starting a company; he just wanted to share his creation with his fellow fanatics. So he wrote to the hobbyist magazine *Radio-Electronics*, which always tried to offer its readers at least one construction project every month. The result was that magazine's cover story for July 1974: "Build the Mark-8: Your Personal Minicomputer."

However, the Altair *was* the first commercially successful microcomputer. By no coincidence, it was also the first to be based on the Intel 8080, which really was a major advance over the 8008. The father of the Altair was H. Edward Roberts of Albuquerque, New Mexico, a thirty-four-year-old electrical engineer who was a big man not only physically–he stood six foot four and weighed 250 pounds–but in his intellectual appetites as well. He once described himself as the kind of guy who devoured whole library shelves at a time. Mostly, however, Roberts was fascinated by electronics. In 1969, while he was still an air force captain working at nearby Kirkland Air Force Base, he and several of his colleagues founded a hobbyist shop called Micro Instrumentation Telemetry Systems, or MITS, which mainly sold radio transmitters for model airplanes through the mail. Then, in the early 1970s, after leaving the air force and buying out his partners, Roberts branched out into mail-order electronic-calculator kits. These sold pretty briskly until the rapidly falling prices of commercial models made them pointless. So in early 1974, with some twenty-five employees now on his payroll and a company that was suddenly facing bankruptcy, Roberts elected to try a computer kit. He'd always been interested in digital electronics, he later explained, and he'd been wanting to try his hand at building a minicomputer anyway. He proceeded to rough out a design using the Intel 8080, which he'd decided was by far the best chip available. And by midyear he was ready to approach the editors of *Popular Electronics,* where he'd been an occasional contributor, to ask if they'd like to feature his kit as a construction project.

They would. The magazine was *Radio-Electronics*'s arch rival, and technical editor Les Solomon just loved the idea of an 8080-based computer project that would top the Mark-8 story. The only ground rules were that the end result had to be a *real* computer, not a toy like the Mark-8, and the kit had to cost less than four hundred dollars.

The price was feasible; Roberts had already talked Intel into selling him the 8080 chips for just seventy-five dollars apiece in bulk. So he and his codesigners at MITS went into overdrive to get the Altair ready in time. Solomon, meanwhile, came up with the perfect name for the machine. (Looking for ideas, he asked his daughter, Lauren, what they called the computer on *Star Trek*. "Computer," she replied. But she added that the starship *Enterprise* was headed for the star Altair that night, so why not call it that?)

The now famous January 1975 issue of *Popular Electronics* hit the newsstands in early December—ironically, just a month before George Heilmeier arrived at ARPA to make J. C. R. Licklider's life miserable. Since the only real, working Altair had gotten lost in transit before reaching the photographers in New York (it would turn up a year later), the cover showed the best mockup that MITS could manage on short notice: a pale-blue Altair shell with an impressive array of switches and diodes across the front that did absolutely nothing. But it certainly looked like the real thing. And right there in the box's upper-left-hand corner was the name: Altair 8800. "Project Breakthrough!" proclaimed the headline. "World's First Minicomputer Kit to Rival Commercial Models." Inside, readers learned that the kits could be had from MITS for just $397.

MITS would be a year or more digging out from under the avalanche. Having expected maybe a few hundred orders, Roberts and his crew ultimately received more than ten thousand. And why not? The kit was a steal, considering that the retail price of the 8080 chip alone was $360. Besides, that $397 really bought something. If the Altair was one step up from a programmable calculator, it was also just one step down from a minicomputer—a message that Roberts and his codesigner William Yates hammered home again and again in their *Popular Electronics* article. The Altair, they insisted, was "a full-blown computer that can hold its own against sophisticated minicomputers."

Indeed, except for the 8080 chip, the Altair *was* a minicomputer, right down to those switches and diodes on its front face; Roberts had copied the layout from a Data General Nova 2 he'd bought for the MITS office (the company was selling time-sharing services with it). The minicomputer heritage was equally obvious on the inside, where the design was identical in spirit to the "Unibus" architecture of DEC's PDP-11. Basically, it was just one big array of slots for add-on cards: everything in the Altair was modular and replaceable.

Even Roberts's later choice for an official programming language was reminiscent of the minis. Created in the spring of 1975 by two young men who had been inspired by the *Popular Electronics* article—Bill Gates, now a Harvard undergrad, and his high school buddy Paul Allen, a programmer working outside Boston—Altair BASIC took a number of key features from DEC's BASIC for the PDP-11. (The language also owed its existence to the Harvard PDP-10, interestingly enough. Since Gates and Allen didn't have access to an Intel 8080 at the time, they used Gates's student account on the big machine to create a simulation of the microprocessor—in the process burning up some forty thousand dollars' worth of computer time that was *not* supposed to be used for commercial purposes. But no matter: once the language was ready, Allen quit his job, Gates dropped out of school, they both moved to Albuquerque to be near MITS, and together they formed a little company called Micro Soft to market it.)

Of course, the Altair did have its share of design quirks and reliability problems, not to mention being about as bare-bones as a computer could get. The only way to enter a program was by flipping tiny switches on the front panel (which tended to tear up your fingers), and the only way to get the output was by deciphering the lit-up diodes. The kit included no keyboard, no monitor, no disk drives, no software—nothing but the box itself, which contained little more than a central processing unit, 256 bytes of memory, and all those empty slots for add-on cards. Paradoxically, however, that very spartan simplicity proved to be another big reason for the Altair's success. The Altair was an open system: Roberts, like DEC before him, had decided to make the specifications for the data bus freely available so anyone could build an add-on card to fit in. And when his own add-on cards for monitors, memory, keyboards, and the like were delayed by the struggle to fill all those back orders for the kit, hundreds of hobbyists and young computer professionals rushed in to do just that. The result was an explosion of entrepreneurial energy, as tiny garage-scale start-ups sprang into being by the hundreds. Roberts's decision to open up his machine had given the ever-expanding Altair community a direct personal stake in the Altair architecture, just as had happened with the PDP-8 a decade earlier.

Unfortunately for Roberts, however, that wasn't quite the same thing as giving that community a direct personal stake in the Altair itself. Among the start-ups were any number of garage shops building microcomputers of their own, including several—with names like IMSAI, SOL, Morrow, Godbout/Compupro, Dynabyte, Cromemco, and Vector Graphic—that were producing outright clones. Since MITS was still laboring to fill those back orders, its rivals found a ready market. (Another open-architecture machine, the IBM PC, would later experience a similar fate.)

Naturally, Roberts fought back. And indeed, once he caught up with the back orders, his company did rather well. By 1976 MITS was offering its own line of expansion boards and peripherals, as well as an upgraded Altair 8800b. Roberts can remember selling equipment to the Secret Service, the FBI, and the CIA, as well as to a fellow working on the special effects for a new science-fiction movie—something called *Star Wars*. But by that point he was also getting tired of the "soap opera," his term for the endless round of product upgrades, frustrated customers, panicked dealers, personnel hassles (MITS now had hundreds of employees), and internal politics. He hadn't had a decent night's sleep in more than a year. So in mid-1976 he announced that he was selling out to a hard-disk manufacturer, Pertec. By year's end he was headed off to medical school; he now works as a doctor in Cochran, Georgia. And Pertec, alas, proceeded to run his company into the ground: both MITS and the Altair had vanished by 1979.

If Roberts and his computer were gone, however, the movement they had done so much to create was already self-sustaining. Not only were new brands of machines flooding the market by the dozens, if not by the hundreds, there were also users' groups for every microcomputer imaginable, as well as pan-micro groups such as the legendary Homebrew Computer Club, which held its first meeting in a Palo Alto garage in March 1975. There were magazines such as *Byte*, which debuted in August 1975, and the software periodical *Dr. Dobbs Journal of Computer Calisthenics and Orthodontia* (motto: "Running Light without Overbyte"), which published its first issue in 1976. There were specialty stores like the Byte Shop and ComputerLand, the latter soon to be a nationwide chain.

And increasingly there were young entrepreneurs who were beginning to imagine that *consumers* might want to use these machines—people who weren't hobbyists, who weren't technically sophisticated, and who had no desire to pick up a soldering iron. Maybe, just maybe, these micros might appeal to the mass market.

At this late date it's impossible to say exactly where this notion came from, though a big part of it was undoubtedly the MIT vision of information utilities, plus the resulting computer-utilities boom. Among the computer cognoscenti, at least, the 1960s had been rife with speculations about "home information systems" linked to data centers via telephone lines or antennas. Other possible influences included the hobbyists' personal experience with minicomputers and pocket calculators, as well as the rumors they'd heard about the wonders at PARC, and the presence of all those thousands of students who had graduated from top computer schools such as MIT, Stanford, Carnegie Mellon, and Berkeley—which also happened to be the top ARPA sites.

But wherever the notion came from, it was obviously in the air by January 1975. As *Popular Electronics* put it in the editorial that introduced the Altair, "we've been reading and hearing about how computers will one day be a household item . . . for many years." By 1976, moreover, some of the start-up companies were visibly taking this prospect seriously. A different kind of customer meant a different kind of product, they realized. Instead of just providing a kit, a bag of parts, they would have to offer something much more like an appliance: a finished system that would work as soon as you plugged it in. Keyboard, monitor, disk drives, operating system, software—everything had to be right there in the box, or else be very easy to add.

On the hardware side, this challenge was taken up most famously by the Apple Computer Company, founded in 1976 by Homebrew Computer Club members Steve Wozniak and Steve Jobs, longtime buddies from the Silicon Valley town of Cupertino. After some encouraging success with their first computer, which they marketed through local hobby shops—it was actually just a single circuit board using the new, 8-bit 6502 microprocessor from MOS Technology, plus 4 kilobytes of RAM—Jobs and Wozniak were joined by the thirty-four-year-old A. C. Markkula, formerly the marketing manager for Intel. Markkula, who had retired from that company two years earlier after earning more than a million dollars in stock options, bought a one-third partnership in Apple for $91,000 and began working his contacts to bring in venture capital and management expertise. The result was the Apple II, a much-upgraded, 6502-based micro that was introduced in April 1977 at the first West Coast Computer Faire in San Francisco. Apple's new machine was a beautiful thing to behold, with a built-in keyboard and a professionally designed beige-colored case. It was comparatively cheap, in that a configuration with 16 kilobytes of RAM and no monitor cost just $1,195. It was easily expandable, with plenty of empty slots for add-on cards. It was the focus of an imaginative ad campaign masterminded by the Palo Alto firm of Regis McKenna (an early ad called it "The home computer that's ready to work, play, and grow with you"). And perhaps most important of all, it was great for playing video games; Wozniak, the technical wizard of the team and a video-game addict himself, had designed it with precisely that use in mind.

Of course, the Apple II had plenty of competition in the consumer market, notably from the Commodore PET, which debuted at the same West Coast Computer Faire, and from the Tandy–Radio Shack TRS-80, which was introduced the following August. In the beginning, moreover, the promises made in the Regis McKenna ad copy–"You'll be able to organize, index and store data on household finances, income taxes, recipes, your biorhythms, balance your checking account, even control your home environment"–were little more than fantasy; the applications software that would work such magic didn't exist yet. Nonetheless, the Apple II was an instant hit. By decade's end Apple itself had become one of the fastest-growing companies in American history. By the time it went public, at the end of 1980–just six years after that *Popular Electronics* cover story on the Altair–the company had sold more than 130,000 Apple IIs and proved to have a market value of $1.2 billion.

On the software side, meanwhile, one of the seminal events had come in April 1976, when *Dr. Dobbs Journal* informed its readers that they could now equip their micros with a real, grown-up operating system for just seventy-five dollars. Known as CP/M–for Control Program for Microcomputers–it had been created by thirty-three-year-old Gary Kildall, a computer-science teacher at the Naval Postgraduate School in Monterey, California. It was quite similar to DECSYSTEM 10, DEC's original software for the PDP-10–indeed, Kildall had carried over many DECSYSTEM features without change, including the use of letters to specify particular disk drives, the use of file names containing a period and a three-letter extension, and the use of commands such as "DIR" to get the contents of a directory. But much more important for the hobbyists was that CP/M could handle drives for "floppy" disks, an IBM invention of the early 1970s that was now emerging as a relatively inexpensive mass-storage medium for the microcomputer world.

Sales were predictably brisk. Within a few months Kildall had resigned from his teaching job and, together with his wife, founded a company called Digital Research to market CP/M. And by the following year, 1977, with more than a hundred companies wanting to license CP/M for their new machines, Kildall was hurriedly rewriting the system in a way that he hoped would keep him sane. His basic idea was to collect all the code that had to be customized for each new computer or disk drive and put it into a small Unix-like kernel that he called the Basic Input/Output System, or BIOS. Getting that right for a given machine would then be the responsibility of the licensee, not Kildall. And once that was done, the rest of CP/M would run without change.

Now, if this all sounds familiar, no wonder: when Bill Gates and his crew at Microsoft were later asked to write a "Disk Operating System" for the new IBM PC, they responded with a clone—MS-DOS—that was just barely different enough from CP/M to avoid legal action. With IBM's imprimatur, moreover, MS-DOS would soon dominate the market, leaving CP/M to play a long and ultimately futile game of catch-up. In 1977, however, the crucial thing was that Kildall had established a clean separation between a microcomputer's hardware and software. In effect, he was proposing a new open-interface standard that would allow hardware and software to evolve independently. In the process, moreover, he had cleared the way for CP/M to become a more or less standard operating system for the non-Apple micros, of which there were many. (Ironically, one of Microsoft's best-selling products in this era was an add-on card that would allow CP/M to run on the Apple II as well.)

The result was the first full flowering of applications software, the shrink-wrapped packages that would become so familiar in the 1980s. By the time Lick's "Multinet" article came out, in 1979, the software industry had already taken on a recognizably modern form. Games were still the most popular choice, of course, but educational software had a large market, too. And business users were beginning to find office-oriented products such as the WordStar word-processing package, the dBase database program, and the first of the must-have killer apps, VisiCalc, an electronic-spreadsheet program for the Apple II that automated the tedious chores of bookkeeping and financial analysis. Businesspeople were soon buying Apple IIs just so they could run the program; Apple later estimated that VisiCalc was responsible for the sale of some twenty-five thousand units in the package's first year alone.

Meanwhile, Big Blue itself was gearing up to enter the fray. Its IBM Personal Computer, based on Intel's new 8088 chip, would be among the first of the next-generation 16-bit micros. When it debuted in August 1981, moreover, the IBM PC immediately legitimized the microcomputer as a *really* serious business machine: retailers couldn't keep it on their shelves. Within weeks IBM was obliged to quadruple its planned production run. Within a year the IBM PC architecture had become a de facto industry standard, microcomputer sales as a whole were going exponential, and so were the fortunes of little Microsoft: Gates had not only provided IBM with its DOS operating system for the PC, but had cannily retained the rights for his company to sell that same operating system to other manufacturers.

By the end of 1982, microcomputers had become such a phenomenon that even the editors of *Time* took notice. On January 3, 1983, having passed over such milestones as the Falklands War, Israel's invasion of Lebanon, the death of Leonid Brezhnev, and the release of *E.T.,* they announced that *Time*'s Man of the Year for 1982–"the greatest influence for good or evil"–was not a human being at all but a machine: the computer. "In 1982," declared the magazine's cover story, "a cascade of computers beeped and blipped their way into the American office, the American school, the American home. The 'information revolution' that futurists have long predicted has arrived, bringing with it the promise of dramatic changes in the way people live and work, perhaps even in the way they think. America will never be the same."

Indeed she would not. Nor, for that matter, would the thousands of techno-geeks who suddenly and miraculously found themselves becoming rich young entrepreneurs. In the fall of 1979, to take just one example, motivated largely by a desire to stay together after they began to graduate, eight current and former students in the MIT Dynamic Modeling group had formed a company called Infocom. Lick and Al Vezza had joined as well, serving as advisers and investors while staying at MIT. Infocom's first product was a microcomputer version of *Zork,* an adventure-style role-playing game that the students had earlier created for fun on the group's PDP-10. It drew rave reviews–"Entertaining, eloquent, witty, and precisely written," said *Byte*–and was soon followed by *Zork II, Zork III, Starcross* (a science-fiction adventure that was packaged in its own flying saucer), *Deadline,* and a slew of other titles. By 1983 Infocom was racking up some $6 million in annual sales and had become a major player in the fast-growing gaming market. The firm had an elegant suite of offices at 55 Wheeler Street in Cambridge. And the former Dynamic Modelers had started work on their first "serious" software product, Cornerstone, a sophisticated database system that (they hoped) would boost them into the same orbit occupied by their Cambridge neighbor Lotus Development Corporation, creator of the megahit spreadsheet Lotus 1-2-3.*

Even Lick was pleased by Infocom's success, says Tracy Licklider, who himself worked at VisiCorp (the maker of VisiCalc) during this period and later served as president of the Boston Computer Society. It wasn't that his father ever cared about money per se, he says, but Lick was certainly happy for his former students' sake. And he was enthralled by the sight of his "self-motivating exhilaration" going mass-market at a speed and on a scale that he had never thought possible. Millions of people had grown up thinking of computers as big machines, mysterious machines, omniscient, cold, and relentless machines. But

---

* Unfortunately, Cornerstone would prove to be the beginning of the end for Infocom. The development process would be extravagantly expensive, not to mention a tremendous distraction from the firm's more lucrative gaming market. In February 1986 the financially troubled company would be acquired by a firm named Activision, which would proceed to run it into the ground. Infocom now exists only as a memory.

now those same people could walk into their local ComputerLand and see all these *little* machines. It was as if the computer had been reinvented before their eyes: the pitiless HAL 9000 of *2001: A Space Odyssey* had become the Computer for the Rest of Us, an instrument of individual empowerment. Indeed, the new micros were not only friendly but downright intimate. They were "mirrors of the mind," as the MIT sociologist Sherry Turkle put it in her 1984 book, *The Second Self.* "In talking to personal computer owners," she explained, "I found that for them the computer is important not just for what it does but for how it makes you feel. It is described as a machine that lets you see yourself differently, as in control, as 'smart enough to do science,' as more fully participant in the future."[11]

Lick couldn't have said it better himself. In fact, he seems to have found only one thing to complain about in the personal-computer revolution: the machines themselves. "He got an IBM PC," remembers Tracy Licklider, "but it never had the resources to do what he wanted." Yes, Lick knew, these talented little micros had been good enough to reinvent the computer in the public mind, which was no small thing. But so far, at least, they had shown people only the faintest hint of what was possible. Before his vision of a free and open information commons could be a reality, the computer would have to be reinvented several more times yet, becoming not just an instrument for individual empowerment but a communications device, an expressive medium, and, ultimately, a window into online cyberspace.

In short, the mass market would have to give the public something much closer to the system that had been created a decade before at Xerox PARC.

LOVING NOT WISELY, BUT TOO WELL

Certainly that was how the wizards out on Coyote Hill Road saw the situation—and why they found it so galling to watch the mass media gushing over the likes of Bill Gates and Steve Jobs.

"It had never occurred to us that people would buy crap," declares Alan Kay, who considered the hobbyists in their garages down the hill to be very bright and very creative ignoramuses—undisciplined kids who didn't read and didn't have a clue about what had already been done. They were successful only because their customers were just as unsophisticated. "What none of us was thinking was that there would be millions of people out there who would be perfectly happy with the McDonald's hamburger approach: they didn't know it wasn't real food."

The PARC-ites had known about the micros from the beginning, of course. They read the hobbyist magazines, too. They shopped at Radio Shack. And at least a few of them attended the twice-weekly meetings of the Homebrew Computer Club, held just a mile or so away, in an auditorium at the Stanford Linear Accelerator Center. Yet like Lick and most of the rest of the ARPA community, they found it hard to see the micros as anything more than toys. Imagine what it was like to go from the Alto's WYSIWYG text editing to an Apple II, says Jerry Elkind: "You had to hook it up to a television, and you got only forty characters per line, with no lowercase letters. It was terrible." "It was a different level of aspiration," agrees Chuck Thacker. "I did buy an Apple II and look at it early on. It turned out to be based on a 6502 microprocessor, which was the same chip we used as a controller in the Alto keyboard!"

PARC, they were confident, had something much more exciting to offer—*if* they could ever get that something out the door.

This was actually getting to be a matter of some urgency. In 1974 PARC's technology had been light-years beyond anything else in the world, but by the late 1970s that lead was dwindling fast. PARC hadn't exactly kept itself a secret, after all. The wizards had published papers, given seminars, recruited students, and proselytized at every opportunity. The lab had likewise kept its door open: just about everybody in the computer field had already visited the place, or knew somebody who was working there, or was angling for a way to work there himself. So PARC's ideas were *known* by the late 1970s.

As early as 1974, for example, MIT's AI Lab had started experimenting with personal machines that were specialized for running Lisp, but with graphics inspired by the bit-mapped display on the Alto. By 1977, hand-built models were in routine use at Tech Square. And by 1980, commercial versions were being produced under license by two new Cambridge-area start-ups: Symbolics and Lisp Machine, Inc. At Carnegie Mellon, likewise, a group that included long-time PARC consultant Allen Newell, PARC veteran Bob Sproull, and MIT AI Lab veteran Scott Fahlman had convinced the computer-science department to embrace a version of personal distributed computing that they called the Scientific Personal Integrated Computing Environment, or SPICE. "The era of time-sharing is ended," they declared in the department's 1979 invitation for vendors to bid on the SPICE computer. "New possibilities [include] high resolution color graphics, 1 million instructions per second, . . . 1 megabyte primary memory, 100 megabyte secondary memory [hard disk]. . . . We would expect that by the mid-1980s such systems could be priced around $10,000." They were right. And while the department's eventual choice in the competition would be a disappointment—the new PERQ computer from start-up Three Rivers, Inc., conceived as a next-generation derivative of the Alto, proved to contain a number of dead-end design features—the Carnegie Mellon effort was nonetheless a wakeup call to the community as a whole: high-end, graphics-based machines of the Alto class and above, a group that people were beginning to call workstations, might actually find a market.*

---

* In the spring of 1982, to take just one example, in the Silicon Valley town of Mountain View, California, a workstation company was founded by four twenty-seven-year-olds. Stanford University grad student Andreas Bechtolsheim contributed the initial product, an elegant but

Meanwhile, PARC's Ethernet was also moving out into the world—carried by Bob Metcalfe himself.

He had been one of the first of the major players to resign from PARC, Metcalfe explains. A certain restlessness had set in. So in 1975, with Ethernet pretty well completed and, sadly, a divorce in its final stages, he had abruptly taken a job with a subsidiary of Citibank in Los Angeles. Then, just seven months later, having accomplished what he was hired for, which was to modernize Citibank's electronic funds-transfer system, and having conceived a loathing for the bank's march-in-step corporate culture, he headed north again. By mid-1976 he was back in Palo Alto, working at Xerox's Systems Development Division. "[But] in the middle of nineteen seventy-eight I decided that I wanted to start my own company," he says, "because that's what you did in Silicon Valley. I didn't even know what I wanted to do. So I just left, and for five months I became a consultant. Then I got the call from Gordon Bell," the chief engineer for the Digital Equipment Corporation.

Bell, it developed, had persuaded DEC's leadership to adopt a long-term "VAX strategy," which called for that machine to replace all the company's earlier, incompatible product lines. There would be mainframe-size VAXes for enterprise-wide functions such as corporate accounting and E-mail distribution; minicomputer-size VAXes for group- or department-level functions such as typesetting and computer-aided design; and desktop-size VAXes for personal-computing needs such as word processing. Moreover, every one of these VAXes would run the same operating system and application software, just as IBM had promised more than a decade earlier with its System/360 computers. However, Bell emphasized, in this case there would be one critical difference. Whereas the System/360 mainframes had operated as stand-alone machines, he wanted the VAXes to be connected the way the Altos were at PARC, so they could function as nodes in an integrated, enterprise-wide work environment.

Thus his call to Metcalfe. "Basically," Metcalfe says, "Gordon wanted me to help DEC develop a local-area network—although it wasn't called that yet—that wouldn't fall afoul of Xerox's Ethernet patents." The idea was to create an open networking standard that would work with any brand of computer, on the theory that customers would balk if DEC tried to lock them into using its machines exclusively. In the long run, the company would do much better if it allowed everyone to play.

Hmm—tricky. Also dangerous: Metcalfe reflected that his former employer

---

comparatively cheap Alto-style workstation that he'd designed using off-the-shelf parts. Berkeley's Bill Joy contributed his world-class expertise in Unix. And Stanford MBA classmates Vinod Khosla and Scott McNealy brought their management acumen to the mix. Since Bechtolsheim had originally designed his workstation to run on the campus-wide Stanford University Network, moreover, the foursome decided to take the name of their firm from the network's initials—thus Sun Microsystems.

could afford a lot more lawyers than he could. "So instead," he says, "I thought, Why not just ask Xerox to cooperate?"

A few discreet inquiries showed that Xerox was game, as was DEC. And so was the semiconductor giant Intel, which was brought into the deal when Metcalfe realized that DEC would need a special chip to implement Ethernet on the VAX. By the summer of 1979, with Metcalfe's continuing to serve as a consultant, representatives of all three companies were hammering out a proposal to make Ethernet an official, open standard for local-area networking. ("Legally," notes Metcalfe, "for them to even talk together in the same room, it *had* to be an open standard.") And in May 1980 Gordon Bell stood up with representatives of Intel and Xerox to announce the plan publicly: for a onetime licensing fee of a thousand dollars, a company could manufacture as many Ethernet cards, cables, transceivers, and peripherals as it wanted. All it had to do was abide by the specifications, which were now published freely.

This was a highly satisfactory deal for Bell. DEC had an inside track on the emerging standard, which would be duly ratified in 1982. IBM was now eating DEC's dust, for a change, at least in this market: Big Blue's "Token Ring" system, which wouldn't be introduced until October 1985, would forever be a networking also-ran. And most important of all, the VAX had a network. Indeed, the VAX strategy would prove to be a masterstroke for DEC: by the mid-1980s, after some admittedly tumultuous years in the development phase, its integrated line of VAX computers would dominate the scientific and engineering markets, and be making real advances in the business market. The company's annual revenue from computers would be measured in the billions of dollars, second only to IBM's—and for a time there would be some who wondered if DEC might one day even *surpass* IBM.*

Finally, the Ethernet agreement was a very good deal for Metcalfe himself, whose post-PARC career had at last gotten a little focus. Early on in the negotiations he had realized that if Ethernet became a formal, open standard, then the technology might have a market—which meant it was *definitely* time to start that little company of his own. So in June 1979 he did exactly that, taking the new firm's name from the three words *Computer, Communication,* and *Compatibility—* in short, 3Com.

So there it was: the "personal" part of the personal-distributed-computing paradigm was diffusing outward in the form of workstations, while the "distributed" part was doing the same in the form of Ethernet. That left the Smalltalk graphical user interface, together with all the other Alto software—the embodiment of computing as a medium of expression and exploration. But this idea was making

---

* Alas, DEC's subsequent history would not be so happy, largely because of its weakness in small computers. It's true that CEO Ken Olsen did change his mind about microcomputers back in the summer of 1980, apparently because an attractive young woman reporter stung

its way into the world, too, thanks to a fateful show-and-tell at the very end of the 1970s.

By now, of course, Steve Jobs's visit to PARC in December of 1979 has long since passed into legend. "Apple's daylight raid," as one writer called it, has become one of the founding myths of the personal-computer revolution, the moment when the nimble young Jason snatched the Golden Fleece from the sleeping dragon. Indeed, as reporter Michael Hiltzik pointed out in his 1999 book about PARC, *Dealers of Lightning,* the mythology has accumulated to the point of burying the reality almost beyond recovery. "The collective memory of the Jobs visit and of its aftermath is so vivid that some former PARC scientists are no longer sure whether they were there themselves, or just heard about it later," as he put it.[15]

Nonetheless, Hiltzik's book includes perhaps the most careful and complete reconstruction of the event to date, and it makes a number of key points. First, Jobs and his crew didn't need a special presentation to learn about graphical user interfaces. That idea was already in the air by 1979, along with everything else PARC had done. Since October 1978, in fact, Apple had been working in secret on a next-generation machine called the Lisa, after Jobs's out-of-wedlock daughter. The 16-bit Lisa featured a bit-mapped graphics screen and, by December 1979, a windowing interface that could be controlled by a mouse. The project was chaotic and unfocused, to be sure, and the interface was clumsy. But it was there.

Second, the dragon wasn't asleep, or even hostile. Quite the opposite: Xerox had courted Apple shamelessly. Or more precisely, the Xerox Development Corporation had. XDC was yet another Xerox subsidiary, recently founded in West Hollywood to make venture-capital investments in promising young firms. Moreover, XDC's director, Abe Zarem, knew exactly what PARC had—and how ineptly its parent company was handling it. His plan was to license PARC's technology to one of these lean and hungry start-ups instead, and then let the kids commercialize it. So in the spring of 1979, when he learned that Apple was raising $7 million with a private stock sale to major institutional investors (the company hadn't gone public yet), Zarem wanted in, and badly.

---

his pride by asking why his company was so far behind in that market. But the multimillion-dollar catch-up effort that resulted proved to be poorly planned and poorly coordinated. By the time DEC's first microcomputers appeared, in mid-1983, they were widely perceived as late, overpriced, and disappointing. Within a few years, having lost some $900 million, DEC effectively abandoned the effort, and Olsen went back to bad-mouthing microcomputers as cheap toys.

By the early 1990s, however, as high-end PCs and workstations began to make serious inroads among former VAX users, the company was paying dearly for that attitude. In 1992 the DEC board of directors regretfully, respectfully, but firmly asked Ken Olsen to resign. But even after he did so, the company was never able to do better than hold its own, despite heroic efforts; the world was moving too fast, thanks to the Internet, and rivals such as Sun had too much of a head start. In January 1998 DEC agreed to be bought by the PC manufacturer Compaq. In effect, it no longer exists.

Third, the hero wasn't quite as nimble as the legends would have it. Initially, at least, Jobs greeted Zarem's overture with contempt, not an unusual response from him. What could a sclerotic bureaucracy like Xerox do that his own team couldn't do better and faster? Partly, this reaction was a function of his counter-cultural disdain for large corporations in general. (The scruffily bearded Jobs wore his T-shirts, jeans, and sandals like a badge of honor.) And partly it was attributable to his memory of Steve Wozniak's former employer, Hewlett-Packard, which had once rebuffed Woz's proposal for a microcomputer. But in any case, Jobs changed his mind after repeated urging by Apple engineer Jef Raskin, who had joined the company to help design the Apple II. Raskin had visited PARC, as it happened, and his friends there had shown him its wonders.

So on April 2, 1979, Jobs and his team met with the XDC people and struck a deal that could make sense only in the go-go world of Silicon Valley: Xerox would be allowed to invest $1.05 million in Apple's private stock sale, and in return it would allow Apple full access to PARC's technology. Jobs had only the vaguest idea of what that involved, evidently. But his engineers were specifically interested in the inner workings of Smalltalk, which they hoped would help them with the Lisa interface.

Fourth, the presentation wasn't a naive giveaway by the Smalltalk team. Granted, Alan Kay himself might have embraced the Apple group as saviors if he'd been there; certainly he had abandoned any hope of seeing Xerox bring PARC's technology to market. But Kay had recently gone off to Los Angeles on a long-promised sabbatical, partly to be with a new girlfriend and partly to "take organ lessons," and the interim team leader, Adele Goldberg, still harbored some hope that Xerox would come through. To her mind, this "alliance" with Apple was potential suicide for PARC. So when Jobs and his top engineers finally showed up for an afternoon visit in December 1979, the presentation was as minimal as Goldberg could make it. She and her Smalltalk colleagues gave Jobs and company the standard visitor's tour: the Alto, the mouse, Bravo word processing, some drawing programs—nothing that the whole world hadn't seen before. And afterward their guests departed, apparently satisfied.

Two days later, having realized almost at once that they'd been shortchanged, Jobs and his crew showed up in the PARC lobby with no advance warning. They wanted to see the good stuff, they said—*now*.

There ensued several hours of argument between the Smalltalk team and its bosses. But in the end, after a direct order, with Xerox headquarters' backing up XDC, a red-faced Goldberg did indeed show Jobs and his people the good stuff. This included education applications written by Goldberg herself, programming tools created by Larry Tesler, and animation tools cooked up by Diana Merry for combining pictures and text in a single document—all of it hitherto top-secret material that showed off Smalltalk's capabilities to the fullest. Just as Goldberg had feared, moreover, Jobs's engineers were asking *very* detailed questions; they seemed to have read everything the Smalltalk team had ever published.

The climactic moment came when Jobs pointed out that Smalltalk scrolled through text one full line at a time, which made its action rather jerky; he wondered aloud if it couldn't scroll continuously, one bit-mapped pixel at a time. Sure, said Dan Ingalls, Smalltalk's lead programmer. He opened a window, changed a few lines of the scrolling code, then closed the window again, and voilà! Through the modular magic of object-oriented programming, Smalltalk now scrolled continuously.

The visitors were blown away. "Why hasn't this company brought this to market?" Jobs famously shouted, waving his arms around while his engineers did their best to ignore him and focus on how the system worked. "What's going on here? I don't get it!"

By the time the Apple team finally departed, according to Jobs's later account, he was a "raving maniac": he had seen the future. According to legend, moreover, he immediately ordered that the Lisa be reconfigured to match the Alto display.

But in reality, Hiltzik concluded in his book, the visit didn't have all that much impact on the Lisa—or not directly, anyway. Apple's partnership with Xerox fell apart soon afterward, the victim of a culture clash that was just too extreme. Lisa's chief programmer, Bill Atkinson, had to re-create most of what he'd seen on his own.

Indirectly, though, the visit did give energy and focus to a project that badly needed both of those things. More important, it was an epiphany for Jobs and his whole team: from now on Apple would follow the gospel according to Smalltalk. "Lisa must be fun to use," declared a project design manifesto written a month or so after the show-and-tell. "Special attention must be paid to the friendliness of the user interaction and the subtleties that make using the Lisa rewarding and job-enriching."[16]

In short, PARC's lead was dwindling fast by the end of the 1970s. Still, things *were* looking up a bit. Slowly, painfully, but inexorably, the hegemony of the former Ford accountants was coming to an end at Xerox, and the company was at last beginning to move.

The turning point, ironically enough, had been the infamous Boca Raton meeting in November 1977. However futile PARC's "Futures Day" demonstration may have seemed at the time, Boca was also where the higher-ups had admitted publicly what by now was obvious to all: the strict bottom-line mentality was a disaster, even in the company's core business in copiers.

Xerox's newest offerings were widely considered unimaginative, unreliable, overpriced, and mediocre in terms of performance. Upstart Japanese firms were offering better machines for less, thereby taking over the low-end copier market. And IBM and Kodak were likewise coming on strong in the high-end market, introducing fast new machines of their own. A few months before, in fact, when Xerox president Archie McCardell had left to head up International Harvester, the board had decided that his successor would be David Kearns—not a former

Ford man but a former IBM marketing executive who seemed to have a much better feel for real technology, and a genuine determination to transform Xerox's corporate culture from the inside out. The company's top management had become remote, closed-minded, indecisive, and out of touch, Kearns had declared in his speech at Boca, where many of his listeners were hearing him for the first time. "We have to change," he added, "and fast."[17]

True, the company's transformation would be anything but "fast" by Silicon Valley standards. Moreover, the effort would keep the Xerox higher-ups tightly focused on copiers and Japan, not computers and PARC; the wizards in Palo Alto would still have a very hard time getting their undivided attention. But by the mid-1980s Xerox would indeed be a new company: slimmed down, decentralized, reinvigorated, armed with a whole new generation of advanced copiers, and fighting back the Japanese challenge on every front. And along the way, whether by accident or by design, the new regime would slowly begin to remove the obstacles that kept PARC's technology imprisoned in the lab.

Only a short time after the Boca Raton meeting, for example, Kearns and board chairman Peter McColough heeded the entreaties of their research chief, Jack Goldman, and approved an attack on one of the most formidable of those barriers: the incomprehension and mistrust of PARC's accomplishments within the rest of Xerox itself. Responsibility for this initiative went to PARC's Jerry Elkind, who would head up yet another new organization, called the Advanced Systems Division. (Elkind's erstwhile deputy, Bob Taylor, at last got a title to match what he had tacitly been all along: director of PARC's Computer Science Laboratory.)*
"The division's charter was to get some market experience with 'beta test' versions of the technology at customer sites," explains Elkind. And indeed, within a year or so of their start-up in January 1978, he and his colleagues had installed Altos, laser printers, and all the necessary Ethernet connections at a major test site in the Santa Clara offices of Xerox's copier sales force—as well as at the Atlantic Richfield Company, in the U.S. Senate, and in Jimmy Carter's White House. True, Elkind's group still had no brief to go beyond "beta test" marketing, a constraint they found increasingly irksome. (At one point in late 1978, they prepared a formal proposal to market the Alto for real, but Kearns, distracted by the Japanese challenge, turned it down.) Nonetheless, they proved that ordinary customers really would pay for what PARC had created. They obtained invaluable feedback on the software from a host of nontechnical, real-world users, which improved the system greatly. And along the way, by letting people directly experience what PARC had wrought, they began to enlist some strong allies inside Xerox. "Despite all the mis-

---

* Elkind's departure from CSL had not been a happy one. Certain top researchers there had simply refused to work with him anymore, declaring that his habit of injecting his own technical judgments into the discussion had alienated too many people to allow him to continue as leader. Elkind himself refuses to discuss the matter, but George Pake, for one, has always believed that the revolt was a coup secretly orchestrated by Taylor himself. Taylor swears he was not involved, insisting that the rebellion was as much of a surprise to him as to anyone

takes at the top, there was actually a lot of support for the Alto at the grass-roots level in the company," notes George Pake. "A lot of people could see that something important was happening. For example, the public-relations people thought it was great for our public image. They felt it built up interest in the company and helped in recruiting."

In 1979, meanwhile, an even tougher barrier crumbled when Office Systems chief Bob Potter followed his former boss, Archie McCardell, to International Harvester. Happily for PARC, Kearns's choice to replace Potter as head of the Dallas operation was the thirty-four-year-old Don Massaro, an energetic Silicon Valley veteran who definitely understood digital technology: he had been a co-founder of the computer disk-drive maker Shugart Associates, which Xerox had purchased in 1977. Brash, cocky, profane, and willing to take risks, Massaro drove his abruptly reenergized division hard, to the point where it would ultimately bring out seven new products in less than three years. And in November 1979, when David Liddle came to Dallas to show him what PARC had been up to, he was ready to listen.

Liddle (pronounced Li-DELL) was an equally hard-driving young PARC veteran who had taken over the northern California office of Steve Lukasik's old System Development Division several years earlier. Liddle was actually rather skeptical about slogging down to Dallas yet again, having made this same pitch to Bob Potter any number of times before, with no success. Then, too, his organization had been whipsawed through the Xeroxian labyrinth mercilessly— now part of this group, now part of that—and some of his best people were leaving in exasperation.

Still, this new guy was supposed to be a techie for a change, and Liddle had plenty to show him. As soon as he took the System Development Division job, Liddle had enticed Chuck Thacker and a number of others to follow him across Coyote Hill Road and start development on a prototype commercial Alto that they had code-named the Star. And now, despite some struggles with the hardware, that project was going quite well. Indeed, the Star software was a work of art. The system had a Smalltalk-like user interface, with windows, icons, and menus that could be manipulated by a mouse. It had a fully integrated suite of applications for everything a professional might need, including documents, business graphics, tables, personal databases, and electronic mail. And perhaps most important, considering the Star's intended, nontechnical audience, the software made the "computerness" of it all as invisible as possible. The icons pictured real-world objects such as file folders, in- and out-boxes, a wastepaper basket, and a telephone, the idea being to make the meaning of each icon obvious at a glance. Moreover, the applications were fixed, always loaded, and, through the magic of object-oriented programming, automatically associated with data files; in principle, at least, users could focus on what they were doing without

---

(and indeed, other accounts support that contention). Either way, the business left a sour taste in everyone's mouth.

having to know anything about such concepts as operating systems, applications, or even software. Meanwhile, of course, every machine would also have a built-in Ethernet connection, giving its user instant access to file servers, laser printers, and the like. The system would be a *system:* personal distributed computing in full.

Thirty minutes into Liddle's demonstration, Massaro told him he could turn off his machine: "I'm already convinced."[18] Or as he later told an interviewer, "I said, 'Fuck it! This is incredible technology, and we're going to bring it to the marketplace!' "[19]

They did. On April 27, 1981, at the National Computer Conference in Chicago, the Star was introduced to the world as the Xerox 8010 Information System. It was a sensation; many of the attendees had never even seen a bit-mapped screen, let alone one that measured seventeen inches diagonally. Demonstrations were given every hour, with people overflowing into the aisles to marvel at the graphical user interface—among them a contingent from Apple, who watched every demo and then retreated into a corner to discuss what they had seen.

And yet within a depressingly short time after this introduction, it became clear that all the brilliance wasn't going to be good enough. The Star wasn't a failure, exactly; its sales figures were reasonably good, especially in the Fortune 500 market it was targeted to. But it was shaping up to be a niche product at best, losing out to those ridiculous micros. For Chuck Thacker, that realization hit home just four months after the Star's introduction when he saw the IBM PC, in August 1981. "I knew we'd lost because the PC's price was right," he says. "Here was a reasonably capable machine that wouldn't take a substantial fraction of my month's salary." For Butler Lampson, acceptance didn't come until early 1983, when Apple finally came out with its Lisa: "It was clear by then that Xerox had blown it," he says. For Jerry Elkind, the moment of truth came a year or so after that, when Xerox was designing a major software upgrade for the Star, called ViewPoint. "I was heading up the planning for that product," he explains. "We made the decision that there had to be an option for it to run IBM PC software. We also decided to put a VisiCalc clone on that machine. And that was the admission that we were not going to be dominant in the marketplace."

Whenever the realization came, however, this was easily the saddest page in the whole PARC story. The young visionaries who had gathered there had participated in one of the great sagas of the information age, the creation of the first modern personal computer. And yet that saga had now become one of the great cautionary tales. Despite a decade's head start and more, history was passing them by. The business writers Douglas K. Smith and Robert C. Alexander said it all in the title of their 1988 account, *Fumbling the Future: How Xerox Invented, Then Ignored, the First Personal Computer.* Indeed, in the pop mythology of the personal-computer era, the story of PARC and the Alto has become an endlessly repeated classic: visionary genius stifled by leaden bureaucracy.

And yet many of the PARC veterans themselves now see the situation a bit differently. The fact is, they believe, *they* finally had a chance, and *they* blew it.

Remember, says Butler Lampson: despite all the incomprehension that head-quarters showed in the early days, and despite all the mistakes that the higher-ups undoubtedly made with the Star (perhaps the most notable being that the copier sales force was turned loose to sell the system without any real training in computers), "the fact is that headquarters had the vision for PARC in the begin-ning. And they sank a lot of money into the development organization." In the end, moreover, the higher-ups let Massaro, Liddle, and the research team at the System Development Division shape the Star as they saw fit. "They knew they didn't understand computers, but they knew they had a great research lab," says Lampson. "So they figured that the engineers would take care of it."

And that was the essential tragedy, he says: whereas headquarters understood marketing but not computers, the wizards out on Coyote Hill Road understood computers but not marketing. They had gone into the design process believing fervently in the PARC vision, in personal distributed computing as a whole, in-tegrated *system*. They knew that it far transcended anything else on the market. And they couldn't imagine that customers would want anything less. They'd also been given to understand that cost was no constraint: their target was the corporate world, the Fortune 500–scale customers that were used to shelling out hundreds of thousands of dollars for computer equipment.

The result was an unfettered exuberance that led the Star's designers into at least three critical errors. The first and most obvious was a rampant case of feature-itis: they loaded up the Star software with every neat thing they could think of, until it had grown to roughly a million lines of code and was at the ragged edge of what the 1981-vintage hardware could support. Indeed, that was the first thing users noticed: the Star was painfully, maddeningly *sloooow*.

A second and more fundamental error was the designers' decision to make the Star a closed system, meaning that all the hardware and all the software had to come from Xerox. Now, in fairness, this was how most major corporations had al-ways operated. "People thought that that was how you became a serious com-puter company," says Jerry Elkind. "Get the customers hooked, and they were yours for life." In the corporate market that the Star was aiming for, moreover, this was what the customers preferred: they were looking for "turnkey" installa-tions like IBM mainframes (or Xerox copiers), which would be fully functional as soon as they were installed and would forever after be serviced by the vendor.

However, if being closed meant that Xerox could lock in its customers, it also meant that it had to guess right about what those customers wanted; it had no way to tap into the creativity of the wider community. And in at least one in-stance, the Star's software designers spectacularly failed to guess right: the Star had no spreadsheet program. Nobody at PARC had ever thought to write one because nobody in a research lab ever needed one. What they *did* need was a good way to write technical papers—"so we got all caught up with WYSIWYG word processing and printing," says Elkind. It was precisely the inverse of priorities in the business world. In the executive suites of 1981, word processing was widely considered to be low-level stenographers' work, whereas a spreadsheet was just the kind of application a manager might use. And the result was another sale for Apple, not for Xerox.

The closed-system approach likewise left potential customers without an inexpensive, low-risk entry path. The Coyote Hill Road designers had implemented the Star system as a seamless whole, and that was the only way Xerox would sell it. So even leaving aside the fact that a single Star workstation cost more than $16,500, or about ten times as much as an Apple II, customers couldn't buy just a single Star to see what it could do. They had to buy Stars for everyone, along with all the Ethernet cabling, laser printers, and such that went with them; a minimal installation cost at least $100,000. And while there were many Fortune 500 executives who could easily afford that, they were understandably leery of betting their entire operation on a brand-new system, especially when they still weren't too sure what a personal computer *was*. From the users' perspective, it was far safer to wait and see—and in the meantime maybe buy a couple of Apple IIs to kick around.

Finally, the third and perhaps most distressing error was that the designers passed up several chances to do something simpler. "The option was open all through the nineteen-seventies," says Lampson. "We could have shipped something much like the IBM PC, and a short time later shipped something very much like the original Alto." Starting around 1979, in fact, when the Star development was going through a rough phase, senior hardware developer Bob Belleville had agitated for a fresh start: Let's forget about the Star platform, he said, and instead get the technology out there quickly by creating a 16-bit microcomputer based on the Motorola 68000 chip, or maybe an Intel 8086. He even built a prototype in the lab to show that it could be done. But the answer was no. "I was one of the people involved in deciding against it," admits Elkind, "largely because all the software would have had to be rewritten, and we didn't have the stomach for it." It would have been too painful, agrees Lampson: "We would have had to scale back too much. [Belleville's machine] wouldn't have supported the wonderful things you could do with the Star."

So there it was: with the best intentions in the world, the researchers on Coyote Hill Road created a product that was lovely to behold—and an also-ran in the marketplace. "We certainly had the idea that computers like the Star would be mass-market items within a decade," says Lampson. "But we just didn't pay attention to a business model of how to do it." And in the end, he says, "the problem wasn't a shortage of vision at headquarters. If anything, it was an excess of vision at PARC."

Fittingly enough, Larry Tesler was among the first to walk. In the spring of 1980, a few months after the famous visit, Steve Jobs tried to lease Smalltalk for the

Lisa. And when Xerox refused (the partnership was already dead by that point), Jobs simply offered to hire Tesler, a Smalltalk team member who had done much of the presentation for the visit. Tesler was receptive, having sensed something stirring in the hobbyist community almost from the beginning. Yes, the garage computers were ridiculously primitive, but people were *buying* them. And yes, the hobbyists tended to be incredibly naive about what had already been accomplished in computing. But if you actually went to the Homebrew Computer Club meetings—and Tesler was one of the few PARC-ites who regularly attended—you could feel the energy there. The world was full of bright people who didn't work for PARC, Tesler realized—people who weren't even part of the ARPA community. And they were learning fast. So he accepted Jobs's offer and went to Apple, where he would eventually become the company's chief scientist.

He was soon followed by Bob Belleville, who was still smarting over Xerox's decision not to pursue his 68000-based alternative to the Star. "Everything you've ever done in your life is terrible," Steve Jobs shouted at him over the telephone not long after the Star introduction, "so why don't you come work for me?"[20] Belleville went on to become chief engineer of Apple's Macintosh project, the Mac being a perky little micro that would feature a *very* PARC-like graphical user interface—and, on the inside, the Motorola 68000.

Even Mr. Dynabook himself came over. Alan Kay never returned to PARC after his one-year sabbatical; instead, he had Apple set him up in his own lab in Los Angeles, where he would happily study kids, computers, and learning for more than a decade.

In that same year of 1981, meanwhile, Charles Simonyi headed north to Microsoft, taking with him all his ideas on WYSIWYG word processing; in 1983 the Bravo editor he'd helped create for the Alto reemerged as Microsoft Word. The contrast with the older company was amazing, says Simonyi, who had interviewed for an executive job at Xerox headquarters just before his first visit to the still-tiny empire in Washington State: "It was like going into the graveyard or retirement home before going into the maternity ward. You could see that Microsoft could do things one hundred times faster, literally, I'm not kidding. Six years from that point, we overtook Xerox in market valuation."[21]

While Simonyi and others had been refining WYSISYG, as it happened, years of work on laser printing had been leading to a truly elegant idea: instead of representing a document as just a collection of bits, why not represent it as a program? The printer could then execute the program to generate the final image. This "page-description" language would contain commands to produce text in any font and size, of course, but it would also have instructions for generating lines, curves, textures, color, and the like anywhere on the page. A few lines of code could thus enable the creation of marvelously complex images. In 1980 this idea was embodied in the Interpress system, designed by Butler Lampson and Bob Sproull with the assistance of John Warnock, chief scientist at PARC's new Imaging Science Lab. Warnock and his boss at that lab, Charles Geschke, would leave PARC in 1982 to found a company called Adobe Systems. There the Interpress idea would be transmuted into a proprietary page-description language named Postscript, which would go on to become a cornerstone of the desktop-publishing industry.

Also in 1982, an exasperated Don Massaro would resign as head of Xerox's Office Systems Division in Dallas after a dispute with Kearns over his sales team. David Liddle would soon resign as well, and together they would found Metaphor Computer Systems, hoping to market applications for workstations—that is, personal-distributed-computer systems like the Star.

And so it would go. For a period in the early 1980s, colleagues seemed to be walking out the door almost daily. And the steady hemorrhage of talent only fueled Bob Taylor's rage. The Jobs visit had been infuriating enough, he says. He'd been out of town at the time, which was regrettable, "because if I'd been in town, I would have told him to get out. And if he hadn't, I would have beat the shit out of him. I had no respect for him. Then they [Xerox] would have fired me—and it would have been good for me and for them."

But far, far worse than that missed opportunity was the knowledge that his group could have changed the world—and that Xerox had let the moment slip by. Most of all, Taylor was convinced, *George Pake* had let it slip by. The director was their spokesman, says Taylor, their pipeline to all the rest of Xerox, their one real hope of conveying to the higher-ups what they had at PARC. And Pake screwed it up utterly, because he didn't understand it himself. No, insists Taylor, the best you could say about George Pake was that he was just . . . out of it.

Pake, for his part, does concede one kernel of truth in that indictment: he *was* out of it, literally. Back in 1978, he explains, in an effort to stop the endless bickering between research chief Jack Goldman and engineering chief Jim O'Neill, Xerox president David Kearns had briefly suggested putting Goldman and all the research centers under O'Neill's direct authority—PARC included. Since O'Neill's numbers-obsessed management would have snuffed out the labs' creativity in very short order, a horrified Goldman had volunteered to fall on his sword: he would let himself be named to the newly created post of "chief scientist," he told Kearns, while George Pake moved up to take his place. Kearns had bought it; Pake had been the corporate research chief since June 1978.

However, says Pake—and this was the out-of-it part—he had taken the corporate job only on condition that he could do it from California; he couldn't stand the thought of doing another tour at Xerox headquarters. "Well, the problem of being out here is that you don't know what the corporate politics are back east until it's too late," he says. So yes, despite an ungodly amount of time spent on airplanes, that probably did hamper his effectiveness somewhat.

As for Taylor and his following, though, Pake had his own opinion. "Fairly early on, it felt to me, and the other managers, that CSL was a cult. Bob thought that he understood where computing was going better than anybody else did; he'd come in with the attitude, 'I understand this, this is how it's going to be.' Plus, his management style was to cultivate a siege mentality; he viewed every interaction with another human being as a win-lose situation, and he was determined to be the winner. So it was almost impossible to deflect CSL from the attitude that 'We're right and you're wrong.' " Indeed, he feared that Taylor's "vision" had become a stultifying orthodoxy, crushing any creative impulse that didn't fit in with it. And even worse, from Pake's standpoint, was that Taylor was constantly squabbling with PARC's other section heads over money. "Taylor wanted their budgets transplanted to him," says Pake. "Well, I believed in diversity. I didn't think we should have one point of view. But it was becoming very divisive."

Pake's assessment is echoed by many others, both inside and outside PARC. "Bob Taylor is one of the most complex human beings on earth," says Bob Kahn, who came through PARC fairly often in those years in his capacity as IPTO director. "He saw the world in black-and-white terms: if you were with him, if you were going in a direction he thought worthwhile, then he supported you all the way. But if you weren't, then whatever you were doing was utterly worthless and contemptible. And if he thought you were actually against him, then he went in with a flame-thrower. No quarter asked or given." Taylor lived in a world of ones and zeros, agrees SSL member Danny Bobrow, who had done his best over the years to maintain good relations with Taylor's people. "When I played tennis with him, he'd play every shot as a killer. If you hit a drop shot, he'd get mad; you were supposed to smash it back."

The inevitable showdown came in late August 1983, when the physicist Bill Spencer, who had been named the director of PARC earlier that year, presented Taylor with a memo detailing a dozen ways in which he was to modify his behavior—along with a demand that he report to the director's office every Monday morning at nine to discuss his progress toward these goals. Taylor, grossly insulted, refused. A mediation effort by the office of David Kearns himself proved fruitless. On Monday, September 19, Taylor called a CSL meeting in the beanbag room, with Spencer in attendance. After an emotional farewell, he announced his resignation and walked out of the room, to be followed a few moments later by Chuck ("This is *bullshit!*") Thacker. A shaken Spencer was left to face the wrath of fifty people who had just lost the only leader many of them had ever known, along with their top hardware wizard ("You can fucking *resign!*" he was told by one anonymous shouter).

Spencer himself did not resign, but many others did. By January 1984 Taylor had arranged for DEC to launch a new DEC Systems Research Center in downtown Palo Alto, with himself as director. Many of PARC's finest soon joined him, a group that included Butler Lampson, Chuck Thacker, and thirteen others. PARC's computer-research capabilities were devastated and would take years to recover.

And the rest, as they say, was history. That magical, integrated vision of personal distributed computing did indeed make its way into the mass market, but only in pieces. And slowly. And with very little of the benefit going to Xerox.

True, when Apple's Lisa debuted, in 1983, it met with a fate not unlike the Star's: audiences loved its Smalltalk-inspired interface but were turned off by its pricey hardware and slow performance. Only one year later, however, Apple's Macintosh hit the sweet spot. It was compact, affordable, friendly, and even sort of . . . cute. The little machines flew out of the stores and developed a fanatically loyal following that persists to the present day. Then, in 1985, Microsoft, never a company to be left behind, started shipping its Smalltalk-inspired Windows 1.0 for IBM PC compatibles. Two massive revisions later, with the release of Windows 3.0, in 1990, Microsoft also hit the sweet spot, and famously began to make up for lost time. By the end of the 1990s, Windows would overwhelmingly dominate the operating-system market, to the point where the Justice Department's Antitrust Division would feel compelled to take an interest in Microsoft.

On the communications front, meanwhile, the proliferation of Macs and PC clones was very good news for networking start-ups such as 3Com, which had already begun to prosper by selling Ethernet products for VAXes and workstations. Indeed, the personal-computer revolution was what transformed the Bob Metcalfe of 1979 into the Bob Metcalfe of today: the proud possessor of a farm in Maine, a town house suitable for an embassy in Boston's tony Back Bay neighborhood, a prestigious second career in technology journalism, and a personal fortune that amounts to a noticeable fraction of Bill Gates's. "I tell people I didn't get rich inventing Ethernet," he says with a laugh. "I got rich selling it!"

Just so—though in truth, the microcomputer market was something of a tough sell in those early days. Partly it was a matter of technology, since the early micros were too feeble to handle network communications very effectively. But much of it was also cultural: back in the hobbyist days of micros, in the 1970s, the emphasis had always been on the individual user and the individual machine. Lots of people had modems, of course, but a permanent network connection, as with Ethernet, was exotic to the point of being incomprehensible. That was why Steve Jobs and his engineers had come away from their infamous visit to PARC having totally missed the significance of the Alto's Ethernet connection—as Jobs himself would later admit. Indeed, in those days Jobs absolutely despised the idea of networking. Like many of his contemporaries, he felt that tying your machine to someone else's meant giving up *your* autonomy—an unthinkable choice for any child of the Me Decade, and a violation of everything that personal computing seemed to stand for. (Legend has it that when Jobs was later asked why the Lisa didn't have a networking port, he threw a floppy disk at the questioner and snarled, "There's my fucking network.")

Such attitudes lingered, with the result's being that local-area networking didn't begin to make much headway in the microcomputer market until the micros *themselves* began to make a serious penetration into the business world—a process that might be called slaughter of the whitecoats, part two.

Back in the 1970s, it seems, computer users in corporate America had been facing much the same frustration faced by researchers such as Steve Wolff: the "whitecoats," or management information service (MIS) departments, dictated the terms. Big companies relied on mainframes, a few minis here and there, and, for "desktop" applications, lots of dedicated word-processing machines from vendors such as Wang and IBM. You played by the MIS rules or not at all.

Ah, but then the micros came along, the high-tech embodiment of individual autonomy, hands-on control, and *freedom*—not to mention the source of a spreadsheet application, VisiCalc, that you couldn't get any other way. Naturally, the MIS types tried to outlaw the things. But it was hopeless; the micros were cheap enough that a manager with any imagination could easily hide them in the petty-cash budget. Within a very few years, the steady trickle of surreptitious Apple IIs had become a flood of IBM PCs, PC clones, and Macs, and the MIS departments had effectively given up. In fairness, they did have a legitimate gripe. With the flood of micros had come a flood of new software, some of it very good—WordPerfect, dBase, and Lotus 1-2-3, for example—but much of it flaky and bug-ridden, and all of it incompatible with most companies existing applications. Too often, the resulting chaos wasted more of a company's time than it saved. But there was no way to keep personal computers out of the office, especially not when more and more employees were using them at home. So instead the MIS managers bowed to the inevitable and started linking the various PCs together into local-area networks, which let them at least get some of the data moved onto a central server where they could be shared. 3Com had helped jump-start the process in 1982 by being among the first to offer an Ethernet card for the new IBM PC. And the company had prospered accordingly throughout the 1980s and 1990s, as the office norm increasingly came to resemble what Bob Taylor and his crew had first imagined two decades earlier: individual computers, Ethernet networking, digital printing, WYSIWYG text editing and graphics programs, a windowing interface—personal distributed computing in full.

JUST DO IT

As a rule, says Al Vezza, retirement parties at MIT are low-key affairs. But not this one. On October 4, 1985, when he and Mike Dertouzos threw a retirement dinner for Professor J. C. R. Licklider—Lick had turned seventy that year—some three hundred people showed up and had a blast, with jokes, memories, and testimonials flowing nonstop. Bob Fano told the story of his train trip with Lick in November of 1962, when they had cooked up the whole idea of Project MAC on the fly. Bob Taylor talked about the Arpanet, and PARC, and Lick's hilarious struggles with fractured Greek when he tried to buy a Coke in Athens. Bob Kahn reminisced about Lick's second tour at ARPA. Mike Dertouzos cited his kindness and good advice. It went on and on, says Vezza; the acrimony that had

followed Lick's unhappy directorship of Project MAC had long since been forgotten. "Lick was accepted for who he was: an extremely important person in
the computer field."

That was news to Stuart Malone, who had been listening to the tributes with
growing astonishment as he waited to give his own testimonial, slated to be the
last of the evening. "When I was an undergraduate," says Malone, meaning the
previous two or three years, "Lick was just a nice guy in a corner office who gave
us all a wonderful chance to become involved with computers."

Their relationship had started in the summer of 1982, he explains. "Al Vezza,
Lick, and the older generation of hackers there felt that something was being lost
in MIT's electrical-engineering and computer-science curricula. Students didn't
have hands-on experience with computers anymore; the courses were all theoretical." At the same time, he says, MIT had launched Project Athena, a massive
effort to provide network connections all over campus, along with easy access to
high-end workstations for everyone. "But because of the central system management, and the high-level security to keep things under control—well, Athena
meant more computers, but paradoxically, it made computers more remote."

Thus the "Hacker Farm," a program that funded undergraduates to come and
get their hands dirty alongside the professionals in Tech Square. "There was a
core group of about six of us who were in the first round and who were actually
living in the same dorm," says Malone. "We wound up in the part of the Hacker
Farm that was located with the Dynamic Modeling group on the second floor,
right near Lick's office. Around our second year, in fact, when LCS got money
for a major remodeling of the second floor, Lick insisted that the undergraduates
get a space of our own *and* that we get to design it ourselves. So we chose a large
open area in the center of the building and did it up with green carpets and blue
walls, so that it would look as much as possible like the outdoors. We called it
the Meadow, after the 'Bloom County' comic strip. Lick's office was right across
from that.

"Meanwhile," Malone continues, "about the same time that Athena was
being set up, there was a parallel effort by DEC to experiment with personal
computing, to learn what it would be like to have one computer per person. So
they donated twenty VAX/750s. We helped uncrate them; they were about the
size of washing machines. Lick stretched the rules and got one assigned to the
undergraduates. The staffers and professors were naming their machines after
beers: Heineken, Molson, XX—Dos Equis—and so on. But since we were all
under the drinking age in Massachusetts, we named ours Grape Nehi. Lick later
got a MicroVAX and named it ClassiCola, which we felt was a vote of solidarity.
Also, since we were sharing a computer, we needed a group name so we could
share files. It had to be one word in lower case because we were using Unix, so
someone came up with lixkids.

"Anyway, as we started doing serious development on Grape Nehi, we started
running out of disk space. Lick had no hesitation: he pulled out a catalog and

bought the best Misubishi disk drive available. Also the fastest controller. The first we heard of it was the day he walked in and said, 'Here it is!'—and went home. Well, the thing weighed forty pounds. There was no case, just a raw drive. It was supposed to be rack-mounted, and we had no rack. The documentation was obviously translated from the Japanese. So we hauled it up to the ninth floor and got an old rack from somewhere. But the holes on the rack didn't line up with the holes on the drive. So we had to drill new holes. Then the controller was a different brand from the drive; we had to rejumper it with raw wires. At any time we could have fried the thing—we could have fried Grape Nehi! It was a miracle we didn't. I don't know how many days it was before we got it working. But we did, and Lick was impressed. He told us much later that he'd fully expected to have to hire an expert to install this thing. But he'd decided to let us try first because he knew it would be a learning experience. He figured that the money was better spent in training us to use the equipment, and to gain confidence, than in bringing in an outside expert. Lick considered that that was what we were there for."

The lixkids group stayed together until they all began to graduate, says Malone, which in his case was in 1984; since then he'd been working at Infocom. But none of that experience had prepared him for what he was now seeing at Lick's retirement party. "What amazed me was how many generations of people he influenced," says Malone. "There were people there from Hewlett-Packard, from DEC—from all over the place, all standing up and crediting Lick with giving them a chance to do their best work. It was a universal feeling."

That it was, agrees David Burmaster, who had been watching the proceedings with much the same feeling of astonishment. "I'd felt I was the only one," he says, thinking of his days as Lick's assistant during the Project MAC years. "That somehow Lick and I had had this mystical bond, and nobody else. Yet during that evening I saw that there were two hundred people in the room, three hundred people, and that all of them felt that way. Everybody Lick touched felt that he was their hero and that he had been an extraordinarily important person in their life."

It was a magical moment, adds Burmaster—and never more so than when Stuart Malone finally got his chance to speak. "They'd started with some of the older group, sort of the Bob Taylor types, and then worked down until the last speaker, an undergraduate [Malone] who stood up and said, 'I'm one of Lick's Kids!' Then the other kids in the room—there were about twenty—got up and cheered. It just knocked the socks off everybody. That's when we all realized what had drawn us to that evening and what we all felt."

Lick, for his part, was deeply touched by the testimonials, even as he enjoyed the party hugely. But he was rather abashed as well. Praise always embarrassed him. And besides, as many people there could attest, Lick had never been one to dwell on the past—not when he could have so much fun exploring the future. After his retirement, Lick still kept a lab at Tech Square (as did many other emeritus faculty members), and he still went in as often as he could to play around with graphical programming. He likewise made a point of attending the faculty lunches on Thursdays, where he would often have an extremely penetrating and/or very funny remark to make. And of course, Lick still let the undergraduates coming through the Hacker Farm program have the run of his office. "Those kids literally lived in those rooms," remembers Louise Licklider. "He made visitors from Washington sit in a little cubbyhole so that the lixkids wouldn't have to leave his office."

And yet for all the joy he took in his work, not even Lick could slow the calendar. "He'd lost energy," says Tracy Licklider. "I don't remember exactly when he was diagnosed with Parkinson's, but the tremors were becoming more obvious in the late nineteen-eighties. He'd had a mentor at Washington University who'd had Parkinson's and had become helpless. Dad had gone to see him a lot and was devastated by the experience." At the time Lick had been terrified that the same thing might one day happen to him—and now that fear had come true.

His asthma was growing worse as well. "He realized that time was running out," says his son. "And yet the generation of new ideas, of things that seemed worthwhile to work on, showed no signs of stopping. So it was very frustrating. He especially felt sad that he didn't have the energy to get involved with this new PC world. Clearly the marketplace momentum was unstoppable." Indeed, the computer was slowly, painfully, but unmistakably being reinvented yet again, this time as a window into cyberspace. "Dad looked at CompuServe, Prodigy, and all these bulletin-board sites as being a manifestation of the vision [of the Multinet], at least in some demonstration sense."

The question was whether Lick himself would live to see the Promised Land become anything more tangible than a blur on the distant horizon. But here the news was also heartening, if a bit disconcerting. Down in Washington, the leadership that Lick had once despaired of was finally beginning to emerge—though once again, it was coming from a direction he never would have predicted.

Actually, if anyone at Lick's retirement party had asked him for odds on the emergence of the Internet/Multinet/whatever as of October 1985, he probably would have said that the prospects looked as dismal as ever.

True, the Internet now existed in reality, not just on paper: the Kahn-Cerf internetworking protocols had become the official standard of the Defense Department in 1980, and the Arpanet itself had switched over to TCP/IP on January 1, 1983—an event that many would call the actual birth of the Internet. Moreover, Vint Cerf had arranged to have the protocols implemented in as many operating systems as possible, Unix most definitely included. But by 1985 the department had also lost its two strongest champions of networking: Cerf had departed for MCI several years earlier—among other things, to create a service called MCI Mail—and Bob Kahn himself was in the process of leaving to form his own consulting firm, the Corporation for National Research Initiatives. And in their absence, ARPA seemed to be cutting back sharply on network research. In 1983, the Reagan administration's new ARPA director, Robert Cooper, had taken Kahn's ideas for ramping up basic research on artificial intelligence and machine architectures and turned them into a massive, agency-wide "Strategic Computing Initiative" that would ultimately cost almost $1 billion. Every office in the agency had been given a piece of it (much to Kahn's disgust; that was one of the reasons he was leaving). And in the face of all that, networking research per se just wasn't a priority anymore; the Defense Department would continue to operate the Arpanet as a service to the ARPA research community, but that was it.

Outside the DoD, meanwhile, the on-line world was more fragmented than ever. Not only were the commercial vendors still pushing their own incompatible networking systems, but there was now a rival *inter*networking protocol as well: the Open Systems Interconnection (OSI) standard, developed by the International Standards Organization starting in 1977. The impetus had come mainly from that organization's European members, which found TCP/IP a little too messy and ad hoc for their tastes—and were loath to go along with a standard defined by the U.S. Department of Defense in any case. If nothing else, such a standard threatened to give American companies too much of a head start on commercial applications. It's true that the full OSI standard as published in 1984 existed only on paper. But it was nonetheless "official," having been blessed with ISO's seal of approval, and many people in the field believed that TCP/IP use would quickly die out as the new standard ramped up, the Defense Department's support notwithstanding. By 1985, in fact, the National Institute of Standards and Technology and several other federal agencies had bought that argument and formally adopted OSI as their internetworking standard.

Back on campus, the disarray was even worse. Inspired by the ARPA community's success with the Arpanet—and by the manifest power of E-mail—other research groups had spent the 1970s lobbying for networks built by their own funding agencies. Thus there was now one network for the researchers working in the Department of Energy's magnetic-fusion energy program, another for the same agency's high-energy physicists, and yet a third for the NASA-funded space physicists. At one point MIT had as many as sixteen different networks coming into campus, none of which could communicate with any of the others. Moreover, not only were these academic networks incompatible in a technical sense—in that some were based on DEC's networking technology, some on IBM's, and so on—but they were also typically restricted to users in one specific discipline. Even the one exception proved the rule: Bitnet, the "Because It's Time" network created by Ira Fuchs and Greydon Freeman at New York University in 1981, had as its goal to be open to everyone in the academic community, regardless of discipline—but it could connect only those sites that had IBM mainframes.

In short, networking was just as balkanized as, half a decade earlier, Lick had feared it would be, with many researchers still lacking network access of any kind. The only bright spot was that unexpected glimmer of leadership from the National Science Foundation.

Now, the irony here was that the hard-core computer mavens had never had much respect for the NSF. Compared to ARPA even in its "defense relevance" era, the foundation seemed to be hopelessly risk-averse and politicized. Unlike their ARPA counterparts, for example, NSF funding officers had to submit every proposal to "peer review" by a panel of working scientists, in a tedious decision-by-committee process that (allegedly) quashed anything but the most conventional ideas. Furthermore, the NSF had a reputation for spreading its funding around to small research groups all over the country, a practice that (again allegedly) kept Congress happy but most definitely made it harder to build up a Project MAC–style critical mass of talent in any one place. And worst of all, from the computerists' point of view, was that the NSF's support for computing had always been anemic at best, particularly in comparison with ARPA's. In the early 1970s, in fact, the agency had even stopped funding campus computer centers, which had left a lot of universities with aging machines and no money to upgrade them.

Still, the NSF *was* the only funding agency chartered to support the research community as a whole. Moreover, it had a long tradition of undertaking projects such as the Kitt Peak National Observatory and the U.S. Antarctic research bases—big infrastructure efforts that served large segments of the community in common. And on those rare occasions when the auspices were good, the right people were in place, and the planets were lined up just so, it was an agency where great things could happen.

Certainly that was the case with the Internet, which not only survived but went on to achieve its stunning growth in the 1990s largely thanks to NSF's efforts in the 1980s.

The impetus for the foundation's first foray into networking had come from computer scientists at the "have-not" universities, who were sick of being shut out of the Arpanet. By the late 1970s it was painfully obvious that the 'Net made it much easier for their colleagues at the "have" schools to share results and collaborate with one another, which in turn made it much easier for them to advance their research. So in 1979, with Larry Landweber of the University of Wisconsin and David Farber of the University of Delaware taking the lead, the have-nots started lobbying the NSF to finance some form of equal access to the Arpanet. And in January 1981 the agency duly launched CSnet, the Computer Science Network. "This was actually a seminal program," says Steve Wolff, who would later become the NSF's networking chief. "Not only did CSnet open up the Arpanet to the entire computer-science community, but it was the first instance of shared use by anyone." In the process, moreover, the TCP/IP protocol—for that was what CSnet used—began to expand beyond its ghetto in the Defense Department.

More significantly, CSnet also set the stage for the NSF's next networking initiative. The impetus in this case came specifically from physicists, who were increasingly finding themselves faced with problems such as condensed-matter phase transitions, black-hole astrophysics, and quantum chromodynamics (the behavior of quarks inside a proton or neutron)—ultramessy situations in which pencil-and-paper analyses were hopeless. Massive computer simulations seemed to be the only way for researchers to make any progress at all. And yet because state-of-the art supercomputers such as the Cray were far too expensive for most universities, scientists who wanted to pursue these kinds of issues were being forced to beg snippets of time from national laboratories such as Los Alamos and Lawrence Livermore, for which the Department of Energy was buying supercomputers by the dozens for nuclear-weapons development. It was humiliating, not to mention incredibly inefficient. So in the early 1980s, led by Larry Smarr of the University of Illinois and Nobel laureate Kenneth Wilson of Cornell, the physicists started agitating for the NSF to set up a system of national supercomputer centers that would function like the national observatories—that is, individual scientists would be awarded blocks of computer time on the machines in much the same way that individual astronomers were awarded blocks of observing time on the Kitt Peak telescopes. In 1985 the NSF duly launched five such supercomputer centers, at Illinois, Cornell, Princeton, Carnegie Mellon, and the University of California at San Diego. And with those national centers came a national network.

The planners had penciled it in from the beginning, explains Steve Wolff: after all, why make the users travel hundreds or thousands of miles just to submit their programs when they could do it far more quickly and cheaply on-line? But then as soon as anyone thought about it, he says, "the idea of using the network *only* to contact the supercomputer centers lasted about fifteen microseconds. It was instantly apparent that the network could be a tool for general scientific communication. So the NSF network from the beginning was designed for access by the general community of scholars. Certainly it would include everyone at the universities. And around the periphery, it would try to include K-through-Twelve education, museums—anything having to do with science education."

True, says Wolff, the creation of such a network exceeded the foundation's mandate by several light-years. But since nobody actually said no—well, they just did it.

Next, he says, in the process of implementing the scheme, the foundation's new Networking Office made two critical decisions. The first came from the office's founder, Dennis Jennings, who ruled in 1985 that the NSF system would be based on TCP/IP. The federal agencies that had already adopted the OSI "standard" had pressured Jennings hard to go that route as well, even though OSI hadn't yet been implemented anywhere. "But a lot of us in the TCP/IP community were jawboning, too," says Wolff, who was then still at Aberdeen. "We were successful because we could demonstrate that our specification was open, that it had been implemented by any number of manufacturers, and that it was compatible with the most popular operating system of the time, Unix." (Indeed, TCP/IP was now incorporated into virtually all the commercial operating systems, even though IBM, DEC, and the other vendors had had to grit their corporate teeth to make it so; they still wanted to push their own proprietary networks, but the Defense Department was just too important a customer to ignore.) Besides, TCP/IP meant instant access to the Arpanet-CSnet combination, which was where all the computer scientists were already; it would just be a matter of linking through a standard gateway.

So anyway, says Wolff, NSFnet became operational in 1986, the same year he replaced Jennings as head of the Networking Office. (Like the Arpanet, NSFnet ran over high-capacity transmission lines leased from the telephone companies.) And the immediate result was an explosion of on-campus networking, with growth rates that were like nothing anyone had imagined. Across the country, those colleges, universities, and research laboratories that hadn't already taken the plunge began to install local-area networks on a massive scale, almost all of them compatible with (and having connections to) the long-distance NSFnet. And for the first time, large numbers of researchers outside the computer-science departments began to experience the addictive joys of electronic mail, remote file transfer, and data access in general.

"When the original NSFnet went up, in nineteen eighty-six, it could carry fifty-six kilobits per second, like Arpanet," says Wolff. "By nineteen eighty-seven it had collapsed from congestion." A scant two years earlier there had been maybe a thousand host computers on the whole Internet, Arpanet included. Now the number was more like ten thousand, and climbing rapidly. With the packets mired in gridlock and users complaining loudly, Wolff and his colleagues hurriedly laid plans to boost their network's capacity by a factor of 30, to 1.5 megabits per second. (In AT&T parlance it would be a T1 line.) Again, they didn't really have the authority; they just did it. The upgraded service took effect on July 1, 1988—and usage exploded again, soaring from seventy-five million packets a day toward a billion packets a day. In 1991 Wolff and company would have to boost the NSFnet's capacity by yet another factor of 30, to "T3" speeds of 45 megabits per second.

In the midst of all this, meanwhile, NSF was getting penciled in for an even wider role, courtesy of the junior senator from Tennessee. As chairman of the Senate Subcommittee on Science, Technology, and Space, Albert Gore, Jr., was widely acknowledged as one of the very few American politicians who took the time to understand technology; he had even been known to show up at meetings of the National Academy of Sciences from time to time, just to listen to the technical sessions. He had been an enthusiastic supporter of NSF's supercomputer centers, and now he was even more intrigued by this notion of digital networks as a continent-spanning infrastructure. (He'd been mulling over that notion for some time, apparently: earlier in the decade he had coined the phrase "Information Superhighway," as an analogy to the Interstate Highway System that his father had helped create back in the 1950s, when *he* was in the Senate.) So in 1986 Gore wrote legislation asking the administration to study the possibility of networking the supercomputer centers with fiber optics.

The request couldn't have come at a better time. The NSF was just then consolidating all its computer-related efforts, including Wolff's Networking Office and the supercomputer centers, into a brand-new directorate headed by one Gordon Bell, the former chief engineer of DEC. And Bell, as hyperkinetic as ever, saw this as a priceless opportunity to promote his vision of the Internet as the VAX strategy writ large: a "fractal" network of networks that would be self-similar from the scale of the planet down to the scale of a microchip. ARPA officials, for their part, likewise saw Gore's request as a priceless opportunity to advance their own Strategic Computing Initiative. Then the Energy Department and NASA saw—well, anyway, by the time the dust had settled in late 1987, Gore found himself looking at a thick, multiagency report advocating a government-wide assault on computer technology across the board. First, said the report, Congress should fund a multiyear, billion-dollar initiative in high-performance computing, with the goal of creating machines in the 1990s whose performance would be measured in teraflops, or trillions of operations per second. When it came to "grand challenges" such as modeling the dynamics of protein molecules or global climate changes, nothing less would do. Second, said the report, Congress should fund an equally ambitious National Research and Education Network, with the goal of moving packets around the country in the 1990s at speeds measured in gigabits, or billions of bits per second.

Gore was dumbfounded. All he'd asked for was a short study on the use of fiber optics to connect the supercomputer centers. This was so much . . . *more*. Then, too, he was a tad distracted at the time by his first, abortive run for the 1988 Democratic presidential nomination. Nonetheless, after he held hearings on the report in August 1988, and heard spokesmen such as Bob Kahn, Len Kleinrock, and many others make the case, he was a convert. Before the year was out he had introduced a bill to fund both initiatives fully. With respect to networking, in particular, the Gore bill gave ARPA the leading role in developing the next-generation gigabit technology, while it put NSF in charge of deploying and maintaining a network based on the best current-generation technology. That is, the NSF would keep on doing exactly what it was doing already, except that now it would be doing it on behalf of the entire federal government and the entire university system.

Very quickly, it's true, the Gore bill ran into flak from the White House, where Republican budget officials argued that such initiatives ought to be left to the private sector. The high-performance-computing half of the proposal wouldn't be passed and signed into law until December 1991. And the networking half wouldn't be added until early 1993, when a newly inaugurated President Bill Clinton would sign a *second* Gore bill turning the effort into the High Performance Computing and Communication program, or HPCC. (By that point, of course, Gore himself had gone on to higher office.) But then, the delay hard mattered: down at the working level, the agencies had quietly been setting up a de facto national network on their own.

"We just found it more useful to cooperate than to fight," explains Chuck Brownstein, who was Gordon Bell's deputy at the NSF computer directorate. Nobody could afford it otherwise. "The networking part of our programs kept getting bigger and bigger," he says. "So finally the program people from ARPA, NSF, NASA, and the Energy Department—the Steve Wolff types—got together and formed a shadow organization: the Federal Research Internet Coordinating Committee." The FRICC would later be reconstituted more officially as the Federal Networking Council, but in the mid-1980s it was a wonderfully ad hoc affair, reminiscent of the beer-and-pretzel get-togethers that Lick had organized back in his first ARPA term. "This FRICC group was like a virtual agency: they didn't agree on everything, but if they came up with a good idea, they would go out and find the money in somebody's budget to do it. So we started doing things like sharing the cost of overseas lines, or putting in exchanges to intercon-nect our networks." And then when the first Gore bill kept falling off the table, he says, the networking chiefs just kept on coordinating their activities and draft-ing their budgets as if the program were already official. After all, they were the ones who had mostly written the report that had gone to Gore in the first place.

Along the way the agencies had also begun to consolidate their various net-works around TCP/IP. Why should each agency maintain a specialized system for its own researchers when the NSFnet was open to everybody? Indeed, in one of the more poignant ironies of the story, the offspring quickly displaced its par-ent. Recognizing that the megabit speeds of the upgraded NSFnet had turned the 56-kilobit Arpanet into a dinosaur, ARPA officials started systematically de-commissioning their venerable IMPs and transferring their users to the faster network. By 1990 the Arpanet was history.

The upshot, says Wolff, was that by the beginning of the 1990s the de facto national research network was in place, and TCP/IP ruled: the Internet had ex-panded from its original base in the Defense Department to the research com-munity as a whole. But of course, the expansion hardly stopped there; Wolff was already taking steps to extend the Internet into the commercial arena as well.

"I pushed the Internet as hard as I did because I thought it was capable of be-coming a vital part of the social fabric of the country—and the world," he says. "But it was also clear to me that having the government provide the network in-definitely wasn't going to fly. A network isn't something you can just *buy;* it's a long-term, continuing expense. And government doesn't do that well. Govern-ment runs by fad and fashion. Sooner or later funding for NSFnet was going to dry up, just as funding for the Arpanet was drying up. So from the time I got here, I grappled with how to get the network out of the government and instead make it part of the telecommunications business."

Unfortunately, Wolff says, the telecommunications companies themselves ill had minimal interest in the Internet: from their point of view, every E-mail t got sent was a telephone call that didn't get made. "As late as nineteen eighty-nine or 'ninety, I had people from AT&T come in to me and say—apologetically—'Steve, we've done the business plan, and we just can't see us making any money.' "

In retrospect, at least, Wolff's solution to this conundrum had a subtlety that seems almost Machiavellian: in what would prove to be the *second* critical choice made in implementing the NSFnet, he decentralized its actual management.

"Starting with the inauguration of the NSFnet program in nineteen eighty-five," he explains, "we had the hope that it would grow to include every college and university in the country. But the notion of trying to administer a three-thousand-node network from Washington—well, there wasn't that much hubris inside the Beltway. So we had to go to a hierarchical structure. We invited groups of universities, research laboratories, companies, and so forth to band together into consortia for regional networks."

The basic idea was to create a three-tiered structure similar to that of the telephone system. At the lowest level would be individual campus-scale networks operated by research laboratories, colleges, and universities, analogous to private branch telephone exchanges run by individual firms. At the middle level would be the new regional networks connecting the local networks, analogous to the telephone systems' regional "Baby Bells." And at the highest level would be a high-capacity nationwide "backbone" operated directly by the NSF, analogous to the long-distance telephone lines.

Now, this scheme didn't get fully implemented until they were planning the 1988 upgrade, says Wolff. But in due course, with NSF's providing the seed money, the regional networks took shape, all operating as not-for-profit Internet access providers to the research community. However, he says, "we told them, 'You guys will eventually have to go out and find other customers. We don't have enough money to support the regionals forever.' So they did—*commercial* customers. We tried to implement an NSF Acceptable Use Policy to ensure that the regionals kept their books straight and to make sure that the taxpayers weren't directly subsidizing commercial activities. But out of necessity, we forced the regionals to become general-purpose network providers." Or as they would soon be called, Internet Service Providers.

Vint Cerf, for one, is still lost in admiration for this little gambit. "Brilliant," he calls it. "The creation of those regional nets and the requirement that they become self-funding was the key to the evolution of the current Internet."

Certainly the gambit was effective. While it's true that many of the regionals themselves faltered rather badly in the marketplace, some of them did quite well. New York's NYSERnet, for example, spun off its operational activities into a very successful commercial subsidiary, Performance Systems International (PSInet) in Reston, Virginia. And in any case, there were any number of independent access providers springing up as well. "So by nineteen-ninety," says Wolff, "we had all these young but extremely vigorous commercial operators who had no commitment to academia. They would just sell Internet access to business, industry, individuals, whoever would buy it."

Indeed, this new wave of sellers and buyers soon became something of a self-exciting system. On the supply side, there were national carriers offering business-class service in the major cities, small-town PC retailers offering dial-in access to local customers as a sideline, and everything in between: entry into the market was comparatively cheap and easy by the early 1990s, if only because all the necessary Internet hardware was now available off the shelf. Routers, connectors, cables, modems—entrepreneurial firms such as Cisco Systems had been churning them out ever since the Defense Department first embraced TCP/IP, in the early 1980s. And then with the coming of the NSFnet in 1986, their market had been virtually guaranteed. The first Interop trade show attracted five thousand engineers, who browsed through exhibits from fifty companies—and that was in September 1988, when there were "only" some sixty thousand host computers on the Internet. By 1991, when the NSF did its second backbone upgrade, the Internet was connecting ten times that many host computers, and the provider business was booming.

It was also becoming obstreperous. During the NSFnet upgrade of 1991, most notably, a group of commercial access providers led by PSI alleged that the upgrade was being managed in a manner that was anticompetitive. The networking industry had matured to the point where NSF (or the regional networks) could simply buy long-distance backbone services on the open market. And yet, they said, Wolff and company had arbitrarily awarded the backbone contract to a single vendor known as ANS—a consortium of IBM, MCI, and Merit, a Michigan firm that had run the earlier backbone. "They called it a partnership," fumed PSI president William Schrader. "I called it a conspiracy."

Back at headquarters, Wolff enjoyed this about as much as any parent enjoys listening to a rebellious teenager. Nonetheless, he insists that he was delighted with the outcome: after some scathing congressional hearings and a two-year investigation by the NSF Inspector General (which found no wrongdoing), his office was "forced" to do what he'd always been aiming toward anyway, which was to push the Internet into the private sector. In 1992 Congress passed a bill that formally allowed for-profit access providers to use the NSFnet backbone. And on April 30, 1995, the NSFnet ceased to exist: Wolff's office would henceforth take the backbone's $12 million annual operating cost and start distributing it to the regional networks to buy long-distance service on the open market from vendors such as MCI. (That funding, in turn, would decline to zero over the next four years.) The foundation would continue to operate a much smaller, very-high-speed system connecting its supercomputer centers at 155 megabits per second. But beyond that, the Internet was at last on its own, and self-sustaining.

On the demand side, meanwhile, all these new, nonacademic customers were discovering that entry into the Internet was considerably less intimidating than it had once been. Thanks to the Usenet newsgroups, for example, as well as the ListServ mailing lists that had originated on Bitnet, the sense of community was stronger than ever: you could always find someone on the Internet who shared your interests. Then, too, E-mail had been standardized and greatly streamlined, first in January 1983, when ARPA adopted the simple mail-transfer protocol (SMTP), and then in January 1986, when a summit of network representatives sorted out the ad hoc chaos of Internet addressing. (They had to do something, said one participant: every host computer on the Internet needed a name, and everybody wanted to claim "Frodo.") The solution was DNS, the now-famous domain-name system that assigns each host computer a sequence of names separated by periods, as in lcs.mit.edu.

New users could likewise find a variety of new tools for navigating the 'Net. In 1990, for instance, programmers at the University of Minnesota created a popular utility known as Gopher, which allowed a user to search for information on the Internet via a series of hierarchical menus. Shortly thereafter, another group at Thinking Machines, Inc., came up with WAIS, the Wide Area Information Service, which enabled users to search for Internet files based on their content. And then around Christmastime 1990, at CERN, the European Center for Particle Physics in Geneva, Switzerland, an English physicist named Tim Berners-Lee finished the initial coding of a system in which Internet files could be linked via hypertext.

Actually, Berners-Lee had already been playing with the idea of hypertext for a full decade by that point, having independently reinvented the idea long before he ever heard of Vannevar Bush, Doug Engelbart, Ted Nelson, or, for that matter, the Internet itself; his first implementation, in 1980, had been a kind of free-form database that simply linked files within a single computer. But having a program follow hyperlinks across the network was an obvious extension of the idea, especially after CERN joined the Internet in the late 1980s. Thus the 1990 implementation, which also included Berners-Lee's notion of "browsing": the program had a word-processor-like interface that displayed the links in a file as underlined text; the user would just click on a link with the mouse, and the program would automatically make the leap, display whatever files or pictures it found at the other end, and then be ready to leap again.

In retrospect, of course, Berners-Lee's combination of hypertext and browsing would prove to be one of those brilliantly simple ideas that change everything. By giving users something to *see,* it set off a subtle but powerful shift in the psychology of the Internet. No longer would it be just an abstract communication channel, like the telephone or the TV; instead, it would become a *place,* an almost tangible reality that you could enter into, explore, and even share with the other people you found there. It would become the agora, the electronic commons, the information infrastructure, cyberspace. Because of Berners-Lee's hypertext browsing, users would finally begin to *get* it about the Internet. And they would want more. . . .

But then, that was still in the future. During the Christmas season of 1990, Berners-Lee was concerned mainly with getting the program finished, not to mention the impending birth of his first child. He was also acutely aware that this initial version of the program would run only on a NeXT computer—his personal favorite, but certainly not a brand that was in widespread use. It would be the better part of a year before others began to use his system in significant numbers, and even longer before browsers became available for Unix machines, PCs, and Macs.* But at least he had found a decent name for the thing—a process that had proved to be surprisingly tricky. What he wanted was something that conveyed his intent to create a new kind of global data structure. But Mesh, or Information Mesh, sounded too much like "mess." Mine Of Information had the unfortunate acronym MOI, French for "me," which he felt was too egocentric. And The Information Mine turned into TIM, which was even worse.

So in the end Berners-Lee settled for a name that was still less than ideal, since its acronym was nine syllables long when spoken. Nonetheless, it did seem to say what he wanted it to say, so he went with it: World Wide Web.

LUCK

"I think he knew something was going wrong with him all that day," says Louise Licklider, thinking back to the beginning of June 1990.

"We were driving back from a vacation trip through the Midwest," she explains. "He loved to drive. But we both knew that this would be our last vacation. Robin [still her pet name for her husband] had asthma, which was rapidly getting worse. He had Parkinson's that was rapidly getting worse. And now he had prostate cancer that had metastasized. So on the trip coming back, I had been steeling myself for rapid changes in our life-styles."

On the last leg toward home, Lick seemed to be in a nostalgic, sentimental mood, as if he were very far away. "It was unusual for him to be that far back in

---

* Writing a new browser wasn't especially hard, once you knew how. And by 1992 or so, as the Web became more popular, computer-science grad students all over the world were doing it as a way of exercising their programming skills. Take Marc Andreessen, for example, a grad student at the University of Illinois's National Center for Supercomputing Applications, one of the NSF's supercomputer centers. Along with staff member Eric Bina, he created a browser for X-Windows, the Unix graphical interface. Mosaic, as they called it, wasn't really any better than anybody else's browser, at least at first. But Andreessen, far more than any other of his contemporaries, listened to the users. Fueled by espresso, he haunted the Internet newsgroups incessantly, trying to figure out what people wanted in a browser and then incorporating new features practically overnight. When Mosaic was finally made available to the public, in January 1993—free for the downloading, in fact, since NCSA was a government-funded operation and not allowed to charge a fee—its combination of high quality and zero cost quickly made it the browser of choice. In December 1993, Andreessen left NCSA and formed a partnership with Silicon Graphics founder Jim Clark; their company soon became known as Netscape Communications Corporation.

the past," she says. "It was different enough that it caught my attention. Then when we got to the end of the Massachusetts Turnpike, there was an inn out in the countryside where we often liked to stop. So he said, 'Let's go in for a martini.' We did, and we sat there on the porch for a while. Finally, he said, 'What do you say we just sit here all afternoon?' It was a beautiful day, and he didn't want to leave. But I said, 'Oh, come on, Robin! You need to get home and rest.' So we went home. And all the way in, he hummed and sang all the old love songs that were popular back when we were dating. So looking back on it later, I had to feel that he had some premonition that something was going on in his body.

"When we got home," Louise continues, "he went upstairs. I was picking up the mail when suddenly he called down to me very urgently, 'Louise, bring me a pen!' "

Louise?

"I knew something was wrong because he never called me Louise—always Sugar, or Shuggie. So I grabbed a pen out of the kitchen—one of those silly little things you keep with a notepad—and ran upstairs."

Linda Licklider Smith, who lived with her own family just a block away, can vividly remember being awakened by her mother's frantic phone call, throwing on some clothes, and then running down the dark street to her parents' house. "The night was beautiful, and the silence of the street was deafening," she says. "I immediately went upstairs to my parents' bedroom and found my father collapsed on the floor. I tried to find a pulse, but I could not."

The ambulance crew were able to keep Lick alive using CPR until they could get him to Symmes Hospital in Arlington, where the emergency-room doctors got his heart beating again. But his wife and daughter both knew that too many minutes had passed in the interim. "He's brain-dead, isn't he?" Louise asked the doctors. And they said, "Yes."

Louise looked at the forest of tubes and pumps forcing oxygen into and out of her husband's lungs, the IV bags dripping nutrients into his bloodstream, the electrodes and wires monitoring his heart.

"Take all that stuff off!" she demanded. She tried somehow to make them understand that this was a man who lived by his brains, whose greatest joy was the playful workings of his imagination, who . . .

And as gently as possible, the doctors explained that they couldn't remove Lick's life support, that the tiny bit of neural activity remaining in the brain stem kept him from meeting the strict definition of brain death prescribed by Massachusetts law. There was no hope whatsoever that he would regain consciousness or have any further meaningful existence, but unless and until he started breathing on his own, they were legally obligated to let the respirator keep on doing it for him.

"It was the law," says Louise, remembering her continued, fruitless demands. "They had to keep it on. Eventually they took him into the intensive-care unit."

Both Tracy and Linda returned with their mother the next morning. "Rob was still on all the gear," says Louise. "They had assigned a young doctor from MIT to the case. The kids accosted him with pictures and newspaper clippings about their father. They started talking about him and told the doctor, 'This is a man who would have jumped out a window if he'd known what you were doing. How can we get this stuff off of him?' "

The doctor understood and was as sympathetic as anyone could be. "But the law says we can't remove the life support until he starts breathing on his own," he told them. "Then we can get him off within the minute."

And there things stood, says Louise. "Then, about the third day, I went to the ICU, and Robin was gone. I was frantic; I just assumed that he had died. But they said no, they had taken him to room such and such." Miraculously, Lick had indeed started breathing on his own, which meant that the hospital staff could begin to take him off life support. "But Robin's heart just kept on beating," she says. "They had given me permission to stay in the room overnight, so I stayed there night and day. Sometimes I'd go downstairs to take care of the grandchildren while Tracy and Linda went up to see him. But that was all.

"After several weeks," she goes on, "I was just pulling up a chair for a night's sleep, when Robin's knee came up and went back down. In all that time he hadn't moved a finger. But I knew there is usually a death spasm. So I told the nurse, 'I think he just died.' And he had."

The date was Tuesday, June 26, 1990. The doctors never knew exactly why Lick had collapsed and stopped breathing, according to Louise. "Since he called for a pen, they thought it might have been a choking reflex from an asthma attack," she says. "But it seemed senseless to do an autopsy. Besides, Robin had donated his corneas and anything else that was harvestable for transplant, and to use those organs they have to take them within just a few hours after death."

Anyway, she says, "the next morning when the kids and I went in to get his things, the young doctor who had been so good to us came up and said how glad he was for us that it was over. And you know, I was devastated when Robin died. It was a horrid loss. But nevertheless I did feel thankful. I thought about how Robin had always felt so lucky in his life: every time he was feeling bored, something new would come up, and off he would go. He loved new experiences: travel, driving, eating in new restaurants—even if the place was bad, at least it was *different*. But his health was rapidly disintegrating at the end. He would have been miserable in another year, when he couldn't learn, couldn't experiment, couldn't teach. And then I was so thankful that Robin had gone quickly that way—because he was indisputably brain-dead from the time he finished calling me. He went out with dignity, still doing what he loved to do.

"So once I could accept the fact that he was gone," she says, "I thought to myself, Well, Robin, you lucked out again."

Certainly he had been a profoundly lucky man in one respect: far more than most of us, J. C. R. Licklider lived to see his dreams become reality. By June 26,

1990, the visions that he had begun to articulate more than thirty years before—Man-Computer Symbiosis, the Intergalactic Network, the global commons of the Multinet—were already taking on concrete form as personal computers, graphical user interfaces, local-area networks, the Internet, and much, much more. Lick's conception of the future would practically define the high-tech 1990s and would seem to grow only more prescient at the dawn of the new millennium.

And yet it's fair to ask, did Lick himself really have all that much to do with these developments? If he had never spun his dreams of symbiosis and all the rest, or if he hadn't accepted Jack Ruina's offer to start ARPA's command-and-control program, wouldn't the history of computing still have unfolded pretty much the way it did?

Maybe. On the one hand, as Lick himself pointed out in his 1988 interview with the Charles Babbage Institute, the pieces were already in place in 1962. Time-sharing experiments were already under way at MIT, BBN, Dartmouth, and elsewhere. DEC was already doing its best to spread the gospel of interactive computing in the commercial arena; its breakthrough success with mini-computers was only a few years away. Enrollment in computer-science courses was already burgeoning, thanks to the explosive growth of IBM and the Seven Dwarfs. And companies such as Texas Instruments and Fairchild were already pushing hard on integrated-circuit technology. There is every reason to believe that the Santa Clara Valley would have become Silicon Valley right on schedule; that computer-on-a-chip microprocessors would have emerged just as they did in the 1970s; and that hobbyists would have started building home computers in their garages soon thereafter. So it's easy to imagine scenarios in which the history of computers would have played out pretty much as it actually did, even without Lick or ARPA.

On the other hand, it's even easier to imagine scenarios in which that history might have played out very differently indeed—scenarios in which those hobbyist computers would stay in their garages for a very long time, for example, while microchips mainly went into building bigger and more powerful centralized machines. Technology isn't destiny, no matter how inexorable its evolution may seem; the way its capabilities are *used* is as much a matter of cultural choice and historical accident as politics is, or fashion. And in the early 1960s history still seemed to be on the side of batch processing, centralization, and regimentation. In the commercial world, for example, DEC was still a tiny niche player, a minor exception that proved the rule. Almost every other company in the computer industry was following Big Blue's lead—and IBM had just made an unshakable commitment to batch processing and mainframes, a.k.a. System/360. In the telecommunications world, meanwhile, AT&T was equally committed to telephone-style circuit switching; its engineers would scoff at the idea of packet switching when Paul Baran suggested it a few years later, and they would keep on scoffing well into the 1980s. And in the academic world, no other agency was pushing computer research in the directions Lick would, or funding it at anything like the ARPA levels. Remember, says Fernando Corbató, "this was at a time when the National Science Foundation was handing out money with eye-droppers—and then only after excruciating peer review. Compared to that, Lick had a *lot* of money. Furthermore, he was initially giving umbrella grants, which allowed us to fund the whole program. So there was this tremendous pump priming, which freed us from having to think small. The contrast was so dramatic that most places gravitated to ARPA. So that opening allowed a huge amount of research to get done." Without that pump priming—or more precisely, without an ARPA animated by J. C. R. Licklider's vision—there would have been no ARPA community, no Arpanet, no TCP/IP, and no Internet. There would have been no Project MAC–style experiments in time-sharing, and no computer-utilities boom to inflame the imagination of hobbyists with wild speculations about "home information centers." There would have been no life-giving river of cash flowing into DEC from the PDP-10 time-sharing machines it sold to the ARPA community. There would have been no windows-icons-mouse interface à la Doug Engelbart. And there would have been no creative explosion at Xerox PARC.

So in the end, about all anyone can really say is that it's hard to play "What if?" with history. What we do know is that in *our* history, J. C. R. Licklider had the vision. He was given the opportunity to realize that vision. He seized that opportunity. And he succeeded beyond anything he could have hoped for. Indeed, nothing testifies more eloquently to his success than one simple fact: in the late 1960s and early 1970s, at the height of the Vietnam debacle, when governments and institutions of all kinds were viewed as hated instruments of oppression, and when punch card–belching mainframes were seen as the most potent symbol of tyranny, a rising generation of computer-science students began to think of computers as *liberating*. This was the generation that would build the Arpanet. This was the generation that would gather at PARC. And this was the generation—together with the students *it* would teach—that would engineer the personal-computer revolution of the 1980s and the networking revolution of the 1990s. Even now, people who never heard of "Man-Computer Symbiosis" or J. C. R. Licklider still fervently believe in his dream, because it is in the very air they breathe.

"I think Lick's being at that place at that time is a testament to the tenuousness of it all," says Bob Taylor. "It was really a fortunate circumstance. You could never get him to admit that, because he was modest to a fault. But he was a very knowledgeable person about scientific endeavor in general, and about the philosophy of science. He was a very good judge of people. And I don't think that Ivan, or I, or anyone who's been in that ARPA position since, has had the vision that Lick had. Most of the significant advances in computer technology—including the work that my group did at Xerox PARC—were simply extrapolations of Lick's vision. They were not really new visions of their own. So he was really the father of it all."

Al Vezza was insistent, remembers Louise Licklider. "Lick had said that he didn't want any kind of to-do when he died," she says. "He wasn't religious himself, even though his father had been a Southern Baptist minister, so it would seem totally phony if he'd had a big religious service." But Vezza kept reminding her of the retirement party. It had been one of the best, he said: just look at all the people who'd shown up, all the stories that had gotten told. This memorial service wouldn't be for Lick, he said. It would be for all the people who had known him.

So she relented. And the result was . . . interesting "All the ushers were undergraduates from the lixkids group," she explains. "There were wonderful speakers, almost all of them with wonderful stories. Tracy gave a delightful discussion of growing up as J. C. R. Licklider's son. I later heard from a lot of people—even some of the more hard-nosed theatrical people from my theater group—who said that they had never felt so much love in a room at any service in their memory. Everyone who spoke genuinely loved the man. But it wasn't sentimental. It was just very, very warm. And very different. I think several people were a little stunned by it."

With reason. The tradition at memorial services was for the MIT string quartet to play a dignified selection of Bach and Brahms. But when Vezza had asked Louise what kind of music her husband had liked, she said Dixieland. And Broadway show tunes. So he got a band to play Lick's favorite songs—"Impossible Dream," "Sunrise, Sunset," and so on—in Dixieland style. "It sort of shook people up the first time they heard it," remembers Vezza. "But then they realized, that was Lick."

Much later, says Louise Licklider, when she decided to move out of the house in suburban Arlington, Massachusetts, where they'd lived for the past twenty-five years, she had to confront the space in the basement where her husband had stored his "papers."

"The MIT Archives had been desperate for his papers," she says. "They kept saying, 'We'll send someone to take it all out.' And I kept saying, 'That just isn't possible.' They thought I was simply being emotional and difficult. But I wasn't. Robin was a very specific, accurate man. But he was not a filer. He had a natural inclination to save everything, even erasers that had worn down. So he had all these personal things in the middle that I wasn't about to sort out.

"Well, Archives got very irritated and finally quit bugging me. But when I moved, a young man named Keith, who loved and adored Lick, came and stood beside me as we went through the basement, just shaking his head. You'd pick up a paper, and it would be a test from a course he taught twenty years ago. Under that would be a flier advertising a lecture. There would be maps of towns he'd been in that had no historical significance, because he liked old maps. I'd throw a few things into a box for MIT, but most of it we threw out.

"Well," she says, "at one point I heard Keith just burst out laughing. Anything Lick loved, he *loved*—chocolate, for one thing. And Keith had come across a folder full of chocolate-bar wrappers that Lick had collected, all neatly folded. Beautiful wrappers from all over the world . . ."

ACKNOWLEDGMENTS

First, a few words of background.

In the summer of 1992, as I was going over the galleys of my previous book, *Complexity*, I learned that the Alfred P. Sloan Foundation was commissioning a new series of books on technologies of the twentieth century. I applied for support, and in October 1992, just as *Complexity* was reaching the bookstores, I got the go-ahead from Sloan to write a book about "software."

Which meant what? I asked the series managers.

Just "software," they told me. The book could be a biography of Bill Gates, say, or it could be about the Soul of a New Shrink-Wrapped Package, or anything else you decide. As long as the book ends up in the right ballpark, we'll be happy.

Thinking that this was too good a deal to pass up, I plunged into my research—and soon began to realize that this kind of freedom was a decidedly mixed blessing. Given the avalanche of books and articles that had come out since the beginning of the PC revolution, what could I possibly say about "software" that hadn't been said a million times before?

There ensued nearly two full years of fits, starts, and blind alleys, as I tried to figure out what story I actually wanted to tell and how to tell it in a compelling way. I talked to Young Turks in the research labs, old Turks in the funding agencies, and everyone else I could think of. Nothing they suggested seemed quite right. And yet, by the summer of 1994, I had begun to notice how often the name J. C. R. Licklider came up in those conversations. Back in 1962, I learned, Lick had founded a little computer-research funding office at the Pentagon's Advanced Research Projects Agency and then had used his position to forge scattered groups of far-thinking computer scientists into a nationwide "ARPA community," which endured through the 1960s and 1970s, and which seemed to evoke a powerful sense of nostalgia from everyone who had experienced it.

Meanwhile, as it happened, I had also been reading about the history of software. And somewhere during that same summer of 1994, I finally began to realize that windows, icons, mice, word processors, spreadsheets, and all the other things that seem so important to the modern software industry actually come at the end of the story, not at the beginning. Historically, in fact, people weren't even able to think about such things until they had grasped a more elemental notion: that a computer could *interact* with a user in real time and give an instant response—that a computer could actually help its users think, in other words, as opposed to just crunching numbers for them. Furthermore, I discovered that this insight had been far from obvious in the early days; it hadn't really begun to emerge until the late 1950s, when one of its earliest and most articulate champions had been—J. C. R. Licklider.

Clearly, I had to learn more about this guy. Sadly, I discovered, Lick himself had died in June 1990; I wasn't going to meet him personally. But in fall 1994 I made several visits to his home turf, MIT, where I talked to many people who had known and worked with him. The undercurrent of affection was palpable; Lick had been not only an admired and respected colleague but a beloved teacher, mentor, and friend.

Finally, on my second or third visit to Cambridge that fall, I put in a call to Lick's widow, Louise, who was living in suburban Arlington, Massachusetts. I hadn't called her earlier because I had somehow gotten the impression that she had also passed away, but she was in fact very much alive. I introduced myself, explaining that I might want to mention her husband in a book I was working on, and asked if I could come out to talk with her about him. Certainly, she said. "But I'm not sure how much I can help you; we never talked much about the technical side of his work." That's fine, I assured her.

So I went out to her house. Louise Licklider proved to be a tall, striking woman in her seventies, given to old-style hospitality. She made sure that there was a cup of hot tea by my elbow, with a wafer or two sitting on the saucer. And then, as I was unpacking my laptop to take notes, she began talking about her life with Lick—how they had met during World War II and gotten married; how they had talked every evening over cocktails, trading stories about the day's events; how he had felt about his work, the people he knew, and the dreams he had dreamed. . . .

Four hours later, my tea was stone cold and her stories were still coming. I finally had to call a halt for the day, simply because my fingers were getting too tired to type. But I knew I would be back. And I knew that I had finally found the story I wanted to tell in this book.

Today, almost seven years later (and almost nine years after I started this project), you have the result before you. I hope you enjoy it. And here I can finally begin to repay some of the many debts I have incurred over those years—debts that the foregoing account can only suggest. So, as inadequate as I know it must be, I would like to offer my thanks . . .

- to the Alfred P. Sloan Foundation, for providing the initial motivation for this book, and for continuing its generous financial support during the book's long creation.
- to my agent, Peter Matson, for finding a publisher for my not-very-coherent proposal and for being a steady voice of encouragement and reassurance when the process of writing seemed to stretch toward infinity.
- to my editor at Viking, Rick Kot, for taking a chance on this book (not to mention its author) when it was still little more than a nebulous idea, for having enough faith in the outcome to bring the book with him as he moved from Hyperion to Little, Brown and finally to Viking, and for exercising his formidable powers of patience and insight as he helped me shape the final manuscript.
- to Bob Kahn, for being the very first to put me on the trail of J. C. R. Licklider and the ARPA community and for showing so much hospitality at the Corporation for National Research Initiatives, where I was able to interview many of the key players in the Arpanet/Internet saga.

- to the Licklider family—Tracy, Linda, and most especially Louise—for sharing their memories of the man they knew as father and husband.
- to the dozens, if not hundreds, of people whom I interviewed for this book—in some cases many times—for giving so generously of their experience and insight.
- to Chigusa Ishikawa of Kyoto University, for perusing the MIT Archives more diligently than I had and then alerting me to several of Lick's papers that I had missed.
- to Tim Anderson, Gordon Bell, Leo Beranek, Marjory Blumenthal, David Burmaster, Vint Cerf, Paul Ceruzzi, Wes Clark, Fernando Corbató, Steve Crocker, Mike Dertouzos, Jerry Elkind, Doug Engelbart, Bill English, Bob Fano, Ed Feigenbaum, Jack Goldman, Charlie Herzfeld, Bob Kahn, Alan Kay, Len Kleinrock, Karl Kryter, Butler Lampson, Louise Licklider, Steve Lukasik, Stuart Malone, Tom Marill, John McCarthy, Victor McElheny, Bob Metcalfe, George Miller, Severo Ornstein, George Pake, Alex Roland, Jack Ruina, Linda Licklider Smith, Bob Taylor, Chuck Thacker, Al Vezza, Barry Wessler, and Steve Wolff, for reading the draft manuscript in whole or in part. Their comments and suggestions have saved me from many errors of fact and interpretation—though, of course, they bear no responsibility for any errors that remain. That responsibility is mine alone.
- and finally, to my wife, Amy Friedlander, for repaying her spouse's seemingly endless preoccupation with years of patience and devotion. An adequate recounting would require a separate volume. So for now, once again, I would like to say thank you. For everything.

NOTES

CHAPTER 1: MISSOURI BOYS

1. Karl Kryter, "In Memoriam of J. C. R. Licklider, 1915–1990" (eulogy delivered during "A Ceremony in Remembrance of J. C. R. Licklider," October 15, 1990).
2. George A. Miller, "Stanley Smith Stevens, 1906–1973," in *Biographical Memoirs*, vol. 47 (Washington, D.C.: National Academy of Sciences, 1975).
3. Ibid.
4. Ibid.
5. Ibid.
6. George A. Miller, "J. C. R. Licklider, Psychologist" (unpublished address given before the Acoustical Society of America, 1991).
7. Steve Heims, *John von Neumann and Norbert Wiener: From Mathematics to the Technologies of Life and Death* (Cambridge, Mass.: MIT Press, 1980), 379.
8. Jerome B. Wiesner, "The Communications Sciences—Those Early Days," in *R. L. E.: 1946+20* (Cambridge, Mass.: Research Laboratory for Electronics, MIT, 1966), 13.
9. Pesi R. Masani, *Norbert Wiener* (Basel: Birkhäuser, 1990), 16.
10. Wiesner, "The Communications Sciences—Those Early Days," 13.
11. Norbert Wiener, *Cybernetics, or Control and Communication in the Animal and the Machine*, 2d ed. (Cambridge, Mass.: MIT Press, 1961), 43.

CHAPTER 2: THE LAST TRANSITION

1. Norbert Wiener, *I Am a Mathematician: The Later Life of a Prodigy* (Cambridge, Mass.: MIT Press, 1956), 112.
2. Vannevar Bush, "The Inscrutable 'Thirties" (1933), in *From Memex to Hypertext: Vannevar Bush and the Mind's Machine*, ed. James M. Nyce and Paul Kahn (San Diego: Academic Press, 1991), 74.
3. Vannevar Bush, "As We May Think" (1945), in Nyce and Kahn, eds., *From Memex to Hypertext*, 89.
4. Quoted in James M. Nyce and Paul Kahn, "A Machine for the Mind: Vannevar Bush's Memex," in *From Memex to Hypertext*, 53–54.
5. Bush, "As We May Think," 101–2.
6. Norbert Wiener, "Memorandum on the Mechanical Solution of Partial Differential Equations"

(1940), in *Norbert Wiener: Collected Works*, ed. Pesi R. Masani (Cambridge, Mass.: MIT Press, 1985), 4: 134.

7. Wiener, *I Am a Mathematician*, 239.

8. Quoted in Larry Owens, "Vannevar Bush and the Differential Analyzer: The Text and Context of an Early Computer," in Nyce and Kahn, eds., *From Memex to Hypertext*, 23–24.

9. Claude Shannon, "A Symbolic Analysis of Relay and Switching Circuits" (1938), in *Claude Elwood Shannon: Collected Papers*, ed. N. J. A. Sloane and Aaron D. Wyner (New York: IEEE Press, 1993), 492.

10. George R. Stibitz, "Early Computers," in *A History of Computing in the Twentieth Century*, ed. N. Metropolis, J. Howlett, and Gian-Carlo Rota (New York: Academic Press, 1980), 481.

11. Herman H. Goldstine, *The Computer from Pascal to von Neumann* (Princeton, N.J.: Princeton University Press, 1972), 182.

12. Ibid., 149

13. Norbert Wiener, *Cybernetics, or Control and Communication in the Animal and the Machine*, 2d ed. (Cambridge, Mass.: MIT Press, 1961), 6.

14. William Aspray, "The Scientific Conceptualization of Information: A Survey," *Annals of the History of Computing* 7 (1985): 130.

CHAPTER 3: NEW KINDS OF PEOPLE

1. Jeremy Bernstein, "Profiles: A.I. (Profile of Marvin Minsky)," *New Yorker*, December 14, 1981.

2. Henry S. Tropp, "Origin of the Term Bit," *Annals of the History of Computing* 6 (1984).

3. Jerome B. Wiesner, "The Communications Sciences–Those Early Days," in *R.L.E.: 1946+20* (Cambridge, Mass.: Research Laboratory for Electronics, MIT, 1966), 12.

4. Steve Heims, *John von Neumann and Norbert Wiener: From Mathematics to the Technologies of Life and Death* (Cambridge, Mass.: MIT Press, 1980), 206.

5. Norbert Wiener, *Cybernetics, or Control and Communication in the Animal and the Machine*, 2d ed. (Cambridge, Mass.: MIT Press, 1961), 23.

6. Heims, *Von Neumann/Wiener*, 189.

7. Norbert Wiener, "A Scientist Rebels," *Atlantic Monthly*, January 1947, and *Bulletin of the Atomic Scientists*, January 1947.

8. Heims, *Von Neumann/Wiener*, 334–35.

9. John von Neumann and Oskar Morgenstern, *Theory of Games and Economic Behavior* (Princeton, N.J.: Princeton University Press, 1944).

10. Heims, *Von Neumann/Wiener*, 359.

11. Richard Rhodes, *Dark Sun: The Making of the Hydrogen Bomb* (New York: Simon & Schuster, 1995), 389.

12. C. Blair, "The Passing of a Great Mind," *Life*, February 25, 1957.

13. Wiener, *Cybernetics*, 159.

14. Ibid., 27.

15. Norbert Wiener, *I Am a Mathematician: The Later Life of a Prodigy* (Cambridge, Mass.: MIT Press, 1956), 325.

16. Ibid., 327–28.

17. Norbert Wiener, "Some Moral and Technical Consequences of Automation," *Science* 131 (1960).

18. Claude Shannon, "The Bandwagon," *IEEE Transactions Information Theory* 2 (1956).

19. *Book Review Digest*, 1948, 920–21; *Book Review Digest*, 1949, 986.

20. Heims, *Von Neumann/Wiener*, 220.

21. Wiener, *Cybernetics*, back cover.

22. Ibid.

23. Heims, *Von Neumann/Weiner*, 335–36.

24. J. C. R. Licklider, "Interactive Dynamic Modeling," in *Prospects for Simulation and Simulators of Dynamic Modeling*, ed. George Shapiro and Milton Rogers (New York: Spartan Books, 1967), 282.

CHAPTER 4: THE FREEDOM TO MAKE MISTAKES

1. Henry S. Tropp, "A Perspective on SAGE: Discussion," *Annals of the History of Computing* 5 (1983): 376.
2. Robert R. Everett, "Whirlwind," in *A History of Computing in the Twentieth Century,* ed. N. Metropolis, J. Howlett, and Gian-Carlo Rota (New York: Academic Press, 1980), 365–66.
3. Tropp, "Perspective on SAGE," 376.
4. Quoted ibid., 380.
5. Everett, "Whirlwind," 375.
6. George A. Miller, "J. C. R. Licklider, Psychologist" (unpublished address given before the Acoustical Society of America, 1991).
7. Quoted in Tropp, "Perspective on SAGE," 379–80.
8. Quoted in Henry S. Tropp, "Reliability of Components (Interview with Jay W. Forrester)," *Annals of the History of Computing* 5 (1983): 399.
9. Quoted in Christopher Evans, "Conversation: Jay W. Forrester," *Annals of the History of Computing* 5 (1983).
10. Quoted in Tropp, "Perspective on SAGE," 378–79.
11. Quoted ibid., 384.
12. Henry S. Tropp, "History of the Design of the SAGE Computer—The AN/FSQ-7," *Annals of the History of Computing* 5 (1983): 340.
13. Norbert Wiener, *Cybernetics, or Control and Communication in the Animal and the Machine,* 2d ed. (Cambridge, Mass.: MIT Press, 1961), vii.
14. Alan M. Turing, "Computing Machinery and Intelligence," *Mind* 59, no. 236 (1950). Reprinted in *The Mind's I: Fantasies and Reflections on Self & Soul,* ed. Douglas R. Hofstadter and Daniel C. Dennett (New York: Basic Books, 1981), 53–67.
15. Ibid.
16. Quoted in Steve Heims, *John von Neumann and Norbert Wiener: From Mathematics to the Technologies of Life and Death* (Cambridge, Mass.: MIT Press, 1980), 276.
17. Quoted ibid., 370.
18. Quoted ibid.
19. Quoted ibid., 371.
20. George A. Miller, "The Magical Number Seven, Plus or Minus Two: Some Limits on Our Capacity for Processing Information," *Psychological Review* 63, no. 2 (1956): 81; http://www.well.com/user/smalin/miller.html.
21. Quoted in Bernard J. Baars, *The Cognitive Revolution in Psychology* (New York: Guilford Press, 1986), 342.
22. Allen Newell and Herbert A. Simon, *Human Problem Solving* (Englewood Cliffs, N.J.: Prentice-Hall, 1972), 109.
23. Quoted in Pamela McCorduck, *Machines Who Think* (San Francisco: W. H. Freeman & Co., 1979), 139.
24. Quoted ibid., 142.
25. Quoted in Baars, *Cognitive Revolution,* 213.
26. Ibid., 198.

CHAPTER 5: THE TALE OF THE FIG TREE AND THE WASP

1. Kenneth Olsen, oral history for the National Museum of American History, Smithsonian Institution, September 28–29, 1988.
2. Wesley Clark, "The LINC Was Early and Small," in *A History of Personal Workstations,* ed. Adele Goldberg (New York: ACM Press, 1988), 353.
3. Olsen, NMAH oral history.
4. Clark, "The LINC Was Early and Small," 357.
5. J. C. R. Licklider, "Man-Computer Partnership," *International Science and Technology,* May 1965, 18.
6. J. C. R. Licklider, "Man-Computer Symbiosis," *IRE Transactions on Human Factors in Electronics,* 1960. Reprinted in *In Memoriam: J. C. R. Licklider, 1915–1990,* ed. Robert W. Taylor, Digital Systems Research Center Reports, vol. 61 (Palo Alto, Calif., 1990).
7. Ibid.

8. Martin Greenberger, ed., *Computers and the World of the Future* (Cambridge, Mass.: MIT Press, 1962), 214.

9. Quoted in Jaime Parker Pearson, ed., *Digital at Work: Snapshots from the First Thirty-five Years* (Burlington, Mass.: Digital Press, 1992), 13.

10. Quoted ibid., 143.

11. Edward Fredkin, "A Machine Remembered: The PDP-1," in Pearson, ed., *Digital at Work,* 37.

12. Greenberger, ed., *Computers and the World of the Future,* 208–9.

13. "The Project MAC Interviews," *IEEE Annals of the History of Computing* 14, no. 2 (1992): 22–23.

14. John McCarthy et al., "A Proposal for the Dartmouth Summer Research Project on Artificial Intelligence," submitted to the Rockefeller Foundation, August 31, 1955.

15. John McCarthy, "Reminiscences on the History of Time-Sharing," *IEEE Annals of the History of Computing* 14, no. 2 (1992): 19–20.

16. Quoted in Richard T. Wexelblatt, ed., *History of Programming Languages* (New York: Academic Press, 1981) 192.

17. John McCarthy, "John McCarthy's 1959 Memorandum," *IEEE Annals of the History of Computing* 14, no. 1 (1992): 20–23.

18. "The CTSS Interviews," *IEEE Annals of the History of Computing* 14, no. 1 (1992): 46.

19. John McCarthy, "Time-Sharing Computer Systems," in Greenberger, ed., *Computers and the World of the Future,* 224–25.

20. "The CTSS Interviews," 41.

21. Wexelblatt, ed., *History of Programming Languages,* 183.

22. Licklider, "Man-Computer Symbiosis."

23. J. C. R. Licklider, "The Truly SAGE System, or Toward a Man-Machine System for Thinking," Box 6, Folder "1957," in the Licklider Papers at MIT Archives (MC 499).

24. J. C. R. Licklider, "Man-Computer Symbiosis: Part of the Oral Report of the 1958 NAS-ARDC Special Study, Presented on Behalf of the Committee on the Roles of Men in Future Air Force Systems, 20–21 November 1958," Box 6, Folder "1958," in the Licklider Papers at MIT Archives (MC 499).

25. Greenberger, ed., *Computers and the World of the Future,* 207.

26. "The Project MAC Interviews," 18.

27. Greenberger, ed., *Computers and the World of the Future.*

28. J. C. R. Licklider, "Interactive Dynamic Modeling," in *Prospects for Simulation and Simulators of Dynamic Modeling,* ed. George Shapiro and Milton Rogers (New York: Spartan Books, 1967), 281.

29. J. C. R. Licklider, "Dynamic Modeling," in *Models for the Perception of Speech and Visual Form,* ed. Weiant Wathen-Dunn (Cambridge, Mass.: MIT Press, 1967), 15.

30. J. C. R. Licklider, "Interactive Information Processing, Retrieval and Transfer," paper presented at "Storage and Retrieval of Information: A User-Supplier Dialog," Neuilly-sur-Seine, France, 1968.

31. J. C. R. Licklider and Robert W. Taylor, "The Computer as a Communication Device," *Science & Technology* 76 (1968): 21–31. Reprinted in *In Memoriam: J. C. R. Licklider, 1915–1990,* ed. Robert W. Taylor, Digital Systems Research Center Reports, vol. 61 (Palo Alto, Calif., 1990).

32. Licklider, "Interactive Information Processing."

33. J. C. R. Licklider, "The System System," in *Human Factors in Technology,* ed. E. Bennett, J. Degan, and J. Spiegel (New York: McGraw-Hill, 1963), 627–28.

34. Steven Levy, *Hackers: Heroes of the Computer Revolution* (Garden City, N.Y.: Anchor Press/Doubleday, 1984), 28–29.

35. Ibid., 47.

36. Olsen, NMAH oral history.

37. Ibid.

38. Quoted in Clark, "The LINC Was Early and Small," 368.

39. "The Project MAC Interviews," 42.

40. "The CTSS Interviews," 44, 46.

41. Ibid., 42.

42. McCarthy, "Time-Sharing Computer Systems," 236.

43. Martin Greenberger, "The Computers of Tomorrow," *Atlantic Monthly,* May 1964.

44. McCarthy, "Reminiscences on the History of Time-Sharing," 23.

CHAPTER 6: THE PHENOMENA SURROUNDING COMPUTERS

1. C. M. Green and M. Lomask, *Vanguard: A History* (Washington, D.C.: Smithsonian Institution, 1971), 186–87.
2. The Rockefeller Panel Reports, *Prospect for America* (New York: Doubleday, 1961), 96–98.
3. "The Project MAC Interviews," *IEEE Annals of the History of Computing* 14, no. 2 (1992): 19.
4. Ibid., 18.
5. Quoted in Howard Rheingold, *Tools for Thought* (New York: Simon & Schuster, 1985), chapter 9.
6. Douglas Engelbart, "The Augmented Knowledge Workshop," in *A History of Personal Workstations*, ed. Adele Goldberg (New York: ACM Press, 1988), 189.
7. Quoted in Rheingold, *Tools for Thought*, chapter 9.
8. Quoted ibid.
9. Engelbart, "Augmented Knowledge Workshop," 190.
10. Quoted ibid., 191.
11. Ibid.
12. "The Project MAC Interviews," 21.
13. "Proposal for a Research and Development Program on Computer Systems to the Advanced Research Projects Agency from the Massachusetts Institute of Technology," January 1963, MIT Institute Archives.
14. "The Project MAC Interviews," 23.
15. Martin Greenberger, "The Computers of Tomorrow," *Atlantic Monthly,* May 1964.
16. Ibid., 33.
17. Robert M. Fano, "Project MAC," in *Encyclopedia of Computer Science and Technology* (New York and Basel: Marcel Dekker, 1979), 12: 343.
18. "The Project MAC Interviews," 25.
19. Ibid.
20. Ibid., 26.
21. Ibid.
22. Ibid., 33.
23. Ibid.
24. Ibid., 30.
25. Ibid.
26. Ibid.
27. Allen Newell, Alan J. Perlis, and Herbert A. Simon, "What Is Computer Science?" *Science* 157, no. 3795 (September 22, 1967): 1373–74.
28. Allen Newell and Herbert A. Simon, *Human Problem Solving* (Englewood Cliffs, N.J.: Prentice-Hall, 1972).
29. J. C. R. Licklider, "Memorandum for Dr. Eugene G. Fubini, Assistant Secretary of Defense (Deputy Director, DR&E), November 26, 1963. Subject: Centers of Excellence in the Information Sciences."
30. Engelbart, "Augmented Knowledge Workshop," 192.
31. Ibid., 195.
32. "Proposal for a Research and Development Program on Computer Systems to the Advanced Research Projects Agency from the Massachusetts Institute of Technology," January 1963, MIT Institute Archives, 7.
33. Robert M. Fano, letter to Dr. J. A. Stratton, June 29, 1964.
34. F. J. Corbató, et al., "Memorandum to Professor R. M. Fano on Technical Evaluation of the IBM and GE Computer Systems Proposed to Project MAC."
35. Quoted in Jaime Parker Pearson, ed., *Digital at Work: Snapshots from the First Thirty-five Years* (Burlington, Mass.: Digital Press, 1992), 82.
36. Quoted ibid., 83.
37. Quoted ibid.
38. "IBM Announces System/360," *IBM News,* 7 April 1964.
39. Fano to Stratton, June 29, 1964.
40. Ibid.
41. Frederick P. Brooks, Jr., *The Mythical Man-Month* (Reading, Mass.: Addison-Wesley, 1975).
42. J. C. R. Licklider, "Graphic Input—A Survey of Techniques," in *Computer Graphics: Utility–Produc-*

tion–Art, ed. F. Gruenberger (London: Academic Press; Washington, D.C.: Thompson Book Company, 1967), 44.

CHAPTER 7: THE INTERGALACTIC NETWORK

1. J. C. R. Licklider and Robert W. Taylor, "The Computer as a Communication Device," *Science & Technology* 76 (1968): 21–31. Reprinted in *In Memoriam: J. C. R. Licklider, 1915–1990,* ed. Robert W. Taylor, Digital Systems Research Center Reports, vol. 61 (Palo Alto, Calif., 1990).
2. Ibid.
3. David E. Weisberg, "Computer Characteristics Quarterly–Recent Trends," *Datamation* 12 (January 1966): 55–56.
4. Martin Campbell-Kelly and William Aspray, *Computer: A History of the Information Machine* (New York: Basic Books, 1996), 218.
5. J. C. R. Licklider, "Social Prospects of Information Utilities," *The Information Utility and Social Choice,* ed. Harold Sackman and Norman H. Nie (Montvale, N.J.: AFIPS Press, 1970), 3–24.
6. J. C. R. Licklider, "Underestimates and Overexpectations," *Computers and Automation,* August 1969, 50. Reprinted from *ABM: An Evaluation of the Decision to Deploy an Antiballistic Missile System,* ed. Abram Chayes and Jerome B. Wiesner (New York: Harper & Row, 1969).
7. J. C. R. Licklider, "Man-Computer Partnership," *International Science and Technology,* May 1965, 20.
8. Robert M. Fano, "Joseph Carl Robnett Licklider, 1915-1990," *Biographical Memoirs,* vol. 75 (Washington, D.C.: National Academy Press, 1998), 18.

CHAPTER 8: LIVING IN THE FUTURE

1. Douglas K. Smith and Robert C. Alexander, *Fumbling the Future: How Xerox Invented, Then Ignored, the First Personal Computer* (New York: William Morrow, 1988), 79.
2. Quoted ibid., 77.
3. Quoted in Tekla S. Perry and Paul Wallich, "Inside the PARC: The 'Information Architects,' " *IEEE Spectrum,* October 1985, 66.
4. Quoted in Smith and Alexander, *Fumbling the Future,* 79.
5. Quoted in Dennis Shasha and Cathy Lazere, *Out of Their Minds: The Lives and Discoveries of 15 Great Computer Scientists* (New York: Copernicus, 1995), 39.
6. Quoted ibid.
7. Quoted ibid., 39–40.
8. Quoted ibid., 41.
9. Alan C. Kay, "The Early History of Smalltalk," in *History of Programming Languages,* ed. Thomas J. Bergin and Richard G. Gibson (New York: ACM Press, 1996), 516.
10. Ibid., 518.
11. Ibid., 522–23.
12. Alan C. Kay and Adele Goldberg, "Personal Dynamic Media," *Computer,* vol. 10 (March 1977), 31–41. Reprinted in *A History of Personal Workstations,* ed. Adele Goldberg (New York: ACM Press, 1988).
13. Kay, "Early History of Smalltalk," 523.
14. Ibid., 527.
15. Butler Lampson, "Personal Distributed Computing: The Alto and Ethernet Software," in *A History of Personal Workstations,* ed. Adele Goldberg (New York: ACM Press, 1988), 295.
16. Vinton G. Cerf and Robert E. Kahn, "A Protocol for Packet Network Interconnection," *IEEE Transactions on Communications Technology* COM-22, no. 5 (May 1974), 627–41.
17. Kay, "Early History of Smalltalk," 554.
18. Ibid., 559.
19. Chuck Thacker, "Personal Distributed Computing: The Alto and Ethernet Hardware," in Goldberg, ed., *History of Personal Workstations,* 285.
20. Quoted in Perry and Wallich, "Inside the PARC," 68.
21. Quoted in Smith and Alexander, *Fumbling the Future,* 168.
22. Quoted ibid.
23. Quoted ibid., 136.

24. J. C. R. Licklider to John McCarthy, et al. (E-mail), April 2, 1975, University Archives, Carnegie Mellon University.

CHAPTER 9: LICK'S KIDS

1. J. C. R. Licklider, "Computers and Government," in *The Computer Age: A Twenty-Year View,* ed. Michael L. Dertouzos and Joel Moses (Cambridge, Mass.: MIT Press, 1979), 91.
2. Ibid., 117.
3. Ibid., 89–90.
4. Ivan E. Sutherland, Carver A. Mead, and Thomas E. Everhart, "Basic Limitations in Microcircuit Fabrication Technology," RAND report R-1956-ARPA, November 1976, 2.
5. Licklider, "Computers and Government," 126.
6. Ibid.
7. Jaime Parker Pearson, ed., *Digital at Work: Snapshots from the First Thirty-five Years* (Burlington, Mass.: Digital Press, 1992), 143.
8. Digital Equipment Corporation, "PDP-8 Typesetting System," advertising brochure ca. 1966, Computer Museum, Boston, Archives, "Milestones of a Revolution" File.
9. Glen Rifkin and George Harrar, *The Ultimate Entrepreneur: The Story of Ken Olsen and the Digital Equipment Corporation* (Chicago: Contemporary Books, 1988), 72.
10. Tom Truscott, "Invitation to a General-Access Unix Network," Duke University, North Carolina.
11. J. C. R. Licklider and Robert W. Taylor, "The Computer as a Communication Device," *Science & Technology* 76 (1968), 21–23. Reprinted in *In Memoriam: J. C. R. Licklider, 1915–1990,* ed. Robert W. Taylor, Digital Systems Research Center Reports, vol. 61 (Palo Alto, Calif., 1990).
12. Paul E. Ceruzzi, *A History of Modern Computing* (Cambridge, Mass.: MIT Press, 1998), 215.
13. Ibid., 222.
14. Sherry Turkle, *The Second Self: Computers and the Human Spirit* (New York: Simon & Schuster, 1984), 20.
15. Michael Hiltzik, *Dealers of Lightning: Xerox PARC and the Dawn of the Computer Age* (New York: HarperBusiness, 1999), 329.
16. Quoted ibid., 343.
17. Douglas K. Smith and Robert C. Alexander, *Fumbling the Future: How Xerox Invented, Then Ignored, the First Personal Computer* (New York: William Morrow, 1988), 200.
18. Ibid., 228.
19. Ibid., 229.
20. Hiltzik, *Dealers of Lightning,* 369.
21. Ibid., 359.

BIBLIOGRAPHY

ORAL HISTORIES FROM THE CHARLES BABBAGE INSTITUTE

A priceless resource for any historian of computing is the collection of oral history interviews gathered over the past two decades by the University of Minnesota's Charles Babbage Institute, a center devoted to the history of information processing. The memories recounted in these conversations are what makes history come alive. The CBI's oral histories may be ordered through its Web site at http://www.cbi.umn.edu/. The ones listed below—constituting only a fraction of the total—are those that proved most helpful in writing this book. (Indeed, most of the oral histories listed here were collected as part of an explicit effort, supported by DARPA itself, to document the history of that agency's Information Processing Techniques Office. The result of that effort—the first full-fledged academic study of IPTO—was the book referenced in this bibliography as *Transforming Computer Technology*, by Babbage Institute director Arthur Norberg and Babbage Institute researcher Judy O'Neill.)

Saul Amarel, OH 176; Paul Baran, OH 182; Allan Blue, OH 173; Alice R. Burks, OH 75; Arthur W. Burks, OH 75, OH 78, OH 136; Vinton G. Cerf, OH 191; Wesley Clark, OH 195; Robert S. Cooper, OH 255; Fernando J. Corbató, OH 162; Stephen Crocker, OH 233; William Crowther, OH 184; Donald Watts Davies, OH 8, OH 189; Jack Bonnell Dennis, OH 177; Michael L. Dertouzos, OH 164; J. Presper Eckert, OH 11, OH 13, OH 193; Robert M. Fano, OH 165; Edward Feigenbaum, OH 14, OH 157; Jay Forrester, OH 16; Howard Frank, OH 188; Bernard A. Galler, OH 236; Herman H. Goldstine, OH 18, OH 19; Frank Heart, OH 186; George H. Heilmeier, OH 226; Charles Herzfeld, OH 208; Cuthbert C. Hurd, OH 261; Robert E. Kahn, OH 158, OH 192; Leonard Kleinrock, OH 190; J. C. R. Licklider, OH 150; Stephen Lukasik, OH 232; John William Mauchly, OH 26, OH 44; Kathleen Mauchly, OH 11; John McCarthy, OH 156; Alexander A. McKenzie, OH 185; Marvin L. Minsky, OH 179; Allen Newell, OH 227; Bernard More Oliver, OH 097; Severo Ornstein, OH 183, OH 258; Raj Reddy, OH 231; Dennis Ritchie, OH 239; Lawrence G. Roberts, OH 159; Douglas T. Ross, OH 65, OH 178; Jack P. Ruina, OH 163; Jules I. Schwartz, OH 161; Ivan Sutherland, OH 171; Robert A. Taylor, OH 154; Ken Thompson, OH 239; Joseph F. Traub, OH 70, OH 89, OH 94; Keith W. Uncapher, OH 174; Frank M. Verzuh, OH 63; David A. Walden, OH 181; Willis H. Ware, OH 37; Frederick W. Weingarten, OH 212; David J. Wheeler, OH 132; Terry Allen Winograd, OH 237; Patrick H. Winston, OH 196.

BOOKS AND ARTICLES

Again, the written materials listed below are only a tiny fraction of what's available on the history of computing, but they were particularly helpful to me in telling the story of Lick and the ARPA community.

Aspray, William. "The Scientific Conceptualization of Information: A Survey." *Annals of the History of Computing* 7 (1985).
——. "John von Neumann's Contributions to Computing and Computer Science." *Annals of the History of Computing* 11, no. 3 (1989).
——. *John von Neumann and the Origins of Modern Computing.* Cambridge, Mass.: MIT Press, 1990.
——. "The Intel 4004 Microprocessor: What Constituted Invention?" *IEEE Annals of the History of Computing* 19, no. 3 (1997).
Augarten, Stan. *Bit by Bit: An Illustrated History of Computers.* New York: Ticknor & Fields, 1984.
Baars, Bernard J. *The Cognitive Revolution in Psychology.* New York: Guilford Press, 1986.
Berners-Lee, Tim. *Weaving the Web: The Original Design and Ultimate Destiny of the World Wide Web by Its Inventor.* San Francisco: HarperSanFrancisco, 1999.
Bernstein, Jeremy. "Profiles: A.I. (Profile of Marvin Minsky)." *New Yorker,* December 14, 1981.
——. *Three Degrees Above Zero: Bell Labs in the Information Age.* New York: Charles Scribner's Sons, 1984.
Birkhoff, Garrett. "Computing Developments 1935–1955, as Seen from Cambridge, USA." In *A History of Computing in the Twentieth Century,* edited by N. Metropolis, J. Howlett, and Gian-Carlo Rota. New York: Academic Press, 1980.
Blair, C. "The Passing of a Great Mind." *Life,* February 25, 1957.
Brink, Jean R. "John von Neumann Reconsidered." *Annals of the History of Computing* 11, no. 3 (1989).
Brooks, Frederick P., Jr. *The Mythical Man-Month.* Reading, Mass.: Addison-Wesley, 1975.
Burke, Colin. *Information and Secrecy: Vannevar Bush, Ultra, and the Other Memex.* Metuchen, N.J.: Scarecrow Press, 1994.
Bush, Vannevar. "Science: The Endless Frontier." Washington, D.C.: Office of Scientific Research and Development, 1945.
——. *Pieces of the Action.* New York: William Morrow, 1970.
——. "The Inscrutable 'Thirties" (1933). In *From Memex to Hypertext: Vannevar Bush and the Mind's Machine,* edited by James M. Nyce and Paul Kahn. San Diego: Academic Press, 1991.
——. "As We May Think" (1945). In *From Memex to Hypertext: Vannevar Bush and the Mind's Machine,* edited by James M. Nyce and Paul Kahn. San Diego: Academic Press, 1991.
Campbell-Kelly, Martin, and William Aspray. *Computer: A History of the Information Machine.* New York: Basic Books, 1996.
Cerf, Vinton G., and Robert E. Kahn. "A Protocol for Packet Network Interconnection." *IEEE Transactions on Communications Technology* COM-22, no. 5 (May 1974).
Ceruzzi, Paul E. *Reckoners: The Prehistory of the Digital Computer, from Relays to the Stored Program Concept, 1935–1945.* Westport, Conn.: Greenwood Press, 1983.
——. "Electronics Technology and Computer Science, 1940–1975: A Coevolution." *Annals of the History of Computing* 10, no. 4 (1989).
——. *A History of Modern Computing.* Cambridge, Mass.: MIT Press, 1998.
Chposky, James, and Ted Leonsis. *Blue Magic: The People, Power and Politics Behind the IBM Personal Computer.* New York: Facts on File, 1988.
Clark, Wesley. "The LINC Was Early and Small." In *A History of Personal Workstations,* edited by Adele Goldberg. New York: ACM Press, 1988.
Committee on Innovations in Computing and Communications: Lessons from History. *Funding a Revolution: Government Support for Computing Research.* Washington, D.C.: National Academy of Sciences, 1999.
Corbató, Fernando J. "On Building Systems That Will Fail." *Communications of the ACM* 34, no. 9 (1991).
Cortada, James W. *The Computer in the United States: From Laboratory to Market, 1930 to 1960.* Armonk, N.Y.: M. E. Sharpe, 1993.
——. "Commercial Applications of the Digital Computer in American Corporations, 1945–1995." *IEEE Annals of the History of Computing* 18, no. 2 (1996).
——. "Economic Preconditions That Made Possible Application of Commercial Computing in the United States." *IEEE Annals of the History of Computing* 19, no. 3 (1997).

Cringely, Robert X. *Accidental Empires: How the Boys of Silicon Valley Make Their Millions, Battle Foreign Competition, and Still Can't Get a Date.* New York: Addison-Wesley, 1992.

"The CTSS Interviews." *IEEE Annals of the History of Computing* 14, no. 1 (1992).

Engelbart, Douglas. "The Augmented Knowledge Workshop." In *A History of Personal Workstations,* edited by Adele Goldberg. New York: ACM Press, 1988.

Evans, Christopher. "Conversation: Jay W. Forrester." *Annals of the History of Computing* 5 (1983).

Everett, Robert R. "Whirlwind." In *A History of Computing in the Twentieth Century,* edited by N. Metropolis, J. Howlett, and Gian-Carlo Rota. New York: Academic Press, 1980.

Fano, Robert M. "Project MAC." In *Encyclopedia of Computer Science and Technology.* New York and Basel: Marcel Dekker, 1979.

——. "Joseph Carl Robnett Licklider, 1915–1990." In *Biographical Memoirs,* vol. 75. Washington, D.C.: National Academy Press, 1998.

Fano, Robert M., et al. "Proposal for a Research and Development Program on Computer Systems to the Advanced Research Projects Agency from the Massachusetts Institute of Technology." Cambridge, Mass.: MIT Institute Archives, 1963.

Flamm, Kenneth. *Creating the Computer: Government, Industry, and High Technology.* Washington, D.C.: The Brookings Institution, 1988.

Fredkin, Edward. "A Machine Remembered: The PDP-1." In *Digital at Work: Snapshots from the First Thirty-five Years,* edited by Jaime Parker Pearson. Burlington, Mass.: Digital Press, 1992.

Galison, Peter. "The Ontology of the Enemy: Norbert Wiener and the Cybernetics Vision." *Critical Inquiry,* Autumn 1994.

Garfinkel, Simson L. *Architects of the Information Society: Thirty-five Years of the Laboratory for Computer Science at MIT.* Cambridge, Mass.: MIT Press, 1999.

Goldstine, Herman H. *The Computer from Pascal to von Neumann.* Princeton, N.J.: Princeton University Press, 1972.

Good, I. J. "Pioneering Work on Computers at Bletchley." In A *History of Computing in the Twentieth Century,* edited by N. Metropolis, J. Howlett, and Gian-Carlo Rota. New York: Academic Press, 1980.

Greenberger, Martin. "The Computers of Tomorrow." *Atlantic Monthly,* May 1964.

——, ed. *Computers and the World of the Future.* Cambridge, Mass.: MIT Press, 1962.

Hafner, Katie, and Matthew Lyon. *Where Wizards Stay Up Late: The Origins of the Internet.* New York: Simon & Schuster, 1996.

Hall, Mark, and John Barry. *Sunburst: The Ascent of Sun Microsystems.* Chicago: Contemporary Books, 1990.

Hauben, Michael, and Ronda Hauben. *Netizens: On the History and Impact of Usenet and the Internet.* Los Alamitos, Calif.: IEEE Computer Society Press, 1997.

Heims, Steve. *John von Neumann and Norbert Wiener: From Mathematics to the Technologies of Life and Death.* Cambridge, Mass.: MIT Press, 1980.

——. *Constructing a Social Science for Postwar America: The Cybernetics Group, 1946–1953.* Cambridge, Mass.: MIT Press, 1993.

Hiltzik, Michael. *Dealers of Lightning: Xerox PARC and the Dawn of the Computer Age.* New York: Harper-Business, 1999.

Hodges, Andrew. *Alan Turing: The Enigma.* New York: Simon & Schuster, 1983.

Ishikawa, Chigusa. "Genesis of J. C. R. Licklider's 'Man-Computer Symbiosis.' " Forthcoming.

Johnson, Luanne (James). "A View from the 1960s: How the Software Industry Began." *IEEE Annals of the History of Computing* 20, no. 1 (1998).

Kay, Alan C. "The Early History of Smalltalk." In *History of Programming Languages,* edited by Thomas J. Bergin and Richard G. Gibson. New York: ACM Press, 1996.

Kay, Alan C., and Adele Goldberg. "Personal Dynamic Media." *Computer* 10 (March 1977). Reprinted in *A History of Personal Workstations,* edited by Adele Goldberg. New York: ACM Press, 1988.

Kidwell, Peggy A., and Paul E. Ceruzzi. *Landmarks in Digital Computing: A Smithsonian Pictoral History.* Washington, D.C.: Smithsonian Institution, 1994.

Kryter, Karl. "In Memoriam of J. C. R. Licklider, 1915–1990." Eulogy delivered during "A Ceremony in Remembrance of J. C. R. Licklider," October 15, 1990.

Lampson, Butler. "Personal Distributed Computing: The Alto and Ethernet Software." In *A History of Personal Workstations,* edited by Adele Goldberg. New York: ACM Press, 1988.

Lee, J. A. N. *Computer Pioneers.* Los Alamitos, Calif.: IEEE Computer Society Press, 1995.

——. " 'Those Who Forget the Lessons of History Are Doomed to Repeat It,' or Why I Study the History of Computing." *IEEE Annals of the History of Computing* 18, no. 2 (1996).

Levy, Steven. *Hackers: Heroes of the Computer Revolution.* Garden City, N.Y.: Anchor Press/Doubleday, 1984.

——. *Insanely Great: The Life and Times of the Macintosh, the Computer That Changed Everything.* New York: Viking, 1994.

Licklider, J. C. R. "The Truly SAGE System, or Toward a Man-Machine System for Thinking." In the Licklider Papers at MIT Archives (MC 499), Box 6, Folder "1957."

——. "Man-Computer Symbiosis: Part of the Oral Report of the 1958 NAS-ARDC Special Study, Presented on Behalf of the Committee on the Roles of Men in Future Air Force Systems, 20–21 November 1958." In the Licklider Papers at MIT Archives (MC 499), Box 6, Folder "1958."

——. "Man-Computer Symbiosis." *IRE Transactions on Human Factors in Electronics* (1960). Reprinted in *In Memoriam: J. C. R. Licklider, 1915–1990,* edited by Robert W. Taylor. Digital Systems Research Center Reports, vol. 61. Palo Alto, Calif., 1990.

——. "The System System." In *Human Factors in Technology,* edited by E. Bennett, J. Degan, and J. Spiegel. New York: McGraw-Hill, 1963.

——. *Libraries of the Future.* Cambridge, Mass.: MIT Press, 1965.

——. "Man-Computer Partnership." *International Science and Technology,* May 1965.

——. "Dynamic Modeling." In *Models for the Perception of Speech and Visual Form,* edited by Weiant Wathen-Dunn. Cambridge, Mass.: MIT Press, 1967.

——. "Graphic Input–A Survey of Techniques." In *Computer Graphics: Utility–Production–Art,* edited by F. Gruenberger. London: Academic Press; Washington, D.C.: Thompson Book Company, 1967.

——. "Interactive Dynamic Modeling." In *Prospects for Simulation and Simulators of Dynamic Modeling,* edited by George Shapiro and Milton Rogers. New York: Spartan Books, 1967.

——. "Interactive Information Processing, Retrieval and Transfer." Paper presented at "Storage and Retrieval of Information: A User-Supplier Dialog," Neuilly-sur-Seine, France, 1968.

——. "Underestimates and Overexpectations." In *ABM: An Evaluation of the Decision to Deploy an Antiballistic Missile System,* edited by Abram Chayes and Jerome B. Wiesner. New York: Harper & Row, 1969.

——. "Social Prospects of Information Utilities." In *The Information Utility and Social Choice,* edited by Harold Sackman and Norman H. Nie. Montvale, N.J.: AFIPS Press, 1970.

——. "Computers and Government." In *The Computer Age: A Twenty-Year View,* edited by Michael L. Dertouzos and Joel Moses. Cambridge, Mass.: MIT Press, 1979.

Licklider, J. C. R., and Robert W. Taylor. "The Computer as a Communication Device." *Science & Technology* 76 (1968). Reprinted in *In Memoriam: J. C. R. Licklider, 1915–1990,* edited by Robert W. Taylor. Digital Systems Research Center Reports, vol. 61. Palo Alto, Calif., 1990.

Machlup, Fritz, and Una Mansfield, eds. *The Study of Information: Interdisciplinary Messages.* New York: Wiley, 1983.

Mahoney, Michael. "The History of Computing in the History of Technology." *Annals of the History of Computing* 10 (1988).

Malone, Michael S. *The Microprocessor: A Biography.* New York: Springer-Verlag, 1995.

Manes, Stephen, and Paul Andrews. *Gates: How Microsoft's Mogul Reinvented an Industry–and Made Himself the Richest Man in America.* New York: Simon & Schuster, 1993.

Masani, Pesi R. *Norbert Wiener.* Basel: Birkhäuser, 1990.

McCarthy, John. "Time-Sharing Computer Systems." In *Computers and the World of the Future,* edited by Martin Greenberger. Cambridge, Mass.: MIT Press, 1962.

——. "John McCarthy's 1959 Memorandum." *IEEE Annals of the History of Computing* 14, no. 1 (1992).

——. "Reminiscences on the History of Time-Sharing." *IEEE Annals of the History of Computing* 14, no. 2 (1992).

McCarthy, John, et al. "A Proposal for the Dartmouth Summer Research Project on Artificial Intelligence." Proposal to the Rockefeller Foundation, August 31, 1955.

McCorduck, Pamela. *Machines Who Think.* San Francisco: W. H. Freeman & Co., 1979.

McDougall, Walter A. . . . *the Heavens and the Earth: A Political History of the Space Age.* New York: Basic Books, 1985.

McGill, William J. "George A. Miller and the Origins of Mathematical Psychology." In *The Making of Cognitive Science,* edited by William Hirst. Cambridge: Cambridge University Press, 1988.

Miller, George A. "The Magical Number Seven, Plus or Minus Two: Some Limits on Our Capacity for Processing Information." *Psychological Review* 63, no. 2 (1956).

——. "Stanley Smith Stevens, 1906–1973." In *Biographical Memoirs,* vol. 47. Washington, D.C.: National Academy of Sciences, 1975.

——. "J. C. R. Licklider, Psychologist." Unpublished address given before the Acoustical Society of America, 1991.

Moritz, Michael. *The Little Kingdom: The Private Story of Apple Computer.* New York: William Morrow, 1984.

Naughton, John. *A Brief History of the Future: The Origins of the Internet.* London: Weidenfeld & Nicolson, 1999.

Newell, Allen; and Herbert A. Simon. *Human Problem Solving.* Englewood Cliffs, N.J.: Prentice-Hall, 1972.

Newell, Allen; Alan J. Perlis; and Herbert A. Simon. "What Is Computer Science?" *Science* 157, no. 3795 (1967).

Norberg, Arthur L. "Changing Computing: The Computing Community and DARPA." *IEEE Annals of the History of Computing* 18, no. 2 (1996): 10–53.

Norberg, Arthur L., and Judy E. O'Neill. *Transforming Computer Technology: Information Processing and the Pentagon, 1962–1986.* Baltimore: Johns Hopkins University Press, 1996.

NRENAISSANCE Committee. *Realizing the Information Future.* Washington, D.C.: National Academy of Sciences, 1994.

Nyce, James M., and Paul Kahn. "A Machine for the Mind: Vannevar Bush's Memex." In *From Memex to Hypertext: Vannevar Bush and the Mind's Machine,* edited by James M. Nyce and Paul Kahn. San Diego: Academic Press, 1991.

Olsen, Kenneth. Oral History. National Museum of American History, Smithsonian Institution, September 28–29, 1988.

O'Neill, Judy E. "The Evolution of Interactive Computing Through Time-Sharing and Networking." Ph.D. dissertation, University of Minnesota, 1992.

——. " 'Prestige Luster' and 'Snow-Balling Effects': IBM's Development of Computer Time-Sharing." *IEEE Annals of the History of Computing* 17, no. 2 (1995).

Owens, Larry. "Vannevar Bush and the Differential Analyzer: The Text and Context of an Early Computer." In *From Memex to Hypertext: Vannevar Bush and the Mind's Machine,* edited by James M. Nyce and Paul Kahn. San Diego: Academic Press, 1991.

Papert, Seymour. *Mindstorms: Computers, Children, and Powerful Ideas.* New York: Basic Books, 1980.

Pearson, Jaime Parker, ed. *Digital at Work: Snapshots from the First Thirty-five Years.* Burlington, Mass.: Digital Press, 1992.

Perry, Tekla S., and Paul Wallich. "Inside the PARC: The 'Information Architects.' " *IEEE Spectrum* (October 1985).

"The Project MAC Interviews." *IEEE Annals of the History of Computing* 14, no. 2 (1992).

Pugh, Emerson W. *Building IBM: Shaping an Industry and Its Technology.* Cambridge, Mass.: MIT Press, 1995.

Pugh, Emerson W., and William Aspray. "Creating the Computer Industry." *IEEE Annals of the History of Computing* 18, no. 2 (1996).

Ralston, Anthony, and Edwin D. Reilly, eds. *Encyclopedia of Computer Science.* 3rd ed. New York: Van Nostrand Reinhold, 1993.

Ranade, Jay, and Alan Nash, eds. *The Best of Byte: Two Decades on the Leading Edge.* New York: McGraw-Hill, 1994.

Randell, Brian. "The COLOSSUS." In *A History of Computing in the Twentieth Century,* edited by N. Metropolis, J. Howlett, and Gian-Carlo Rota. New York: Academic Press, 1980.

Redmond, Kent C., and Thomas M. Smith. *Project Whirlwind: The History of a Pioneer Computer.* Bedford, Mass.: Digital Press, 1980.

Reid, T. R. *The Chip.* New York: Simon & Schuster, 1985.

Rheingold, Howard. *Tools for Thought.* New York: Simon & Schuster, 1985.

Rhodes, Richard. *The Making of the Atomic Bomb.* New York: Simon & Schuster, 1986.

——. *Dark Sun: The Making of the Hydrogen Bomb.* New York: Simon & Schuster, 1995.

Richard J. Barber Associates. *The Advanced Research Projects Agency, 1958–1974.* Washington, D.C.: Barber Associates, 1975.

Rifkin, Glen, and George Harrar. *The Ultimate Entrepreneur: The Story of Ken Olsen and the Digital Equipment Corporation.* Chicago: Contemporary Books, 1988.

Riordan, Michael, and Lillian Hoddeson. *Crystal Fire: The Birth of the Information Age.* New York: W. W. Norton, 1997.

Salus, Peter H. *A Quarter Century of UNIX.* Reading, Mass.: Addison-Wesley, 1994.

———. *Casting the Net: From ARPANET to INTERNET and Beyond . . .* Reading, Mass.: Addison-Wesley, 1995.

Segal, Irving Ezra. "Norbert Wiener, November 26, 1894–March 18, 1964." In *Biographical Memoirs*, vol. 62. Washington, D.C.: National Academy of Sciences, 1992.

Shannon, Claude. "A Symbolic Analysis of Relay and Switching Circuits" (1938). In *Claude Elwood Shannon: Collected Papers,* edited by N. J. A. Sloane and Aaron D. Wyner. New York: IEEE Press, 1993.

———. "The Bandwagon." *IEEE Transactions Information Theory* 2 (1956).

Shannon, Claude, and Warren Weaver. *The Mathematical Theory of Communication.* Urbana: University of Illinois Press, 1949.

Shasha, Dennis, and Cathy Lazere. *Out of Their Minds: The Lives and Discoveries of 15 Great Computer Scientists.* New York: Copernicus, 1995.

Slater, Robert. *Portraits in Silicon.* Cambridge, Mass.: MIT Press, 1987.

Smith, Douglas K., and Robert C. Alexander. *Fumbling the Future: How Xerox Invented, Then Ignored, the First Personal Computer.* New York: William Morrow, 1988.

Smith, Richard E. "A Historical Overview of Computer Architecture." *Annals of the History of Computing* 10, no. 4 (1989).

Southwick, Karen. *High Noon: The Inside Story of Scott McNealy and the Rise of Sun Microsystems.* New York: John Wiley, 1999.

Stern, Nancy. "John von Neumann's Influence on Electronic Digital Computing, 1944–1946." *Annals of the History of Computing* 2, no. 4 (1980).

Stibitz, George R. "Early Computers." In *A History of Computing in the Twentieth Century,* edited by N. Metropolis, J. Howlett, and Gian-Carlo Rota. New York: Academic Press, 1980.

Thacker, Chuck. "Personal Distributed Computing: The Alto and Ethernet Hardware." In *A History of Personal Workstations,* edited by Adele Goldberg. New York: ACM Press, 1988.

Tropp, Henry S. "The Smithsonian Computer History Project, and Some Personal Recollections." In *A History of Computing in the Twentieth Century,* edited by N. Metropolis, J. Howlett, and Gian-Carlo Rota. New York: Academic Press, 1980.

———. "History of the Design of the SAGE Computer–The AN/FSQ-7." *Annals of the History of Computing* 5 (1983).

———. "A Perspective on SAGE: Discussion." *Annals of the History of Computing* 5 (1983).

———. "Reliability of Components (Interview with Jay W. Forrester)." *Annals of the History of Computing* 5 (1983).

———. "Origin of the Term Bit." *Annals of the History of Computing* 6 (1984).

Turing, Alan M. "On Computable Numbers, with an Application to the Entschidungsproblem." *Proceedings of the London Mathematical Society* 2, no. 42 (1937).

———. "Computing Machinery and Intelligence." *Mind* 59, no. 236 (1950). Reprinted in *The Mind's I: Fantasies and Reflections on Self & Soul,* edited by Douglas R. Hofstadter and Daniel C. Dennett. New York: Basic Books, 1981.

Turkle, Sherry. *The Second Self: Computers and the Human Spirit.* New York: Simon & Schuster, 1984.

Umpleby, Stuart A., and Eric B. Dent. "The Origins and Purposes of Several Traditions in Systems Theory and Cybernetics." *Cybernetics and Systems: An International Journal,* no. 30 (1999).

Usselman, Steven W. "Fostering a Capacity for Compromise: Business, Government, and the Stages of Innovation in American Computing." *IEEE Annals of the History of Computing* 18, no. 2 (1996).

von Neumann, John. "The Principles of Large-Scale Computing Machines." *Annals of the History of Computing* 3, no. 3 (1981).

———. "First Draft of a Report on the EDVAC." *IEEE Annals of the History of Computing* 15, no. 4 (1993).

von Neumann, John, and Oskar Morgenstern. *Theory of Games and Economic Behavior.* Princeton, N.J.: Princeton University Press, 1944.

Wexelblatt, Richard L., ed. *History of Programming Languages.* New York: Academic Press, 1981.

Wiener, Norbert. "Memorandum on the Mechanical Solution of Partial Differential Equations" (1940). In *Norbert Wiener: Collected Works,* vol. 4, edited by Pesi R. Masani. Cambridge, Mass.: MIT Press, 1985.

———. "A Scientist Rebels." *Bulletin of the Atomic Scientists,* January 1947. *Atlantic Monthly,* January 1947.

——. *The Human Use of Human Beings: Cybernetics and Society.* Boston: Houghton Mifflin Company, 1950.

——. *I Am a Mathematician: The Later Life of a Prodigy.* Cambridge, Mass.: MIT Press, 1956.

——. "Some Moral and Technical Consequences of Automation." *Science* 131 (1960).

——. *Cybernetics, or Control and Communication in the Animal and the Machine.* 2d ed. Cambridge, Mass.: MIT Press, 1961.

Wiesner, Jerome B. "The Communications Sciences—Those Early Days." In *R. L. E.: 1946 + 20.* Cambridge, Mass.: Research Laboratory for Electronics, MIT, 1966.

——. "Vannevar Bush, March 11, 1890–June 28, 1974." In *Biographical Memoirs,* vol. 50. Washington, D.C.: National Academy of Sciences, 1979.

——. "Memories of Lick." Speech delivered at "A Ceremony in Remembrance of J. C. R. Licklider," October 15, 1990.

Wright, Robert. *Three Scientists and Their Gods: Looking for Meaning in an Age of Information.* New York: Harper & Row, 1988.

Wylie, Francis E. *M. I. T. in Perspective: A Pictorial History of the Massachusetts Institute of Technology.* Boston: Little, Brown, 1975.

Zachary, G. Pascal. "Vannevar Bush Backs the Bomb." *Bulletin of the Atomic Scientists,* December 1992.

——. *Endless Frontier: Vannevar Bush, Engineer of the American Century.* Cambridge, Mass.: MIT Press, 1999.

Zuse, Konrad. "Some Remarks on the History of Computing in Germany." In *A History of Computing in the Twentieth Century,* edited by N. Metropolis, J. Howlett, and Gian-Carlo Rota. New York: Academic Press, 1980.

INDEX